Powell-Smith and Furmston's

Building Contract Casebook

Powell-Smith and Furmston's
Building Contract Casebook

Fifth Edition

Michael Furmston
TD, BCL, MA, LLM

WILEY-BLACKWELL
A John Wiley & Sons, Ltd., Publication

This edition first published 2012

Wiley-Blackwell is an imprint of John Wiley & Sons, formed by the merger of Wiley's global Scientific, Technical and Medical business with Blackwell Publishing.

Registered office:
John Wiley & Sons, Ltd, The Atrium, Southern Gate, Chichester, West Sussex, PO19 8SQ, UK

Editorial offices:
9600 Garsington Road, Oxford, OX4 2DQ, UK
The Atrium, Southern Gate, Chichester, West Sussex, PO19 8SQ, UK
2121 State Avenue, Ames, Iowa 50014-8300, USA

For details of our global editorial offices, for customer services and for information about how to apply for permission to reuse the copyright material in this book please see our website at www.wiley.com/wiley-blackwell.

First edition published 1984 by Granada Publishing
First edition (revised) published 1987 by BSP Professional Books
Second edition published 1990
Third edition published 2000 by Blackwell Science Ltd
Fourth edition published 2006 by Blackwell Publishing
Fifth edition published 2012 by John Wiley & Sons, Ltd

Library of Congress Cataloging-in-Publication Data

Furmston, M. P.
Powell-Smith and Furmston's building contract casebook / Michael Furmston. – 5th ed.
p. cm.
Includes bibliographical references and index.
ISBN 978-0-470-65592-4 (hard cover : alk. paper) 1. Construction contracts–England–Cases. 2. Construction contracts–Wales–Cases. I. Powell-Smith, Vincent. Powell-Smith and Furmston's building contract casebook. II. Title. III. Title: Powell-Smith and Furmston's building contract casebook. IV. Title: Building contract casebook.
KD1641.F87 2012
343.41'07869–dc23
2011035017

A catalogue record for this book is available from the British Library.

Wiley also publishes its books in a variety of electronic formats. Some content that appears in print may not be available in electronic books.

Set in 10/12.5 pt Minion by Toppan Best-set Premedia Limited
Printed and bound in Singapore by Markono Print Media Pte Ltd

1 2012

Contents

Preface

The last edition of this book appeared in 2005. The flow of new and interesting cases on Building Contracts has continued unabated. I have tried to keep the book to the same length though a change in typeface has actually meant a reduction in the number of pages. I have included 14 new cases and taken out altogether 8 cases as well as reducing somewhat the quotations from a few other cases.

In the first four editions Julia Burden provided invaluable editorial support (while working for three different publishers). Her successors at Wiley-Blackwell (as it now is) have been most helpful. The support of Angie Choo Kum Woon in Singapore has been invaluable.

Michael Furmston
October 2011

PART I

FORMATION OF THE CONTRACT

Chapter 1
Letters of Intent

Usually, a letter of intent merely expresses an intention to enter into a contract in the future, and creates no contractual liability. It is a question upon the facts of each case whether a letter of intent gives rise to any and, if so, what liability.

British Steel Corporation v. Cleveland Bridge & Engineering Co Ltd

QUEEN'S BENCH DIVISION

(1981) 24 BLR 94

CBE were involved in the construction of a bank in Saudi Arabia, and needed to be supplied with cast steel nodes, for which supply they negotiated with BSC. On 21 February 1979, CBE sent the following letter of intent to BSC:

> 'Samma Bank: Damman
> We are pleased to advise you that it is [our] intention to enter into a subcontract with your company, for the supply and delivery of the steel castings which form the roof nodes on this project . . . We understand that you are already in possession of a complete set of our node detail drawings and we request that you proceed immediately with the works pending the preparation and issuing to you of the official form of subcontract.'

Consequent upon this letter BSC started work. On 27 February 1979 CBE sent a telex to BSC listing the sequence in which they required the nodes to be delivered, this being the first intimation that CBE required the nodes to be manufactured in a particular order. Despite the problems this caused, BSC continued to manufacture the nodes.

CBE did not send 'the official form of sub-contract' to BSC, nor did the parties agree on price or delivery dates. Ultimately, CBE did not pay BSC for the delivered nodes.

BSC claimed £229,832.70 from CBE as the price of the nodes, the claim being based in contract and alternatively upon a quantum meruit. CBE admitted that the goods were sold and delivered to them, and admitted liability in part. This admission was subject to a plea of set-off totalling £867,735.68, which CBE counter-claimed on the ground that BSC, in breach of contract, delivered the nodes late and out of sequence.

Powell-Smith and Furmston's Building Contract Casebook, Fifth Edition. Michael Furmston.

BSC's primary contention was that they were entitled to be paid on a quantum meruit because, if there was no binding contract between the parties, there was no legal basis for CBE's counter-claim. CBE argued that the agreement between the parties was comprised in the letter of 21 February, the telex concerning the delivery sequence, and BSC's conduct in proceeding with the manufacture of the nodes.

Held: No contract had come into existence between the parties on the basis of the letter of intent and BSC's performance, but BSC were entitled to be paid upon a quantum meruit.

ROBERT GOFF J: [The] question whether in a case such as the present any contract has come into existence must depend on a true construction of the relevant communications which have passed between the parties, and the effect (if any) of their actions pursuant to those communications. There can be no hard and fast answer to the question whether a letter of intent will give rise to a binding agreement; everything must depend on the circumstances of the particular case. In most cases where work is done pursuant to a request contained in a letter of intent, it will not matter whether a contract did or did not come into existence; because if the party who has acted on the request is simply claiming payment, his claim will usually be based upon a quantum meruit, and it will make no difference whether that claim is contractual or quasi-contractual. Of course, a quantum meruit claim (like the old actions for money had and received, and for money paid) straddles the boundaries of what we now call contract and restitution; so the mere framing of a claim as a quantum meruit claim, or a claim for a reasonable sum, does not assist in classifying the claim as contractual or quasi-contractual. But where, as here, one party is seeking to claim damages for breach of contract, the question whether any contract came into existence is of crucial importance.

As a matter of analysis the contract (if any) which may come into existence following a letter of intent may take one of two forms – either there may be an ordinary executory contract, under which each party assumes reciprocal obligations to the other; or there may be what is sometimes called an 'if' contract, i.e. a contract under which A requests B to carry out a certain performance and promises B that, if he does so, he will receive a certain performance in return, usually remuneration for his performance. The latter transaction is really no more than a standing offer which, if acted upon before it lapses or is lawfully withdrawn, will result in a binding contract.

The former type of contract was held to exist by Judge Fay in Turriff Construction Ltd v. Regalia Knitting Mills Ltd (1971), and it is the type of contract for which [CBE] contended in the present case. Of course, as I have already said, everything must depend on the facts of the particular case; but certainly, on the facts of the present case – and, as I imagine, on the facts of most cases – this must be a very difficult submission to maintain. It is only necessary to look at the terms of CBE's letter of intent in the present case to appreciate the difficulties. In that letter, the request to BSC to proceed immediately with the work was stated to be 'pending the preparation and issuing to you of the official form of sub-contract', being a subcontract which was plainly in a state of negotiation, not least on the issues of price, delivery dates, and the applicable terms and conditions. In these circumstances, it is very difficult to see how BSC, by starting work, bound themselves to any contractual performance. No doubt it was envisaged by CBE at the time they sent the letter that negotiations had reached an advanced stage, and that a formal contract would soon be signed; but since the parties were still in a state of negotiation, it is impossible to say with any degree of certainty what the material terms of the contract would be. I find myself quite unable to conclude that, by starting work in these circumstances, BSC bound themselves to complete the work. In the course of argument, I put to [counsel] the question whether BSC were free at any time, after starting work, to cease work; his submission was that they were not free to do so, even if negotiations on the terms of the formal contract broke down completely. I find this submission to be so repugnant to common sense and the commercial realities that I am unable to accept it. It is perhaps revealing that, on 4 April 1979, BSC did indeed state that they were not prepared to proceed with the contract until they had an agreed specification – a reaction which, in my judgment, reflected not only the commercial, but also the legal, realities of the situation.

I therefore reject CBE's submission that a binding executory contract came into existence in this case. There remains the question whether, by reason of BSC carrying out work pursuant to the request contained in CBE's letter of intent, there came into existence a contract by virtue of which BSC were entitled to claim reasonable remuneration; i.e., whether there was an 'if' contract of the kind I have described. In the course of argument, I was attracted by this alternative – really on the basis that, not only was it analytically possible, but also that it could provide a vehicle for certain contractual obligations of BSC concerning their performance – e.g., implied terms as to the quality of goods supplied by them. But the more I have considered the case, the less attractive I have found this alternative. The real difficulty is to be found in the factual matrix of the transaction, and in particular the fact that the work was being done pending a formal sub-contract the terms of which were still in a state of negotiation. It is, of course, a notorious fact that, when a contract is made for the supply of goods on a scale and in circumstances such as the present, it will in all probability be subject to standard terms, usually the standard terms of the supplier. Such standard terms will frequently legislate, not only for the liability of the seller for defects, but also for the damages (if any) for which the seller will be liable in the event not only of defects in the goods but also of late delivery. It is a commonplace that a seller of goods may exclude liability for consequential loss, and may agree liquidated damages for delay. In the present case, an unresolved dispute broke out between the parties on the question whether CBE's or BSC's standard terms were to apply – the former providing no limit to the seller's liability for delay, and the latter excluding such liability altogether. Accordingly when in a case such as the present the parties are still in a state of negotiation, it is impossible to predicate what liability (if any) will be assumed by the seller, e.g., for defective goods or late delivery, if a formal contract should be entered into. In these circumstances, if the buyer asks the seller to commence work 'pending' the parties entering into a formal contract, it is difficult to infer from the seller acting upon that request that he is assuming any responsibility for his performance, except such responsibility as will rest upon him under the terms of the contract which both parties confidently anticipate they will shortly enter into. It would be an extraordinary result if, by acting on such a request in such circumstances, the seller were to assume an unlimited liability for his contractual performance, when he would never assume such liability under any contract which he entered into.

For these reasons, I reject the solution of the 'if' contract. In my judgment, the true analysis of the situation is simply this. Both parties confidently expected a formal contract to eventuate. In these circumstances, to expedite performance under that anticipated contract one requested the other to commence the contract work, and the other complied with that request. If thereafter – as anticipated – a contract was entered into, the work done as requested will be treated as having been performed under that contract; if, contrary to their expectation, no contract was entered into, then the performance of the work is not referable to any contract of which the terms can be ascertained, and the law simply imposes an obligation on the party who made the request to pay a reasonable sum for such work as has been done pursuant to that request, such an obligation sounding in quasi-contract or, as we now say, in restitution. Consistently with that solution, the party making the request may find himself liable to pay for work which he would not have had to pay for as such if the anticipated contract had come into existence, e.g. preparatory work which will, if the contract is made, be allowed for in the price of the finished work: cf. William Lacey (Hounslow) Ltd v. Davis (1957). This solution moreover accords with authority, e.g. the decision in Lacey v. Davis (above); the decision of the Court of Appeal in the unreported case of Sanders & Forster Ltd v. A. Monk & Co. Ltd (1980), though that decision rested in part on a concession; and the crisp dictum of Parker J in O.T.M. Ltd v. Hydranautics (1981), when he said of a letter of intent that 'its only effect would be to enable the defendants to recover on a quantum meruit for work done pursuant to the direction' contained in the letter. I only wish to add to this part of my judgment the footnote that, even if I had concluded that in the circumstances of the present case there was a contract between the parties and that that contract was of the kind I have described as an 'if' contract, then I would still have concluded that there was no obligation under that contract on the part of BSC to continue with or complete the contract work, and therefore no obligation on their part to complete the work within a reasonable time. However, my conclusion in the present case is that the parties never entered into any contract at all.

In the course of his argument, [Counsel] submitted that, in a contract of this kind, the price is always an essential term in the sense that, if it is not agreed, no contract can come into existence. [He] relied upon a dictum of Lord Denning MR in Courtney & Fairbairn Ltd v. Tolaini Brothers (Hotels) Ltd (1975), to the effect that the price in a building contract is of fundamental importance. I do not however read the Master of the Rolls' dictum as stating that in every building contract the price is invariably an essential term, particularly as he expressly referred to the substantial size of the contract then before the Court. No doubt in the vast majority of business transactions, particularly those of substantial size, the price will indeed be an essential term, but in the final analysis it must be a question of construction of the particular transaction whether it is so. This is plain from the familiar trilogy of cases – May & Butcher Ltd v. The King (1929); W.N. Hillas and Co. Ltd v. Arcos Ltd (1932); and Foley v. Classique Coaches Ltd (1934) – which show that no hard and fast rule can be laid down but that the question in each case is whether, on a true construction of the relevant transaction, it was consistent with the intention of the parties that even though no price had been agreed a reasonable price should be paid. In the present case, however, I have no doubt whatsoever that, consistently with the view expressed by the Master of the Rolls in Courtney v. Tolaini Brothers, the price was indeed an essential term, upon which (among other essential terms) no final agreement was ever reached.

It follows that BSC are entitled to succeed on their claim; and that CBE's set-off and counterclaim must fail.

RTS Flexible Systems v Molkerei Alois Mŭller GmbH & Co KG

SUPREME COURT

(2010) 129 Con LR 1; [2010] UKSC 14

The defendant (Mŭller) was a leading dairy product supplier. At its premises in Market Drayton it produced different brands of Yoghurt and dairy rice products. The claimant (RTS) specialised in the supply of automated machines for packaging and product handling in the foods industry. There were negotiations between the parties from 2000 about plans to automate the process of packaging multi-packs of yoghurt. In April 2004 RTS provided quotation A and subsequently ten further quotations (down to quotation K), the later ones all in a format requested by Mŭller.

The work which RTS undertook was the design, manufacture, assembly, works testing, delivery, installation and commissioning at Mŭller's factory at Market Drayton of equipment described as line 1 and line 2. Mŭller were to supply the packaging equipment for the lines.

On 13 January 2005 RTS issued its quotation I. It indicated that it was prepared to enter into a contract based on MF/1 conditions and expressed concern about the time it would take to reach agreement on the contract conditions. On 13 February RTS were told that they had been awarded the contract.

On 15 February RTS issued quotation J. This specified a fixed price of £1,682,000. It provided for liquidated damages for delay of 0.5% of total contract value for a delay up to five weeks. There were many other detailed conditions and appendices.

On 27 February Mŭller sent RTS a letter of intent agreeing the contract price, instructing a start to work and providing for a full contract based on Mŭller's amended version of MF/1 within four weeks. Prior to completion of the contract only Mŭller were to have the right to terminate. On termination Mŭller were to be liable for reasonable out-of-pocket expenses but not for loss of profit.

On 1 March RTS wrote to Mŭller confirming that they had started work and raising points about the letter of intent. Negotiations as to the terms commenced. In April the four-week period for negotiating the contract was extended by agreement until 16 May and later until 27 May.

On 16 March Mŭller sent RTS a first draft of MF/1 amended and including provisions as to liquidated damages and limitation on RTS's liability. There were further exchanges and by 25 May the parties were agreed as to the wording of the amended MF/1 conditions save for the definition of force majeure which was agreed by an exchange of e-mails on 5 July. Meanwhile there were negotiations as to the schedules of which there were 17. No written contract was ever signed. Mŭller paid RTS 30% of £1,682,000 on about 28 April 2005, a further 30% in September 2005 and 10% in January 2006.

A dispute arose. Both parties argued before Christopher Clarke J as their primary position that there was a contract. Christopher Clarke J held that there was a contract but that it did not include the final draft version of the MF/1 terms. Before the Court of Appeal RTS argued that there was no contract. The Court of Appeal held that it was open to RTS so to argue and that since the parties were negotiating around the MF/1 conditions clause 48 of those made it clear that for a contract on those conditions to come into existence there had to be an agreement in writing. It followed that there was no contract.

Mŭller appealed.

Held: (allowing the appeal) – an objective analysis of the behaviour of the parties led to the conclusion that:

(1) they intended their relationship to be contractually binding and (2) after the expiry of the letter of intent contract, they intended to be bound by the terms they had agreed on or before 5 July as subsequently varied on 25 August, except (3) that the agreement (though incorporating the agreed versions of the MF/1 conditions) did not

include cl 48 and was not therefore dependent on the completion of a written contract.

LORD CLARKE SCJ:

The first point remains. Had the parties agreed to be bound by the agreed terms without the necessity of a formal written contract or, put another way, had they agreed to waive that requirement and thus cl 48? We have reached the conclusion that they had. The circumstances point to the fact that there was a binding agreement and that it was not on the limited terms held by the judge. The price had been agreed, a significant amount of work had been carried out, agreement had been reached on 5 July and the subsequent agreement to vary the contract so that RTS agreed to provide line 1 before line 2 was reached without any suggestion that the variation was agreed subject to contract. The clear inference is that the parties had agreed to waive the subject to contract clause, viz cl 48. Any other conclusion makes no commercial sense. RTS could surely not have refused to perform the contract as varied pending a formal contract being signed and exchanged. Nobody suggested that it could and, of course, it did not. If one applies the standard of the reasonable, honest businessman suggested by Steyn LJ, we conclude that, whether he was an RTS man or a Mŭller man, he would have concluded that the parties intended that the work should be carried out for the agreed price on the agreed terms, including the terms as varied by the agreement of 25 August, without the necessity for a formal written agreement, which had been overtaken by events.

Chapter 2
Costs of Tendering

The cost of tendering is generally borne by the contractor, but if the contractor performs additional services at the employer's request, he may be entitled to reasonable payment.

William Lacey (Hounslow) Ltd v. Davis

QUEEN'S BENCH DIVISION
[1957] 2 All ER 712

The plaintiff tendered for the re-construction of war-damaged premises belonging to the defendant, who led the plaintiff to believe that it would receive the contract. At the defendant's request, the plaintiff calculated the timber and steel requirements for the building, and prepared various further schedules and estimates which the defendant made use of in negotiations with the War Damage Commission. Eventually, the plaintiff was informed that the defendant intended to employ another builder to do the work. In fact, the defendant sold the premises.

The plaintiff claimed damages for breach of contract, and, alternatively, remuneration on a *quantum meruit* in respect of the work done by it in connection with the reconstruction scheme.

Held: Although no binding contract had been concluded between the parties, a promise should be implied that the defendant would pay a reasonable sum to the plaintiff in respect of the services rendered. Judgment was entered for the plaintiff in the sum of £250 13s 5d.

> BARRY J: I am unable to see any valid distinction between work done which has to be paid for under the terms of a contract erroneously believed to be in existence, and work which was to be paid for out of the proceeds of a contract which both parties erroneously believed was about to be made . . . In both cases, when the beliefs of the parties were falsified, the law implies an obligation . . . to pay a reasonable price for the services which had been obtained.

Powell-Smith and Furmston's Building Contract Casebook, Fifth Edition. Michael Furmston.

There is no implication that the costs incurred by the contractor in preparing his tender will be reimbursed by the employer, but the contractor may be entitled to reasonable payment if, at the employer's request, he does substantial preparatory work over and above normal tender preparation and no contract is ever placed.

Marston Ltd v. Kigass Ltd

QUEEN'S BENCH DIVISION
(1989) 15 Con LR 116

The plaintiffs claimed £25,946.23 for preparatory works over and above the costs and work of preparing their tender for a design and build contract to provide a factory at Warwick to replace one which had burnt to the ground. The contract for the work was never placed because the proceeds of the insurance money from the old factory were insufficient to cover the cost of rebuilding to the local authority's standards.

The plaintiffs' tender was the best value for money but because there was a tight time scale it needed to be supplemented with further details. The tender was followed by a vital meeting on 18 December 1986, when the plaintiffs were the only tenderers invited to discuss their tender. The defendants' chairman made it clear that no contract to rebuild the factory would be entered into unless and until he had obtained the insurance money to pay for the rebuilding. Nothing was said which indicated that in the event of the insurance money not being forthcoming all preparatory work done by the plaintiffs up to that point would be at the contractors' risk. The defendants were well aware that the plaintiffs would have to start preparatory work before the contract was signed, and it was understood if not spelt out that there should be a four-week lead-in period. It was expressly agreed that the contract would be signed (if the insurance money had been received) on 16 January 1987 and that work would start on site on 2 February 1987. It was also found that it was made plain to the plaintiffs at the meeting that subject to payment of the insurance money and the employer's surveyor being satisfied on points of detail normal to a tender appraisal the contract would be given to the plaintiffs.

Held: There was an express request made by the defendants to the plaintiffs to carry out a small quantity of design work and there was an implied request to carry out preparatory works in general. Both the express and implied requests gave rise to a right to payment of a reasonable sum.

JUDGE PETER BOWSHER QC: I find that the facts of the present case, although different in important respects are similar in kind to the facts in *William Lacey* v. *Davis* (1957). There was a request to do the work, though the request in respect of the bulk of the work was implied rather than express. It was contemplated that the work would be paid for out of the contemplated contract. Both parties believed that the contract was about to be made despite the fact that there was a very clear condition which had to be met by a third party if the contract was to be made. The defendants obtained the benefit of the work in my judgment, though [counsel for the defendants] submitted that they did not. That submission led into an interesting consideration of academic writings which in this sphere have gone ahead of the decisions of the courts.

At one stage it appeared that [counsel for the plaintiffs] on behalf of the plaintiffs would pray in aid the theory of 'free acceptance', a concept put forward by Lord Goff and Professor Gareth Jones in their book *The Law of Restitution* and also propounded by Mr Peter Birks in his book *Introduction to the Law of Restitution*. In a recent article, 'Free acceptance and the Law of Restitution' [1988] *LQR* 576, Mr A.S. Burrows has put forward the thesis that neither in principle nor on authority does free acceptance have a place within the law of restitution. It is unnecessary for me to consider this large and interesting debate because [counsel for the plaintiffs] later made it clear that her submission was limited to saying that later free acceptance of services from the plaintiffs was evidence that the defendants had previously requested those services on 18 December. [Counsel for the defendants] on behalf of the defendants rightly submitted that on the pleadings and on the formulation of the preliminary issues it was not open to the plaintiffs to advance a case of 'free acceptance'. But the writings on the subjects of free acceptance are of interest in considering the meaning of 'benefit' in the law of restitution.

When considering benefit, a distinction has to be made between the delivery of money and goods on the one hand and the provision of services on the other. [Counsel for the defendants] submits that for there to be any recovery by the plaintiffs in this case they must show that the defendants have received an actual benefit, not a potential benefit that is never realised. [Counsel for the defendants] relies in particular on a passage in para 1943 of the chapter on Restitution in *Chitty on Contracts* (25th edn) written by Mr J. Beatson:

> The nature of the enrichment. This may take the form of a direct addition to the recipient's wealth, such as by the receipt of money, or an indirect one, for instance where an inevitable expense has been saved. The most common example of the second type of benefit is the discharge of an obligation of the defendant, whether by paying his creditor or abating a nuisance or performing some other service for which he is primarily responsible.
>
> In the case of the rendering of services as opposed to the payment of money 'the identity and value of the resulting benefit to the recipient may be debatable'. Services may take many forms and while some result in an indirect accretion to the defendant's wealth, for instance by improving his property, other 'pure' services do not. The fact that services cannot be restored and the influence of the implied contract theory has meant that they were often not regarded as beneficial so as to give rise to a *quantum meruit* unless the defendant had requested the services or, knowing that they were to be paid for, had freely accepted them. Many but not all such cases are capable of analysis as a genuine implied contract. These cases do not depend upon the service adding to the defendant's wealth, the service *per se* is treated as a benefit. There is, however, also some authority that treats a service as beneficial where it results in an 'incontrovertible benefit' to the defendant (*Craven-Ellis* v. *Canons Ltd* (1936), *Greenwood* v. *Bennett* (1973)). With the possible exception of necessitous intervention to preserve life or health, only services that result in an accretion to the defendant's wealth can constitute an incontrovertible benefit. Goff and Jones state that incontrovertible benefit is established where the defendant has made 'an immediate and realisable financial gain or has been saved an expense which he otherwise would have incurred'.

Although Goff and Jones refer to the gain as being 'realisable' (pp. 19 and 148) Birks argues that it must not be merely realisable but realised. But I am not convinced by his argument despite [counsel for the defendants'] contention that there is no reported case in quasi-contract where there has not been actual benefit.

> It seems to me that in appropriate cases, the benefit may consist in a service which gives a realisable and not necessarily realised gain to the defendant particularly when, as here, the service is a part of what was impliedly requested.
>
> The preliminary works requested were undoubtedly done for the benefit of the defendants and were only done for the benefit of the plaintiffs in the sense that they hoped to make a profit out of them. As a result of the works some progress was made towards getting consents and in the end the defendants had in the hands of their agent some designs and working drawings (though not a complete set) together with an implied licence to build to those drawings even though that licence be limited as I think (without having heard argument) to a licence to have the factory built by the plaintiffs. Whether the defendants decide ultimately to build a factory or to sell the land, they have a benefit which is realisable.
>
> . . . I therefore conclude that there was no agreement as alleged in the statement of claim. I find that there was an express request made by the defendants to the plaintiffs to carry out a small quantity of design works and that there was an implied request to carry out preparatory works in general and that both the express and the implied requests gave rise to a right of payment of a reasonable sum.

The reasoning in the above two cases and in particular the result in Marston Ltd v. Kigass Ltd were doubted in the next case.

Regalian Properties plc v. London Dockland Development Corporation

CHANCERY DIVISION

(1994) 45 Con LR 37

The plaintiffs offered to buy a licence for the residential development of land in Wapping from the defendant on terms that the defendant would grant them a building lease. In July 1986 this offer was accepted in a letter headed:

> SUBJECT TO CONTRACT
>
> Further to your submission of11 June 1986 in respect of the above I am pleased to inform you that the Corporation's Board has accepted your company's offer for this site, subject to: (1) Contract; (2) The District Valuer's certificate of market value; (3) Your scheme achieving the desired design quality and the obtaining of detailed planning consent.

In fact the building lease was never granted because of delays caused partly by the defendants' requests for further designs and partly by delay in the defendants becoming owners of all the land. By the time these difficulties had been overcome, the value of the land had fallen dramatically.

The plaintiffs sought to recover from the defendant nearly £3 million in professional fees which they had incurred in connection with the proposed development.

Held: By using the words 'subject to contract' each party had taken the risk that the transaction would not come to fruition.

RATTEE J: Although I have to say with respect that I do not find the reasoning of Barry J entirely easy to follow, the result seems to me to make perfectly good sense on the facts of that case. At the request of the defendant the plaintiff had done work which had clearly benefited the defendant quite outside the ambit of the anticipated contract and had only not charged for it separately, as one would otherwise have expected him to do, because he thought he would be sufficiently recompensed by what he would be paid by the plaintiff under the contract. In those circumstances it is not surprising that the law of restitution found a remedy for the plaintiff when the contract did not materialise. I do not consider that the decision lends any real support to the claim made by Regalian in the present case for compensation for expenditure incurred by it for the purpose of enabling itself to obtain and perform the intended contract at a time when the parties had in effect expressly agreed by the use of the words 'subject to contract' that there should be no legal obligation by either party to the other unless and until a formal contract had been entered into. It was frankly accepted by Mr Goldstone of Regalian that he knew and intended that this should be the effect of the use of the phrase 'subject to contract', and indeed Regalian admits in its pleadings that those words were not intended to have any unusual meaning in the present case.

I have to say, with all respect for the judge, that I find this a surprising decision, not least because, as I have recited from the judge's findings of fact, the plaintiff had earlier requested and been refused an assurance that it would be compensated for the preparatory work concerned. In this respect I agree with the critical commentary on the decision by the editor of the report of the decision that is to be found in 46 BLR 109. However, whether the decision be right or wrong, I do not feel obliged to apply it in the present case, which is distinguishable on the facts in two particular respects. First, in the present case, unlike *Marston Construction Co* v. *Kigass Ltd*, even if a contract had materialised no part of any costs incurred or work done by Regalian in connection with the contract would have been paid for by LDDC. The only obligation on LDDC would have been to grant the building lease. Second, as I have already said, I am not satisfied in the present case that the preparatory works resulted in any benefit to LDDC.

Chapter 3
Tenders and Estimates

The contractor's offer to carry out the works is called a tender. It must be definite and unambiguous in its terms if its acceptance is to conclude an agreement enforceable by the law as a contract.

Peter Lind & Co Ltd v. Mersey Docks & Harbour Board

QUEEN'S BENCH DIVISION

[1972] 2 Lloyds Rep 234

The plaintiffs submitted to the defendants alternative tenders for the construction of a freight terminal, one on a fixed-price basis, and the other on a 'cost-plus' basis. The defendants purported to accept 'your tender', but did not specify which one. The plaintiffs carried out the work and claimed payment on a *quantum meruit.*

Held: There was no concluded contract because the defendants' purported acceptance did not specify which tender was being accepted. The plaintiffs were entitled to payment on a *quantum meruit.*

Normally an estimate will be held to be a firm offer in law, and its unequivocal acceptance will result in a binding contract.

Crowshaw v. Pritchard and Renwick

QUEEN'S BENCH DIVISION

(1899) 16 TLR 45

The plaintiff wrote to the defendant enclosing a drawing and specification, and inviting a tender for some alteration work. The defendant replied:

> ESTIMATE – Our estimate to carry out the sundry alterations to the above premises, according to the drawings and specifications, amounts to £1,230.

Powell-Smith and Furmston's Building Contract Casebook, Fifth Edition. Michael Furmston.

The plaintiff wrote next day accepting the defendant's 'offer to execute' the works, but later the defendant refused to go ahead. He contended that by using the word 'estimate' he did not intend it as an offer to do the work, and also that there was a trade custom that a letter in this form was not to be treated as an offer.

Held: The 'estimate' was an offer which had been accepted by the plaintiff and the defendant was liable in contract.

> BIGHAM J: It has been suggested that there is some custom or well-known understanding that a letter in this form is not to be treated as an offer. There is no such custom, and, if there is, it is contrary to law.

Normally there will be no contract before the tender is accepted, but in some circumstances the invitation to tender may create contractual obligations.

Blackpool and Fylde Aero Club Ltd v. Blackpool Borough Council

COURT OF APPEAL
[1990] 3 All ER 25

The Council owned an airport. It raised money by granting a concession to operate pleasure flights from the airport. In 1983 it invited the plaintiff and six other parties to tender for the next three-year concession. The invitation stated that the tender was to be in the envelope provided, was to bear no mark which would identify the sender and that tenders received after 12 noon on 17 March 1983 would not be considered. The plaintiff placed its tender in the Town Hall letter box at 11am on 17 March 1983, but the letter box was not cleared at 12 noon by Council staff. As a result, the plaintiff's tender was classified as late and excluded from consideration.

Held: In the circumstances the Council had promised that any tender properly submitted by the deadline would be considered and the plaintiff's submission of a tender created a binding contract to that effect.

> BINGHAM LJ: A tendering procedure of this kind is, in many respects, heavily weighted in favour of the invitor. He can invite tenders from as many or as few parties as he chooses. He need not tell any of them who else, or how many others, he has invited. The invitee may often, although not here, be put to considerable labour and expense in preparing a tender, ordinarily without recompense if he is unsuccessful. The invitation to tender may itself, in a complex case, although again not here, involve time and expense to prepare, but the invitor does not commit himself to proceed with the project, whatever it is; he need not accept the highest tender; he need not accept any tender; he need not give reasons to justify his acceptance or rejection of any tender received. The risk to which the tenderer is exposed does not end with the risk that his tender may not be the highest (or, as the case may be, lowest). But where, as here, tenders are solicited from selected parties all of them known to the invitor, and where a local authority's invitation prescribes a clear, orderly and familiar procedure (draft contract conditions available for inspection and plainly not open to negotiation, a prescribed common form of tender, the supply of envelopes designed to preserve the absolute anonymity of tenderers and clearly to identify the tender in question and an absolute deadline) the invitee is in

my judgment protected at least to this extent: if he submits a conforming tender before the deadline he is entitled, not as a matter of mere expectation but of contractual right, to be sure that his tender will after the deadline be opened and considered in conjunction with all other conforming tenders or at least that his tender will be considered if others are. Had the club, before tendering, inquired of the council whether it could rely on any timely and conforming tender being considered along with others, I feel quite sure that the answer would have been 'of course'. The law would, I think, be defective if it did not give effect to that.

It is of course true that the invitation to tender does not explicitly state that the council will consider timely and conforming tenders. That is why one is concerned with implication. But the council does not either say that it does not bind itself to do so, and in the context a reasonable invitee would understand the invitation to be saying, quite clearly, that if he submitted a timely and conforming tender it would be considered, at least if any other such tender were considered.

At least in the public sector, the letting of contracts by competitive tender gives rise to an implicit contract to conduct the tendering process fairly.

Harmon CFEM Facades (UK) Ltd v. Corporate Officer of the House of Commons

QUEEN'S BENCH DIVISION

(1999) 67 Con LR 1

This dispute arose out of the letting of the fenestration contract for the new House of Commons (H of C) building in Bridge Street, Westminster. In the original tendering process, the claimant was the lowest tenderer at £40,479,469. All the tenders were higher than had been expected and the specification was altered, with the claimant again the lowest tenderer. There was a further round of tendering in which the tenderers were in effect tendering for different options. Those managing the tendering process were anxious for the contract to go to a British firm. The contract was awarded to a consortium with a British element.

Held: The defendant was in breach of its obligation to conduct the tendering process fairly and openly.

The claimant was entitled to recover the wasted costs of tendering. Further, since the defendant could not lawfully let the contract to the successful tenderer and since in practice restarting the tendering process was not a practical option, the claimant was entitled to recover damages for its failure to be awarded the contract.

[Note that the defendant was also in breach of the European Union rules for the procurement of public contractors.]

JUDGE HUMPHREY LLOYD QC:

214. In my judgment it is clear from the *Blackpool* case and from the other authorities that there must be something more than a request for a tender which is to be submitted competitively along with others. An invitation to tender is by its nature not normally an offer; it solicits offers. It does not carry with it an obligation to accept any offer that is made in response to it, even if the customary disclaimer is not made. It would be quite a change if the very fact

that tenderers were informed that competitive tenders were being sought was treated in law as an offer that any tenderer who submitted a tender would expect to be treated fairly. It would intrude into the ordinary commercial freedom or discretion to accept or reject a tender or to negotiate with whoever seemed best in the eyes of the person seeking tenders. There must therefore be some good reason why obligations of the kind suggested by Harmon can arise.

215. In this instance tenders were sought using the restricted procedure under the 1991 Regulations which provides tenderers with some protection in as much as they should not be tendering along with anyone who was not considered to be qualified to carry out the work in question and had satisfied other requirements. The restricted procedure is intended both to provide the contracting authority with a mechanism whereby tenders are only sought from a selected short list and to provide the tenderers with the knowledge that the competition will come from true competitors. In addition tenderers who had not been selected can challenge that decision. The 1991 Regulations are seemingly comprehensive. They give disappointed tenderers rights to prevent the procedure from being abused and to obtain reasons and to question the award. The requirement to give reasons itself imports an obligation of fairness even if it were not grafted on by the decisions of the European Court of Justice governing the interpretation of the parent legislation. I therefore consider that H of C is right in its primary submission.

216. On the other hand, the procedure of the 1991 Regulations was not followed. H of C did not inform the other tenderers it would be considering Seele/Alvis' option B2 alongside tenders based on the new specification. If I am right in my conclusion that H of C only sought alternative details then can Harmon complain of the decision to consider Seele/Alvis's tender on the basis that it offered an alternative design? The discussions in the autumn of 1995 deviated from the 1991 Regulations procedures in that they went beyond mere negotiations and clarifications. They entertained an alternative design, of which the other tenderers were unaware (at least formally). If Harmon cannot complain of the result under the 1991 Regulations does this not mean that the 1991 Regulations are not as comprehensive as they might first appear – or that they do not effectively control deviations of the kind adopted by H of C? In my judgment by repeating the offer to consider alternatives, on 1[illegible] September and on 2 and 30 October 1995, it was to be implied in that offer that by submitting a tender any alternatives would be equivalent to the schemes or schemes for which revised tenders were being sought and would be options only in terms of refinements of detail design which would reduce cost, albeit confidential to the tenderer but falling short of different proposals which were more than matters of detail but ones of changes of design, of which tenderers were not informed and therefore were entitled to assume were not matters which they needed to take into account. In my judgment even though all tenderers accepted that they would not be entitled to see alternatives of detail which were considered to be commercially confidential to a given tenderer, H of C in soliciting new or revised tenders under the European public works regime (to which effect is given by the Regulations) impliedly undertook towards any tenderer which submitted a tender that its submission would be treated as an acceptance of that offer or undertaking and: (a) that the alternative submitted by any tenderer would be considered alongside a compliant revised tender from that tenderer; (b) that any alternative would be one of detail and not design; and (c) that tenderers who responded to that invitation would be treated equally and fairly.

These contractual obligations derive from a contract to be implied from the procurement regime required by the European directives, as interpreted by the European Court, whereby the principles of fairness and equality form part of a preliminary contract of the kind that I have indicated. Emery shows that such a contract may exist at common law against a statutory background which might otherwise provide the exclusive remedy. I consider that it is now clear in English law that in the public sector where competitive tenders are sought and responded to, a contract comes into existence whereby the prospective employer impliedly agrees to consider all tenderers fairly (see the *Blackpool* and *Fairclough* cases).

217. H of C was in breach of all these obligations. In my judgment it also broke the implied duty to treat all tenderers fairly and equally by considering an alternative design without giving any other tenderer the opportunity of competing with it on its terms.

Pratt Contractors Ltd v. Transit New Zealand

JUDICIAL COMMITTEE OF PRIVY COUNCIL

[2004] BLR 143

The defendant (Transit) was a Crown entity set up by statute with responsibility for maintenance and improvement of state highways. It was required by statute to employ approved competitive pricing procedures and it had drawn up and approved such procedures (CPP). It also had internal rules contained in *The State Highway Evaluation Manual* (SHM) and *Contract Administration Manual* (CAM). CPP provided for two methods, one based simply on lowest price and one (the weighted attribute method) in which marks (70%) were given for price and other marks for relevant experience (4%), track record (10%), technical skills (4%), resources (4%), management skills (4%) and methodology (4%).

Transit wished to invite tenders for the realignment of state highway 1 at Vinegar Hill and to use the weighted attribute method. The claimant (Pratt) was one of those invited to tender and in the initial tenders its price was the lowest by a substantial margin. The marking of the other attributes involved the use of consultants who regarded Pratt with suspicion as 'lowballers' (that is, contractors who seek to get contracts by quoting low prices and then pursue aggressive claims policies).

The contract was awarded to another tenderer and Pratt claimed damages from Transit for breach of a preliminary contract that the tendering process would be conducted fairly and in good faith. It was accepted that there was such a contract and that it did require Transit to act fairly and in good faith.

Held: Transit were not in breach of the contract. The requirement of good faith did not incorporate the internal procedures set out in SHM and CAM. The requirement of fairness did not require the appointment of decision makers or advisers with no previous knowledge of the tenderers, or that tenderers should be given a hearing.

LORD HOFFMANN:

The contractual terms

44. Their Lordships agree with the Court of Appeal that the RFT [request for tenders] did not incorporate the terms of the SHM or CAM. The RFT said only that tenders would be evaluated in accordance with the weighted attributes method prescribed by the CPP. The detailed procedures prescribed by the other manuals are no doubt intended in part to ensure that Transit does comply with the terms of the CPP and in part for its own administrative convenience. But they are not something upon which an outsider can rely. The claim based on the single envelope procedure must therefore fail. Their Lordships also agree that the adoption of the sub-attribute methodology does not mean that the TET [tender evaluation team] were assessing attributes other than those specified in the RFT. As a matter of construction

the sub-attributes, including financial viability, could legitimately be regarded as aspects of the principal specified attribute.

45. The nature of the implied duty to act fairly and in good faith has been the subject of a good deal of discussion in Commonwealth authorities. In *Pratt Contractors Ltd* v. *Palmerston North City Council* [1995] 1 NZLR 469, 483, Gallen J said that fairness was a 'rather indefinable term'. In *Hughes Aircraft Systems International* v. *Airservices Australia* (1997) 146 ALR 1, 36–37, Finn J said that the duty in cases of preliminary procedural contracts for dealing with tenders is a manifestation of a more general obligation to perform any contract fairly and in good faith. That is a somewhat controversial question into which it is unnecessary for their Lordships to enter because it is accepted that, in general terms, such a duty existed in this case. The issue is rather as to its specific content in relation to the particular acts required to be performed by Transit in evaluating the tenders.

46. In relation to this question, Finn J emphasised that such an implied term –

does not as such impose on [the employer] under the guise of contract law, the obligation to avoid making its decision or otherwise conducting itself in ways which would render it amenable to judicial review of administrative action.

47. This observation was adopted by McGrath J in the Court of Appeal and their Lordships agree with it. The judge's findings of apparent bias were therefore no ground for holding Transit to have been in breach of contract. It is nevertheless necessary to identify exactly what standard of conduct was required of the TET in making its assessment. In their Lordships' opinion, the duty of good faith and fair dealing as applied to that particular function required that the evaluation ought to express the views honestly held by the members of the TET. The duty to act fairly meant that all the tenderers had to be treated equally. One tenderer could not be given a higher mark than another if their attributes were the same. But Transit was not obliged to give tenderers the same mark if it honestly thought that their attributes were different. Nor did the duty of fairness mean that Transit were obliged to appoint people who came to the task without any views about the tenderers, whether favourable or adverse. It would have been impossible to have a TET competent to perform its function unless it consisted of people with enough experience to have already formed opinions about the merits and demerits of roading contractors. The obligation of good faith and fair dealing also did not mean that the TET had to act judicially. It did not have to accord Mr Pratt a hearing or enter into debate with him about the rights and wrongs of, for example, the Pipiriki contract. It would no doubt have been bad faith for a member of the TET to take steps to avoid receiving information because he strongly suspected that it might show that his opinion on some point was wrong. But that is all.

48. Their Lordships do not consider that the judge made any finding of bad faith on the part of . . . any member of the two TETs. There is no doubt that [one member of the TET] was strongly of the view that Pratt's business methods and lack of competence made it unwise for Transit to engage it as a contractor. But that was no reason for him to disqualify himself from the TET. Transit had paid for his expert opinion and were entitled to pay attention to it. Nor is there anything to show that the marks on which he and his colleagues agreed did not reflect a true consensus of their honestly held opinions.

49. It follows that their Lordships do not think that the findings of fact justify a conclusion that there was a breach of the express or implied terms of the preliminary procedural contract at either of the tender rounds. They also agree with the Court of Appeal that even if there was such a breach in the first round, it would have had no causative effect on Pratt's failure to obtain the contract. They will therefore humbly advise Her Majesty that the appeal should be dismissed. The appellants must pay the costs before their Lordships' Board.

Chapter 4
Incorporation of Documents

A reference in a contractual document to the contract being subject to conditions 'available on request' and brought to the notice of the other party is sufficient to incorporate the current edition of those conditions into the contract.

Smith v. South Wales Switchgear Ltd

HOUSE OF LORDS

[1978] 1 All ER 18

Smith had overhauled SWS's electrical equipment for some years. On 18 March 1970 SWS wrote to Smith asking him to carry out an overhaul of the equipment in July 1970 and Smith replied that he would do so. On 27 May, SWS sent Smith a 'purchase note' requesting the overhaul 'subject to . . . our General Conditions of Contract 24001, obtainable on request'. Smith replied stating that he had given instructions for the work to be carried out. He did not request a copy of the conditions, although on 1 July 1970 a copy of the 1969 conditions was sent to him. There were two other versions of the conditions, including one dated March 1970.

Held: The reference in the purchase order to 'General Conditions . . . available on request' was sufficient to incorporate the general conditions as revised in March 1970. The reference clearly indicated how the terms might be ascertained by Smith. The conditions referred to were those current at the date of the contract, because it was common knowledge that such conditions were revised from time to time. Had Smith asked for a copy of the general conditions, he could reasonably expect to have received the latest edition.

LORD KEITH: I consider that the contract was concluded by the appellants receiving the respondents' original purchase order dated 27 May 1970, coupled with their letter to the respondents dated 16 July 1970 saying that they had instructed the work to be carried out . . . The purchase order referred to the respondents' 'General Conditions of Contract 24001 obtainable on request'. It thus clearly indicated the manner in which the terms of these conditions could be ascertained, and that was sufficient in law, unconditional acceptance having followed, for their incorporation into the contract. A question however arises as to

Powell-Smith and Furmston's Building Contract Casebook, Fifth Edition. Michael Furmston.

what in the circumstances is the meaning to be attributed to 'General Conditions of Contract 24001, obtainable on request', it having emerged that the original terms of these conditions were revised by the respondents in January 1969 and again in March 1970. In my opinion the meaning reasonably to be attributed is that the conditions referred to are the current conditions, that is to say, the conditions as revised in March 1970. It is common experience that the general conditions of contract of various undertakers are revised from time to time, and anyone requesting a copy of such conditions would reasonably expect to receive the current up-to-date edition. Further, that is the edition which a responsible employee of the undertaker clearly ought to send to an inquirer. Viewing the matter objectively therefore, I am of opinion that the parties contracted on the basis of the respondents' General Conditions of Contract 24001 as revised in March 1970. I do not consider that the reference in the purchase order to 'General Conditions of Contract 24001', without any mention of revision, precludes the relevant conditions from being the latest revised version. All three versions in fact bore the number 24001. The situation does not differ from that which would exist had the reference simply been to 'General Conditions of Contract' without any number. It would be unreasonable to suggest that wherever such conditions are revised the fact of revision must necessarily be indicated on any contractual reference to them. Any reference to conditions can only be understood, in the mind of an ordinary reasonable man, as a reference to the conditions currently in force. The circumstance that on 1 July 1970 the respondents sent to the appellants a purchase amendment order having printed on the reverse a copy of the general conditions as revised in January 1969 does not, in my view, affect the matter. This occurred at a time when both parties were erroneously attributing the number NQ492122 to contractual work which had been completed at the end of 1969, and engaged in fixing the price to be paid for that work. In that situation the January 1969 version of the general conditions was quite rightly attributed to the particular work in question.

The terms of a standard form contract may be incorporated into the contract by reference.

Killby & Gayford Ltd v. Selincourt Ltd

COURT OF APPEAL

(1973) 3 BLR 104

A contract was made by an exchange of letters. In a letter dated 11 February 1972, the architects asked the contractors to price alteration work. The letter concluded:

> Assuming that we can agree a satisfactory contract price between us, the general conditions and terms will be subject to the normal standard form of RIBA contract.

The contractors submitted a written estimate on 6 March 1972, to which the architect replied that he was accepting 'on behalf of the client . . . your estimate dated 6 March 1972 and I shall be obliged if you will accept this letter as your formal instruction to proceed with the work'. No JCT contract was ever signed, but the contractors proceeded with the work.

Held: The exchange of letters incorporated the current JCT form.

Even where the parties use loose or inappropriate language when referring to standard form contracts the court will endeavour to make sense of vague references to them. However, the court will not create a contract for the parties, but will seek to apply their presumed intention objectively ascertained.

Brightside Kilpatrick Engineering Services v. Mitchell Construction (1973) Ltd

COURT OF APPEAL

(1975) 1 BLR 64

The plaintiffs were nominated as sub-contractors under JCT 63. The subcontract was placed by the defendants on an order form dated 24 March 1972, which stipulated that the sub-contract documents should consist of a 'Standard Form of Tender dated 9.11.71', a specification, 'Conditions of Contract' and a 'Schedule of Facilities'. The order continued that 'The Form of Contract with the Employer is RIBA 1963 Edition, July 1971 Revised', and concluded 'The Conditions applicable to the Sub-Contract shall be those embodied in RIBA as above agreement'. Printed references to the 'green' form of sub-contract had been deleted.

Held: The words should be read as referring to those clauses in the main contract which referred to matters concerning nominated sub-contractors. Clause 27 of JCT 63 regulated the nominated sub-contract relationship, and the only way to give a sensible meaning to the relationship was to read into it the terms of the 'green' form.

BUCKLEY LJ: It is not an easy point, for, as so often happens when parties try to import into their contractual relationships terms in other documents which do not easily fit those relationships, it is difficult to know what is the proper construction to put upon the documents in the case.

The contention for the defendants has been that the statement in the document of 24 March 1972 that 'The Conditions applicable to the Sub-Contract with you shall be those embodied in RIBA as above' is a clear reference to the main contract, which was the RIBA 1963 form of contract, and that the effect of that is to import into the contractual relationship between the contractors and the sub-contractors all the terms of the head contract *mutatis mutandis* substituting for references to the building owner references to the contractor, and substituting for references to the contractor references to the sub-contractor, and that that contract as so modified rules the relationship between the contractor and the sub-contractor accordingly, at any rate so far as the terms of that contract as so modified are capable of applying as a practical matter to the relationship between a contractor and sub-contractor. I say 'capable of applying as a practical matter' because . . . when one comes to read the head contract adapted *mutatis mutandis* in the way I have indicated, many of its clauses are already quite inappropriate and incapable of being given a sensible effect. That that is the position, to my mind, is a strong indication that one should approach interpreting the order form of 24 March 1972 in this respect in this way with the greatest circumspection, particularly as so construing the document brings it into conflict with the terms of the tender document, which, as I have said, contemplated that the parties would enter into a contract between contractor and sub-contractor in the green form and not in a form produced by modifying the RIBA contract.

As I have said, it is not easy to know precisely what these words were intended to mean, but I am much inclined to the view that upon their true construction what they say, and all that they say, is this: that the sub-contractual relationship between the contractor and the sub-contractor shall be such as to be consistent with all those terms in the head contract that specifically deal with matters relating to sub-contractors, so that there shall be no conflict between the head contract and the sub-contract.

It is to be observed that the words, 'The Conditions applicable to the Sub-Contract with you shall be those embodied in RIBA as above' do not say, in terms at any rate, that the sub-contract between the contractor and the sub-contractor shall be in the RIBA form, or the RIBA form suitably adapted, or anything of that kind, but that some conditions applicable to the sub-contract shall be found in the RIBA contract. In my judgment, the best way to give a sensible interpretation to those words is to read them as referring to those clauses in the head contract which relate to matters concerning sub-contractors.

Modern Building (Wales) Ltd v. Limmer & Trinidad Co. Ltd

COURT OF APPEAL

(1975) 14 BLR 101

The plaintiffs were nominated sub-contractors to the defendants, the subcontract having been placed by an order which read:

> To supply adequate labour, plant and machinery to carry out [the sub-contract works] . . . in full accordance with the appropriate form for nominated sub-contractors (RIBA 1965 edition) . . . All as your quotation.

There was no RIBA sub-contract form, nor was there any RIBA contract in a 1965 edition. The main contract was in JCT Standard Form, 1963 edition.

Held: The words used in the order should be read as describing the current 'green' form of sub-contract, which was thereby incorporated. The words in brackets in the order were inappropriate and could be disregarded.

BUCKLEY LJ: I think we have to consider whether, on the language of the order of 18 December, the arbitration clause in the 'green' form was or was not imported into the contract between the parties . . . [Counsel] has said that there are various ways in which that order can be construed which would result in its being held that the arbitration clause was not so imported. First of all, he says that the words within the brackets, '(RIBA 1965 edition) are really non-sensical, and the whole of the importation clause in the order form, if I can so describe it, is therefore of no effect at all, upon the basis, I think, that it must be read as a whole, and if it purports to import a non-existent form of contract, then it has no effect. In my judgment, the answer to that argument is this: that this is a clear case of *falsa demonstratio* The words 'the appropriate form for nominated sub-contractors', if read in the light of the evidence . . . are clearly capable of identifying the document that was intended to be referred to . . . There is language perfectly appropriate to describe the 'green' form. It is not language which ought to be read as intended to cut down and restrict the identifying operation of the preceding words. It is just an added description which on investigation turns out to be factually inaccurate . . . I think the right way to construe this order form is to ignore the words in brackets altogether . . . and to accept that the reference to the appropriate form for sub-contractors is a reference to the 'green' form, that being the only form to which it is suggested that those words could apply and the form to which it is said that anybody in the trade would understand them as applying . . .

[Counsel] has suggested that the written contract contains insufficient indication as to how various matters that are left blank in that 'green' form contract, and in particular the appendix to that contract, which relates to such matters as the completion period, the retention money and so forth, should be filled in. Where the parties by an agreement import the terms of some other document as part of their agreement those terms must be imported in their entirety, in my judgment, but subject to this that if any of the imported terms in any way conflict with the expressly agreed terms, the latter must prevail over what would otherwise be imported. Here it is not disputed that the written contract between the parties, consisting of the quotation and the order, contains all the essential terms of the contract, and, in my judgment, the 'green' form of contract must be treated as forming part of the written contract, subject to any modifications that may be necessary to make the clauses in the 'green' form accord in all respects with the express terms agreed between the parties.

Terms can be incorporated into a contract by a course of dealing between the parties as well as by express reference. Terms can also be incorporated on the basis of the common understanding of the parties.

British Crane Hire Corporation Ltd v. Ipswich Plant Hire Ltd

COURT OF APPEAL

[1974] 1 All ER 1059

The parties were engaged in the plant hire business. A drag line crane was hired by telephone. Hiring charges were agreed, but nothing was said about the other terms. After delivery, the plaintiff owners sent to the defendants the then current 'General Conditions for the Hiring of Plant', but the defendants failed to sign and return the acceptance note.

Held: As both parties were in the plant hire industry, and both used the same standard conditions when hiring plant, they must be treated as having assumed that the contract was made on those terms.

LORD DENNING MR: In support of the course of dealing, the plaintiffs relied on two previous transactions in which the defendants had hired cranes from the plaintiffs. Each was on a printed form which set out the hiring of a crane, the price, the site, and so forth; and also setting out the conditions the same as those here. There were thus only two transactions many months before and they were not known to the defendants' manager who ordered this crane. In the circumstances I doubt whether those two transactions would be sufficient to show a course of dealing.

In *Hollier* v. *Rambler Motors (AMC) Ltd* (1972), Salmon LJ said he knew of no case 'in which it has been decided or even argued that a term could be implied into an oral contract on the strength of a course of dealing (if it can be so called) which consisted at the most of three or four transactions over a period of five years'. That was a case of a private individual who had had his car repaired by the defendants and had signed forms with conditions on three or four occasions. The plaintiff there was not of equal bargaining power with the garage company which repaired the car. The conditions were not incorporated.

But here the parties were both in the trade and were of equal bargaining power. Each was a firm of plant hirers who hired out plant. The defendants themselves knew that firms in the plant-hiring trade always imposed conditions in regard to the hiring of plant; and that their conditions were on much the same lines. The defendants' manager (who knew the crane), was asked about it. He agreed that he had seen these conditions or similar ones in regard to the hiring of plant. He said that most of them were, to one extent or another, variations of a form which he called 'the Contractors' Plant Association form'. The defendants themselves (when they let out cranes) used the conditions of that form. The conditions on the plaintiffs' form were in rather different words, but nevertheless to much the same effect . . .

From [the] evidence it is clear that both parties knew quite well that conditions were habitually imposed by the supplier of these machines: and both parties knew the substance of those conditions. In particular that, if the crane sank in soft ground, it was the hirer's job to recover it; and that there was an indemnity clause. In these circumstances, I think the conditions on the form should be regarded as incorporated into the contract. I would not put it so much on the course of dealing, but rather on the common understanding which is to be derived from the conduct of the parties, namely, that the hiring was to be on the terms of the plaintiffs' usual conditions.

As Lord Reid said in *McCutcheon* v. *David MacBrayne Ltd* (1964). . . . 'The judicial task is not to discover the actual intentions of each party: it is to decide what each was reasonably entitled to conclude from the attitude of the other.' It seems to me that, in view of the relationship of the parties, when the defendants requested this crane urgently and it was supplied at once – before the usual form was received – the plaintiffs were entitled to conclude that the defendants were accepting it on the terms of the plaintiffs' own printed conditions – which would follow in a day or two. It was just as if the plaintiffs had said, 'We will supply it on our usual conditions,' and the defendants said, 'Of course, that is quite understood.'

Special problems may attend the incorporation of arbitration clauses.

Aughton Ltd v. M.F. Kent Services Ltd

COURT OF APPEAL

(1991) 31 Con LR 60

Aughton were sub-sub-contractors to Kent under six sub-sub-contracts. Kent were sub-contractors to Press Construction Ltd who were main contractors to the Ministry of Defence. The main contract was on GC/Works/1, second edition and the sub-contract between Press and Kent was based on those terms. The question was whether the sub-sub-contract incorporated an arbitration clause.

The sub-sub-contract was contained in a letter from Kent to Aughton dated 3 June 1988. This included '(11) Our previous correspondence and the documentation encompassed in our inquiry forms part of our agreement'. This previous correspondence included reference to the main contract being on GC/Works/1 Second Edition terms, to Press conditions of contract and Kent conditions of sub-contract.

Condition 61 of GC/Works/1 as incorporated in the sub-contract between Press and Kent contained an arbitration agreement in the following terms:

> 61.1 All disputes, differences or questions between the parties to the Sub-Contract with respect to any matter or thing arising out of or relating to the Sub-Contract other than a matter to which the decision or report of the Contractor or of any other person is by the Sub-Contract expressed to be final and conclusive shall after notice by either party to the Sub-Contract to the other of them be referred to a single Arbitrator agreed for that purpose, or in default of such agreement within twenty eight days of service of such notice to be appointed at the request of the Contractor by the President of such one of the undermentioned as the Contractor may decide, viz, the Law Society . . . the Royal Institute of British Architects, the Royal Institution of Chartered Surveyors . . . the Institutions of Civil Engineers, Mechanical Engineers, Heating and Ventilating Engineers, Electrical Engineers or Structural Engineers.
>
> 61.2 Unless the parties otherwise agree, such reference shall not take place until after the completion, alleged completion or abandonment of the Works or the determination of the Sub-Contract . . .
>
> 61.4 If the dispute, difference or question between the Contractor and the Sub-Contractor is substantially the same as or connected with a matter or thing which has, under the Main Contract, already been referred to an Arbitrator, the Contractor and Sub-Contractor may, with the consent of the Authority, refer that dispute, difference or question to that Arbitrator. If the Authority so consents, the relevant reference under the Sub-Contract shall take place at such time as may have been agreed between the Contractor and the Authority.

That provision could not, without at least some verbal amendment, be applied to the contract between Kent and Aughton, since it referred to contractor and sub-contractor.

A two judge Court of Appeal, though for very different reasons, held that the arbitration clause had not been incorporated in the contract.

> RALPH GIBSON LJ: For my part, however, I have no doubt that the propositions stated by Brandon J are authoritative guides to construction in any case where the court is considering whether an arbitration clause has been incorporated by reference. If the particular case is concerned with contracts and circumstances different from those normally found in charterparty/bills of lading cases, the relevant differences will or may affect the answer which the court will be constrained to give, and in particular to the question whether, in the circumstances, the 'precise words alleged to do the incorporating' can properly be given the effect contended for as disclosing the intention of the parties. Thus, in this case, in my judgment, the issue turns upon the proper construction of the words of incorporation which, put shortly, were that sub-sub-contract no. 2 would be 'a sub-sub-contract based on Press/Kent'.

The distinction between conditions of a contract which define the rights and obligations of the parties with reference to the subject matter, (e.g. the standard and extent of the work, the time for completion, and the power to vary the work by instructions, together with the terms for payment and the control of the relationship of the parties on the site), on the one hand, and on the other hand, those which control or affect the rights of the parties to enforce those rights and obligations by proceedings at law was, as stated above, emphasised in *Thomas* v. *Portsea*. That distinction is as relevant, in my judgment, in a case of this nature about engineering work as it is in a case of charterparty and bill of lading. It provides good reason for requiring that an alleged intention of the parties to exclude the ordinary right of access to the court by an arbitration agreement, which may well include special terms of limitation, be clearly demonstrated from the terms of the contract.

The general words of incorporation in this case were, as I have said, that the sub-sub-contract no. 2 was to be based upon the Press/Kent conditions. Those of the Press/Kent conditions which control the relationship of Kent as sub-contractor and of Aughton as sub-sub-contractor, so far as concern the nature and extent of the work to be done, and payment for it, together with those conditions which dealt with access to the site, security and secrecy, are, in my judgment, examples of conditions which are apt for effective incorporation by such general words on the ground that, as it was put by Oliver LJ in *Skips Nordheim, The Varenna* [1983] 3 All ER 645 at 650, [1984] QB 599 at 619, the intention to incorporate such conditions, save as excluded or varied by terms expressly agreed, can properly be held to have been thus so clearly expressed as to require or justify by necessary implication sufficient modification of those conditions to make them fit. We have not heard full submissions upon those conditions generally, and Mr Williamson made no concessions with reference to them. I think, however, that it is clear that many, if not all, of those conditions must properly be treated as incorporated as suitably modified on the facts of this case. No argument was advanced that any particular conditions, apart from the arbitration clause, were from their terms not within the intention of the parties. The clear intention was, I think, that the work should be done on the site and payments made in respect of the work in accordance with the relevant terms of Press/Kent subject to the terms expressly agreed, and that the parties must be taken to have known and intended that, if necessary, for the application of a particular condition, the words would be modified to make them fit by substituting 'sub-sub-contractor' for 'sub-contractor' etc.

> A contract is complete as a contract as soon as the parties have reached agreement as to what each of its essential terms is or can with certainty be ascertained: for it is an elementary principle of the English law of contract *id certum est quod certum reddi potest*.
>
> in *Sudbrook Trading Estate Ltd* v. *Eggleton* [1982] 3 All ER 1 at 6, [1983] 1 AC 444 at 478, *per* Lord Diplock.

I come, then, to the arbitration clause itself in condition 61. Unless it is modified by changing the words 'contractor', 'sub-contractor', 'authority', as considered by the judge and found by him to be practicable, it plainly does not apply to the present disputes. There was no express reference to the arbitration clause in sub-sub-contract no. 2. The judge found those circumstances to be conclusive against the defendant. The question is whether, if I am right in holding that there is no special rule of construction which prevents effective incorporation of the arbitration clause by these general words, the court should, on ordinary rules of construction, hold those words to be effective for that purpose.

SIR JOHN MEGAW: There are, in my opinion, three important inter-related factors peculiar to arbitration agreements. First, an arbitration agreement may preclude the parties to it from bringing a dispute before a court of law. That, of course, is something which is not only permissible, but may also be a very desirable way of settling disputes. But, as was said by Lord Gorell in *Thomas* v. *Portsea* [1912] AC 1 at 9:

> Now I think, broadly speaking, that very clear language should be introduced into any contract which is to have that effect.

Secondly, it has been laid down by statute (section 32 of the Arbitration Act 1950 as re-enacted in section 7(1)(e) of the Arbitration Act 1979) that an arbitration agreement has to be 'a written agreement'. (There was a corresponding requirement of 'a written agreement', in the Arbitration Act 1889, in force when *Thomas* v. *Portsea* was decided.) The object, or the effect, of that statutory requirement must be to emphasise, and seek to ensure, that one is not to be deprived of his right to have a dispute decided by a court of law, unless he has consciously and deliberately agreed that it should be so. Thus, if one party to a dispute under a contract wishes to litigate it in a court of law, it would be no answer for the other party to prove, by the clearest evidence, or even an admission, that there had been an oral agreement to arbitrate, which had not been reduced into writing or specifically evidenced by writing. If, as is the position by statute, an oral agreement will not suffice, it must surely follow that an agreement depending, in any essential part, in inference will not suffice.

Thirdly, the status of a so-called 'arbitration clause' included in a contract of any nature is different from other types of clause because it constitutes a 'self-contained contract collateral or ancillary to' the substantive contract. These are the words of Lord Diplock in *Bremer Vulkan* v. *South India Shipping* [1981] 1 All ER 289, [1981] AC 909. It is a self-contained contract, even though it is, by common usage, described as an 'arbitration clause'. It can, for example, have a different proper law from the proper law of the contract to which it is collateral. This status of 'self-contained contract' exists irrespective of the type of substantive contract to which it is collateral. In *Bremer Vulkan* it was a shipbuilding contract. It appears to me that this consideration (which I believe has not infrequently been overlooked) is another important reason why arbitration clauses are to be treated as being in a category of their own, as was the arbitration clause in the charterparty, which the House of Lords declined to permit to be incorporated into the bill of lading contract in *Thomas* v. *Portsea*. If this self-contained contract is to be incorporated, it must be expressly referred to in the document which is relied on as the incorporating writing. It is not incorporated by a mere reference to the terms and conditions of the contract to which the arbitration clause constitutes a collateral contract.

Each of the three factors which I have mentioned above applies equally to an incorporation from an engineering sub-contract into a sub-sub-contract as it applies to the incorporation from a charterparty into a bill of lading. The distinction which *Thomas* v. *Portsea* drew between the arbitration clause and other clauses of the charterparty applies equally to the arbitration clause of the engineering sub-contract in the present case. That, in my opinion, is decisive of the appeal.

I should add that I agree with Ralph Gibson LJ that the appellants have failed to show that the statutory requirement for a written agreement has been complied with. I should, however, add that I find it difficult to see how the arbitration clause in the agreement between Press and Kent can be a relevant arbitration agreement in writing between the appellants and the respondents.

The parties may attempt to contract by exchanging letters, estimates, orders, and acknowledgments, each of which purports to be on the usual terms of the party producing it. This gives rise to the 'battle of the forms'. Different views have been expressed as to who wins this 'battle'. Probably the correct approach is to see whether one party has by words or conduct accepted the other's terms. In the absence of such acceptance there is probably no contract.

Butler Machine Tool Co Ltd v. Ex-Cell-O Corporation (England) Ltd

COURT OF APPEAL

[1979] 1 All ER 965

On 23 May 1969 the sellers offered to sell machine tools to the buyers for £75,535, delivery ten months forward. The offer took the form of a written quotation subject to terms which 'shall prevail over any terms and conditions in the buyer's order'. These terms included a price variation clause. The buyers replied on 27 May 1969 placing an order for the machinery, subject to their own terms, which did not include a price variation clause. There was a tear-off acknowledgment of receipt of order attached. This said: 'We accept your order on the terms and conditions stated thereon'. The sellers completed and signed the acknowledgment and returned it to the buyers with a letter saying that the order was 'being entered in accordance with our revised quotation of 23 May for delivery in 10/11 months'. After delivery, the sellers claimed increased costs, which the buyers refused to pay.

Held: The buyers' order of 27 May was a counter-offer which destroyed the offer made in the seller's quotation of 23 May. By completing and returning the acknowledgment slip, which was said to be on the buyers' terms and conditions, the sellers had accepted the counter-offer on those terms, which did not include a price variation clause. The sellers' accompanying letter referring to the quotation of 23 May was irrelevant: it merely referred to the price and identity of the machine-tools.

LORD DENNING MR: In most cases where there is a 'battle of forms' there is a contract as soon as the last of the forms is sent and received without objection being taken to it . . . The difficulty is to decide which form, or which part of which form, is a term or condition of the contract. In some cases the battle is won by the man who fires the last shot He is the man who puts forward the latest terms and conditions: and, if they are not objected to by the other party, he may be taken to have agreed to them . . . In some cases, however, the battle is won by the man who gets the blow in first. If he offers to sell at a named price on the terms and conditions stated on the back and the buyer orders the goods purporting to accept the offer on an order form with his own different terms and conditions on the back, then, if the difference is so material that it would affect the price, the buyer ought not to be allowed to take advantage of the difference unless he draws it specifically to the attention of the seller. There are yet other cases where the battle depends on the shots fired on both sides. There is a concluded contract but the forms vary. The terms and conditions of both parties are to be construed together. If they can be reconciled so as to give a harmonious result, all well and good. If differences are irreconcilable, so that they are mutually contradictory, then the conflicting terms may have to be scrapped and replaced by reasonable implication.

LAWTON LJ: The modern commercial practice of making quotations and placing orders with conditions attached, usually in small print, is indeed likely, as in this case, to produce a battle of forms. The problem is how should that battle be conducted? The view taken by the judge was that the battle should extend over a wide area and the court should do its best to look into the minds of the parties and make certain assumptions. In my judgment, the battle has to be conducted in accordance with set rules. It is a battle more on classical 18th century lines when convention decided who had the right to open fire first rather than in accordance with the modern concept of attrition.

The rules relating to a battle of this kind have been known for the past 130-odd years. They were set out by the then Master of the Rolls, Lord Langdale, in *Hyde* v. *Wrench* (1840), and Lord Denning MR has already referred to them; and, if anyone should have thought they were obsolescent, Megaw J in *Trollope & Colls Ltd* v. *Atomic Power Construction Ltd* (1962) called attention to the fact that those rules are still in force.

When those rules are applied to this case, in my judgment, the answer is obvious. The sellers started by making an offer. That was in their quotation. The small print was headed by the following words: 'General. All orders are accepted only upon and subject to the terms set out in our quotation and the following conditions. These terms and conditions shall prevail over any terms and conditions in the buyer's order.' That offer was not accepted. The buyers were only prepared to have one of these very expensive machines on their own terms. Their terms had very material differences in them from the terms put forward by the sellers. They could not be reconciled in any way. In the language of article 7 of the Uniform Law on the Formation of Contracts for the International Sale of Goods they did materially alter the terms set out in the offer made by the sellers.

As I understand *Hyde* v. *Wrench* (1840) and the cases which have followed, the consequence of placing the order in that way, if I may adopt Megaw J's words, was 'to kill the quotation'. It follows that the court has to look at what happened after the buyers made their counter-offer. By letter dated 4 June 1969 the sellers acknowledged receipt of the counter-offer, and they went on in this way: 'Details of this order have been passed to our Halifax works for attention and a formal acknowledgment of order will follow in due course'. That is clearly a reference to the printed tear-off slip which was at the bottom of the buyers' counter-offer. By letter dated 5 June 1969 the sales office manager at the sellers' Halifax factory completed that tear-off slip and sent it back to the buyers.

It is true, as counsel for the sellers has reminded us, that the return of that printed slip was accompanied by a letter which had this sentence in it: 'This is being entered in accordance with our revised quotation of 23 May for delivery in 10/11 months'. I agree with Lord Denning MR that, in a business sense, that refers to the quotation as to the price and the identity of the machine, and it does not bring into the contract the small print conditions on the back of the quotation. Those small print conditions had disappeared from the story. That was when the contract was made. At that date it was a fixed price contract without a price escalation clause.

As I pointed out in the course of argument to counsel for the sellers, if the letter of 5 June which accompanied the form acknowledging the terms which the buyers had specified had amounted to a counter-offer, then in my judgment the parties never were ad idem. It cannot be said that the buyers accepted the counter-offer by reason of the fact that ultimately they took physical delivery of the machine. By the time they took physical delivery of the machine, they had made it clear by correspondence that they were not accepting that there was any price escalation clause in any contract which they had made with the plaintiffs.

Chichester Joinery Ltd v. John Mowlem & Co plc

QUEEN'S BENCH DIVISION

(1987) 42 BLR 100

Chichester submitted a quotation to Mowlem in November 1984, accompanied by their standard terms and conditions. In January 1985 Mowlem sent Chichester a pro-forma enquiry form. Although the face of this document referred to conditions on its reverse there were no conditions printed on the original. The parties discussed the proposed sub-contract at two subsequent meetings. On 14 March 1985 Mowlem sent Chichester their purchase order. This said:

> THE TERMS AND CONDITIONS OF PURCHASE NOS 1–18 INCLUSIVE AS SET OUT ON THE REVERSE HEREOF ARE EXPRESSLY DECLARED TO APPLY TO THE PURCHASE ORDER.
>
> You are requested to sign the acceptance of order and return the same within seven days. Any delivery made and accepted will constitute an acceptance of this Order.

Chichester did not sign and return the acceptance of order. On 30 April 1985, before any delivery had been made, they sent to Mowlem a printed form headed acknowledgement of order. This purported to accept Mowlem's order 'subject to the conditions overleaf'. The question arose as to whether the sub-contract was governed by Mowlem's terms and conditions as set out in their purchase order of 14 March 1985 or by Chichester's terms and conditions as set out in the acknowledgment of order dated 30 April 1985.

Held: The sub-contract was governed by Chichester's terms and conditions. Mowlem's purchase order of 14 March 1985 had been a counter-offer to Chichester's quotation and had killed that quotation. However, Mowlem's counter-offer had in turn been killed by Chichester's acknowledgment of 30 April 1985 which had been a counter-offer accepted by Mowlem by conduct, namely by accepting the joinery later delivered to site by Chichester.

JUDGE JAMES FOX-ANDREWS, QC: On 14 March 1985 Mowlem sent two documents in identical terms, save that at the bottom of the fronting page of the first appeared the word 'Original', and on the second the words 'Acceptance (Return to John Mowlem)'. The forms bore the words:

> Please supply the Materials/Equipments/Goods set out below. This order is placed subject to the Terms and Conditions set out on the reverse side of this form, and to any Terms and Conditions set out below. Your quotation if any is taken only as a basis for pricing.

The works were described in relation to Mowlem's enquiry of 28 January; the rates in Chichester's quotation of 31 August 1984; Chichester's letter of 14 February 1985 and relevant architect's detail. There were six notes, to two of which I shall refer shortly.

The common wording at the bottom of each of the two fronting pages of the forms were:

> THE TERMS AND CONDITIONS OF PURCHASE NOS 1–18 INCLUSIVE AS SET OUT ON THE REVERSE HEREOF ARE EXPRESSLY DECLARED TO APPLY TO THE PURCHASE ORDER.
>
> You are requested to sign the acceptance of order and return same within seven days. Any delivery made and accepted will constitute an acceptance of this order.

The acceptance form provided a space for a signature and date.

It is unnecessary to consider in detail Mowlem's terms and conditions which in this case did appear on their purchase order. I find that there are significant material differences between Chichester and Mowlem's sets of conditions. They materially altered the terms. Note 2 of the purchase order provided that February 1985 would be the base date for the NEDO formula rules. But there was added that there was a 10% non-adjustable element. It was accepted that there had been no previous reference to this. It was, however, argued that this was an insignificant term. I do not accept that. It is a term which can be of considerable significance to a supplier.

Note 3 did not set out accurately what had been 'agreed'. Different dates from those in the programme had been 'agreed' orally. But this I regard as a minor matter which could have been easily rectified.

This clearly was not an acceptance of the offer made by Chichester on 8 February and confirmed by them by their letter of 14 February. It killed Chichester's earlier offer. It was, I find, a counter-offer.

It was argued that the conduct of Chichester thereafter in preparing for the manufacture of the joinery constituted an acceptance. If this submission had been open to Mowlem on the pleadings, I would have had no hesitation in rejecting the argument on two quite separate grounds. First, the terms of their Purchase Order only contemplated two forms of acceptance: (1) written acceptance; (2) delivery and acceptance of the goods. Further, the evidence falls far short of that necessary to establish acceptance by conduct. Chichester neither orally nor in writing either within seven days or at all accepted Mowlem's order prior to 30 April 1985.

On 30 April 1985 – a substantial time before any deliveries were made – Chichester sent a printed form headed 'ACKNOWLEDGEMENT OF ORDER'. It read:

> Dear Sirs, We acknowledge with thanks receipt of your valued order of – and there were then inserted the words 'recent date' – 'which will receive our best attention. Your order is, unless expressly agreed otherwise in writing, accepted subject to the conditions overleaf.
>
> Please check the following specification and inform us within two days in the event of any errors or omissions. If we are not advised of any errors we will assume all details are correctly shown and manufacture will proceed on that basis.

There then appeared in capitals the words:

> a copy of this acknowledgement is also a specification to our works so please read this carefully.

On the back were set out the identical conditions to those on their original 1984 quotation. There was typed at the foot of the acknowledgement:

> re: Royal Holloway & Bedford College – Egham Joinery Phase 1 to Maths Arts Building. Curtain Walling panels all as shown on elevation drawings . . .

Although it is headed 'Acknowledgement' and not 'Acceptance', some of the sentences I have read are much more consistent with acceptance than mere acknowledgement. It fails to indicate the contrary to Chichester's condition 2 that discount was 5%. It does not deal at all with the non-adjustable 10% element unless condition 1(ii) is relevant.

But I am unable to spell out of the relevant facts any intention by Chichester to be bound by Mowlem's conditions. The facts are significantly different from those in *Butler Machine Tool Co Ltd* v. *Ex-Cell-O Corporation (England) Ltd* (1979). A necessary conclusion therefore is that this acknowledgment constituted a counter-offer thus killing Mowlem's 14 March 1985 offer.

The question for determination is whether the subsequent acceptance by Mowlem of joinery when it came to be delivered on the site constituted an acceptance of Chichester's counter-offer of 30 April and therefore subject to their conditions, or whether the parties were never *ad idem*.

These cases are never easy. After some hesitation I have reached the conclusion that Mowlem did accept Chichester's terms. Mowlem at an earlier stage were specifically offering to treat a contract as complete if, on joinery being delivered to the site, it was accepted by them. In all those circumstances, I find that this preliminary issue is decided in favour of Chichester.

Sauter Automation Ltd v. Goodman (Mechanical Services) Ltd

CHANCERY DIVISION

(1986) 34 BLR 81

The plaintiffs were sub-contractors to the defendants for the supply, delivery and commissioning of equipment required as part of a boiler installation which the defendants were providing. The defendants' contract incorporated the Government standard conditions GC/Works/1, Edition 2. The plaintiffs had sent a quotation to the amendments making their offer subject to their standard conditions, clause 7 of which read:

> Property at Risk
>
> The Purchaser shall be liable for all risks in the goods from the time of delivery to the Purchaser but the title in all goods in each order/contract shall remain with the Company until payment for that order/contract be made in full. Where the Purchaser has sold the goods to a third party before payment in full has been received by the Company the Purchaser shall hold all sums received for such goods as trustees for the Company.

The defendants sent the plaintiffs an order which provided inter alia, 'terms and conditions in accordance with Main Contract . . .'.

The plaintiffs supplied most of the equipment to the defendants before the defendants went into liquidation. The employer determined the main contract.

The plaintiffs sought a declaration that certain equipment delivered to the defendants was their property and orders for delivery up of that equipment together with payment of the proceeds of the sale of the equipment to the employer as well as the price of the equipment and damages.

Held: The defendants' order was a counter-offer which killed the plaintiffs' offer and which had been accepted by the plaintiffs in proceeding to carry out the works. Accordingly there was no incorporation in the contract of clause 7 or any other provision of the plaintiffs' quotation.

MERVYN DAVIES J: The plaintiffs submitted that the Goodman order dated 20 June 1985 was not such a counter-offer as killed the original offer made by Sauter on 10 April 1985. He referred to *Butler Machine Tool* v. *Ex-Cell-O Corporation* (1979), suggesting that on an examination of the 'offer' and 'counter-offer' there was no great discrepancy between them; the principal differences being that the offer to install (rather than to supply and to install) was refused and that the price was put at £32,715 instead of £33,555. He pointed, too, to the fact that the specification for the main contract with PSA referred expressly to another sub-contract and the terms which were to apply to such other sub-contract, whereas there was in the specification no such reference to the Sauter sub-contract. He relied as well on some unchallenged evidence which reads:

> The terms and conditions of the Main Works Contract are wholly inappropriate in the context of the Supply Contract with which the plaintiffs were concerned.

I see the force of [the plaintiffs'] submissions. Nevertheless, I see that the approach to the problem before me has been set out by Lawton LJ in the Butler case. Adopting that approach, I see that the Sauter offer dated 10 April 1985 is met on 20 June 1985 with an 'order' from Goodmans which desires Sauter to supply (but not to install) certain equipment at a price 2½% below the Sauter price. Those points are, in my view, quite neutral. But then there are the words 'Terms and Conditions in accordance with Main Contract GC/Works/1 Edition 2 plus Amendment No 4 (see Form of Contract attached)'. The 'Form of Contract' refers to the general conditions of the main contract and to certain other matters. To my mind the words quoted plainly intimated to the plaintiffs that the 'order' was an offer to contract on the terms and conditions mentioned, i.e. not on the Sauter terms and conditions. What was offered was a sub-contract with a relationship between Goodmans and Sauters such as to be consistent with the terms of the main contract; so that there should be no conflict between the main contract and the sub-contract: see *Brightside* v. *Mitchell Construction* (1975). I conclude that Goodmans were making a counter-offer which killed the Sauter offer. Sauters implicitly accepted the counter-offer by proceeding to execute the Goodmans' order. Accordingly, there was no incorporation in the contracts of clause 7 or any other Sauter condition.

Both counsel made submissions on the footing that a possible construction of the documents was that while GC/Works/1 ('the GC conditions') was incorporated into the Sauter/Goodman contract, there was also incorporated the Sauter conditions insofar as the Sauter conditions were not inconsistent with the GC conditions. In that event [counsel] for the plaintiffs conceded that the GC conditions would prevail. I proceed to consider the matter on that footing, i.e. that both sets of conditions are incorporated with the GC conditions prevailing in case of inconsistency. It seems to me that on this footing condition 7 cannot have any operation. I have already set out condition 30(2)(a) of the GC conditions. Sauter knew of this condition, or had an opportunity of knowing the condition, at least from the time when they received the Goodman order dated 20 June 1985; that order expressly referring to the GC conditions. Thus Sauter must be taken to have known that Goodmans were under an obligation to the PSA to procure a vesting in themselves of such materials as they (Sauter), the sub-contractor, delivered to the site. True it is that Goodmans did not expressly exact from

Sauter such a provision as is mentioned in 30(2)(a). Nevertheless Sauter must be taken to have known that the PSA and Goodmans desired sub-contractor's materials to vest in Goodman once on site. In that situation I do not see how clause 7 of the Sauter conditions can have any operation, given that the GC conditions prevail over the Sauter conditions. There is then the further consideration that condition 3(1) of the GC conditions operates to nullify the Sauter condition 7.

General condition 3(1) reads as follows:

> 3(1). From the commencement to the completion of the Works, the Works and any things (whether or not for incorporation) brought on the Site in connection with the Contract and which are owned by the Contractor or vest in him under any contract shall become the property of and vest in the Authority subject to his right of rejection of any things for incorporation which are not approved . . .

If condition 3 is read, as it may be, as a term of the Sauter/Goodman contract, one takes the reference to 'Contractor' as a reference to Sauter and a reference to the 'Authority' as a reference to Goodman. Thus mutatis mutandis general condition 3(1) is a term subsisting between Sauter and Goodman. That was, as I understand, the course followed by Buckley LJ in the Brightside case.

Where the express terms of the contract are clear and unambiguous the courts will not imply a term simply to extricate a party from difficulties.

Trollope & Colls Ltd v. North-West Metropolitan Regional Hospital Board

HOUSE OF LORDS

(1973) 9 BLR 60

Building works were to be carried out in three phases, for each of which there was a separate contract sum and set of conditions, all based on JCT terms. The date of commencement of Phase III was to be fixed by reference to the completion date of Phase I. Phase III was to be completed by a specified date. Phase I was delayed by 59 weeks, 57 of which were allowed under the extension of time clause, and in effect the period for completing Phase III was reduced from 30 to 16 months. The contractors contended that a term for extending the date for completion of Phase III should be implied.

Held: No such term could be implied. The contract was clear and free from ambiguity in giving the date for completion of Phase III.

LORD PEARSON: The basic principle [is] that the court does not make a contract for the parties. The court will not even improve the contract which the parties have made for themselves, however desirable the improvements might be. The court's function is to interpret and apply the contract which the parties have made for themselves. If the express terms are perfectly clear and free from ambiguity, there is no choice to be made between different possible meanings: the clear terms must be applied even if the court thinks some other terms would have been more suitable. An unexpressed term can be implied if and only if the court finds that the parties must have intended that term to form part of their contract: it is not enough for the court to find that such a term would have been adopted by the parties as reasonable men if it had been suggested to them: it must have been a term that went without saying, a term necessary to give business efficacy to the contract, a term which, though tacit, formed part of the contract which the parties made for themselves.

The relevant express term is entirely clear and free from ambiguity: the date for completion of Phase III is the date stated in the appendix to [the] conditions . . . That term in itself can have only one meaning.

There is no general rule of law implying into a building sub-contract terms that the main contractor will make sufficient work available to the sub-contractor to enable the subcontractor to maintain reasonable progress and execute the sub-contract works in an efficient and economic manner or that the main contractor will not hinder or prevent the sub-contractor in the execution of the sub-contract works.

Martin Grant & Co Ltd v. Sir Lindsay Parkinson & Co Ltd

COURT OF APPEAL

(1985) 3 Con LR 12

Grant were sub-contractors to the respondents (Parkinson) for formwork on a number of local authority housing projects. The main contracts were in JCT 63 form. The sub-contracts were in a non-standard form, and contained no provision for the risk of delay. There were substantial delays in the performance of the main contracts, and as a result the appellants incurred substantial losses since they had to carry out the work several years later than contemplated. Grant issued a writ against Parkinson claiming damages for breach of contract and/or alternatively for various sums due under the sub-contracts. Grant contended, *inter alia*, that the following term was to be implied in the sub-contracts:

> That (a) [the main contractor] would make sufficient work available to the [sub-contractors] to enable them to maintain reasonable progress and to execute their work in an efficient and economic manner; and (b) [the main contractor] should not hinder or prevent the [sub-contractors] in the execution of the sub-contract works.

The trial judge decided that the express terms of the sub-contracts left no room for the kind of implied term for which Grant contended. Grant appealed.

Held: The appeal would be dismissed. The express terms of the sub-contract left no room for the term contended for to be implied. There is no general rule of law implying such a term into a building sub-contract by reason of the relationship of the parties or otherwise.

LAWTON LJ: It is necessary now to go to the terms of the sub-contract. It provided by way of recital that the respondents had an agreement with the London Borough of Islington, that the sub-contractor had submitted a tender for that portion of the contract which related to form-work, and it went on as follows:

> . . . which tender the Contractor is desirous of accepting and whereas the Sub-Contractor has read the Principal Contract and the Specification and Bills of Quantities appertaining thereto and forming part thereof and the terms and conditions thereof are hereby embodied in this Agreement.

It follows that the appellants must be taken to have known what was in the principal contract and, in particular, they must be taken to have known of the risks against which the principal contract made provision. In any building contract work there is always the possibility of risk and, over the years, those who have been concerned with negotiating the kind of contracts with which this case is concerned have learned to identify the risks and make provision for them.

It is necessary now to consider what risks were in the contemplation of the London Borough of Islington and the respondents and, by implication, the risks of which the appellants were aware. Those risks were set out in clause 23 of the principal contract. That clause provided for delays which would not frustrate the contract and, in the event of these various delays occurring, provided the architect was satisfied, certificates could be given for extending the contract. Those risks were divided into those in respect of which there should be no compensation payable to the respondents and those in respect of which there would be compensation payable to them. For example, there is always in every building scheme the risk of delays by *force majeure*, delays by reason of exceptionally inclement weather, delays by the kind of matters against which one insures, floods, fires and the like, delays by reason of the nominated sub-contractors not performing their work in due time. In respect of those delays the London Borough of Islington were not bound to pay any compensation at all. The appellants must be taken to have appreciated that there might be delays certified by the architect as being reasonable which would prolong the operation of the main contract and, by implication, prolong the operation of the sub-contract. There were two risks in respect of which compensation was payable to the respondents. One was by reason of delays on the part of the architect not giving instructions and, secondly, in not making necessary drawings and plans available.

So it is against that background that it is necessary to turn now to the specific clauses in the sub-contract. Clause 1 provided that the respondents should accept the tender of the sub-contractor and then come clauses 2 and 3 to which I must refer in full:

> 2. The Sub-Contractor will provide all materials labour plant scaffolding in addition to that provided by the Contractor for his own requirements haulage and temporary works and do and perform all the obligations and agreements imposed upon or undertaken by the contractor under the Principal Contract in connection with the said works to the satisfaction of the Contractor and of the Architect or Engineer under the Principal Contract (hereinafter called "the Architect")

and then come these important words:

> at such time or times and in such manner as the Contractor shall direct or require and observe and perform the terms and conditions of the Principal Contract so far as the same

> are applicable to the subject matter of this Contract as fully as if the same had been herein set forth at length and as if he were the Contractor under the Principal Contract.
>
> 3. The Sub-Contractor shall proceed with the said works expeditiously and punctually to the requirements of the Contractor and so as not to hinder hamper or delay the work or the portions of the work at such time or times as the Contractor shall require having regard to the requirements of the Contractor in reference to the progress or conditions of the Main Works and shall complete the whole of the said works to the satisfaction of the Contractor and of the Architect and in accordance with the requirements of the local and other authorities.

In my judgment, those sub-clauses mean, and would have been understood by the appellants to mean, that if the main contract was extended, then the sub-contractors' contract would be extended and that, during the time when the contract was in existence, the appellants would do such portions of the work and at such times as might be required by the respondents. In other words, there was a clear risk for the appellants that the principal contract and the sub-contract might go on much longer than was originally contemplated.

The contemplation seems to have been that the sub-contract work would be completed in about 103 weeks more or less. But the appellants must be taken to have envisaged that it might not be completed within 103 weeks more or less, and they took that risk. Because they knew what the risks were as set out in the principal contract, they undertook those risks and they knew that the respondents would not be paid compensation in respect of most of the causes of delay. They themselves made no provision in their sub-contract for compensation to be paid to them in respect of any of the risks covered by clause 23. They may have been unwise in not protecting themselves against the kind of risks which were envisaged in clause 23 but, as Lord Pearson pointed out in the well-known case of *Trollope & Colls Ltd* v. *North West Metropolitan Regional Hospital Board* (1973):

> Faced with the conflict of judicial opinion in this case, I prefer the views of Donaldson J and Cairns LJ as being more orthodox and in conformity with the basic principle that the court does not make a contract for the parties. The court will not even improve the contract which the parties have made for themselves, however desirable the improvement might be. The court's function is to interpret and apply the contract which the parties have made for themselves. If the express terms are perfectly clear and free from ambiguity, there is no choice to be made between different possible meanings: the clear terms must be applied even if the court thinks some other terms would have been more suitable.

Applying that principle to this case, it would have been more suitable from the appellants' point of view if some provision had been made for the risk of delay. There was no such provision. There were considerable delays in the performance of the London Borough of Islington contract. It had been contemplated by the main contract that the building of these 375 houses would be completed within three years. It took much longer than that to complete the work and the main contract subsisted because of various certificates given by the architect. The sub-contract went on too for no less than five years. As a result, it proved to be most uneconomical for the appellants. They say that they have lost a great deal of money. The term of the sub-contract which provided for fluctuations in the price of labour and materials, they claim, does not sufficiently indemnify them against the losses they have sustained as a result of the work taking much longer than they had contemplated . . .

. . . [The] appellants' advisers had decided that the basis of the Marquess Road claim, and probably of the other claims too, was the existence of an implied term in each of the sub-contracts.

> It was an implied term of the said sub-contract that (a) the Defendants would make sufficient work available to the Plaintiffs to enable them to maintain reasonable progress and to execute their work in an efficient and economic manner; and (b) the Defendants should not hinder or prevent the Plaintiffs in the execution of the sub-contract works.

[The appellants' counsel] appreciated that the implied term could not arise on what has been called the *Moorcock* principle, namely it was not necessary for the efficacy of the performance of the contract; and he also seems to have accepted, as far as I have followed his argument, that it could not be said to be implied on the basis that it was the obvious but unexpressed intention of the parties. The way he put his case was that that kind of term was to be implied by reason of the relationship of the parties, namely a contractor making a contract with a sub-contractor. He relied upon the general approach to such implied terms which was discussed in *Lister* v. *Romford Ice and Cold Storage Co Ltd* (1957). The general law is stated concisely and clearly in the speech of Lord Tucker:

> Some contractual terms may be implied by general rules of law. These general rules, some of which are not statutory, for example, Sale of Goods Act, Bills of Exchange Act, etc., derive in the main from the common law by which they have become attached in the course of time to certain classes of contractual relationships, for example, landlord and tenant, innkeeper and guest, contracts of guarantee and contracts of personal service. Contrasted with such cases as these there are those in which from their particular circumstances it is necessary to imply a term to give efficacy to the contract and make it a workable agreement in such manner as the parties would clearly have done if they had applied their minds to the contingency which has arisen.

The problem, therefore, is this. Is there a general rule of law implying a term such as was pleaded? The first observation to make is that building disputes have been regularly tried in this court all my professional lifetime and long before. If there were a general rule . . . it is surprising that it has never been recognised before. [There] is no trace in the reported English cases of any such general rule, nor is there in the standard text books. That, of course, does not conclude the matter by any means because it was accepted [by the respondents] that the relationship between contractor and sub-contractor by implication of law does give rise to obligations on each side. There is, for example, by implication of law, an obligation to cooperate with one another. But the degree of co-operation will depend, in my judgment, upon the express terms of any contract which the parties may have made. It was at this point that there was a marked divergence between [the arguments]. [The appellants'] submission was that, as the law imposes an obligation on contractors and sub-contractors to co-operate with one another in the performance of the contract, any express terms in it must be so construed so as to ensure that co-operation will not be vitiated to such an extent as to destroy its economic purpose. [The respondents] approached the matter the other way round, and had the support of a passage in the speech of Lord Simon in the well-known case of *Luxor (Eastbourne) Ltd* v. *Cooper* (1941). The passage is to this effect:

> There is, I think, considerable difficulty, and no little danger, in trying to formulate general propositions on such a subject, for contracts with commission agents do not follow a single pattern and the primary necessity in each instance is to ascertain with precision what are the express terms of the particular contract under discussion, and then to consider whether these express terms necessitate the addition, by implication, of other terms.

> [The respondents'] argument was that, by the express terms, the appellants took the risk of the contract extending longer than what they had in mind, namely 103 weeks, more or less. They took the risk that, if it were extended, they would have to do such work and such portions of it at such times as the respondents directed. In those circumstances, there is no room at all for the implied term pleaded in paragraph 5 of the re-re-re-amended statement of claim. The point became important in the following circumstances. After the trial had gone on for about six days, it seemed to the learned judge that everything in the London Borough of Islington case was going to turn upon the existence of this term. The reason for that was that it had not been suggested in the voluminous particulars delivered under the statement of claim that the respondents had taken any positive action which could be said to have hindered the appellants in the performance of the sub-contract. The whole case against them was that there had been delays in the performance of the main contract. If that were so, then there was no basis for a claim against the respondents in relation to the London Borough of Islington contract unless it could be shown that there was some kind of implied term as set out in the re-re-re-amended statement of claim. After discussion between the parties it was agreed that the learned judge's decision on this matter of the implied term decided the case in relation to the particular contract . . .
>
> The learned judge gave a judgment on this preliminary issue. He decided that the sub-contract by its express terms left no room for the kind of implied term for which the appellants were contending. It is in those circumstances that the preliminary issue comes before this court. I am of the same opinion as [the trial judge]. It seems to me, having regard to the express terms and for the reasons I have indicated, that there is no room for this implied term. What I have had to consider is whether the term pleaded arises as a matter of law because of the relationship between contractor and sub-contractor in the circumstances of this case. I am firmly of the opinion that it does not.

The above case can usefully be contrasted with the next case.

J. & J. Fee Ltd v. The Express Lift Co Ltd

QUEEN'S BENCH DIVISION

(1993) 34 Con LR 147

In this case, Judge Peter Bowsher QC implied into the contract a term that:

> The plaintiff by itself its servants agents or sub-contractors would provide the defendant with correct information concerning the works in such a manner and at such times as was reasonably necessary for the defendant to have in order for it to fulfil its obligations under the sub-contract.

> JUDGE PETER BOWSHER QC: For the plaintiffs it is submitted that that clause requires that the defendants should specifically apply in writing for information and no term should be implied which requires that they should be entitled to have information without applying in writing. But clause 11.10.6 only refers to an event happening between the employer and the companies employed (the contractor and the sub-contractor) and it does not impinge on the way in which the contractor and the sub-contractor necessarily have to co-operate if the work is to be done.

There is nothing in this contract, DOM/2, which puts on the contractor an express duty to provide necessary information. For that reason it is agreed by the parties that a term should be implied into the contract that information should be supplied. The contract cannot be made to work unless the information is supplied: it is essential. Moreover, there is no point in supplying that information unless it is supplied in such a manner and at such times as is reasonably necessary for the defendants to have it in order for the defendants to fulfil their obligations under the sub-contract. It is necessary and reasonable that the whole of the implied term for which the defendants contend should be implied into the contract. It is not necessary or reasonable that it should be implied that the defendants should only be given essential information if they ask for it in writing. I see no inconsistency between such an implied term and the express terms of DOM/2.

The standard forms contain no technical information about the work to be done. The contract will consist, therefore, of a combination of the contract conditions and other documents such as drawings, specifications and bills of quantities which describe the works. Sometimes the various contractual documents contain provisions which appear inconsistent with each other. Normally the courts in such a situation would attach more weight to documents such as bills which have been produced for the particular contract. The effect of JCT 98, clause 2.2.1 [JCT 63, clause 12(1)1], which provides that 'nothing contained in the contract bills shall override or modify the application or interpretation of that which is contained in the Articles of Agreement, the Conditions or the Appendix' is that the printed provisions prevail over any special clauses in the bills. This reverses the ordinary rule of interpretation that written words prevail over printed words in standard form contracts. Accordingly, the bills cannot modify the interpretation of the contract conditions. See now JCT 2005, clause 1.3.

English Industrial Estates Corporation v. George Wimpey & Co Ltd

COURT OF APPEAL

(1972) 7 BLR 122

For the facts and decision, see p. 97.

STEPHENSON LJ: To apply the general principle that type should prevail over print seems to me to contradict the express provision of clause 12 that the reverse is to be true of this particular contract: the special conditions in type are to give way to the general conditions in print. The words in clause 12 'or affect in any way whatsoever the application or interpretation of that which is contained in these conditions' seem to mean that in this contract the bills of quantities are 'Contract Bills' in so far as they deal with 'the quality and quantity of the work included in the contract sum', but that in so far as they state conditions of the contract they have no effect on the printed conditions. In so far as they repeat or copy printed conditions or amendments which have been added to the printed conditions . . . they add nothing to those conditions. In so far as they introduce further contractual obligations, as they do . . . they may add obligations which are consistent with the obligations imposed by the conditions, but they do not affect them by overriding or modifying them or in any other way whatsoever.

It follows from a literal interpretation of clause 12 that the court must disregard – or even reverse – the ordinary and sensible rules of construction and that the first of the documents comprising the works Wimpeys offered in their tender . . . to carry out expressly prevents the court from looking at the second of those documents to see what the first of them means. But that is because the second document is . . . a hybrid document and part of it deals with matters which should have been incorporated in the first; and that part can only live with clause 16 if it has . . . a different subject matter.

Those building owners who put forward the RIBA contract ought, in my judgment, to see either that clause 12 is struck out or amended or that all contractual terms which should be read together with the printed conditions are not left in the bills of quantities only but are incorporated in the conditions so that they cannot be shut out by clause 12 from being so read. Otherwise building owners have only themselves to blame if they find that they are saddled with such a liability as they have successfully contested in this case.

Chapter 5
Completion of the Contract

An unambiguous acceptance of a tender by or on behalf of the employer and communicated to the contractor results in a binding contract, but there is no binding contract unless the parties are agreed on all essential terms.

Courtney & Fairbairn Ltd v. Tolaini Brothers (Hotels) Ltd

COURT OF APPEAL

(1975) 2 BLR 97

The defendants wished to develop a site by building a motel complex. They approached the plaintiffs, who asked that if they (the plaintiffs) arranged the necessary finance for the project, the defendants would 'be prepared to instruct [their] Quantity Surveyor to negotiate fair and reasonable contract sums in respect of the . . . projects as they arise . . . based upon agreed estimates of the net cost of the work and general overheads with a margin for profit of 5%'. The defendants replied, stating that they 'agreed to the terms specified'. The plaintiffs arranged the finance, but were not awarded the contract. The question was whether there was a binding and enforceable contract between the parties.

Held: There was no enforceable contract, because there was no agreement about the price or any method of ascertaining it. The price was a matter of fundamental importance to the contract.

LORD DENNING MR: I can find no agreement on the price or any method by which the price was to be calculated. The agreement was only an agreement to 'negotiate' fair and reasonable contract sums. The words of the letter are 'your Quantity Surveyor to negotiate fair and reasonable contract sums in respect of each of the three projects as they arise'. Then there are words which show that estimates had not yet been agreed, but were yet to be agreed. The words are: 'These [the contract sums] would, incidentally be based upon agreed estimates of the net cost of work and general overheads with a margin for profit of 5 per cent.' Those words show that there were no estimates agreed and no contract sums agreed. All was left to be agreed in the future. It was to be agreed between the parties themselves. If they had left the price to be agreed by a third person such as an arbitrator, it would have been different. But here it was to be agreed between the parties themselves.

Powell-Smith and Furmston's Building Contract Casebook, Fifth Edition. Michael Furmston.

Now the price in a building contract is of fundamental importance. It is so essential a term that there is no contract unless the price is agreed or there is an agreed method of ascertaining it, not dependent on the negotiations of the two parties themselves. In a building contract both parties must know at the outset, before the work is started, what the price is to be, or, at all events, what agreed estimates are. No builder and no employer would ever dream of entering into a building contract for over £200,000 without there being an estimate of the cost and an agreed means of ascertaining the price.

In the ordinary course of things the architects and the quantity surveyors get out the specification and the bills of quantities. These are submitted to the contractors. They work out the figures and tender for the work at a named price: and there is a specified means of altering it up or down for extras or omissions and so forth, usually by means of an architect's certificate. In the absence of some such machinery, the only contract which you might find is a contract to do the work for a reasonable sum or for a sum to be fixed by a third party. But here there is no such contract at all. There is no machinery for ascertaining the price except by negotiation. In other words, the price is still to be agreed. Seeing that there is no agreement on so fundamental a matter as the price, there is no contract.

But then this point was raised. Even if there was not a contract actually to build, was not there a contract to negotiate? In this case Mr Tolaini did instruct his quantity surveyor to negotiate, but the negotiations broke down. It may be suggested that the quantity surveyor was to blame for the failure of the negotiations. But does that give rise to a cause of action? There is very little guidance in the book about a contract to negotiate. It was touched on by Lord Wright in Hillas & Co Ltd v. Arcos Ltd (1932) where he said:

> 'There is then no bargain except to negotiate, and negotiations may be fruitless and end without any contract ensuing'. Then he went on: '. . . yet even then, in strict theory, there is a contract (if there is good consideration) to negotiate, though in the event of repudiation by one party the damages may be nominal, unless a jury think that the opportunity to negotiate was of some appreciable value to the injured party'. That tentative opinion by Lord Wright does not seem to me to be well founded. If the law does not recognise a contract to enter into a contract (when there is a fundamental term yet to be agreed) it seems to me it cannot recognise a contract to negotiate. The reason is because it is too uncertain to have any binding force. No court could estimate the damages because no one can tell whether the negotiations would be successful or would fall through; or if successful, what the result would be. It seems to me that a contract to negotiate, like a contract to enter into a contract, is not a contract known to the law . . . I think we must apply the general principle that when there is a fundamental matter left undecided and to be the subject of negotiation, there is no contract. So I would hold that there was not any enforceable agreement in the letters between the plaintiff and the defendants. I would allow the appeal accordingly.

However, there may be a contract, even though the parties have left one important clause 'to be discussed and agreed'.

Mitsui Babcock Energy Ltd v. John Brown Engineering Ltd

QUEEN'S BENCH DIVISION

(1996) 51 Con LR 129

The defendants (JBE) were main contractors for the turnkey construction of a 680 MW combined cycle power station. JBE engaged the plaintiffs Mitsui Babcock Energy Ltd

(BEL) to design, manufacture and install two unfired triple pressure heat recovery steam generators (HRSGs) for the power station. The HRSGs were required to generate steam from the hot exhaust gases emitted from two gas turbines to drive a steam turbine.

It was always envisaged that the boilers, when complete, would be tested. The expert evidence was that such performance tests could not be carried out exactly, however accurate and sophisticated the instruments of measurement might be. This raised the possibility that BEL might perform the contract exactly but that the measurements might show a defect. To be sure of passing the tests, BEL would have had to allow for the possible margin of error (±2%) but they were not anxious to do this because of other design constraints.

The result of these considerations was that in the contractual negotiations JBE were pressing for strict compliance with the design requirement of the tests with very substantial liquidated damages if the units failed the tests, and BEL were pressing for an explicit agreement for tolerances.

In June 1993 a contractual document was signed on behalf of both parties. Clause 35 of this document, which dealt with performance tests, was struck out and a marginal note made – 'to be discussed and agreed'. No agreement was in fact ever reached.

In September 1995 BEL wrote to JBE alleging that there was no contract or that if there were a contract the subsequent failure to agree terms for the performance tests made the agreement unworkable or void for uncertainty.

By this time the HRSGs had been installed and had passed the performance tests.

Held: (1) The document signed was a binding contract.

(2) The failure to agree on the performance tests did not render the contract unworkable or void for uncertainty.

JUDGE ESYR LEWIS QC: My review of the authorities leads me to the conclusion that there is no reason in principle why two parties should not enter into a binding agreement, if that was their intention, which is to be objectively determined, even though they have agreed that some proposed terms should be the subject of further discussion and later agreement. That is explicitly recognised in Lloyd LJ's fourth proposition in the passage from his judgment in the *Pagnan SpA* case which I have quoted. This also seems to me to be recognised in the part of Lord Wright's speech in *Hillas & Co Ltd* v. *Arcos Ltd* which I have also quoted. Of course, as Lord Wright said, if proposed terms are left out of an agreement for later resolution, the agreement may fail for want of certainty.

[Counsel for the plaintiff's] submission was founded on the premise that the importance of clause 35 was so fundamental that the parties could not have intended to bind themselves contractually by signing the contract documents with the annotated reservations relating to it and that what was agreed was conditional upon the agreement of further terms. Although the parties did not expressly state that the contract documents were signed conditionally on further terms being agreed that is in fact the effect of the annotation 'To be discussed and agreed'. This is, of course, a different question from that of whether, without such a clause, the parties concluded an agreement which was capable of being performed. He has stressed in his argument the importance attached by both BEL and JBE to a tolerance provision against measurement uncertainties as revealed in the documentary evidence and in the evidence of [an executive director of BEL] and [a general manager within BEL]. It was disagreement over this which led to the parties signing the contract documents on the explicit basis that this clause would be the subject of subsequent discussion and agreement. It was obviously important to JBE since they had a main contract which contained no tolerance provision and firmly and repeatedly resisted BEL's wish for such a clause.

I am left in no doubt by the evidence that clause 35 as a whole and the issue of tolerances were important to both sides. However, it does not necessarily mean that by reserving this clause for further discussion and agreement the parties failed to conclude any binding agreement when the contract documents were signed. As Lloyd LJ said in his sixth statement of principle in the passage from his judgment in *Pagnan SpA* v. *Feed Products Ltd* [1987] 2 Lloyds Rep 601 at 619 which I have quoted: 'It is for the parties to decide whether they wish to be bound and, if so, by what terms whether important or unimportant'. As I have indicated, [Counsel for the plaintiff] has stressed the importance of tolerances and thus the appropriate wording of clause 35 to BEL. Again as I have already indicated, the thrust of [counsel for defendant's] submissions has been that this was not as important to BEL as [an executive director of BEL] and [a general manager within BEL] made out and was not in any event an impediment to an intention by BEL to make a binding agreement with JBE when the contract documents were signed as the details of the performance guarantee did not have to be settled then. The essential question for me to decide is whether, when the documents were signed, there was a common intention to make a binding agreement notwithstanding the reservations contained in the annotations made in the contract documents. This question cannot, in my view, be resolved simply by reference to the language of the annotations.

I conclude that the parties intended to make a contract on the basis of the terms recorded in the contract documents which were not reserved and set aside clause 35 in the expectation that they would be able to reach agreement on it at a later time just as they did with clauses 30, 34 and 35 of section 5. In my view the circumstances here are those described by Lloyd LJ in the fourth and sixth principles summarised by him in his judgment in *Pagnan SpA* v. *Feed Products Ltd* [1987] 2 Lloyds Rep 601 at 619. Undoubtedly clause 35 was important both to BEL and JBE, but it was for them to decide whether it was essential that this clause should be agreed before they made a contract or whether they should leave it for further discussion after making a binding agreement. I find that the circumstances show that the parties did not regard agreement on clause 35 vital to the conclusion of a binding contract.

Where the parties use unclear or ambiguous language, the court will do the best it can to give a sensible meaning to their expressed intention, but it will not make a contract for them where none exists.

Constable Hart & Co Ltd v. Peter Lind & Co Ltd

COURT OF APPEAL

(1978) 9 BLR 1

Constable Hart were sub-contractors to Peter Lind. The sub-contract was in the FCEC standard form and incorporated Constable Hart's quotation which provided that it would 'remain Fixed Priced until 3 June 1975; any work carried out after this date to be negotiated' and this was confirmed by Peter Lind's purchase order which read 'Price fixed until 3 June 1975 – £434,732.29p. less 5 per cent discount'. The works were not completed by 3 June 1975 and no agreement was reached thereafter upon the terms on which Constable Hart were to be paid for work executed after that date.

Held: The sub-contract was subject to an implied term that the rates for work carried out after 3 June 1975 would be reasonable rates.

SIR DOUGLAS FRANK QC (whose judgment was affirmed by the Court of Appeal): [Counsel] for the claimants submitted that where parties have agreed that building work be carried out and have concluded a contract but where the price has not been agreed for part of the work, the law implies that a reasonable sum be paid. He referred me to *Foley* v. *Classique Coaches Ltd* (1934), *F. & G. Sykes (Wessex) Ltd* v. *Fine Fare Ltd* (1967), *Liverpool City Council* v. *Irwin* (1977). However, those cases only go to establish the proposition that where a contract is silent as to the price, the court, in order to save the contract, will say that there is an implied term that the price is to be a reasonable price. They do not assist in determining a formula for a reasonable price or in the case where there is more than one possible formula, for ascertaining a price. He also referred to *Shell UK Ltd* v. *Lostock Garage Ltd* (1976), and in particular to Lord Denning's two broad categories of implied terms. It was accepted by him and by . . . the respondents that this case falls within the second category, that is, it is not of common occurrence and that the implication is based on an imputed intention by the parties from their actual circumstances. These are what he described as the 'officious bystander' type of case, and in such cases a term is not to be implied on the ground that it would be reasonable but only when it is necessary and can be formulated with sufficient degree of precision. Therefore the question I have to decide is what advice would have been given by an officious bystander, albeit one with experience of civil engineering contracts, to and accepted by the parties had they applied their minds to the term which should have been incorporated in the contract . . .

[The respondents' counsel] said that the words in the letter of 8 June 1973 'any works carried out after this date to be negotiated' are too uncertain to be given any meaning unless there is a yardstick by which they are to be agreed, and therefore one is led back to the implied term. That term, he said, must be determined in the light of the expressed terms. He emphasised that this is a lump sum contract, that is, at a fixed price for defined works, and that is the price referred to in clause 5. He pointed out that the contract provides for the valuation of variations in the work at specified rates and there is no similar provision for additional cost for work carried out after 3 June 1975 and so it can only be paid for as an additional sum under sub-clause (5) of clause 15. Thus there is no means of paying for the additional cost other than as an additional sum without modifying the price and that, he argued, cannot be done.

One difficulty I find with [that] argument is that it seems to me only to go to the time at which the additional payment is to be made and not to the method of its quantification. I think I must go back to June 1973 to see what was in the minds of the parties and I may here say that although [counsel] contended to the contrary, I find that the words 'any work carried out after this date to be negotiated' were embodied in the contract. It is clear that the claimants in arriving at their quotation had done so on the footing that the work would be completed by about 3 June 1975. Indeed, the period of completion specified in the contract was approximately 75 per cent of the main carriageway to be constructed in 1974, and the remainder in 1975. It is also clear that they had quoted on the basis of their priced bills and that their quotation had been accepted on that basis, and their quotation was expressed to form part of the contract. Thus, that the quotation was based on the priced bills and on the footing that the price would hold good until 3 June 1975 was recognised by both parties. It seems to me to follow naturally that had the parties then been asked the basis for payment after that date they would have replied that appropriate adjustments either up or down would have to be made to the prices in the bills and I think that it was when the claimants referred to the work to be negotiated they meant that there would have to be negotiations for the adjustment of the prices in the bills. This is a matter mainly of impression from the documents and all the surrounding circumstances. I do not think that the fact that the price is defined in the contract vitiates this conclusion. It seems to me that the terms of the letter of 8 June 1973 are broad enough to include adjustment of the price as defined and that is the conclusion I have reached.

An informal instruction to carry out construction work can give rise to a contract if accepted, even though there is no agreement on the price or the precise extent of the work.

ACT Construction Ltd v. E. Clarke & Sons (Coaches) Ltd

COURT OF APPEAL

(2002) 85 Con LR 1 [2002] EWCA Civ 972

The defendant (Clarke) was a family controlled company which conducted a coach operating business in London. The claimant (ACT) was a company which operated as civil engineers and builders. ACT had successfully carried out several pieces of work for Clarke. In 1990 Clarke wanted to acquire much bigger premises and bought a disused cold store. Adaptation required planning permission and much extensive demolition and reconstruction.

Work started in April 1992. There was no formal contract. From time to time ACT made applications for payment which were met, the latest of these in October 1993 for work up to 28 September 1993. In May 1994, Clarke asked ACT to leave the site, which it did. In October 1994 ACT sent a statement showing £19,367 as outstanding on 28 September 1993 and asking for £25,000 on account for work after 28 September. Clarke paid £19,000. A final account was submitted on 1997 which led to a dispute.

Held: There was a contract arising from the instruction to do work, followed by the doing of the work. Clarke were bound to pay a reasonable sum for the work.

WARD LJ:

28. As the judge found, Mr Clarke and Mr Blake [the person who ran the building side of ACT] conducted their business on an informal basis, each seeming to have a high degree of trust and respect for the other. Mr Clarke wanted a 'state of the art' coach station and Mr Blake was ready, willing and able to provide it. As at April 1992, when the final decision was taken to go ahead with the project, there was, as the judge found, some lack of precision. Mr Blake had prepared a schedule of works for phase 1 but that omitted the roofing work to cover the yard where the coaches would be parked, the paving of that area and the refurbishment of the existing two-storey office block. He found, correctly in my judgment, that the precise content of phase 2 had not been finally agreed. An application to amend the existing planning permission in vital respects required by Mr Clarke had only just been submitted and the scope of the work could not be finalised until that permission had been obtained. Contracts to purchase the property were only exchanged after this first stage of work had commenced and completion did not take place until planning permission had been agreed, in the event, not until June 1992. The judge was entitled to conclude that there was no 'building contract' to design and build, and particularly to complete, a state-of-the-art or top quality coach station. But he also took the view that the actions of the parties belied there being an intention to create legal relations with immediate effect. That led the judge to conclude that ACT started on the site 'on an informal basis without any contractual framework' and that the work would therefore

'be paid for as and when it was carried out on the basis of a reasonable remuneration'. In his May judgment he summarised his conclusion in this way:

> I concluded that there was no contract between the parties. The parties' relationship was not a contractual one, with the consequence that the value of the work carried out by the claimant could be recovered and paid for, but on the basis of a quantum meruit, a reasonable sum, a restitutionary basis in fact.

29. I cannot accept that finding by the judge. As Henry LJ observed when granting permission to appeal, the decision that there was no contract came as a surprise. In my view, the proper conclusion was to find that there was, as [Counsel] for Clarke submits, 'a contractual *quantum meruit*'. In focusing on the essential ingredients for 'a building contract of some complexity' the judge may have lost sight of the fact that even if there is no entire contract, and especially even if there is no 'formal' contract, there may still be an agreement to carry out work, the entire scope of which was not yet agreed, even if a price has not been agreed. Provided there is an instruction to do work and an acceptance of that instruction, then there is a contract and the law will imply into it an obligation to pay a reasonable sum for that work. That is what happened here.

It may be that the parties have made a series of contracts.

Emcor Drake & Scull Ltd v Sir Robert McAlpine Ltd

COURT OF APPEAL

(2004) 98 ConLR 1; [2004] EWCA Civ 1733

The appellant/defendant (McAlpine) was the main contractor on the Dudley Hospitals PFI project. The respondent/claimant (EDS) did much of the sub-contract mechanical and electrical (M&E) work. The principal question in the preliminary issue in the present litigation was the nature of the contractual arrangements for this work. McAlpine contended that EDS had contracted to do all the M&E work for £34.25m and had committed a repudiatory breach by failing to complete the work. EDS's position was that it had only agreed to do work up to a value of £14m.

There had been extensive negotiations before work began and a price of the order of £34.25m had been agreed in principle. It was intended that the parties should enter into a formal sub-contract but this was never done.

The main contract had been signed on 17 May 2001. Thereafter McAlpine sent EDS a number of letters which EDS was invited to sign. There was a meeting on 3 July and a letter was then sent to EDS to allow commencement of design work etc up to a maximum of £1m and EDS started work on 4 July 2001. It was accepted that at this stage EDS was entitled to be paid a reasonable sum up to a maximum of £1m and that it was an implied term that the work would be carried out in a good and workmanlike manner.

There continued to be negotiation about the terms. Further orders were issued by McAlpine which increased the maximum payment by stages to £14m. The last order was issued on 15 August 2002.

Held: Work had been done under a series of contracts under which EDS was entitled to be paid a reasonable sum for work done and agreed to do work of a good and workmanlike quality. EDS had never agreed to do the whole M&E work and was neither bound nor entitled to do work in excess of the value of the currently applicable order.

CLARKE LJ:

In all the circumstances I have reached the conclusion that the judge was right. My reasons may be summarised as follows.

(i) Before the letter of 20 August was sent, EDS was performing work pursuant to the order of 3 July for which it would be paid a reasonable sum up to a maximum of £1m. It was an implied term of the agreement that the work would be carried out in a good and workmanlike manner ([14] above).

(ii) The mere signing and returning of the letter of 20 August did not create a contract subject to the terms of the letter ([23] and [32]).

(iii) At some stage thereafter the letter probably became of contractual effect whereafter the contractual position was as contemplated in the order of 3 July, namely that the order continued to be effective on the acceptance by both parties of the terms of the letter ([24], [32] and [33]).

(iv) Construed together (as they should be), the order and the letter do not on their true construction evidence an agreement that EDS would carry out the whole of the M&E works or that it would do so for £34.25m or any other sum ([28]–[31]).

(v) If regard is had only to the letter, no such agreement was reached because SRM never proffered a sub-contract which conformed to the terms of the letter ([127] and [35]–[38]) and because the letter left a number of important points unspecified or unsettled ([27], [33] and [44]–[47]).

(vi) The work was carried out pursuant to a series of limited orders ([26], [34] and [39]–[43]).

(vii) I agree with the answers given by the judge to the preliminary questions he was asked to determine as set out in [4] above.

A very useful restatement of the appropriate principles is to be found in the following case.

G. Percy Trentham Ltd v. Archital Luxfer Ltd

COURT OF APPEAL

(1992) 63 BLR 44

STEYN LJ: Before I turn to the facts it is important to consider briefly the approach to be adopted to the issue of contract formation in this case. It seems to me that four matters are of importance. The first is the fact that English law generally adopts an objective theory of contract formation. That means that in practice our law generally ignores the subjective expectations and the unexpressed mental reservations of the parties. Instead the governing criterion is the reasonable expectations of honest men. And in the present case that means that the yardstick is the reasonable expectations of sensible businessmen. Secondly, it is true that the coincidence of offer and acceptance will in the vast majority of cases represent the mechanism of contract formation. It is so in the case of a contract alleged to have been made by an exchange of correspondence. But it is not necessarily so in the case of a contract alleged to have come into existence during and as a result of performance. See *Brogden* v. *Metropolitan Railway* (1877) 2 App Cas 666; *New Zealand Shipping Co Ltd* v. *A.M. Satterthwaite & Co Ltd* [1975] AC 154 at page 167 D–E; *Gibson* v. *Manchester City Council* [1979] 1 WLR 294. The third matter is the impact of the fact that the transaction is executed rather than executory. It is a consideration of the first importance on a number of levels. See *British Bank for Foreign Trade Ltd* v. *Novinex* [1949] 1 KB 628, at page 630. The fact that the transaction was performed on both sides will often make it unrealistic to argue that there was no intention to enter into legal relations. It will often make it difficult to submit that the contract is void for vagueness or uncertainty. Specifically, the fact that the transaction is executed makes it easier to imply a term resolving any uncertainty, or, alternatively, it may make it possible to treat a matter not finalised in negotiations as inessential. In this case fully executed transactions are under consideration. Clearly, similar considerations may sometimes be relevant in partly executed transactions. Fourthly, if a contract only comes into existence during and as a result of performance of the transaction it will frequently be possible to hold that the contract impliedly and retrospectively covers pre-contractual performance. See *Trollope & Colls Ltd* v *Atomic Power Construction Ltd* [1963] 1 WLR 333.

On behalf of Archital counsel challenged specific findings of fact regarding successive stages of the dealings between the parties on the ground that there was no evidence to support the findings. I have held that the individual criticisms were not well founded. But ultimately the only issue is whether there is sufficient evidence to support the judge's central finding of fact that a binding contract on phase 1 came into existence. Even if, contrary to my view, a particular finding was not supported by evidence, it would not matter provided that there was sufficient evidence to support the ultimate finding of fact.

In a case where the transaction was fully performed the argument that there was no evidence upon which the judge could find that a contract was proved is implausible. A contract can be concluded by conduct. Thus in *Brogden* v. *Metropolitan Railway*, supra, decided in 1877, the House of Lords concluded in a case where the parties had acted in accordance with an unsigned draft agreement for the delivery of consignments of coal that there was a contract on the basis of the draft. That inference was drawn from the performance in accordance with the terms of the draft agreement. In 1992 we ought not to yield to Victorian times in realism about the practical application of rules of contract formation. The argument that there was insufficient evidence to support a finding that a contract was concluded is wrong. But, in deference to counsel's submissions, I would go further.

One must not lose sight of the commercial character of the transaction. It involved the carrying out of work on one side in return for payment by the other side, the performance by both sides being subject to agreed qualifying stipulations. In the negotiations and during the performance of phase 1 of the work all obstacles to the formation of a contract were removed. It is not a case where there was a continuing stipulation that a contract would only come into existence if a written agreement was concluded. Plainly the parties intended to enter into binding contractual relations. The only question is whether they succeeded in doing so. The contemporary exchanges, and the carrying out of what was agreed in those exchanges, support the view that there was a course of dealing which on Trentham's side created a right to performance of the work by Archital, and on Archital's side it created a right to be paid on an agreed basis. What the parties did in respect of phase 1 is only explicable on the basis of what they had agreed in respect of phase 1. The judge analysed the matter in terms of offer and acceptance. I agree with his conclusion. But I am, in any event, satisfied that in this fully-executed transaction a contract came into existence during performance even if it cannot be precisely analysed in terms of offer and acceptance. And it does not matter that a contract came into existence after part of the work had been carried out and paid for. The conclusion must be that when the contract came into existence it impliedly governed pre-contractual performance. I would therefore hold that a binding contract was concluded in respect of phase 1.

If the contractor makes a unilateral mistake in his tender price, or there are individual errors in pricing, multiplication, or addition, he is bound by such errors unless, before acceptance, the employer or the architect discovers the error and realizes that it is not intentional.

W. Higgins Ltd v. Northampton Corporation

CHANCERY DIVISION

[1927] 1 Ch. 128

Higgins contracted with the local authority to erect 58 houses. Higgins completed the tender incorrectly. The result was that Higgins thought he was tendering for the erection of the houses at £1670 a pair, when in fact it was £1613 because of the way in which Higgins had made up the bills.

Held: Higgins was bound by his mistake and could not claim to have the contract set aside or rectified.

ROMER J: The result is very unfortunate for the plaintiff and an extreme hardship on him. But can I rectify the contract so as to make it give effect to what was undoubtedly his intention when he entered into the contract? It appears to me that I cannot. [In] my opinion the parties were not labouring under a mutual mistake. I think the contract expresses what the defendants intended it to express. At any rate, it has not been shown to me that the defendants were labouring under any mistake, and of course the onus is upon the plaintiff to prove that fact, if fact it were . . .

In these circumstances, with the utmost regret, I come to the conclusion that I cannot give the plaintiff the relief that he desires, and unless, even at the eleventh hour, the defendants can make some concession to the plaintiff, I am afraid that the ratepayers will have benefited by the bona fide and serious mistake made on the part of the plaintiff, which was really brought about by the carelessness of some official of the defendant corporation in drawing up the original bill of quantities. If he had made it clear in that bill of quantities, as he ought to have done, that the 50 yards of chimneys were already included in the 321 yards of wall, Mr Higgins would never have made the mistake that has resulted in this action.

If the employer discovers an error in the contractor's tender he cannot accept the mistaken tender so as to create a contract where the contractor's mistake is as to the terms of the offer itself and not merely as to the motive or underlying assumptions on which the offer is based.

McMaster University v. Wilchar Construction Ltd

ONTARIO HIGH COURT

(1971) 22 DLR (3d) 9

The employer purported to accept a tender, knowing that the contractor had omitted the entire first page of his bid. That page included an intended price fluctuations clause.

Held: The contractor was not liable.

THOMPSON J: There is not the slightest doubt in my mind that the real reason the plaintiff purported to accept Wilchar's tender was in the hope that it might be able to recover the penalty of the bid bond, knowing full well as early as 3 October that Wilchar had made a mistake in its tender and that it would refuse to enter into a contract unless the mistake were remedied. To me this is patently a case where the offeree, for its own advantage, snapped at the offeror's offer well knowing that the offer was made by mistake.

[Counsel for McMaster] concedes that Wilchar had, by mistake, not included the escalator clause as a term of its tender, but he argues that this was not a mistake of a fundamental character such as to vitiate the tender and that it was a mistake merely in the motive or the reason for making the offer.

I am not prepared to accede to such an argument. In a construction contract the price is always a fundamental term of the contract. In fact it is the very quid pro quo of such a contract. In the instant case, for the contractor the mistake meant a loss of thousands of dollars as contrasted with a profit in its absence; for the contractee, it meant an advantage of some $16,000.

Were it not that a provision or stipulation as to price were a fundamental term of a contract, then such decisions of the courts, ranging over a number of years, as in *Webster* v. *Cecil* (1861); *Garrard* v. *Frankel* (1862); *Hartog* v. *Colin and Shields* (1939) and *Bennett* v. *Adams River Lumber Co* (1910) have all been in vain . . .

In my view, this is truly a case of unilateral mistake. While perhaps inadvertent, it was an honest mistake, and one which Wilchar hastened to make known to the plaintiff as soon as it was discovered; one which was made known to the plaintiff long before the tender or offer was accepted. Moreover, I conclude that from the circumstances, the plaintiff must, in any event, be taken to have known of the mistake before acceptance of the offer. In this context, it should be stressed that one is taken to have known that which would have been obvious to a reasonable person in the light of the surrounding circumstances: see *Hartog* v. *Colin and Shields* (1939).

With full knowledge of the mistake, and the evidence as to this is irrefragable, apart from the concession of counsel, the plaintiff purported to accept Wilchar's tender, despite the protest of Wilchar and its endeavour to amend it.

There can be little doubt that there was no real agreement between the parties and that this is but a bold attempt by the plaintiff to force the defendant Wilchar to fulfil a promise in a sense which the plaintiff knew that Wilchar did not intend it and to which its mind did not assent: and so I hold.

To put it simply, this is a case where one party intended to make a contract on one set of terms and the other intended to make it upon another set of terms, with the result that there is lack of consensus. The parties were not ad idem. The existing circumstances prevented the formation of a contract.

I further hold that in any event, the circumstances were such that it would be unconscionable, unfair and unjust to permit the plaintiff to maintain the contract in the light of the conduct of its project manager, conduct which I find was not only fraudulent in that more broad equitable sense of the word, but which went even further.

Wilchar would be entitled to the equitable right of rescission, in the event that a contract had been consummated, on the basis that under the circumstances, the contract was voidable for fundamental mistake in its formation. Under such conditions, such a mistake is a good defence to any action brought to enforce such an alleged contract or to obtain damages for its breach: see *Raffles* v. *Wichelhaus* (1864).

Stent Foundations Ltd v. Carillion Construction (Contracts) Ltd

COURT OF APPEAL

(2000) 78 Con LR 188

The claimants (Stent) did preparatory work by way of building a piled retaining structure around the perimeter of a site at South Quay in London's Docklands between 12 October 1988 and 31 January 1989. Wiggins Waterside Ltd (Wiggins) were the owners of the site. Wimpey Construction Management (WCM) were appointed as management contractors. The contract between Wiggins and WCM was not finally agreed until January 1989 although WCM had tendered in May 1988.

It became necessary to decide whether Stent's contract was with Wiggins or WCM. Stent were originally approached in June 1988 by ECHP (quantity surveyors for Wiggins). On 29 June 1988 ECHP accepted Stent's tender on behalf of Wiggins subject to agreement of a contract programme. By this time WCM was the front runner for appointment as management contractor and became involved in the negotiation with the works contractors (including Stent). On 14 July 1988 at a meeting at which Stent, ECHP and WCM were represented, WCM made it clear that the contract would be on the 1989 Works Management Contract. Stent said that it would proceed on a letter of intent. On 17 August ECHP sent a letter on behalf of Wiggins to Stent instructing them to start work and indicating that they would be required to enter into a sub-contract with WCM and at a meeting on 28 September that it would be signing a JCT Management Contract with all works contractors.

Held: There was a contract between Stent and WCM. The parties were substantially agreed on the terms by September 1988 and although the parties clearly contemplated the completion of a formal contract, they did not intend not to be bound if there was no formal contract. It was the completion of the contract between Wiggins and WCM which was the delaying factor and once the main contract was executed the sub-contract came into existence.

HALE LJ: [Counsel] for WCM, no longer seeks to argue that the parties were not *ad idem*. They were agreed on all the essential terms, and any amendments to the works contract required as a result of the precise terms of the management contract would not in fact have presented a problem. He also accepts that in many cases parties to a building contract may make an agreement either orally or in correspondence which is simply confirmed in the formal written contract. But he argues that this was a case in which their agreement was, in effect, subject to contract.

In support of the argument that formality was a condition precedent of a binding contract between them, [Counsel for WCM] made two main points. First, no negotiations about the precise terms of the works contract could take place until after the terms of the management contract had been finalised. It would be dangerous for a management contractor to bind itself to a works contractor without knowing what those terms were. However, that is no longer the case once he has bound himself to the employer and knows what the terms of the management contract are, and this is particularly so if the work has been completed and no problems have arisen with those contractual terms.

Secondly, [Counsel for WCM] argues that it was Stent's own understanding that the contract would be embodied in formal documentation, and that this meant that it would indeed be subject to contract. He relies particularly on the terms of Stent's own explanatory memorandum which was twice sent to ECHPS, and to the minutes of the meetings which continued as late as 23 November 1988, to record that Stent was working under the terms of the letter of intent. However, the explanatory memorandum was sent in before the work began, and the minutes of the meetings are equally consistent with the condition precedent which the judge did find – that a management contract had to be placed before the formal contract between Stent and WCM came into being.

It also seems to me clear that everyone behaved as if the works contract was in place. Payment was made under that contract. The developing dispute about the ground conditions was being handled by WCM as if it was under that contract.

[Counsel for WCM] accepts that it would not usually be possible to argue that there was no contract once a transaction had been fully performed on both sides (see *G. Percy Trentham Ltd* v. *Archital Luxfer Ltd* (1992) 63 BLR 44 at 52 per Steyn LJ). But in this case there was a letter of intent between Wiggins and Stent which gave sufficient comfort. He relies in particular on *Galliard Homes Ltd* v. *J. Jarvis & Sons plc* (1999) 71 Con LR 219, a decision of this court. But the facts and circumstances in the Jarvis case were very different from these. The pre-contractual arrangements were made between the same parties. The letter of intent in that case promised payment upon a quantum meruit basis 'in the event that we do not enter into a formal contract with you'. All sorts of matters remained in active dispute and variation as the work proceeded. Crucially these included whether there was a fixed-price contract at all, as Galliard wished it to be, or whether, as the contractor contended, a contract for payment on a *quantum meruit* basis. Hence, although the Court of Appeal did find that this was indeed subject to contract, Evans LJ found it impossible to say that a contract had been concluded at a later meeting because:

> First, and more importantly, it is not possible to be certain what the terms of the contract were.

He did, however, think that the correct legal analysis was that there was a contract to pay a quantum meruit based on the letter of intent, rather than simply a quasi-contractual obligation so to do once the work had been done.

If anything, in this case the history of the letter of intent supports rather than undermines the claimant's case. It clearly contemplates a works contract being made; it promises reimbursement of costs but not loss of profit if no such contract is made. That was no longer acceptable to Stent once they started work. They were then paid by WCM according to the agreed tender sums and not in accordance with the letter of intent. Everything else that happened after then was in accordance with a contract between WCM and Stent. This includes the procurement by WCM for Wiggins of the bond and warranty, which would not have been necessary, or at least as necessary, if, as [Counsel for WCM] contends, the letter of intent had been a contract between Wiggins and Stent which was still in existence.

For my part, therefore, it seems to me quite clear that the learned judge was entirely correct in this case to hold that this work had been done under a contract between Stent and WCM which came into being as soon as WCM concluded a contract with Wiggins in January 1989, and I would dismiss this appeal.

Where work has commenced before a contract is concluded, but the parties later agree on terms, the terms so agreed will have retrospective effect.

Trollope & Colls Ltd v. Atomic Power Constructions Ltd

QUEEN'S BENCH DIVISION

[1962] 3 All ER 1035

In February 1959 the plaintiffs submitted to the defendants a tender for carrying out, as sub-contractors, certain civil engineering work. Negotiations then took place and were continuing when, in June 1959, the plaintiffs began work at the defendants' request, and on the basis of a letter of intent. On 11 April 1960, the form of the general conditions was finally agreed between the parties, the judge holding that on that date the parties were then agreed on all the essential terms. Disputes arose as to whether the terms of the contract so agreed, and in particular the provisions as to variations, governed the rights of the parties for work done prior to 11 April 1960.

Held: The contract finally concluded on 11 April 1960 governed the rights of the parties as to prior work.

MEGAW J: So far as I am aware, there is no principle of English law which provides that a contract cannot in any circumstances have retrospective effect, or that, if it purports to have, in fact, retrospective effect, it is in law a nullity. If, indeed, there were such a principle, there would be many important mercantile contracts which would, no doubt to the consternation of the parties, be nullities. Frequently, in large transactions a written contract is expressed to have retrospective effect, sometimes lengthy retrospective effect: and this in cases where the negotiations on some of the terms have continued up to almost, if not quite, the date of the signature of the contract. The parties have meanwhile been conducting their transactions with one another, it may be for many months, on the assumption that a contract would ultimately be agreed on lines known to both the parties, though with the final form of various constituent terms of the proposed contract still under discussion. The parties have assumed that when the contract is made – when all the terms have been agreed in their final form – the contract will apply retrospectively to the preceding transactions. Often, as I say, the ultimate contract expressly so provides. I can see no reason why, if the parties so intend and agree, such a stipulation should be denied legal effect. Take, as an example, a simple case. Suppose that a contract for the sale of goods is under negotiation. The offeror has said: 'I will sell you one thousand tons of coal on such and such terms'. While the detailed terms of sale are still under negotiation, the offeree asks for the delivery of five hundred tons and the offeror makes delivery, both parties intending and anticipating that this delivery will count against the contract quantity if and when the contract is made, or perhaps believing wrongly that the contract has already been made. The final terms are then agreed, the offer is accepted, and the contract is made. Even if, in the actual acceptance by the offeree, no express reference is made to the antecedent delivery of the five hundred tons, I should have thought that there would be little room for doubt that the contract was intended to govern, and in law did govern, that antecedent delivery; and that neither party could successfully assert that there was no contract, or that the five hundred tons was delivered on a *quantum meruit* basis; or that the whole one thousand tons still fell to be delivered after the contract. Of course, the position would be different if no contract were ultimately made, as, for example, by the offeror's withdrawal of the offer before acceptance. So here. If a contract was made on 11 April 1960, and if the contract expressly provided, or should in law be assumed to have provided, that its terms as then agreed were to apply retrospectively to previous acts of the parties done since the date of the tender in anticipation of the making of such a contract at a later date, I see no reason why, in law, effect should not be given to such a provision. It is true that such retrospectivity may give rise to difficulties for one or other or both of the parties, where things have been done at a time when certain of the terms which ultimately and retrospectively become contractual were in a state of flux. Thus, it might turn out that something done by a party before the making of the contract had been done otherwise than in accordance with a term as subsequently agreed. That is a matter which may have to be taken into account as tending to militate against the implication of such an agreement as to retrospective effect. It would not, however, be sufficient to negative the existence of a contract with retrospective effect, if the assent of the parties to such a contract were clearly established.

In the present case, so far as I have seen, a retrospective effect is nowhere expressly stated or stipulated. Is it to be implied into the contract of 11 April 1960, if contract there was? That, I conceive, is the crucial question on this issue.

It has been said on many occasions that terms are not to be implied merely because they are desirable, or merely because the parties, if they had considered the question, would probably or as reasonable men have agreed such terms. Terms can only be implied where, to use the common phrase, they are necessary in order to give 'business efficacy' to the contract. In the present case, in one sense at least, this term of retrospectivity is necessary to give business efficacy to the contract, if contract there was. The submission of counsel for the plaintiffs is in my view right to this extent, that, in the absence of retrospective effect to the variations clauses, an agreement made on 11 April 1960, would not only be devoid of business efficacy but also would not be capable of being a contract at all, because it would pre-suppose a state of affairs (the original specifications and price) which had long since ceased to be realistic or practicable or within the true intentions of either party in relation to the work to be done under an agreement. On the other hand, I do not think that a term such as this can be implied simply for the purpose of upholding the existence of a contract, unless it can clearly be seen that it conforms with what the parties truly intended and with what they both would have accepted as a matter of course had the question been raised in the course of the negotiations or at the moment of the making of the supposed contract. But if those factors are present, in my judgment it is right and necessary that such a term should be treated as being implied.

I am satisfied from all the circumstances that both parties, in all that they did in the course of the negotiations, in the defendants' requests or instructions to the plaintiffs to carry out the work as varied, and in the plaintiffs' acceptance of those instructions, were doing so on the understanding and in the anticipation that, if a contract were made, and whenever it was made, that contract would apply to and govern what was meanwhile being done by the parties. I am satisfied that if, on 11 April 1960 (still assuming that a contract was, otherwise, made on that date) the question had been raised, both parties would have said, as a matter of course: 'This contract is to be treated as applying, not only to our future relations, but also to what has been done by us in the past since the date of the tender in the anticipation of the making of this contract'.

PART II

PERFORMANCE OF THE CONTRACT

Chapter 6
The Contractor's Obligations

Design

In the classic form of building contract, the employer and his professional advisers assume responsibility for the design of the building, and the contractor agrees to build the building as designed. In such a contract, the contractor does not in principle undertake liability for the design.

Bower v. Chapel-en-le-Frith Rural District Council

KING'S BENCH DIVISION

(1911) 9 LGR 339

The plaintiffs were the successful tenderers for the erection of a waterworks for the Council. The contract was for a lump sum and contained specifications and bills of quantities. The plaintiffs were required by the contract to buy a windmill tower and pump from a named supplier at a named price and to fix them. The windmill proved hopelessly defective. The Council claimed that the plaintiffs must replace it with an efficient one. The plaintiffs argued that they were not responsible as they had played no part in the choice of the windmill. It was not argued that the default was in any way due to defective installation.

Held: The plaintiffs were not liable.

LAWRENCE J: The question is: Were the plaintiffs liable for the inefficient working of the mill, or did their liability cease when they properly erected it under the contract? Another way of putting the question is: Was the plaintiffs' contract to make the mill answer its purpose; or was it a contract to do the work in accordance with the specification and plans? I need hardly say that in my judgment the answer to each of these questions should be in favour of the plaintiffs on the ground that I have already given, namely, that the mill was chosen by the defendants, and it was the mill the plaintiffs were bound by their contract to erect, and which they did erect, and that the plaintiffs had no means or power of objecting to the mill, or altering it, or doing anything to it. It was no part of their contract to guarantee in any way that the mill would be efficient.

Powell-Smith and Furmston's Building Contract Casebook, Fifth Edition. Michael Furmston.

The Court of Appeal ordered a new trial on the ground that the judge had misunderstood the way in which the parties intended to present the case, but expressed no opinion on the question of the law involved.

Cable (1956) Ltd v. Hutcherson Brothers Pty. Ltd

HIGH COURT OF AUSTRALIA

(1969) 43 ALJR 321

The respondents successfully tendered for the design, supply, and installation of a bulk storage and handling plant. This was to be built on reclaimed harbour land for the appellant. A formal contract was entered into, under which the drawings of the proposed installations, prepared by the respondents, had been approved by the appellant's consulting engineer. These drawings showed ring beam foundations for storage bins. Subsidence occurred when the bins were erected and filled.

Held: The promise of the respondent was to carry out the work agreed in the drawings.

> BARWICK CJ: The principal matter to be resolved is the effect upon the promise of the respondent of the incorporation in the contract of the terms of the specification. The competing views are that of the appellant that the promise of the respondent was to produce a result, namely, storage bins of given holding capacity which with the ancillary plant would operate and carry out the appellant's scheme or project, that is to say, receive, hold and discharge the volume of the stated mineral to the specified extent at the specified rate: and that of the respondent, that the promise was to do only the work described in the agreed drawings, the appellant being itself satisfied upon its own engineer's advice that work so done would operate and carry out the scheme or project it proposed. In my opinion, the choice between the contentions can be quite securely made without reference to decided cases or indeed any attempt to find an analogy in any of them or in the facts of any of them. No doubt at times the question whether the promise of a builder is to produce a result or merely to do specific work is difficult to answer. In reaching a conclusion, the fact that it can be seen that reliance is placed upon the skill and judgment of the builder may on occasions be an important if not a decisive consideration. Here, there is no doubt that the respondent knew the result which the appellant sought to achieve and it might well be that at the time of tender an area of the design of the installation was left to the tenderer. Indeed, the respondent tendered a price to include 'design' (see the letter of tender) though oddly enough the word 'design' is not found in the basis of tender clause in the specification. The respondent did in fact propose the design of the foundation for the storage bins and prepared the first drawings for them. But, by the time the agreement was made the 'design' of the foundations had been agreed, after participation of the appellant's consulting engineer in the determination of its form. That agreed design, extrapolated in a drawing, was expressed to be part of the contract works . . . But the appellant says that notwithstanding that fact so much may be drawn from the inclusion of the specification in the articles of agreement, that the responsibility for the suitability of the design to the appellant's mutually known purpose remained with the respondent. It is therefore necessary and important to observe what is brought out of the specifications prepared for tendering into the operation of the articles of agreement.

The description of the 'works' in the portion of the articles of agreement which I set out earlier in these reasons includes a reference to the specifications. The respondent's promise in clause 1 of the articles is to do the work shown in the drawings and described by or referred to in the specification and conditions. The conditions refer to the works shown in the drawings and described in the specification: but the contract conditions are paramount (clause 1(c)). However, in any case the specification is only incorporated insofar as it describes the 'works', presumably, the works which are to be found in or which form part of the works found in the drawings. It seems to me that the part of the specification most relied upon by the appellant, namely the basis of tender, does not perform or relate to the function of describing the works. In terms it relates only to the footing on which the tenderer should make his bid. Whether or not the building owner will exact a promise to make good that footing is quite another matter, particularly where as here, it is throughout intended that the mutual rights and obligations of the parties shall be exclusively expressed in a formal manner. So far from the basis of tender clause in the specification being a description of work, it seems to me to have become at the time of the execution of the contract no more than an historical step in a negotiation which preceded the making of the agreement. It is not reflected in the terms of that agreement either expressly or by implication. But out of deference to the submissions which were pressed upon us, let me say that, even if one moved the whole of the clause 'Basis of tender' into the operative part of the articles of agreement, the result would, in my opinion, be the same. Nothing can be made, in my opinion, of the use of the word 'turnkey'. It is not a term of art and, even if it could be taken to mean that the works must be handed over as a going concern, I would not have thought that in the context of these articles the word or expression meant that the builder warranted the efficacy of the works he had agreed to erect. The same reasoning would, in my opinion, deny that the acceptance of all responsibility for supply and erection and efficient operation of the project – itself a very inapt word in any case in the circumstances – involved any undertaking as to the suitability of the agreed design. Lastly, the paramountcy of the contract conditions, which do describe the works agreed to be done, would prevent the clause of the specification having the submitted effect, if in truth it could be construed as imposing the obligations the appellant seeks to draw from it.

In my opinion, for all these reasons I conclude that the respondent promised no more than to carry out the specified work in a workmanlike manner. This, according to the arbitrator it did.

Work under a JCT contract does not normally include 'design' as part of the contractor's obligation and in so far as a performance specification in the bills of quantities purports to impose upon the contractor a responsibility for design it is seeking to modify the interpretation of the contract conditions which JCT 98, clause 2.2.1 [JCT 63 clause 12(1)] does not permit. See now JCT 2005, clause 1.3.

John Mowlem & Co Ltd v. British Insulated Callenders Pension Trust Ltd

QUEEN'S BENCH DIVISION

(1977) 3 Con LR 63

Mowlem contracted with BICP to build a warehouse and office block. The contract was in JCT 63 form, private edition, with quantities. The third party (Jampel) were consultant structural engineers to the project and, inter alia, designed the reinforced concrete floor slab of the basement area. The actual design work was carried out by one of their employees.

The bills of quantities contained the following provision:

> Watertight Construction
> Retaining walls forming external walls to buildings and basement slabs are to be constructed so that they are impervious to water and damp penetration, and the contractor is responsible for maintaining these in this condition.

The basement floor subsequently developed cracks and water penetrated, necessitating expenditure on remedial work and resulting in loss and damage because of the consequent delay in completion and occupation.

Apart from damages, two issues fell to be determined:

(1) Whether the defects were the result of inadequate design and/or faulty workmanship;
(2) If the lack of watertightness was due to design failure, whether Mowlem were nonetheless liable because of the performance specification in the bills of quantities.

Held: On the evidence and in the circumstances of the failure, Jampel, in designing the basement slab, failed to exercise the requisite skill and care and the failure was due to inadequate design. Design was no part of the contractors' obligation and, in so far as the performance specification purported to impose such liability on them, it was ineffective by virtue of the provisions of clause 12(1) of the contract.

JUDGE WILLIAM STABB QC: Having concluded that the failure of this basement slab and its consequent lack of watertightness was due to a design fault on the part of the consulting structural engineer, am I compelled to hold the contractor liable for the resulting loss and damage by reason of the provision in the bill of quantities to which I have already referred, but which I shall read again? That specification, under the heading of 'Watertight construction', read: 'Retaining walls forming external walls to buildings and basement slabs are to be constructed so that they are impervious to water and damp penetration, and the contractor is responsible for maintaining these in this condition.' It goes on to say: 'The following are minimal requirements and shall not be construed as relieving the contractor from any of the responsibilities placed upon him by the foregoing paragraph'. Then there comes a provision with regard to the mix of the concrete to be used, saying 'The Engineer does not require a mix richer than 1–2–4, as specified previously. The position and arrangement of construction joints shall be agreed between the Engineer and the contractor, and how the construction joints shall be constructed, and the watertight work shall be concrete in lengths not exceeding 25ft, and adjoining sections shall only be cast after a time interval of ten days', and it ends by saying 'The contractor's prices for watertight concrete are to include for the provision of water bars at construction and day work joints as specified herein.'

The contentions put forward by [Counsel for the third party] can be summarised as follows: first of all, clause 1(1) of the contract provides that the contractor shall carry out and complete the works shown upon the contract drawings and described in the bill of quantities, in every respect to the reasonable satisfaction of the architect; secondly, clause 12(1) provides that the quality and quantity of the work shall be deemed to be that which is set out in the bill of quantities, but nothing contained in the bill of quantities shall override, modify or affect the conditions of the contract; thirdly, the watertight construction specification in the bill of quantities is a matter of the quality of the work, and work here includes design; otherwise, he says, this specification in the bill of quantities is quite unnecessary, for, if it was referring only to workmanship the contractor would be liable in any event; fourthly, the watertight construction specification in the bill of quantities does not conflict with the conditions of the contract because clause 1(2) and clause 24(1)(c) provide for compensation to the contractor; and, fifthly, any other interpretation, he contends, does violence to the construction of the bill of quantities which imposes, in effect, an absolute liability on the contractor to produce and maintain external retaining walls and basement slabs that are watertight.

[Counsel], on behalf of the plaintiffs, contends, first of all, that clause 12(1) and the bill of quantities are tailor-made in the sense that the additional provision in clause 12(1), referring to the effect of the amended or amplifying comments upon the conditions of the contract in the preliminaries bill, and those comments in the bill of quantities, show that the contract and the bill of quantities were drawn up together; secondly, that the bill of quantities, except for those comments, is not to modify the conditions of the contract; thirdly that 'work' in such a contract is not normally interpreted as including design, which is no part of a contractor's obligation; fourthly, that therefore, in so far as the bill of quantities watertight construction specification purports to impose upon the contractor a responsibility for structural design, it is seeking to modify the interpretation of the contract conditions which clause 12(1) of the contract does not permit. I accept [his] contention, which seems to me to have the added attraction of according with common sense. I should require the clearest possible contractual condition before I should feel driven to find a contractor liable for a fault in the design, design being a matter which a structural engineer is alone qualified to carry out and for which he is paid to undertake, and over which the contractor has no control. I agree that the construction for which [the third party] contends places the contractor in an impossible position. He cannot alter the faulty design without being in breach of contract, for this fault in the design is not, in my view, a discrepancy or divergence between the contract drawings and/or the bill of quantities, and yet if he complies with the design he would still be in breach. I decline to hold that the specification in the bill of quantities makes the contractor liable for the mistakes of the engineer, and, in so far as it may purport to do so, I think it is ineffective by reason of clause 12(1) of the contract.

In practice the contractor may be under a measure of design liability for various reasons. The contractor's obligations as to materials and workmanship often include a measure of liability for what may be called 'second order design', i.e. detailed decisions as to suitability of materials and methods of construction.

Reg. Glass Pty Ltd v. Rivers Locking Systems Pty Ltd

HIGH COURT OF AUSTRALIA

(1968) 120 CLR 576

The appellants employed the respondents, who were specialists in security installations, to fit a number of doors at their shop premises. The quotation for one door read:

> Rear single door: supply, fit and hang core door to suit opening, steel sheet outside area, fit Rivers 4 point non-break-out locking system operated by key from inside, hinge, stops, etc.

The respondents fitted the door upon an existing wooden frame made of Pacific maple, a softwood. Thieves later broke into the shop by forcing the door away from the frame with a jemmy.

Held: Although the door had been installed in compliance with the express terms of the quotation, the respondents were in breach of an implied term that they would supply, fit and hang a door which would provide reasonable protection against persons seeking to break in.

> BARWICK CJ: By common law, however, there was a term to be implied in the contract. The business of the defendant was to provide burglar-proof protection and the plaintiff unquestionably relied upon its skill and ability to supply, fit and hang a door which would provide reasonable protection against persons seeking to break in when the locking devices were in operation . . . It would, for instance, have been a plain breach of the contract if the locking devices had done no more than penetrate the wooden jambs and lintel so that pressure against the door could have split the wood and pushed the door in. Similarly, if the defendant had fixed the steel sheeting to the outside area of the door in such a way as it could have readily been removed, e.g., by exposed screws, although there would have been compliance with the express terms of the contract, the implied term would have been broken. The problem, as we see it therefore, is whether the Supreme Court was correct in setting aside the finding of the learned trial judge that the manner of the hanging of the door was such that the door did not provide reasonable protection against persons seeking to enter the shop.

It is increasingly common for the contractor expressly to assume some liability for design, especially under 'package deal' or 'design and build' contracts.

Greaves & Co (Contractors) Ltd v. Baynham Meikle & Partners

COURT OF APPEAL
(1975) 4 BLR 56

The plaintiff contractors were employed under a 'package deal' contract to build a warehouse which was to be used to store oil drums. The drums were to be kept on the first floor of the warehouse and to be moved about by fork-lift trucks. In fact, the warehouse as constructed was defective in that the movement of the fork-lift trucks caused vibrations which caused the floor to crack. The plaintiffs admitted liability to the employers, and the present action concerned the plaintiffs' claim against the defendant structural engineers, whom they had engaged to design the warehouse.

> LORD DENNING MR: [As] between the owners and the contractors, it is plain that the owners made known to the contractors the purpose for which the building was required, so as to show that they relied on the contractors' skill and judgment. It was therefore the duty of the contractors to see that the finished work was reasonably fit for the purpose for which they knew it

was required. It was not merely an obligation to use reasonable care. The contractors were obliged to ensure that the finished work was reasonably fit for the purpose. That appears from the recent cases in which a man employs a contractor to build a house: *Miller* v. *Cannon Hill Estates Ltd* (1931); *Hancock* v. *B.W. Brazier (Anerley) Ltd* (1966). It is a term implied by law that the builder will do his work in a good and workmanlike manner; that he will supply good and proper materials; and it will be reasonably fit for human habitation. Similarly in this case the contractors undertook an obligation towards the owners that the warehouse should be reasonably fit for the purpose for which, they knew, it was required, that is as a store in which to keep and move barrels of oil. In order to get the warehouse built, the contractors found they needed expert skilled assistance, particularly in regard to the structural steel work . . .

What was the cause of this cracking of the floors? The structural engineers said that it was due to the shrinkage of the concrete for which they were not responsible. There was nothing wrong, they said, with the design which they produced. But the judge did not accept that view. He found that the majority of the cracks were caused by vibration and not by shrinkage. He held that the floors were not designed with sufficient strength to withstand the vibration which was produced by the stacker trucks.

On those findings the first question is: what was the duty of the structural engineers towards the contractors? The judge found that there was an implied term that the design should be fit for the use of loaded stacker trucks, and that it was broken. Alternatively, that the structural engineers owed a duty of care in their design, which was a higher duty than the law in general imposes on a professional man; and that there was a breach of that duty . . .

What then is the position when an architect or an engineer is employed to design a house or a bridge? Is he under an implied warranty that, if the work is carried out to his design, it will be reasonably fit for the purpose? Or is he only under a duty to use reasonable care and skill? This question may require to be answered some day as matter of law. But in the present case I do not think we need answer it. For the evidence shows that both parties were of one mind on the matter. Their common intention was that the engineer should design a warehouse which would be fit for the purpose for which it was required. That common intention gives rise to a term implied in fact.

In a design and build contract for dwellings it will be an implied term that the buildings will be fit for habitation on completion. If for some reason such a term is not to be implied there will be a lesser implied term that the contractors shall have designed the building with skill and care.

Basildon District Council v. J.E. Lesser (Properties)

QUEEN'S BENCH DIVISION

(1984) 8 Con LR 89

In this case the buildings were erected on JCT 1963 terms but the specification and design drawings were prepared by the defendant contractors. The Council's architect's department had commented on the drawings.

JUDGE JOHN NEWEY QC: The implication of a term that a builder should provide a building which is fit for purposes imposes upon him a higher duty than is ordinarily implied in a contract for the design of a building by an architect. The latter, like the engineer, solicitor, doctor and other professionally qualified persons has only to use reasonable care and skill:

Bolam v. *Friern Hospital Management Committee* [1957] 1 WLR 582 and *Greaves (Contractors) Ltd* v. *Baynham Meikle & Partners* [1975] 1 WLR 1095 . . .

In the light of the authorities to which I have referred, if in the present case the council and the contractors had entered into a contract with a minimum of terms providing for the contractors to build system-dwellings for the council it would, I think, have been placing reliance upon the contractors, despite an earlier appraisal of the contractors' system by the National Building Agency.

The conclusion between the council and the contractors of an agreement incorporating standard conditions made the position much less clear. Terms as to workmanship and materials are invariably implied in standard construction contracts and other terms are obviously possible, but an implied term as to design is unheard of; since a principal feature of such contracts is that there is a division of function: the employer's architect undertakes design, while the contractor is responsible for workmanship and materials. The specification A3 and the nine drawings prepared by the contractors became part of the contract drawings, showing the contract works.

Since the third recital stated only that A3 and the drawings had been prepared by or under the direction of a council officer and not that he had prepared them, the council is not precluded from asserting their true authorship. The reality is that the design of the dwellings was, apart from such small contribution as was made by the addendum to A3, the exclusive work of the contractors. As the contractors well knew the council wanted dwellings, the contractors were the experts in their system and the council were relying upon them to apply it properly, so as to produce habitable dwellings. The case was quite unlike the usual standard contract case, in which the architect has as contemplated by the contract designed the works. The contractors had I think a continuing responsibility for the design of the dwellings, even after their drawings had become contract drawings.

The case can I think be distinguished from Lynch v. Thorne, in which the court concentrated its attention upon the agreed defective plan and did not consider whether if the builder was its author he was in breach of an implied term either on account of its faults or in failing to warn against the use of it.

I hold that it was an implied term of the agreement that the buildings designed by the contractors as dwellings should be fit for habitation on completion. There was no comparable implied term in relation to the garages and the civil engineering works which were designed by the council's officers.

If I had not decided that there was an implied term as to fitness for habitation in respect of the dwellings, I should have decided that a lesser term, namely that the contractors should have designed them with the skill and care to be expected of system builders, should have been implied.

Appropriate design will often involve careful consideration of the surrounding circumstances. These will often involve ground conditions which can be difficult to investigate.

Mirant Asia-Pacific Construction (Hong Kong) Ltd and another v. Ove Arup & Partners International Ltd and another (No.2)

COURT OF APPEAL

(2005) 105 Con LR 1, [2005] EWCA Civ 1585

The claimants brought proceedings against the defendants in relation to a contract for the construction of a coal-fired power station at Sual in the Philippines for a loss arising out of the failure of two of the foundations of one of the boiler houses at the power station about April 1997.

The buildings were designed by the defendants, who were both companies in the Arup Group.

MAY LJ:

I reject Arup's grounds of appeal which say that the judge was wrong for these reasons:

(1) Arup were obliged under the design agreement to carry out their design with the skill and care of ordinarily competent engineers for work of this kind. This is not contentious.

(2) A competent foundation design requires a sufficient knowledge of the ground conditions to determine a safe bearing capacity for foundations.

(3) If the designing engineer's knowledge of the ground conditions is insufficient to enable him to determine a safe bearing capacity, he may work initially upon assumptions. But he has an obligation to see to it that the requisite additional information is acquired to verify the assumptions. He does not necessarily have to get the additional information personally, but he must see to it that someone does, and he must see to it that the client knows that the additional information has to be obtained. Absent an explicit warning and disclaimer, it would not be sufficient for a designer, whose initial design is based on an unverified assumption, to leave it to the client alone to obtain and evaluate the additional information. The designing engineer is responsible for the design, and he should normally see to it that the necessary additional information is conveyed back to him, so that he may judge that it is sufficient for the purpose of his design.

I accept Arup's submissions that foundation design is not an exact science, and that perfect and complete knowledge of the ground may be impossible to achieve. That does not, however, mean that a foundation designer does not have to see to it in appropriate circumstances that his assumptions are verified to the extent that a reasonably competent design would require. The judge had ample evidence to determine, as he did, what the required extent was in this case.

Even where it is clear who bears responsibility for design, there may be legitimate grounds for argument as to whether the defect is one of design.

One element of design is whether those likely to be employed to construct are likely to be capable of executing the design: what is sometimes called buildability.

Department of National Heritage v. Steensen Varming Mulcahy

QUEEN'S BENCH DIVISION

(1998) 60 Con LR 33

This action arose out of the construction of the British Library. The plaintiff (DNH) was the emanation of the Crown responsible at the relevant time for the construction of the library. The defendants (SVM) were the consulting mechanical and electrical engineers. The first third parties, Balfour Beatty Ltd (BB), were the electrical contractors.

The action arose out of the discovery of very extensive damage to the low voltage cabling installed in the library. Some 300km were reinstated. DNH and BB both asserted that the damage was caused by faulty design by SVM. (DNH and BB were covered by the same insurance.)

Held: SVM were not in breach of their duty of care in design. BB were under a duty under their contract to employ 'directly employed qualified staff' and were frequently in breach of this duty. If they had employed such staff, the electrical scheme designed by SVM would have been buildable.

JUDGE PETER BOWSHER QC:

129. SVM were entitled to assume that their design would be executed by a workforce composed of directly employed labour provided by a contractor with one of the best reputations in the world, bearing in mind that that contractor had undertaken to work to a high standard (not just a 'reasonable standard') on one of the most prestigious buildings to be built in this country in this century. (In this connection, it is to be noted that Counsel for DNH submitted in his closing submissions that the nature of the project was relevant in considering the standard to be expected of SVM: he submitted that having regard to the nature of the project, 'the highest reasonable competence' was required of SVM.) SVM were also entitled to assume that this highly skilled contractor would, in compliance with its contract, draw to the attention of SVM, through the management contractor, any requirement of the design which was contrary to good practice. Many years after the event, in this trial it was suggested that even if there were good reasons for lid-down trunking in areas where there were ceilings, those reasons did not apply in the plant rooms. There is some merit in that contention, but it does not amount to showing bad practice, and why did BB not mention it when tendering to the specification which distinguished between the trunking in plant rooms and other areas as regards IP41 and IP44 protection only?

130. I return to the submission which DNH now submit as their 'primary case', namely that, given SVM's design it would have been impossible even if special or exceptional care had been taken to avoid substantial cable damage. I find that DNH and BB are urging too low a standard of care as the standard to be applied by BB. 'Exceptional' care was not required of BB, but a high standard was required: I am not sure if that rates as a special care in Counsel's terminology, but it is higher than 'ordinary care'. Even if they were relying on the correct standard of care, both DNH (through the PSA [Property Services Agency]) and BB should have objected to the design at the time of tendering, or in default of such objection being accepted, BB should have refrained from tendering. Having undertaken a task which they now say was too difficult (although the difficulty, if it existed, must have been apparent at the stage of tendering), BB cannot now be heard to say that the task which they undertook with open eyes was impossible of achievement. Equally, DNH, who had skills available to them through the PSA which matched the skills of BB, cannot now be heard to complain that they entered into a contract with BB which was made impossible of fulfilment by virtue of the clear and open requirement of lid-down trunking made by that contract.

131. It does not lie in the mouth of BB to complain that it was not within the capability of their workforce to install cabling to a high standard in lid-down trunking, and it does not lie in the mouth of DNH to rely on and adopt that case against SVM.

132. I should add that I am convinced by the evidence (to which I shall refer later) that the reason why BB did not say when tendering or when doing the work that to require the trunking covers to be lid down was contrary to good practice, was that it was not contrary to good practice.

133. Taking a further point from the passage from the judgment of Judge Newey in *Equitable Debenture Assets Corp Ltd* v. *William Moss Group Ltd* from which I have quoted, DNH submit that the design must be 'supervisable', capable of being executed under supervision so that the quality of work can be checked by the supervisor. As to that, I agree with what was said by Judge Newey, but subject to a qualification. Some work is by its nature difficult to supervise and such that no design can remove the difficulty of supervision. There are obvious difficulties about supervising the work of a steeplejack or of a diver working on an underwater construction or repair. Equally, there are difficulties about supervision of the work of an electrician in places where access for inspection is difficult. The overall method can be supervised, and in many respects it can be seen by the supervisor if the electrician is going about his work in the wrong way, but in one vital respect the quality of the work rests very much on the individual actually doing the work. When a cable is being led or drawn through a confined space, the onlooker can see whether the overall method used is correct, but the man (and oddly enough it does always seem to be a man) actually doing the work is the only one who can feel the tug on the cable which indicates that some damage may be done if extra care is not taken at that point with that cable. If, having felt the tug, the workman carries on regardless, there is little that the supervisor can do about it except by inspection afterwards and inspection is difficult unless spot checks are made on individual cables while the work is in progress. That is one reason why the quality of the workforce is vitally important.

134. On the evidence, I do not see that it was any more difficult to supervise either during the work or by inspection afterwards with lid-down than with lid-up trunking. There are difficulties about inspection inherent in both, but as a general rule one was not more difficult than the other: in specific places one might be more difficult to inspect than the other. For example, where trunking was set tight up close to a ceiling, lid-down trunking would be the only type of trunking which could be inspected.

In exceptional circumstances it may be possible to infer that the employer looked to the contractor to advise him that the design was faulty.

Brunswick Construction Ltd v. Nowlan

SUPREME COURT OF CANADA

(1974) 21 BLR 27

Nowlan engaged an architect to design a house and then contracted with the appellants to build the house according to the architect's design. The design was faulty and made no sufficient provision for ventilation of the roof space and timbers. The result was a serious attack of rot. No architect was engaged to supervise the construction.

Held: The appellants were liable.

RITCHIE J: In my opinion a contractor of this experience should have recognized the defects in the plans which were so obvious to the architect subsequently employed by the respondents, and, knowing of the reliance which was being placed upon it, I think the appellant was under duty to warn the respondents of the danger inherent in executing the architect's plans, having particular regard to the absence therein of any adequate provision for ventilation.

Plant Construction plc v. Clive Adams Associates (No.2)

COURT OF APPEAL

(2000) 69 Con LR 106

The claimant (Plant) was the main contractor in a contract to install two new engine mount rigs and a suspension rig in a laboratory at the Ford Research Engineering Centre. The contract was under Ford terms, which put the risk of any damage to the works caused by Ford or its servants or agents on Plant. JMH Construction Services Ltd (JMH) were the sub-structure sub-contractors and Clive Adams Associates (Adams) were consulting structural engineers.

On 2 January 1994, the roof of the building collapsed. This was because of defects in the temporary support provided to the roof when the steel column which supported the roof had temporarily to be removed. Plant paid Ford £1,313,031 in settlement of Ford's claim and sought to recover its total loss of about £2 million from Adams and JMH. Adams and Plant reached a settlement and the action continued as a trial of Plant's claim against JMH.

It was accepted that the effective reason for adopting the faulty system of support was that it was insisted on by Ford's chief engineer, who was a dominating personality who did not brook dissent. Relevant personnel in Plant, Adams and JMH all realised that the temporary support proposal was defective.

Held: In the circumstances, JMH was under a duty to warn Plant that the support system was defective. The matter should be remitted to decide what would have happened if Plant had given a more effective warning. (At a later hearing, Judge John Hicks QC held that if a more effective warning had been given, the collapse would not have occurred.)

MAY LJ: Any analysis of implied terms in a building contract must start with and take proper account of its express terms. Subject to the express terms, there will normally be an implied term that the contractor will perform his contract with the skill and care of an ordinarily competent contractor in the circumstances of the actual contractor. In my judgment, the factual extent of the performance which this term requires will depend on all relevant circumstances, which may vary enormously. I shall not attempt to make a comprehensive list of possible circumstances. But they may in particular cases include the size, nature and details of the works; the experience and perceived expertise of the contractor; relevant elements of the relationship between the contractor and the employer and of their respective relationships with others, for example, architects, engineers, surveyors, contracts managers, clerks of works, sub-contractors, local authority building inspectors and so forth; and crucially details of the particular parts of the works and other facts which give rise to the question whether the contractor fulfilled the obligation which the implied term imports.

The present appeal concerns (a) temporary works which were (b) designed and specifically instructed by the employer, so that (c) they became part of JMH's sub-contract works, which (d) were obviously dangerous, and which (e) JMH knew to be dangerous. JMH were sub-contractors to Plant under a sub-contract whose terms were, so far as is relevant, unsophisticated. Plant had the services of a consulting civil engineer. Plant and JMH are each to be taken as experienced in their respective roles. In my judgment, of the elements which I have referred to, all are relevant but (d) and (e) are crucial. These temporary works were, to the knowledge of JMH, obviously dangerous to the extent that a risk of serious personal injury or death was apparent. JMH were not mere bystanders and, in my judgment, there is an overwhelming case on the particular facts that their obligation to perform their contract with the skill and care of an ordinarily competent contractor carried with it an obligation to warn of the danger which they perceived. The fact that no one was injured is irrelevant. The question is, not whether JMH owed a duty of care to someone who was injured, but what was the scope of the implied contractual term in their subcontract with Plant. Nor is it relevant whether the loss which Plant claim should be categorised as economic loss, since economic loss is a problem which arises in the analysis of duties of care in tort. The facts that the design and details of the temporary works had been imposed by Ford and that Plant had Mr Adams as their consulting engineer do not, in my view, negative or reduce the extent of performance which the implied term required in this case. The fact that other people were responsible and at fault does not mean, in my judgment, that on the facts of this case JMH were not contractually obliged to warn of a danger. Nor in this case is the extent of performance negated by the fact that JMH were expressly obliged by contract to do what [Ford's Senior Civil Engineer] instructed. JMH, with others, had a duty to guard against the risk of personal injury to a potentially large number of people. That duty extended to giving proper warnings about the risk. It was not itself a contractual duty owed to Plant, but it is a relevant circumstance in determining the extent of performance which JMH's implied duty of skill and care required. In my judgment, the judge in this case came to the correct conclusion about JMH's implied contractual duty and I would reject [Counsel for the appellant's] first submission.

This analysis and conclusion accords, with one gloss, with Judge Newey's decision in *Lindenberg* v. *Canning*. It also, I think, accords with Judge Lewis' analysis in the *Oxford University Press* case, since he was obviously inclined to imply a duty in tort on the contractor to warn of design defects which might give rise to danger to personal safety. He would also, no doubt, have been similarly inclined to imply a duty in contract. I have also noted that Judge Bowsher in the *University of Glasgow* case did not suggest that there were no circumstances in which a term might be implied requiring a contractor to warn a building owner of defects in the design. The gloss relating to *Lindenberg* v. *Canning* is that in the present case JMH knew that the design of the temporary works was defective. This knowledge is not, in my view, to be taken as having been displaced by what Mr Adams said about snow loading nor [Ford's Senior Civil Engineer's] manifestly irresponsible suggestion about the continued support from J11. In *Lindenberg* v. *Canning*, the builder did not apparently know that the 9 inch wall was load bearing; but he ought to have known. I would expressly reserve for future consideration circumstances where (a) the contractor did not know, but arguably ought to have known, that the design was dangerous, and (b) where there was a design defect, of which the contractor knew or ought to have known, which was not dangerous. Neither of these circumstances arises in the present appeal.

As to [Counsel for the appellant's] second submission, the judge was in my view entitled on the evidence to conclude that JMH were in breach of the obligation which I have identified notwithstanding the warning which they did give. The finding in para 69 of the judge's judgment, which I have set out in full, was in my view a justified finding of fact which there is no basis for this court to disturb. It seems to me that there are a number of possible answers to [Counsel for the appellant's] question, What more could JMH have done? Generally speaking, the answer is that they could have protested more vigorously. It is not, I think, appropriate to be more specific in the light of my view about [Counsel for the appellant's] third submission.

As to this third submission, [Counsel for the respondent] accepts that the judge made no express finding about what would have happened if JMH had protested more vigorously. The question of causation was an issue on the pleadings. We are told that it was a live issue during the trial, although my strong impression is that it was well down the list of issues which featured as important. I do not think that the judge can in the circumstances properly be criticised for a judgment which takes as implicitly obvious a conclusion that, if JMH had protested more vigorously and sufficiently, steps would have been taken so that the roof collapse would have been avoided. There is, however, a properly arguable case to the contrary. In my view, this court is in no position to fill the factual gap in the judge's judgment. He had heard extensive evidence, most of it given orally, and we have not. Although we can read the transcripts, I consider that this question is one that can only properly be determined by a judge who has heard the evidence. It is a small omission from a judgment about which no other proper criticism can be advanced. It seems to me that, in the unusual circumstances of this case, the parties are entitled to have further findings of fact explicitly made. The judge's implicit conclusion in his present judgment is perhaps obvious. But, without further findings of fact, I would not assume that any particular conclusion is obvious.

I would therefore allow this appeal to the limited extent only of remitting for further findings of fact the question what would have happened if JMH had protested more vigorously so as to fulfil their contractual obligation. I would hear further submissions about the details of this, but my provisional inclination is that the matter should be remitted to Judge Hicks himself. This appeal has raised no matter critical of him such as would suggest that it ought to go to a different judge. It would save the parties a great deal of money if the matter is dealt with by the same judge. I would leave it to the judge's discretion to direct how the additional hearing should be managed and conducted. He will no doubt wish to hear further submissions from the parties in the light of the decision of this court. The additional findings of fact should, however, in my view be made without additional evidence, except perhaps in the judge's discretion any limited additional evidence which would be admissible upon the hearing of an appeal to the Court of Appeal. The additional finding should determine the causation issue explicitly. I would set aside the present implicit conclusion and would not assume that the judge will upon reconsideration necessarily reach the same conclusion.

To this extent, I would allow this appeal.

Materials and workmanship

Where a building owner has engaged a contractor both to design and construct a building, unless the parties have agreed to the contrary, the builder impliedly agrees:

(1) that he will do his work in a good and workmanlike manner;
(2) that he will supply good and proper materials; and
(3) (where the contract is to build a house) that the house will be reasonably fit for human habitation. In other cases, this obligation is to ensure that the completed structure will be reasonably fit for its intended purpose.

Test Valley Borough Council v. Greater London Council

COURT OF APPEAL

(1979) 13 BLR 63

A dispute arose between the parties as to the interpretation of a Town Development Agreement and the issue of principle before the court related to the standard of duty to be implied into the agreement.

Held: There were implied terms of the agreement that the respondents would provide dwellings which were constructed in a good and workmanlike manner, of materials which were of good quality and reasonably fit for their intended purpose and so as to be fit for human habitation.

PHILLIPS J (whose judgment the Court of Appeal affirmed): Where a house proves defective, the defects may be of materials, of workmanship or of design. The duty of the person supplying the house will depend on the circumstances, and, in particular, upon whether he is responsible for the design. Where he is, the duty is said to be to do the work in a good and workmanlike manner, to supply good and proper materials and to provide a building reasonably fit for human habitation: see *Hancock* v. *B.W. Brazier (Anerley) Ltd* (1966). Of course, the circumstances may be such that the third limb of the duty does not arise; for example, where the client employs his own architect: see *Lynch* v. *Thorne* (1956). Similar duties arise in the case of a 'package deal' or a 'design and build' contract: see *Greaves & Co (Contractors) Ltd* v. *Baynham Meikle & Partners* (1975).

Hancock v. B.W. Brazier (Anerley) Ltd

COURT OF APPEAL

[1966] 2 All ER 901

LORD DENNING MR: In the autumn of 1958 the builders, B.W. Brazier (Anerley) Ltd were proposing to develop an estate of some 40 houses known as the Haven Close Estate in Swanley. They laid it out in plots. Each prospective purchaser could choose his own plot. The three plaintiffs did so. Each of them selected a plot and paid a sum down to get it allotted to him; but no contract was made at that time for the erection of the houses. In the next three months, between October 1958 and January 1959 the defendants did a good deal of work on the foundations. They put in some hardcore and then put a layer of four inch concrete above it. They started to build the houses. All this before any contracts were signed. After the work had been done written contracts were entered into by the purchasers with the builders By clause 1 the purchaser contracted to buy the house for £2,750. Clause 9 said: 'The vendor [Braziers] will prior to completion at its own cost and charge and in a proper and workmanlike manner erect build and complete on the above freehold land a messuage or dwelling-house in accordance with the plan and specification supplied to the purchaser subject only to such variations as may be ordered by the purchaser'. The plan and specification were in the ordinary form providing for the brickwork, the rooms, and so forth. There was nothing in the specification about the foundations, except that the site concrete was to be 4 in thick and laid to required levels. On the plan there was this: 'four-inch site concrete on hardcore'.

The builders went on with their work. By May 1959 the houses were substantially finished. Completion took place on 20 May. The purchasers went into occupation. All appeared well; but two years later there was trouble in one house. Three or four years later in another house. Then in other houses. The trouble was this: the floors were cracking and breaking up. It was investigated. The cause was found in the hardcore which had been put into the foundations. It was not just the ordinary kind of hardcore which comes out of demolished buildings, such as bricks, flints, chalk and earth. There was quite a lot of sodium sulphate in this hardcore. We have been shown some of this sodium sulphate. The blocks of it look very much like flints or chalk. Anyone might easily mistake it for ordinary hardcore. Somebody must have got it from some old dump from a factory, must have thought it was stones and must have used it for hardcore. Yet it turns out to be a dangerous substance to use as hardcore. I say it is a 'dangerous substance' because, when it is dry, it soaks up water whenever it gets the chance, and then it expands with almost irresistible force. If it is already damp, it will soak up more water and expand. If it gets into contact with concrete, it brings about a chemical reaction which may cause the concrete to disintegrate. Hence the damage caused to these houses. It was the sodium sulphate in the hardcore that caused the floors to break up. The purchasers have been put to much expense to get it put right.

Let me say at once that all this trouble was in no way the fault of the builders. They had no reason to suspect the hardcore. They bought it in good faith. The defects were not apparent. No one could tell by looking at this hardcore that it had sodium sulphate mixed in with it. The builders used all reasonable care, skill and judgment. Counsel for the builders says that in these circumstances they were not guilty of any breach of contract. The entire obligation of the builders, was contained in clause 9 of the contract. That was a clause whereby they had to use only all reasonable care and skill and judgment in their work. It did not mean that they warranted the suitability of the materials which they used. He relies on the principle that where a matter is covered by express provision in a contract, it is not for the court to make an implication on the same matter.

It is quite clear from *Lawrence* v. *Cassell* (1930) and *Miller* v. *Cannon Hill Estates Ltd* (1931), that when a purchaser buys a house from a builder who contracts to build it, there is a threefold implication: that the builder will do his work in a good and workmanlike manner; that he will supply good and proper materials; and that it will be reasonably fit for human habitation. Sometimes this implication, or some part of it, may be excluded by an express provision, as for instance in *Lynch* v. *Thorne* (1956). The specification there expressly provided that the walls were to be 9 in brick walls. The work was done with good materials and workmanship and exactly in accordance with the specification; but the walls did not keep out the driving rain. The builder was held not liable. The question in this case is whether the threefold implication is excluded by clause 9. I think that it is not, for this simple reason: clause 9 deals only with workmanship. It does not deal with materials. The quality of the materials is left to be implied; and the necessary implication is that they should be good and suitable for the work. I am quite clear that it is implied in the contract that the hardcore must be good and proper hardcore, in the same way as the bricks must be good and proper bricks. I know that the builders were not at fault themselves. Nevertheless this is a contract: it was their responsibility to see that good and proper hardcore was put in. As it was not put in, they are in breach of their contract. If it is any consolation to them, they can try and get hold of their suppliers and sue them if they can prove it against them; but they have to take responsibility so far as the purchasers are concerned.

In principle implied terms may be varied by agreement so as to impose a higher or lower duty on the contractor. (Any reduction of the contractor's liability might now be subject to the requirement of 'reasonableness' under the Unfair Contract Terms Act 1977 and Supply of Goods and Services Act 1982.) Such an agreement may be arrived at either from express words or as a necessary deduction from the circumstances.

Steel Company of Canada Ltd v. Willand Management Ltd

SUPREME COURT OF CANADA

(1966) 58 DLR(2d) 595

The respondent roofing contractor had bid for, and been awarded, a number of roofing contracts on buildings being constructed for the appellants. The roofs of the buildings were to be covered by a fluted steel deck, and the roofing membrane (a layer of insulation and roofing felt) was to be attached to the steel deck by a compound of fire-resistant material known as 'Curadex'. As part of the contract the respondent guaranteed 'for a period of five years that all work above specified will remain weather tight and that all material and workmanship employed are first-class and without defect'.

The roof suffered damage in high winds and the evidence was that this was due to the 'Curadex' proving an unsatisfactory way of fixing the membrane to the steel deck.

Held: The guarantee extended to damage to the roof arising out of faulty design, and the appellant was therefore entitled to recover the cost of repairing the roof.

RITCHIE J: It accordingly appears to me that the question which lies at the heart of this appeal is whether the responsibility for the results of using 'Curadex' rests upon the appellant who prescribed it or upon the respondent who applied it, and in this regard it seems to me to be of first importance to consider the circumstances under which this adhesive came to be included in the specifications . . .

It therefore appears to me that when [a director] signed the tenders on behalf of the respondent, although he had had no actual experience in the use of 'Curadex' on sloping roofs, [he] was, as the result of lengthy discussions with the appellant's officers and of his having previously used the product on flat roofs, fully aware of the factors necessary to enable him to decide whether or not this adhesive was a first class material for its intended use, and whether it was one which his company was prepared to guarantee to remain 'weather tight' for a period of five years . . .

In construing the guarantee as he did [the trial judge], was clearly influenced by the fact that he did not think that it would have been reasonable for the defendant to have expected, and the plaintiff to have given, an absolute guarantee against the elements when neither had had any experience with the capacity of 'Curadex' to perform properly on the sloping steel deck.

In this regard it is, however, to be remembered that the respondent is an experienced contractor specializing in the roofing business and that it was bidding in competition with several other roofing contractors. Under these circumstances the language employed by Cockburn CJ, in *Stadhard* v. *Lee* (1863), which was quoted with approval in the Exchequer Court of Canada in *Jones* v. *The Queen* (1877), appears to me to be particularly pertinent: 'It frequently happens, in the competition which notoriously exists in the various departments of business, that persons anxious to obtain contracts submit to terms which, when they come to be enforced, appear harsh and oppressive. From the stringency of such terms escape is often sought by endeavouring to read the agreement otherwise than according to its plain meaning. But the duty of a court in such cases is to ascertain and to give effect to the intention of the parties as evidenced by the agreement, and though, where the language of the contract admit of it, it should be presumed that the parties meant only what was reasonable, yet, if the terms are clear and unambiguous the court is bound to give effect to them without stopping to consider how far they may be reasonable or not.'

In construing the guarantee by supplying the words 'in so far as the plans and specifications with which it had to comply would allow' it appears to me that the Courts below have tacitly accepted the proposition that no matter how experienced a contractor may be in a particular field, he nevertheless bears no responsibility for the employment of defective material in the work which he has undertaken, provided that it is a material which has been selected by the owner and included in the specifications. This proposition finds support in the judgment of Vann J, in the *MacKnight* case (1950), in a passage which was expressly adopted by Hughes J, which reads as follows: 'The defendant [i.e., the owner] specifically selected both material and design and ran the risk of a bad result. If there was an implied warranty of sufficiency, it was made by the party who prepared the plan and specifications, because they were its work, and in calling for proposals to produce a specified result by following them, it may fairly be said to have warranted them adequate to produce that result.'

I cannot accept this proposition which appears to me to run contrary to a long line of decisions in England starting with *Thorn* v. *London Corporation* (1876), which have been followed in this country . . . and the effect of which is summarised in part in *Hudson's Building & Engineering Contracts* (1959), 8th edn, page 147 where it is said: 'Sometimes, again, a contractor will expressly undertake to carry out work which will perform a certain duty or function in conformity with plans and specifications, and it turns out that the work constructed in accordance with the plans and specifications will not perform that duty or function. It would appear that generally the express obligation to construct a work capable of carrying out the duty in question overrides the obligation to comply with the plans and specifications, and the contractor will be liable for the failure of the work not-withstanding that it is carried out in accordance with the plans and specifications. Nor will he be entitled to extra payment for amending the work so that it will perform the stipulated duty.'

The agreement to furnish a written guarantee 'that all work above specified will remain weather tight' for five years in my view constitutes at the very least an express undertaking 'to carry out work which will perform a certain . . . function in conformity with plans and specifications' and in accordance with the principles stated in the paragraph last above cited, I think that it follows that when a work so constructed does not perform the function which the contractor agreed that it would perform, the contractor is liable for the failure of the work and is not entitled to extra payment for repairing it 'so that it will perform the stipulated duty'.

In the course of his reasons for judgment, Hughes J expresses the following view: 'It would seem . . . that from this evidence that the defendant corporation was taking a calculated risk in specifying the adhesive designed and required to fasten the roofing membrane to a roof of new design, and it would seem that they knew this to be the case.'

In my opinion the evidence discloses that both parties were fully alerted to any limitations which may have attached to the use of 'Curadex' as an adhesive on these roof decks and in view of the fact that neither of them had had any experience in using it on sloping roofs, I think that some risk was involved. This may have been the reason why the appellant required the contractors who were tendering on the work to provide the guarantee in question, but whatever the reason may have been, it appears to me that any risk involved in the undertaking was accepted by those who were prepared to tender in accordance with specifications which included the requirement of providing a written guarantee that all material employed in the work was first class and without defect, and that 'all work . . . specified' would remain weather tight for a period of five years.

Young and Marten Ltd v. McManus Childs Ltd

HOUSE OF LORDS

(1969) 9 BLR 77

The respondents, builders engaged in developing a housing estate at Gerrards Cross, sub-contracted the roofing of the houses to the appellants. The roofing contract called for the use of 'Somerset 13' tiles, which were manufactured only by JB. The appellants obtained supplies of tiles from their own suppliers in London who in turn obtained them from JB. Some of the tiles supplied had a defect, apparently due to faulty manufacture, which was not discoverable by any reasonable inspection by the appellants. After less than 12 months the tiles began to disintegrate and the respondents claimed the cost of the re-roofing of the houses. The appellants argued that, as they had not chosen the type of tile and could not have discovered the defect, they were not liable.

Held: This argument was rejected.

LORD REID: This is a contract for the supply of work and materials and this case raises a general question as to the nature and extent of the warranties which the law implies in such a contract. As regards the contractor's liability for the work done there is no dispute in this case: admittedly it must be done with all proper skill and care. The question at issue relates to his liability in respect of material supplied by him under the contract. The appellants maintain that the warranty in respect of materials is similar to that in respect of work, so that if the selection of material and of the person to supply it is left to the contractor, he must exercise due skill and care in choosing the material and the person to supply it. But where, as in this case, the material and the supplier were chosen by the respondents, the appellants maintain that there was no warranty as to the fitness or quality of the tiles. The respondents admit that, if it is held that the choice of this type of tile was theirs and theirs alone, there can be no implied warranty that this type of tile was fit for the contract purpose. But they say that there still was a warranty that the tiles would be of good quality and that that warranty must be implied notwithstanding the fact that they left no choice to the appellants in selecting the person who was to supply the tiles. If that is right then the respondents must succeed. The loss was not caused by Somerset 13 tiles being unsuitable for the contract purpose: it was caused by the tiles which were supplied being of defective quality.

There is not very much authority on this matter so it may be well first to consider it as a question of principle. In my view no warranty ought to be implied in a contract unless it is in all the circumstances reasonable. If authority be required for that proposition I find it in the judgment of the Exchequer Chamber in *Readhead* v. *Midland Railway Co* (1869): 'Warranties implied by law are for the most part founded on the presumed intention of the parties, and ought certainly to be founded on reason, and with a just regard to the interests of the party who is supposed to give the warranty, as well as of the party to whom it is supposed to be given.' I take first the general question of the contractor's liability where the material which he is required to use can be obtained from any one of several suppliers and the choice of suppliers is left to him. There is no doubt that in every case he is bound to make a proper inspection of the material before using it, and he will be liable if the loss is caused by the use of material which reasonable inspection would have shown to be defective. The question is whether he warrants the material against latent defects.

There are in my view good reasons for implying such a warranty if it is not excluded by the terms of the contract. If the contractor's employer suffers loss by reason of the emergence of the latent defect, he will generally have no redress if he cannot recover damages from the contractor. But if he can recover damages the contractor will generally not have to bear the loss: he will have bought the defective material from a seller who will be liable under section 14(2) of the Sale of Goods Act because the material was not of merchantable quality. And if that seller had in turn bought from someone else there will again be liability so that there will be a chain of liability from the employer who suffers the damage back to the author of the defect. Of course the chain may be broken because the contractor (or an earlier buyer) may have agreed to enter into a contract under which his supplier excluded or limited his ordinary liability under the Sale of Goods Act. But in general that has nothing to do with the employer and should not deprive him of his remedy. If the contractor chooses to buy on such terms he takes the risk of having to bear the loss himself if the goods prove to be defective.

Moreover many contracts for work and materials closely resemble contracts of sale: where the employer contracts for the supply and installation of a machine or other article, the supply of the machine may be the main element and the work of installation be a comparatively small matter. If the employer had bought the article and installed it himself he would have had a warranty under section 14(2), and it would be strange that the fact that the seller also agreed to install it should make all the difference.

The speciality in the present case is that these tiles were only made by one manufacturer. So the contractor had to buy them from him or from someone who bought from him. Why should that make any difference? It would make a difference if that manufacturer was only willing to sell on terms which excluded or limited his ordinary liability under the Sale of Goods Act, and that fact was known to the employer and the contractor when they made their contract. For it would be unreasonable to put on the contractor a liability for latent defects when the employer had chosen the supplier with knowledge that the contractor could not have recourse against him. If the manufacturer's disclaimer of liability caused him to supply the goods at a cheaper price, as in theory at least it should, the employer ought not to get the benefit of a cheap price as well as a warranty from the contractor.

A more difficult case would be where the employer and contractor had no reason to suppose, when they made their contract, that the manufacturer would refuse to sell subject to a seller's ordinary liabilities in respect of the goods which he sells. But I need not consider that case now because there is no suggestion that [JB] had refused to sell except on terms which limited their ordinary liability in respect of latent defects in their tiles. No doubt there will be some cases where, although the contractor had a right of recourse against the manufacturer, he cannot in fact operate that right. The supplier may have become insolvent, or, as in the present case, the action against the contractor may be so delayed that he has no time left in which to sue his supplier. But these cases must be relatively few and it would seem better that the contractor should occasionally have to suffer than that the employer should very seldom have any remedy at all. It therefore seems to me that general principles point strongly to there being an implied warranty of quality in this case.

Gloucestershire County Council v. Richardson

HOUSE OF LORDS

[1969] 1 AC 480

The council engaged the respondents under a RIBA contract (1939 edition, 1957 revision) as main contractors for the erection of extensions to a technical college. The architect nominated C as supplier of concrete units, and sent to the respondents the quotation from C, instructing the respondents to accept it, which they did. The quotation contained clauses limiting C's liability if the goods supplied proved defective. Under the terms of the main contract the respondents were entitled to make reasonable objection to nominated sub-contractors but not to nominated suppliers.

The concrete units proved defective and this led to great delays. In due course the respondents purported to exercise their right to determine the contract on the ground that the work had been delayed for more than one month 'by reason of architect's instructions', and the appellants in turn claimed damages for wrongful repudiation. It was conceded that the respondents were not entitled to terminate if they were in breach of an implied warranty as to the quality of the concrete units.

Held: The respondents were not in breach of any implied warranty.

LORD PEARCE: My Lords, the contractor in any particular field of business, when he engages to do certain work and supply material, impliedly warrants that the materials will be of good quality, unless the particular circumstances of the case show that the parties intended otherwise. To find the intention one must consider the express terms of the contract and any admissible surrounding circumstances.

Here the parties entered into a Royal Institute of British Architects contract, a complicated and sophisticated document. There is no express acceptance by either party of liability for the quality of the nominated materials. The contractor must comply with the instructions of the architect. He must accept the architect's nomination in respect of certain sub-contractors and nominated suppliers. No nominated sub-contractor, however, can be employed (clause 21) if the contractor makes reasonable objection to him or if, inter alia, the sub-contractor will not enter into a sub-contract indemnifying the contractor against claims for negligence of the sub-contractor and 'against the same obligations in respect of the sub-contract as those for which the contractor is liable in respect of this contract'. These words seem to make it clear that the contractor is accepting liability in respect of work done by the nominated sub-contractor.

The situation with regard to nominated suppliers, however, is noticeably different. The clause [22] which deals with nominated suppliers follows directly on that which deals with nominated sub-contractors. It provides no veto on the ground of the contractor's reasonable objections, nor on the ground of the nominated supplier refusing to indemnify the contractor. This omission cannot, I think, be unintentional. It seems, in contrast to clause 21, to point to an intention that the contractor is not undertaking liability for materials provided by a nominated supplier. Otherwise he must surely have been given, as in the case of a nominated sub-contractor, an opportunity of making reasonable objections, and a right to insist on an indemnity from the supplier.

It would not be unreasonable for the parties to intend that an employer should take the responsibility for materials provided by nominated suppliers. They have been selected, without giving the contractor any right to express views, by the employer's own expert architect who has decided that the nominated goods are suitable for the purpose and who has made the preliminary arrangements with the suppliers either before or during the main contract. The contractor is simply instructed to obtain his supplies from the nominated supplier. It is the employer who, through his architect, alone arranges the price, which is liable to be reflected in the quality and who alone can insist on tests and checks of quality. All the circumstances of the nomination appear actually to exclude any reliance on the contractor's skill and judgment. And, though the contractor receives a profit on the nominated supply, it is a controlled profit and he has certain duties to perform such as co-ordinating the delivery with the work and doing his best to see that there are no delays (see clause 18(vii)).

On the other hand, if the contractor is not liable for material provided by nominated suppliers, the employer is left without a remedy for faulty material. For the contract, by clause 22, indicates clearly that the nominated supplier is in contractual relation with the contractor only and, although the employer is paying for the nominated materials, he pays the contractor for them and the contractor pays the nominated supplier. Thus to hold that the contractor is not liable for nominated supplies is to go against one of the important reasons for the general rule that there is a warranty of good quality in materials supplied under a contract for labour and materials, namely that the employer should have a remedy against the contractor who can in turn enforce it against the supplier with whom the fault lies.

Yet, in spite of this important fact, I think that the contrast between the wording of clause 21 and that of clause 22 persuades one to the view that the contract shows an intention to exclude a warranty by the contractor in respect of nominated supplies. And the particular circumstances of the case fortify this view. The employers (or their architect) employed a skilled engineer to advise in respect of the columns in question. The engineer, without any consultation with the contractor, prepared detailed designs of the columns, chose out [C] as suitable suppliers and gave them such instructions as he considered were necessary to specify the composition of the columns. The architect obtained a quotation from [C] which was lower than the amount quoted by another firm and also substantially lower than the P.C. item, and which contained a substantial limitation of the purchaser's right of recourse in the event of the columns being defective. The architect presumably was satisfied with this and without discussing the matter with the contractor, instructed him to accept the quotation.

The employers, through their architect, having directed the contractor to buy from a manufacturer who had substantially limited his own liability, it would not be reasonable to suppose that the parties were intending the contractor to accept an unlimited liability, which might have been very great, for those columns over whose manufacture he had no control whatsoever. It was suggested in argument by the appellants that one should assume some implied intention that there will be liability on the contractor limited to the curtailed right of recourse provided under [C's] quotation. If so, one would have to imply in respect of each nominated supplier that there is a liability limited to the right of recourse provided in each particular sub-contract of supply. But I see no ground for such a complicated implication which, if it existed, must surely have been embodied in express terms in clause 22. In my opinion, the limitation of recourse against the nominated supplier and the other particular circumstances in respect of the supply of the columns provide confirmation of the intention indicated by the contract itself, namely, that any warranty by the contractor in respect of the quality of the nominated supplies was to be excluded.

LORD WILBERFORCE: The situation thus created was one of a special and complex character, differing greatly from that which arose in Young and Marten Ltd v. McManus Childs Ltd (1969). There the employer nominated a brand article to be supplied by the manufacturer with no limitation on the contractor's freedom to contract with the manufacturer as he thought fit. The contractor could, and it would be the expectation that he would, or at least it would be his responsibility if he did not, deal with the manufacturer on terms attracting the normal conditions or warranties as to quality or fitness.

But here, the design, materials, specification, quality and price were fixed between the employer and the sub-supplier without any reference to the contractor: and so far from being expected to secure conditions or warranties from the sub-supplier, he had imposed upon him special conditions which severely restricted the extent of his remedy. Moreover, as reference to the main contract shows, he had no right to object to the nominated supplier, though, by contrast, the contract does provide a right to object to a nominated sub-contractor if the latter does not agree to indemnify him against his liability under the contract.

In these circumstances, so far from there being a good reason to imply in the contract (which throughout its elaborate 27 clauses makes no express provision as regards this matter) a condition or warranty binding the contractor in respect of latently defective goods, the indications, drawn from the conduct of the contracting parties, are strongly against any such thing. It would, indeed, be most unjust if when the employer has (possibly to his own advantage as reflected in the price) limited the contractor's right or recourse as severely as he has (since consequential damage or loss in such cases as this may be very great and much in excess of the cost of replacement) he should be given by implication an unlimited right to recover damages from the contractor.

Norta Wallpapers (Ireland) Ltd v. John Sisk & Sons (Dublin) Ltd

IRISH SUPREME COURT

(1977) 14 BLR 49

The claimants engaged the respondents as main contractors to build a factory for making wallpaper and nominated H as sub-contractors for the design, supply and erection of the superstructure including the roof and roof lights. H were specialists in 'system built' superstructures.

H completed the work in May 1967 and the respondents completed work under the main contract in August 1967. In February 1968 a major leak occurred in the roof of the factory, which was found by an arbitrator to be due 85 per cent to defective design of the roof lights, 12 per cent to use of inferior materials and 3 per cent to bad workmanship.

The trial judge held that:

(1) The respondents were liable for so much of the damage as was due to the use of inferior material by, and bad workmanship of, the sub-contractor; and
(2) The respondents were not liable for the loss which was due to the defective design of the roof lights.

The claimants appealed against (2), but the appeal was dismissed.

HENCHY J: In all cases of supply and installation by a sub-contractor I conceive the law to be that, unless the particular circumstances give reason for its exclusion, there is implied in the contract a term to the effect that the contractor will be liable to the employer for any loss or damage suffered by him as a result of the goods, materials or installations not being fit for the purpose for which they were supplied. The basis for this rule is that, while the contractor is thus made primarily liable, he will be able, under the sub-contract, to have recourse, by third-party procedure or otherwise, against the sub-contractor for the indemnity in respect of the contractor's liability to the employer. I do not think there is any dispute about the law as stated thus. The cleavage of opinion arises as to whether the circumstances of this case are such as to exclude an implied term in the contract making the contractor liable for the sub-contractor's default.

An implied term in a contract to the effect that goods or installations supplied under it will be suitable for their purpose cannot be held to exist unless it can be said to have come within the presumed intention of the parties: see *G.H. Myers & Co* v. *Brent Cross Service Co* (1934) which was approved by all the judges in the House of Lords in *Young & Marten Ltd* v. *McManus Childs Ltd* (1969).

As I say, the real question in the present case is whether the implied warranty of fitness is excluded in the particular circumstances. It is common ground – at least there is no appeal from the judge's decision to that effect – that Sisk are liable to Norta under an implied term in the contract in respect of the defective materials and the bad workmanship employed in erecting the roof. The judge went on to hold that there was no such implied term in regard to the defective design of the roof lights . . . The modern cases (particularly the decisions of the House of Lords in *Young & Marten Ltd* v. *McManus Childs Ltd* (1969) and *Gloucestershire County Council* v. *Richardson* (1969)) show that the liability of a contractor under an implied term in the contract is approximated to the liability of a seller under section 14 of the Sale of Goods Act, 1893. Thus he may be liable, as Sisk are here, for bad workmanship done or inferior materials supplied by the sub-contractor. But if the subject matter of the employer's complaint is something which under its patent or trade name he has required to be supplied or installed under sub-contract, he cannot hold the contractor liable if, because of bad design, it turns out not to be suitable for a particular purpose. If he nominated it under its patent or trade name he has, in effect, accepted the design as being suitable for the particular purpose. If that be the position here, then the trial judge's reasoning for rejecting Norta's claim against Sisk is unanswerable.

However, I am inclined to think that the defective design of this roof did not simply render it unsuitable for a particular purpose. The arbitrator has held that 'the roof has now deteriorated to the extent that, in my opinion, it is not suitable for its purpose'. By that he meant, presumably, that the roof is not fit for the purpose or purposes for which a roof of that kind is commonly bought, in that it is not rainproof. If that be so, then it might be held to be wanting in merchantable quality: per Lord Denning MR in *Cehave NV* v. *Bremer Handelsgesellschaft mbH* (1976). Without more findings than the arbitrator has set out in the case stated, one would not be justified in firmly reaching such a conclusion. But such a conclusion is not necessary for the purpose of deciding the case. Even if the defect in the design of the roof is such as to justify describing it as lacking merchantable quality, a warranty on the part of the contractor that it will not have such a defect should not be read into the contract unless it would be reasonable to do so.

University of Warwick v. Sir Robert McAlpine

QUEEN'S BENCH DIVISION

(1989) 42 BLR 6

McAlpine were main contractors to the University. The buildings had reinforced concrete frames with a white ceramic tile external cladding. The cladding had been carried out by nominated sub-contractors. The cladding began to fail. The architects blamed bad workmanship and McAlpine blamed the design. Eventually, the University decided to remedy the defects by a resin injection process, which was a recent development. The architects recommended that CCL, the sole British licensees of the process, be employed by McAlpine to perform the remedial works. The main contract was varied accordingly. CCL were not technically nominated sub-contractors but were employed by McAlpine. The remedial works were not successful and, inter alia, the University alleged that McAlpine were in breach of an implied warranty of fitness for purpose.

Held: A term that the resin be fit for its purpose could only be implied in the main contract if the University had relied on McAlpine. As they had not done so, no such term was to be implied.

GARLAND J: The University allege that McAlpine are in breach of an implied warranty of fitness for purpose. They accept that the resin injection process was of good quality (or merchantable). The purpose must be injection into the tiling not only to bond delaminations, but also as far as possible, to fill voids. Since the contract with McAlpine was for 'work and materials' the implication must be at common law and parallel to the statutory implication in the case of sale of goods (Gloucestershire County Council v. Richardson (1969)). There are certain features of the contractual relationship between the University, McAlpine and CCL which must be borne in mind:

(1) The University did not rely on the skill and judgement of McAlpine. McAlpine had considerable reservations about the use of the process.
(2) The University did rely on the skill and judgment of CCL.
(3) CCL were not nominated sub-contractors so that YRM's instructions to McAlpine to enter into the sub-contracts were not within the machinery of the main contract but when accepted by McAlpine constituted variations. McAlpine could have refused to accept CCL as sub-contractors. The terms on which McAlpine contracted with CCL were not prescribed by the University.
(4) I have dealt later in this judgment with the terms of the sub-contracts and have found that they all incorporated the Green Form so that McAlpine were entitled to be indemnified by CCL.
(5) McAlpine were at all material times aware that the University were relying on CCL.
(6) The University's solicitors were worried about the 'guarantee' offered by CCL but at no stage suggested that CCL should enter into a direct warranty with the University. A form of direct warranty was in use in 1974. It was intended for use where a sub-contractor was to be nominated pursuant to clause 27 of [JCT 63], but there would be no difficulty in using it outside any main contract machinery. The impact of such a direct warranty has recently been considered by the Court of Appeal in Greater Nottingham Co-operative Society v. Cementation Piling and Foundations Ltd (1988) in the context of defining any duty in tort owed by a sub-contractor to a building owner.

[The University] submits that in these circumstances, when McAlpine were instructed to enter into the sub-contracts with CCL and did so, it was an implied term of the main contract as varied that the resin injection process would be not only of good quality but fit for the purpose of filling voids. If this were not so, it is submitted, the University's reliance on CCL is contractually stultified and the practicality of 'up and down the line' rights and remedies in chain contracts lost. [The University] starts with the well known passage in Myers & Co. v. Brent Cross Service Co (1934):

> A person contracting to do work and supply materials warrants that the materials which he uses will be of good quality and reasonably fit for the purpose for which he is using them, unless the circumstances of the contract are such as to exclude the warranty.

Lord Pearce in *Young & Marten Ltd* v. *McManus Childs Ltd* (1969) said:

> If it is known to both parties that the manufacturer gives no warranty to the contractor, that fact is a strong indication that no warranty is being given by the contractor. So too, of course, if a contractor advises against a particular material. But the circumstances of contracts are so various that it must be a question of fact and degree whether the circumstances of a particular case suffice to exclude a warranty which the general rule implies.

In *Young & Marten Ltd* the contractor specified in a domestic sub-contract a particular tile made by only one manufacturer. The tiling sub-contractor bought the tiles and fixed them. The tiles were defective. It was held that the sub-contractor was liable under the implied warranty of quality because there were no circumstances pointing to its exclusion; the fact that the manufacturer was only willing to supply tiles on terms excluding liability would have been such a circumstance. As far as the warranty of fitness was concerned, this was excluded because 'the employers (the contractor) chose and ordered the tiles under their trade name of Somerset 13' (per Lord Pearce). This is a reference to the exception in section 14(1) of the 1893 Act which is really no more than an express statutory presumption that when goods were ordered by a 'patent or trade name' there was no reliance on the skill and judgment of the seller.

[The University] relied negatively on the *Gloucestershire County Council* case as demonstrating the extreme circumstances required to exclude the warranty of quality. Under the 1939 RIBA standard form the contractor had no right to object to the nomination of a supplier, in contrast to a sub-contractor. In addition, the nomination was to enter into a sub-contract for the supply of components on terms which substantially limited the supplier's liability for defects. The precise *ratio decidendi* of this case has always been difficult to ascertain. Lord Pearce decided on the 'no right to object' point; Lord Upjohn on the limitation of liability point; Lord Wilberforce on both, Lord Reid agreed with them and Lord Pearson dissented. Lord Upjohn appears to equate both warranties (quality and fitness); Lord Pearce appears to take the view that the warranty of fitness never arose in the first place:

> [the materials] have been selected, without giving the contractor any right to express views, by the employer's own expert architect who has decided that the nominated goods are suitable for the purpose . . . the contractor is simply instructed to obtain his supplies from the nominated supplier. All the circumstances of the nomination appear actually to exclude any reliance on the contractor's skill and judgment.

A constant thread running through the decision is the commercial convenience of chain remedies in contract, for example, Lord Pearce:

> Thus to hold that the contractor is not liable for nominated supplies is to go against one of the important reasons for the general rule that there is a warranty of good quality in materials supplied under a contract for labour and materials, namely that the employer should have a remedy against the contractor who can in turn enforce it against the supplier with whom the fault lies.

[The University] also relied on *IBA* v. *EMI and BICC* (1980). EMI were engaged as contractors to build a television mast and were instructed by IBA to enter into a nominated sub-contract with BICC for the design, supply and erection of the mast. EMI in fact undertook vis-a-vis IBA responsibility for the design so that this was really a 'package deal' contract for design supply and construction: EMI's liability for design was conceded. There was an issue whether the design liability was 'fitness for purpose' or 'professional liability' but since the House of Lords decided that the trial judge's finding of negligence against BICC was correct and that EMI were therefore also liable in negligence, it was not necessary to decide that issue. [The University], however, relied on passages in the speeches of Lord Fraser and Lord Scarman. Lord Fraser said:

> If the terms of the contract alone had left room for doubt above that, I think that in a contract of this nature a condition would have been implied to the effect that EMI had

accepted some responsibility for the quality of the mast, including its design, and possibly also for its fitness for the purpose for which it was intended. The extent of the responsibility was not fully explored in argument, and, having regard to the decision on negligence in the design, it does not require to be decided. It is now well recognised that in a building contract for work and materials a term is normally implied that the main contractor will accept responsibility to his employer for materials provided by nominated sub-contractors. The reason for the presumption is the practical convenience of having a chain of contractual liability from the employer to the main contractor and from the main contractor to the sub-contractor – see *Young & Marten Ltd* v. *McManus Childs Ltd* (1969). Of course, as Lord Reid pointed out in that case, 'No warranty ought to be implied in a contract unless it is in all circumstances reasonable'. In most cases the implication will work reasonably because if the main contractor is liable to the employer for defective material, he will generally have a right of redress against the person from whom he bought the material. In the present case it is accepted by BIC that, if EMI are liable in damages to IBA for the design of the mast, then BIC will be liable in turn to EMI. Accordingly, the principle that was applied in *Young & Marten Ltd* in respect of materials, ought in my opinion to be applied here in respect of the complete structure, including its design. Although EMI had no specialist knowledge of mast design, and although IBA knew that and did not rely on their skill to any extent for the design, I see nothing unreasonable in holding that EMI are responsible to IBA for the design seeing that they can in turn recover from BIC who did the actual designing. On the other hand it would seem to be very improbable that IBA would have entered into a contract of this magnitude and this degree of risk without providing for some right of recourse against the principal contractor or the sub-contractors for defects of design.

We were referred to the Irish case of *Norta Wallpapers Ltd* v. *John Sisk Ltd* (1978), where the facts were in many ways similar to those in the present case. The claimant company ('Norta') had accepted a tender from the respondents ('Sisk'), who were building contractors, to build a new factory. Before receiving tenders Norta had approved a quotation from a German company to provide and erect the superstructure of the factory including the roof, and when they accepted Sisk's tender for erecting the factory as a whole they nominated the German company for the supply and erection of the superstructure. Sisk made a sub-contract with the German company. Later the roof developed a major leak which was found to be caused as to 85 per cent by defective design. McMahon J held, and the Supreme Court agreed, that Sisk were liable to Norta for all loss or damage caused by the sub-contractors' use of defective material or bad workmanship, but not for loss caused by the defective design of the roof. The decision was reached upon its own facts but it is useful to notice how Henchy J, in the Supreme Court, approached the case. He said:

> . . . the real question in the present case is whether the implied warranty of fitness is excluded in the particular circumstances.

He held that it was excluded, and the other learned judges agreed. The question in the instant case is exactly the same, but in my opinion the answer is different. Although the facts in Norta were very like those in the present case, they are distinguishable in respect that, before the contract between Norta and Sisk was made, Norta's engineers had already approved the German company's design and specification and Norta had promised that the order for the superstructure would be given to them at a price which was also agreed. Sisk thus were given no option either as to the identity of the sub-contractor or as to the design or price of the superstructure. In the present case, although EMI had no option but to appoint BIC as sub-contractor for the mast, they were not bound to accept any particular

> design at any particular price. If they had checked BIC's design and had considered it unsatisfactory they would have been entitled to insist on its being improved. That is not a mere theoretical possibility, as is shown by the fact that on a previous occasion, in 1958, when EMI had tendered to IBA for the erection of a lattice mast at Mendlesham, they had expressly accepted 'overall responsibility' for the mast including its design and they had then told BIC that they wished to arrange for a check on the structural design as an additional safeguard. That procedure could I think have been repeated in 1963 if EMI had wanted.

[McAlpine] submitted that this passage, which is obiter, does represent the position which normally obtains, but Lord Fraser plainly did not have in mind the case where the employer or his architect selects a material or process and simply asks the contractor to make the necessary contractual arrangements to supply it without in any way relying on the contractor with regard to fitness for purpose. [They] submit that merely because the sub-contractor may be liable to the contractor on the implied warranty of fitness, it does not follow that when it comes to implying the warranty between employer and contractor, reliance can be dispensed with and the implication based on the existence either of a chain of contracts or the employer's reliance on the sub-contractor. [The University] also relied on Lord Scarman [but McAlpine] says that this (also obiter) passage does not assist the University:

> My Lords, I have had the advantage of reading in draft the speeches delivered by my noble and learned friends, Viscount Dilhorne and Lord Fraser of Tullybelton. For the reasons which they give I also would dismiss the appeal and allow the cross-appeal.
>
> The finding of negligence in the design of the mast makes it unnecessary for the House to determine the extent of the contractual responsibility for the design assumed by EMI to IBA (and by BIC to EMI). As my Lord Fraser of Tullybelton observes,
>
> > it must at the very least have been to ensure that the design would not be made negligently.
>
> But I would not wish it to be thought that I accept that this is the extent of the design obligation assumed in this case. The extent of the obligation is, of course, to be determined as a matter of construction of the contract. But, in the absence of a clear, contractual indication to the contrary, I see no reason why one who in the course of his business contracts to design, supply, and erect a television aerial mast is not under an obligation to ensure that it is reasonably fit for the purpose for which he knows it is intended to be used. The Court of Appeal held that this was the contractual obligation in this case, and I agree with them. The critical question of fact is whether he for whom the mast was designed relied upon the skill of the supplier (i.e. his or his sub-contractor's skill) to design and supply a mast fit for the known purpose for which it was required.
>
> Counsel for the appellants, however, submitted that, where a design, as in this case, requires the exercise of professional skill, the obligation is no more than to exercise the care and skill of the ordinarily competent member of the profession. Although it might be negligence today for a constructional engineer not to realise the danger to a cylindrical mast of the combined forces of vertex shedding (with lock-on) and asymmetric ice loading of the stays, he submitted that it could not have been negligence before the collapse of this mast: for the danger was not then appreciated by the profession. For the purpose of the argument, I will assume (contrary to my view) that there was no negligence in the design of the mast, in that the profession was at that time unaware of the danger. However, I do

not accept that the design obligation of the supplier of an article is to be equated with the obligation of a professional man in the practice of his profession. In *Samuels v. Davis* (1943), the Court of Appeal held that, where a dentist undertakes for reward to make a denture for a patient, it is an implied term of the contract that the denture will be reasonably fit for its intended purpose. I would quote two passages from the judgment of du Parcq LJ. At page 529 he said (omitting immaterial words):

> . . . if someone goes to a professional man . . . and says: 'Will you make me something which will fit a particular part of my body?' . . . and the professional gentleman says: 'Yes', without qualification, he is then warranting that when he has made the article it will fit the part of the body in question.

And he added:

> If a dentist takes out a tooth or a surgeon removes an appendix, he is bound to take reasonable care and to show such skill as may be expected from a qualified practitioner. The case is entirely different where a chattel is ultimately to be delivered.

I believe the distinction drawn by du Parcq LJ to be a sound one. In the absence of any terms (express or to be implied) negativing the obligation, one who contracts to design an article for a purpose made known to him undertakes that the design is reasonably fit for the purpose. Such a design obligation is consistent with the statutory law regulating the sale of goods: see Sale of Goods Act 1893, the original section 14, and its modern substitution enacted by section 3, Supply of Goods (Implied Terms) Act 1973.

I do not think that this passage assists the University's case.

[McAlpine] submitted that there was no clear authority governing the question I have to decide where E (the employer) instructs (M) a main contractor to order goods or services from (S) a sub-contractor relying on the skill and judgement of S but not M. If the basis for the implication of the warranty of fitness is reliance, M should not be liable for fitness but will be liable for quality. If it can be said that the implication should be made as a matter of contractual convenience because E relied on S, and M is a contractual conduit protected by his indemnity against S, then M will be liable for both fitness and quality.

Young & Marten and *Gloucestershire CC* proceeded on the basis once the warranty is prima facie to be implied, the question is whether the circumstances are such that it should be excluded. This appears also to be the approach of Lord Scarman in the passage quoted. In *Young & Marten* fitness was never prima facie implied and the question was whether quality was excluded. In *Gloucestershire CC* I read the report as a whole as indicating (subject to the passages in Lord Upjohn) that only quality was in issue because the columns had been chosen in careful detail by the architect. In the IBA case, EMI were under the terms of the contract responsible for the design and owed a duty of care in respect of it. In *North* v. *Sisk* (1977) where a nominated sub-contractor had undertaken the design of the relevant defective part of the building which had been approved by the employer's engineer and the price agreed, the contractor was held not liable to the employer. The case is not on all fours with one that I have to decide, but Henchy J placed great emphasis on there having been no reliance by the employer on the contractor and consequently no warranty of fitness; Kenny J emphasised the fact that the contractor had not been consulted and was bound to accept the sub-contractor and his design. He thought that the implication of warranties in a contract for work and materials was not a matter of reliance on skill and judgment but of presumed intention based

> on reason and a just regard to the interests of both parties. I venture the thought that his reason for not basing his decision on reliance was incorrect since he appeared to think that where there is no reliance, there is also no warranty of quality. Reliance may be no more than the factual foundation of presumed intention, but in my view the correct approach is to ask, as between E and M (to revert to my abbreviations) what warranties are prima facie to be implied into the contract, (or in the present case, the contracts as varied) and then to ask whether the circumstances are such as to exclude one or both of them. On this basis there can be no implication of the warranty of fitness because there was no reliance on McAlpine.
>
> Can there nevertheless be an implication because the University relied on CCL to McAlpine's knowledge, and McAlpine were not obliged to contract with CCL on particular terms, or indeed, at all? In fact, they obtained an indemnity. The issue is a finely balanced one, but in my view the first approach is correct. I do not accede to the suggestion that if there is no implication the University are left remedyless. They could have taken an express warranty from McAlpine or they could have taken a direct warranty from CCL. Indeed, in the circumstances and given the concern about CCL's guarantee, I am surprised that they did not.
>
> I therefore find McAlpine not liable in contract to the University.

Rotherham Metropolitan Borough Council v. Frank Haslam Milan & Co Ltd

COURT OF APPEAL

(1996) 59 Con LR 33

The plaintiffs employed the two defendants to build phase I and phase II respectively of a new five storey office building to be used as the new civic offices of Rotherham. The contracts were on JCT 1963 terms (1977 revision). Both contracts involved the provision of hardcore. The bills of quantities referred in relation to the first defendant to granular hardcore and in relation to the second defendant to hardcore and further provided:

> [Granular] hardcore shall be graded or uncrushed gravel, stone, rock fill, crushed concrete or slag or natural sand or a combination of any of these. It shall not contain organic material, material susceptible to spontaneous combustion, material in a frozen condition, clays or more than 0.2% of sulphate ions as determined by BS 1377.

In both contracts the hardcore included steel slag. Steel slag is subject to expansion in confined conditions and did expand, causing cracking to the reinforced concrete slabs.

Held: (1) It was appropriate to imply a term that the hardcore should be of merchantable quality but the steel slag was of merchantable quality since it was reasonably saleable under the description 'steel slag' for use as fill material in unconfined spaces; but

(2) In the circumstances it was not appropriate to imply a term as to fitness for purpose since the architect was more expert than the contractors as to the selection of fill material and none of the parties knew that the slag was unfit.

LEGGATT LJ: The critical question is whether the circumstances show that Rotherham did not rely on the contractors' skill and judgment. As a general rule where the efficacy of a building or other object depends upon a designer it is the designer who may be expected to bear the responsibility for ensuring the suitability of the components incorporated into it. Where the designer himself relies on those who have specialist skills his reliance may show or suggest that he is abrogating that responsibility in relation to matters within the purview of the specialist. Here the designers of the building indisputably were Rotherham's architect and engineer. The design made necessary the importation of hardcore as fill material. In the bills of quantities the types of hardcore that would be acceptable were specified, as was the grading and sulphate content of the fill material. All those stipulations were made that the architect thought necessary to ensure the provision of a suitable fill material. To the extent that there were no stipulations and the supplier was free to choose, the freedom was accorded not in order to enable the supplier to exercise some supposed skill and judgment but because the architect believed that no further stipulations were necessary.

In truth the architect would have rightly regarded himself as more expert than the contractors. They did not profess, and cannot have been regarded as enjoying, any special expertise in the selection of fill material. For a permitted product like slag there was no assurance that either contractor would use a specialist supplier or would buy slag from a slag reduction company. Nor did they. It was not like ordering a propeller from a specialist manufacturer who would necessarily have a greater knowledge of the requisite dimensions than anyone else. Here what was called for was hardcore of specified types which then had to be of a stated grading and sulphate content. The product was subject to sampling, testing, inspection and approval by the architect. If he had realised that steel slag was unsuitable he would have said so because he took it upon himself to determine suitability instead of leaving it to whoever the ultimate supplier might prove to be. Had the officious bystander plucked the contracting parties by the sleeves as the contracts were about to be signed, and asked them whether it was left to the contractors to determine the suitability of the fill material, they would all, I believe, have replied, 'Of course not'. Rotherham might have added that that was what they engaged an architect for.

ROCH LJ: In this case the experts agreed and the employers admitted that the fill supplied by the contractors conformed to the requirements particularised in the contract specification for hardcore. The evidence of Mr Ward was to the effect that although he either did not read, or, if he did read, he did not take in the significance of the passage in BRE 222 concerning steel slag, that he and others in the employer's architectural and engineering departments would have had this document in 1979. Mr Ward candidly admitted that the definition of hardcore in the contract bills was such as to embrace steel slag because the design team thought that they knew but did not actually know the true nature of steel slag. In effect the employers are saying that despite the fact that the use of steel slag arose from and was within the wording of their specification of hardcore, and despite the fact that they had the means, of knowledge that the steel slag was not inert (albeit the judge found that they were not at fault in not knowing) that nevertheless the contractors who complied with the contractual specification should pay for the damage because a term as to fitness of purpose should be implied. In my judgment the proposition only has to be stated in that way for it to be seen that the implication of the term in the circumstances of this case would be both unreasonable and unjust.

Progress and completion

Normally, the contractor undertakes to complete the work in its entirety. He will normally be in breach of contract if he fails to complete, unless the contract has come to an end for some reason. (See Part III – Discharge of the Contract.)

If the contract is 'entire', the contractor who does not complete substantially will not be entitled to payment. (See Chapter 7 – Payment.) The practical effect of this rule is often modified by provision for payment against certificates, but the contractor who stops work without reason still does so at his peril.

Ibmac Ltd v. Marshall (Homes) Ltd

COURT OF APPEAL

(1968) 208 Est. Gaz. 852

The defendants, who were building a council housing estate, employed the plaintiffs to build a roadway to serve it. The plaintiffs quoted a figure of £1,494 but also supplied a bill of quantities with rates for the work. (The Court of Appeal thought it unnecessary to decide whether this was a fixed price or a remeasurement contract).

The plaintiffs found the work was more difficult than they had expected, as the site lay at the bottom of a steep hill and there were serious difficulties with water. The plaintiffs abandoned the works and claimed payment for the work they had done.

Held: Their claim must fail.

LORD DENNING MR: After a few weeks, the difficulties which they did not expect arose. The water hazard was indeed serious, and they did not keep to their side of the contract. They did not get a six-ton roller to consolidate the earth of the roadway, and so forth. So the difficulties of the water were enhanced and eventually, in June, they left the work. They said it was all Marshalls' fault. But it was not. On June 9, Ibmac wrote, 'We shall not return to the above site until the problem of surface water has been dealt with' – taking the line, as they did then, that the water was not their responsibility but the fault of Marshalls: Marshalls had not given them a proper site on which to do the work. There was some attempt to get the matter in order for Ibmac to come back. There was no doubt that they had left the work, especially on the trial judge's findings of fact, which he made after a full hearing, that they were not justified in leaving the work. They abandoned it without just cause and excuse when they had about one-third of the work done. In point of law, on that footing they were not entitled to anything, because of the rule in *Sumpter* v. *Hedges* (1898), which was quite clear, that if one had an entire contract where one had to complete the work, if one abandoned it half or a third of the way through without just excuse one could not recover anything.

Note: In the absence of contractual provision, the contractor would only be required to complete by the agreed completion date or, if no date is expressly agreed, within a reasonable time. In practice, however, the contract will usually require the contractor, for example, 'to begin the works forthwith and regularly and diligently proceed with the same', etc. In practice, this obligation is important in relation to disputes concerning liquidated damages and extensions of time (see Chapter 14 – Liquidated damages and extension of time), and determination of the contractor's employment under express terms (see *J.M. Hill & Sons Ltd* v. *London Borough of Camden* (1980), 18 BLR 26).

Where the contract requires the contractor 'regularly and diligently (to) proceed with' the works (JCT 98, clause 23.1; JCT 63, clause 21), whether this standard is achieved is probably to be judged according to the usage of the industry. See now JCT 2005, clause 2.4.

Hounslow Borough Council v. Twickenham Garden Developments Ltd

CHANCERY DIVISION

(1970) 7 BLR 81

MEGARRY J (discussing the phrase 'regularly and diligently' in JCT 63, clause 21): These are elusive words on which the dictionaries help little. The words convey a sense of activity, of orderly progress, and of industry and perseverance; but such language provides little help on the question of how much activity, progress and so on is to be expected. They are words used in a standard form of building contract and in those circumstances it may be that there is evidence that could be given, whether of usage among architects, builders and building owners or otherwise, that would be helpful in construing the words. At present, all that I can say is that I remain somewhat uncertain as to the concept enshrined in these words.

West Faulkner Associates v. Newham London Borough Council

COURT OF APPEAL

(1992) 42 Con LR 144

SIMON BROWN LJ: My approach to the proper construction and application of the clause would be this. Although the contractor must proceed both regularly and diligently with the works, and although each word imports into that obligation certain discrete concepts which would not otherwise inform it, there is a measure of overlap between them and it is thus unhelpful to seek to define two quite separate and distinct obligations.

What particularly is supplied by the word 'regularly' is not least a requirement to attend for work on a regular daily basis with sufficient in the way of men, materials and plant to have the physical capacity to progress the works substantially in accordance with the contractual obligations.

What in particular the word 'diligently' contributes to the concept is the need to apply that physical capacity industriously and efficiently towards that same end.

Taken together the obligation upon the contractor is essentially to proceed continuously, industriously and efficiently with appropriate physical resources so as to progress the works steadily towards completion substantially in accordance with the contractual requirements as to time, sequence and quality of work.

It is a general principle of building contracts that it is for the contractor to plan and perform his work as desired during the contract period. If he does that, he cannot be said to have failed to exercise due diligence and expedition. The contractor's primary obligation is to complete

the work within the contract period. A clause requiring him to proceed 'with due diligence and expedition' is a subordinate and subsidiary obligation.

Greater London Council v. Cleveland Bridge & Engineering Co Ltd

COURT OF APPEAL

(1986) 8 Con LR 30

The respondents agreed with the appellants to manufacture, deliver and erect the gates and gate arms of the Thames Barrier by certain key dates. There was provision in the contract for the adjustment of price in the event of an increase or reduction in the respondents' costs. A dispute arose as to whether a sum was due by reason of an increase in costs affecting the manufacturing part of the contract. The question that fell to be decided was whether the costs notionally incurred by the contractors had been increased by reason of their default or negligence. The appellants claimed that the respondents were under an obligation to perform with due diligence and expedition.

Held: The contract did not as a matter of construction or implication provide that the contractors should carry out the contract works with due diligence and expedition and on the facts assumed there was not shown to be any default or negligence on the part of the respondent contractors. If there is no obligation to do more than complete by a certain date, the contractor is free to conduct his programme in a manner which suits him.

STAUGHTON J (whose judgment was affirmed): [There is] a general principle applicable to building and engineering contracts that in the absence of any indication to the contrary, a contractor is entitled to plan and perform the work as he pleases, provided always that he finishes it by the time fixed in the contract. That is exemplified by the case of *Wells* v. *Army and Navy Co-operative Society* (1902), where Wright J said:

> The plaintiffs must, within reasonable limits, be allowed to decide for themselves at what time they are to be supplied with details. The plaintiffs were entitled to do the work in what order they pleased.

See also Vaughan Williams LJ:

> . . . in the contract one finds the time limited within which the builder is to do this work. That means not only that he is to do it within that time, but it means also that he is to have that time within which to do it. It seems to me that in the construction [counsel] wished us to put upon clause 16 of this contract he was inviting us to put the construction which would really put it in the power of the directors of this company to deprive the builders altogether of the benefit of that limitation of time. To my mind that limitation of time is clearly intended, not only as an obligation, but as a benefit to the builder.
>
> In my judgment, where you have a time clause and a penalty clause, it is always implied in such clauses that the penalties are only to apply if the builder has, as far as the building owner is concerned and his conduct is concerned, that time accorded to him for the execution of the works which the contract contemplates he should have.

> In the present case, the contract specifies dates by which the whole of the contractors' work is to be completed, and clause 39 deals with extension of time for completion and not anything else.
>
> One's first impression, on reading clause 19, is that the contract does, as a matter of construction or by implication, provide that the contractors shall carry out the contract works with due diligence and expedition. But, on further consideration, I do not think that that is the right conclusion. The point is very nicely balanced [but] I conclude that although neglect by the contractors to execute the works with due diligence and expedition would entitle the employers to discharge them, under clause 19, it would not by itself be a breach of contract on the part of the contractors.
>
> If I had not reached that conclusion, I would have held, without hesitation, that due diligence and expedition must be interpreted in the light of the other obligations as to time in the contract. That seems to me to follow from the construction of the contract as a whole and from the considerations I have already mentioned. If there had been a term as to due diligence, I consider that it would have been, when spelt out in full, an obligation on the contractors to execute the works with such diligence and expedition as were reasonably required in order to meet the key dates and completion date in the contract. As I understand it, the employers would not be contending in any further proceedings before the arbitrator, that the contractors had been in breach of such a term.

Note: In this case there was no express term requiring the contractor to proceed with diligence and expedition. Furthermore, the contract was one in which the contract period was much longer than the time needed by the contractor to carry out the contract.

Indemnities and insurance

The contract will normally contain provisions as to which of the parties is to insure against certain risks. These provisions may present serious difficulties of interpretation, as illustrated by the next two cases.

Gold v. Patman & Fotheringham Ltd

COURT OF APPEAL

[1958] 2 All ER 497

The plaintiff engaged the defendants under a standard RIBA Form (1939 edition with quantities, 1952 revision). Clause 14 required the contractors to indemnify the employer in respect of injury to persons and property (subject to certain exceptions). Clause 15 required the contractors to 'effect . . . such insurances . . . as may be specifically required by the bills of quantities'. The risks required to be insured by the bills included 'insurance of adjoining properties against subsidence or collapse'.

A neighbouring owner brought a claim against the plaintiff in respect of subsidence caused by the building operations and it appeared that the defendants had taken out an insurance policy which covered their liability for subsidence but not that of the plaintiff. The plaintiff brought an action claiming that the defendants were in breach of their obligations under clause 15.

Held: The claim must fail.

HODSON LJ: We must deal separately with the question who is to be insured, though it falls for consideration on the question of certainty; but, so far as the other matters are concerned, we agree with the learned judge that there is no uncertainty as to the risk to be covered by insurance. That which was to be insured against was the legal liability to adjoining owners arising out of the subsidence or collapse of the premises. It would be wholly wrong in our judgment to say, in the words of Lord Wright, that the language used was so obscure and so incapable of any definite or precise meaning that the court is unable to attribute to the parties any particular contractual intention.

The only real difficulty which arises is as to the determination of the question who is to be covered by insurance, but the court ought not to be deterred by any difficulty of interpretation which can fairly be surmounted. Looking first at condition 14 of the contract dealing with (a) injury to persons, (b) injury to property, the contractors are bound to indemnify the building owner against loss in the circumstances therein set out. When one comes to condition 15, the contractors are thereby required to effect insurance or cause any sub-contractor to do so as may be specifically required by the bill of quantities. This condition prima facie refers to the indemnity required by condition 14, for it is introduced by the words 'without prejudice to his liability to indemnify the employer under clause 14 hereof', which indicates plainly that the indemnity is to stand independently of the insurance and that the contractors cannot claim to have fulfilled their obligation under the indemnity clause in respect of injury to persons or property by effecting a policy or policies of insurance or causing any sub-contractor to do so.

These policies and their premium receipts must be produced as and when required by the architect. Loss or damage by fire is dealt with in clause 15(b)[A], and there, in contrast to the policies referred to in clause 15(a), the contractors must insure in the joint names of the building owner and themselves with a company or companies approved by the architect, and the contractors are obliged to deposit (not only produce when required) the policies and premium receipts with the architect. It would thus appear that there is a contrast between policies in which the contractors are insuring themselves, no doubt for the additional security of the building owner, and cases where the contractors are insuring in the joint names of the owner and themselves. Thus, when the contract requires the building owner (described in the contract as the employer) to be insured by the contractors, it is expressly so provided. Further, it would appear appropriate in cases where the insurance is for the benefit of the building owner to require approval by his architect and deposit of the policy of insurance with him. Condition 25, which deals with fluctuations, refers also to the burden of insurance falling on the contractors under the heading of expenses, including the cost of employer's liability and third-party insurance, and lends support to the contention of the contractors that these are among the specific insurances contemplated by condition 15(a) of the contract. These considerations derived from the conditions in the main contract are persuasive to suggest that the insurances specifically required by the bill of quantities will be found to be insurances effected by the contractors for themselves and not for the building owner.

When one turns to the bill of quantities itself, nothing is found to lead to a contrary conclusion. In construing the obligation beginning with the words 'The contractor shall insure', one looks at the whole list of matters in respect of which they have to insure or make payments, and nothing is found in that list pointing to an obligation to insure anyone else but themselves.

We cannot find anything in the initial phrase or in the context to lead one to understand that the obligation is to insure not the contractors alone but the building owner either in addition to the contractors or separately from him. The list following the initial phrase is not very illuminating and there is on any view some duplication . . . All the items in the list would appear to be cases where the contractors must insure themselves, at least so far as they are intelligible, bearing in mind that 'Unemployment Insurance Acts' do not exist, and insurances against specific industrial diseases do not add anything to the words above, viz., 'National Insurance Acts and National Insurance (Industrial Injuries) Acts'. Damage by storm etc. would be the contractors' risk for them to cover, for they would not be relieved of their obligation by such damage . . .

We find no sufficient reason for thinking that, when one comes to the last item 'insurance of adjoining properties against subsidence or collapse', the obligation to insure is for the first time in the list an obligation to insure, not the contractors themselves, but the building owner. If there were to be a change of obligation, one would expect some words to be used to show that there is a change, and here there are none; nor would one expect to find such an obligation in this place, having regard to the words of condition 15(a), where reference is made to such insurances to be effected as may be specifically required by the bill of quantities, the policies to be produced when required and not only to be approved by and deposited with the architect. It is true that there is duplication, since the heading 'Employer's liability, common law, third party' no doubt produces the same result; but duplication, as we have said, appears elsewhere in the miscellany and the last item is only a specific identification of one kind of third-party risk.

Counsel for the contractors reinforced his argument by a contention that there would be grave difficulties in the way of a contractor knowing what to tender if he had to insure the building owner, for he would not survey the property before tendering, whereas in the case of insuring himself against the risk in question there is no practical difficulty, because the premium is based on the contractor's past record and the number of men he employs. Counsel sought in this connection to rely on the oral evidence of an insurance expert, to show the practical difficulties involved. We see no ground for admitting evidence of this character to show the surrounding circumstances of this particular contract. No trade custom was sought to be proved and it is not contended that both parties were contracting in the light of some special insurance background which could be established by evidence. In any event, we should not have been deterred by the difficulty of effecting an insurance from finding that the obligation to insure the building owner had been placed on the contractors by the contract if the words used had produced that result. The evidence of [the expert] could, one would assume (although no argument was addressed to the court on this aspect of his evidence) be used to show uncertainty if there were a latent ambiguity in the use of the word 'insurance' in its context. We find nothing in his evidence which was read *de bene esse* to lead to the conclusion that the word 'insurance' as here used is uncertain in its meaning.

English Industrial Estates Corporation v. George Wimpey & Co Ltd

COURT OF APPEAL

(1972) 7 BLR 122

LORD DENNING MR: This case arises out of a big fire at Hartlepool on 18 January 1970. Much of a new factory was gutted. The damage is estimated at £250,000. Who is to bear the loss?

The factory was occupied by Reeds Corrugated Cases Ltd, who make corrugated cardboard. The factory and its site are owned by a statutory corporation called English Industrial Estates Corporation. The corporation let it on lease to Reeds.

In 1969 Reeds wanted to extend the factory greatly. It was to be three times as large as before. The Corporation agreed to make the extension and to let the whole to Reeds on a new lease. But Reeds wanted to continue making their corrugated cardboard all the time whilst the extension work was being done. In order to make the cardboard they intended to install a great new machine. It was to be fed with reels of paper weighing over a ton each. The machine could eat up these reels at such a rate that Reeds needed storage space for hundreds of these big reels of paper.

On 17 December 1968, George Wimpey & Co Ltd, the big contractors, put in a tender. They offered to build the extension for £687,860. In making the tender they had before them the form of agreement, the RIBA conditions, and the bills of quantities. They agreed that, if the tender was accepted, they would enter into a formal contract on those terms. Their tender was accepted.

On 7 February 1969, the Corporation and Wimpeys duly entered into articles of agreement. It was on the standard RIBA form and incorporated the bills of quantities. The most important provision for present purposes was clause 20A(1) of the RIBA form. It provided that the contractors, Wimpeys, should insure the works against fire and other perils until they were practically completed. Wimpeys had a floating policy which covered this.

In 1969 Wimpeys started on the work. By January 1970 they had done a great deal of it. Reeds had installed their great new machine and started producing corrugated cardboard. They had stored over 1500 big reels of paper in the new reel warehouse. But the contractors had not finished their work when on 18 January 1970, there was this disastrous fire. Much of the new extension was gutted. The loss is said to be £250,000. On whom should this loss fall? The Corporation say that the works had not been completed: that it was the duty of Wimpeys to insure: and that the loss should fall on them or their insurance company. But Wimpeys (or rather, I expect, their insurers) said that the Corporation, through Reeds, had taken possession of several parts of the works: and that, under clause 16 of the RIBA form, the risk had passed from Wimpeys to the Corporation so far as those parts were concerned.

To decide the case, I must set out some parts of the RIBA form, and also of the bills of quantities.

1. *The RIBA form*

The most material clauses were as follows:

20A(1) The Contractor shall . . . insure against loss and damage by fire . . . all work executed and all unfixed materials and goods . . . and shall keep such work, material and goods so insured until Practical Completion of the Works, and until the employer shall authorise in writing the cancellation of such insurance.

16 If at any time . . . before Practical Completion of the Works, the Employer, with the consent of the Contractor, shall take possession of any part or parts of the same (any such part being hereinafter in this clause referred to as "the relevant part") then notwithstanding anything expressed or implied elsewhere in this Contract . . .

(a) Within seven days from the date on which the Employer shall have taken possession of the relevant part, the Architect shall issue a certificate stating his estimate of the approximate total value of the said part . . .

(d) The Contractor shall reduce the value insured under Clause 20(A) of these Conditions by the full value of the relevant part, and the said relevant part shall, as from the date on which the Employer shall have taken possession thereof, be at the sole risk of the Employer.

It is plain that, in applying clause 16, the great question is: Had the employers, before the date of the fire, 18 January 1970, 'taken possession' of any part or parts of the works? If they had done so, that part was at the risk of the employers. If they had not done so, it was at the risk of the contractors. The contractors say that 'taking possession' is a question of fact which is answered here by the fact that the employers were using and occupying the factory at the time of the fire. The employers say that 'taking possession' only took place when the relevant part was handed over to them and accepted by them: and that never took place.

2. *The facts as to 'taking possession'*

(i) The car park

The car park was ready for use by September 1969, but there was some work yet to be done, such as painting the white lines to show the spaces for the cars. The contractors asked the architect to inspect it prior to handover. The architect did inspect it. On 23 September 1969, he accepted the handover and issued a certificate on a printed form issued by the RIBA. It said:

> I certify that, subject to the making good of any outstanding items, or of any defects, shrinkages and other faults which appear during the defects liability period . . . a part or section of the works, namely, Car Park, the value of which I estimate to be £10,000 was completed to my satisfaction and taken into possession on 22 September 1969, and that in relation to the said part or section of the works, the said defects liablity period will end on 23 March 1970.
>
> I declare that one moiety of the retention moneys deducted under previous certificate in respect of the said section is to be released.

The architect gave evidence that that form of certificate was in general use, and was the normal certificate which was used by architects for sectional completion. He said that, when issuing that certificate, he understood in principle that clause 16 required him to certify that a section or part of the works had been completed to his satisfaction and taken into possession.

Since this case started (and I expect because of it) the RIBA have issued an amended form in which the words 'completed to my satisfaction' have been struck out: but have left the rest of the form as it was.

That certificate as to the car park was the only certificate of the kind issued by the architect in respect of any part of the works. No such certificate was issued for any other part.

(ii) The board machine house

At the date of the fire – 18 January 1970 – the board machine house was virtually complete. The new 87-inch board machine had been installed and begun working. It was in production.

120 tons of boarding had been produced, about 80 per cent of which was suitable for marketing. There were some things still to be done by the contractors. For instance, a sliding fire door had yet to be fixed. The architect had inspected the floor but not the rest of the machine house.

On 12 December 1969, the architect had written to Wimpeys:

> We have inspected the floor of the Board Machine House and accepted this on behalf of our clients . . .
>
> This letter is in no way a Practical Completion Certificate: this will be issued at the appropriate time for the whole of the Board Machine House.

The position was, therefore, that the architect had not inspected the machine house itself, nor agreed to accept it.

(iii) The reel warehouse

The reel warehouse has three bays, namely, the eastern, central and western bays. At the date of the fire the whole was roofed and walled in, but the outside doors were not completed. In the central and western bays the sprinkler works and permanent lighting had been installed; but not in the eastern bay. There was some painting and glazing work to be done. The emergency exit doors had still to be fixed. Reeds had for some weeks used the warehouse for storing reels of paper. They had stored 1500 tons on the floors of the central and western bays, and 300 to 400 tons in the eastern bay. The floors had been inspected by the architect, but not the rest of the reel warehouse. On 8 December 1969, the architects had written to Wimpeys:

> We inspected the floor only of the central bay of the reel warehouse and accepted this on behalf of our client . . .
>
> This letter is in no way a Practical Completion Certificate: this will be issued at the appropriate time for the whole of the reel warehouse.

On 9 January 1970, the architects wrote a letter in the same terms as to the floor of the western bay of the reel warehouse.

The position was, therefore, that the central and western bays were approaching completion, save for the doors: but the eastern bay had still much to be done to it. The architects had not inspected the reel warehouse itself, nor agreed to accept it.

(iv) Factory extensions south and east

At the time of the fire, these were well advanced, 90 reels were stored in the southern extension and 45 in the eastern extension. But work remained to be done. The architects had not inspected these extensions, nor accepted them. Nor had they written any letter about them.

(v) The respective contentions

The contractors, of course, relied greatly on the use which Reeds had made of the works. Reeds had installed a great new machine in the board machine room. They had stored nearly 2,000 tons of paper in the reel warehouse. This use by Reeds hampered Wimpeys very much in their work. So much so, that Wimpeys said that the use and occupation by Reeds amounted to 'taking possession' of those parts.

But the Corporation replied that the parties themselves did not so regard it. They said that there was no 'taking possession' until the relevant part or section had been handed over by the contractors to the employers. They referred especially to a letter of 22 December 1969, in which Wimpeys wrote to the architect:

> We write to inform you that sectional handovers of completed works are being delayed due to a number of impediments which are beyond our control. The major disruptions to our Contract are:
> (a) Tenants' reel movement and storage;
> (b) Excessive occupation of areas by plant erectors . . .
> The effect of those factors has been to retard our performance and, in consequence, we apply for an extension of section completion dates . . .

3. *Provisions 'C' and 'D'*

One of the most important points in this case is whether the employers can rely on two special provisions (C and D) in the bills of quantities. These show that the parties contemplated that Reeds would install plant and equipment, and would occupy and use part of the works: but that the contractors were still to keep the works covered by insurance: and they could charge the increased cost of insurance to the employers. The two provisions were in these words:

> 'C' Employer may install equipment as the work proceeds
> As the work proceeds, the contractors shall permit the Employer or any other person authorised by him to place and install as much plant and equipment during the progress of the building as is possible before the completion of the various parts of the works, and such placing and installing shall not in any way alter the period fixed for completion of the works or the liability of the Contractor hereunder provided that:
> (a) in making arrangements for such placing and installation of plant and equipment, due regard shall be made to the requirements of the Contractor;
> (b) if by reason of such installation, the cost of any insurance which the Contractor is required to maintain under this contract shall be increased, then the amount of such increased cost shall be in addition to the contract sum.
>
> 'D' Occupation of completed parts of the Works
> The Contractor shall allow the Employer or any person authorised by him to occupy and use any part or parts of the Works as soon as, in the opinion of the Architect or Employer, such part or parts can be occupied and used without impeding the progress of completion of the Works, and such occupation shall not in any way alter the period fixed for completion of the Works or the liability of the Contractor hereunder, provided that if, by any reason of such occupation, the cost of any insurance which the Contractor is required to maintain under this contract shall be increased, then the amount of such increased cost shall be an addition to the contract sum.

Wimpeys contended that it was not permissible for the court to have regard to these special provisions 'C' and 'D'. They relied on clause 12(1) of the RIBA form, which says:

> . . . Nothing in the Contract Bills shall override, modify, or affect in any way whatsoever the application or interpretation of that which is contained in these Conditions.

Mocatta J left to himself would have had regard to provisions 'C' and 'D', despite clause 12. But he felt bound by some observations by Lord Hodson in *Gold* v. *Patman & Fotheringham Ltd* (1958); *North-West Metropolitan Regional Hospital Board* v. *T.A. Bickerton & Son Ltd* (1970); just as he had in *M. J. Gleeson (Contractors) Ltd* v. *London Borough of Hillingdon* (1970). So he thought it was best not to place any reliance on clauses C and D in interpreting clause 16.

4. *Type versus print*

I must say that I think that, in construing this contract, we should have regard to provisions C and D. They were carefully drafted and inserted in type in the bills of quantities. They were put in specially so as to enable the contractors to make their calculations. It was on the basis of these that the contractors made their tender and the employers accepted it. They were incorporated into the formal contract just as much as the conditions in the RIBA form. In contrast, conditions 12 and 16 were not specially inserted at all. They were two printed conditions in the middle of 23 pages of small print. It was in quite general terms. On settled principles they should have taken second place to the special insertion – see the passages collected by Scrutton LJ in *Sutro & Co* v. *Heilbut, Symons & Co* (1917). If and in so far as conditions 12 and 16 contain anything inconsistent with provisions C and D, they should be rejected, just as the printed form was in *Love and Stewart Ltd* v. *Rowtor S.S. Co Ltd* (1916).

5. *The construction of 'C' and 'D'*

In my opinion, however, the various provisions can all be reconciled in this way: The general printed condition 16 is designed to deal with the usual building contract in which the employer does not take possession of the works, or any particular part of the same, until it is practically complete. Whereas the special provisions C and D are designed to deal with this special contract in which it is contemplated that Reeds will be installing plant and equipment whilst the works are still in progress and will be using and occupying parts before they are fully completed. These special provisions apply to the situation which they expressly cover and should be given their proper effect without being restricted to condition 16.

Applying conditions C and D they show that the employers do not 'take possession' of a part of the works simply by installing plant and equipment in that part or by occupying and using that part. That is all the employers have done here, in regard to the board machine house, the reel warehouse and the factory extensions south and east. So the contractors remain liable to insure it and could charge the increased cost to the employers.

6. *Apart from 'C' and 'D'*

I would, however, not leave the matter there. I am quite prepared to consider clause 16 without placing any reliance on provisions 'C' and 'D'. In my opinion the words 'take possession' of a part of the works must be so interpreted as to give precision to the time of taking possession and accuracy in defining the part. I say this because of the consequences which follow on it. [We were given] a list of seven. They include these:

(i) That part is deemed to be practically complete and the defects liability period begins to run

(ii) A moiety of the retention money, in respect of that part, becomes payable

(iii) The architect is, within seven days, to issue a certificate of value of that part
(iv) That part is at the sole risk of the employer and not of the contractor; and the contractor is to reduce the value insured, accordingly.

In order to achieve the requisite precision and accuracy, the parties themselves, as I read the evidence, evolved a suitable machinery to determine it. It was by way of a definite handing over of the part by the contractors to the employers. The practice was for the contractors to tell the architect that a part was ready for handover. The architect would then inspect it to see if the part was in a condition to be accepted. If it was, he would accept it on behalf of the employers. He would give a certificate which defined the part, and would give its value and date of taking possession, and went on to state the consequences. So the handover was precise and definite. I regard that conduct as admissible either as explaining an ambiguity or uncertainty in the contract as it originally stood, or, alternatively, as an agreed gloss or variation on it made subsequently to the contract. I realise that the original form of certificate issued by the architect was inaccurate in speaking of 'practical completion' of the part. But that does not alter the fact that it was the accepted means of defining the handover. It should be given effect, accordingly.

7. *Conclusion*

In my opinion the contractors, at the time of the fire, had not handed over to the employers the responsibility for the board machine house, the reel warehouse, or the factory extensions east and south. Reeds were using and occupying those places but, until actual handover, it was the responsibility of the contractors to insure them. The risk had not passed to the employers. So it remained with the contractors. They, or their insurers, must bear the loss.

Note: Edmund Davies and Stephenson LJJ delivered judgments reaching the same conclusion, but differing as to the effect of clause 12. See p. 41.

On its true interpretation, clause 20[C] of JCT 63 imposes on the employer the whole risk of damage by fire, whether or not that fire is caused by the contractor's negligence. Fire caused by the contractor's negligence falls within the exception to the indemnity which he gives to the employer under clause 18. The insurance provisions in JCT 80 and JCT 98, clauses 21 and 22, are rather different. See now JCT 2005, section 6.

Wimpey Construction UK Ltd v. Scottish Special Housing Association

HOUSE OF LORDS

(1986) 9 Con LR 19

In 1980 the appellants contracted with the respondents on JCT 63 terms to modernise 128 houses in Edinburgh. In the course of carrying out the works one of the houses was damaged by fire and, for the purposes of Scottish legal procedure, it was assumed that the fire was caused by the appellants' negligence. The parties presented a special case to the Scottish Court of Session as to whether, on the true construction of clauses 18(2) and 20[C] of the contract, the appellants were responsible for the damage caused by the fire. The Court of Session held that the appellants were so liable. The appellants appealed.

Held: On the true interpretation of the contract, the contractor's liability set out in clause 18(2) was subject to the provisions of clause 20[C] which made existing structures and their contents at the sole risk of the employer as regards damage by fire and other specified perils. Clause 20[C] does not distinguish between fire due to the contractor's negligence and other causes. Consequently, fire caused by a contractor's own negligence falls within the exception to the contractor's liability set out in clause 18(2), and the appellants were not liable for the damage.

LORD KEITH OF KINKEL: The provisions of the contract which principally bear on the issue are to be found in clauses 18, 19(1)(a) and 20[C] of the standard form, which are in these terms:

> Clause 18(1) The contractor shall be liable for, and shall indemnify the Employer against, any liability, loss, claim or proceedings whatsoever arising under any statute or at common law in respect of personal injury to or the death of any person whomsoever arising out of or in the course of or caused by the carrying out of the Works, unless due to any act or neglect of the Employer or of any person for whom the Employer is responsible. (2) Except for such loss or damage as is at the risk of the Employer under . . . clause 20[C] of these conditions . . . the Contractor shall be liable for, and shall indemnify the Employer against, any expense, liability, loss, claim or proceedings in respect of any tempest, flood, bursting or overflowing of water tanks, apparatus or pipes, earthquake, aircraft and other aerial devices or articles dropped therefrom, riot and civil commotion (excluding any loss or damage caused by ionising radiations or contamination by radioactivity from any nuclear fuel or from any nuclear waste from the combustion of nuclear fuel, radioactive toxic explosive or other hazardous properties of any explosive nuclear assembly or nuclear component thereof, pressure waves caused by aircraft or other aerial devices travelling at sonic or supersonic speeds), and the Employer shall maintain adequate insurance against those risks. If any loss or damage affecting the Works or any part thereof or any such unfixed materials or goods is occasioned by any one or more of the said contingencies then, upon discovering the said loss or damage the Contractor shall forthwith give notice in writing both to the Architect/Supervising Officer and to the Employer of the extent, nature and location thereof and (a) The occurrence of such loss or damage shall be disregarded in computing any amounts payable to the Contractor under or by virtue of this contract . . . (c) If no notice of determination is served as aforesaid, or, where a reference to arbitration is made as aforesaid, if the arbitrator decides against the notice of determination, then (i) the Contractor with due diligence shall reinstate or make good such loss or damage, and proceed with the carrying out and completion of the Works (ii) the Architect/Supervising Officer may issue instructions requiring the Contractor to remove and dispose of any debris and (iii) the reinstatement and making good of such loss or damage injury or damage whatsoever to any property real or personal in so far as such injury or damage arises out of or in the course of or by reason of the carrying out of the Works, and provided always that the same is due to any negligence, omission or default of the Contractor, his servant or agents or of any sub-contractor his servants or agents.
>
> Clause 19(1)(a) Without prejudice to his liability to indemnify the Employer under Clause 18 of these conditions the Contractor shall maintain and shall cause any sub-contractor to maintain such insurances as are necessary to cover the liability of the

Contractor or, as the case may be, of such sub-contractor in respect of personal injury or death arising out of or in the course of or caused by the carrying out of the Works not due to any act or neglect of the Employer or of any person for whom the Employer is responsible and in respect of injury or damage to property, real or personal, arising out of or in the course of or by reason of the carrying out of the Works and caused by any negligence, omission or default of the contractor, his servants or agents or, as the case may be, of such sub-contractor, his servants or agents . . .

Clause 20[C] The existing structures together with the contents thereof owned by him or for which he is responsible and the works and all unfixed materials and goods, delivered to, placed on or adjacent to the Works and intended therefore (except temporary buildings, plant, tools and equipment owned or hired by the Contractor or any sub-contractor) shall be at the sole risk of the Employer as regards loss or damage by fire, lightning, explosion, storm, and (when required) the removal and disposal of debris shall be deemed to be a variation required by the Architect/Supervising Officer.

The opening words of clause 18(2) make it clear that the liability of the contractor for damage to property caused by his negligence or that of a sub-contractor or of anyone for whom either of them is responsible is subject to an exception. The ambit of the exception is to be found in clause 20[C]. Clause 19(1)(a), dealing with the contractor's obligation to insure against, inter alia, damage to property, does not shed any light on that matter, since the insurance is to cover only the contractor's liability for such damage, whatever that liability may be. Clause 20[C] provides that the existing structures and contents owned by the employer are to be at his sole risk as regards damage by, inter alia, fire. No differentiation is made between fire due to the negligence of the contractor and that due to other causes. The remainder of the catalogue of perils includes some which could not possibly be caused by the negligence of the contractor, such as storm, tempest and earthquake, but others might be, such as explosion, flood and the bursting or overflowing of water pipes. There is imposed on the employer an obligation to insure against loss or damage by all these perils, in quite general terms. I have found it impossible to resist the conclusion that it is intended that the employer shall bear the whole risk of damage by fire, including fire caused by the negligence of the contractor or that of sub-contractors. The exception introduced by the opening words of clause 18(2) must have the effect that certain damage caused by the negligence of the contractor or of sub-contractors, for which in the absence of these words the contractor would be liable, is not to result in liability on his part. The nature of such damage is to be found in clause 20[C], which refers in general terms to damage by fire to the existing structures. No sensible content can be found for the words of exception in clause 18(2) if they are not read as referring to damage of the nature described in clause 20[C]. Counsel for the respondents strove valiantly to indicate some such alternative content but was unable, in my view, to do so convincingly.

A similar conclusion was arrived at by the Court of Appeal in England in *James Archdale & Co Ltd* v. *Comservices Ltd* (1964) on the construction of similarly but not identically worded corresponding clauses in a predecessor of the standard form. I consider that case to have been correctly decided and to be indistinguishable from the present one.

The judges of the First Division were much impressed by what Lord Cameron described as a bizarre consequence of the construction contended for by the appellants, namely that if correct it would result in their being remunerated, assuming the contract was not terminated under clause 20[C](b), for putting right damage caused by their own negligence. The result, however, does not appear bizarre when it is kept in view that [the employer] would have received policy monies representing the cost of putting right the damage under the insurance which clause 20[C] required them to effect. In substance, the question at issue comes to be one as to which party has the obligation to insure against damage to existing structures due to fire caused by the negligence of the contractors or of sub-contractors.

In each case it is necessary to analyse the wording of the contract carefully to see whether it operates to produce this result.

Dorset County Council v. Southern Felt Roofing Ltd

COURT OF APPEAL

(1989) 29 Con LR 61

The defendants undertook works of repair and renewal to a flat roof of the plaintiff's primary school at Bourton. During the course of the works the school caught fire and the building and contents were seriously damaged. It was assumed for the purpose of the proceedings that this was due to the defendant's negligence. The plaintiff claimed for the loss caused by the fire.

The contract was on the county architect's conditions, which were derived from JCT 1963 (1977 revision) but had significant differences. The relevant clauses were 1.7, 1.9 and 2.1, which provided:

The Contractor Shall

1.7 Indemnify the Council against any liability, loss claim or proceedings in respect of injury or death to persons or damage to property and shall, without prejudice to his liability to indemnify the Council, or cause any sub-contractor to insure against the above risks for a minimum of £500,000 in respect of any one occurrence, the cover to be unlimited in amount, except that the insurance in respect of claims for personal injury or death from the Contractor's own employees arising out of and in the course of their employment shall comply with the Employer's Liability (Compulsory Insurance) Act 1969 or any amendment or re-enactment thereof.

1.9 Insure or cause any sub-contractor to insure against loss or damage by fire, etc, to any temporary buildings, plant, tools and equipment owned or hired.

The Council Shall

2.1 Bear the risk of loss or damage in respect of the Works and (where appropriate) the existing structure and contents thereof (excepting temporary buildings, plant, tools and equipment owned or hired by the Contractor or any sub-contractor) by fire, lightning, explosion, aircraft and other aerial devices or articles dropped therefrom.

Held: These clauses did not put on the plaintiff the risk of fire caused by the Contractor's negligence.

SLADE LJ: The conclusion that the operation of clause 1.7 is limited to claims etc. by third parties is, in my judgment, reinforced by reference to the pattern of the conditions, when read as a whole. They contain comprehensive provisions for the apportionment of risk. Clause 1.9

places squarely on the contractor the risk in respect of loss or damage by fire etc to 'any temporary buildings, plant, tools and equipment owned or hired' and obliges him to insure against loss or damage of this nature. Clause 2.1, while not requiring any insurance, places squarely on the council the risk in respect of loss or damage by fire etc. in respect of the works and the existing structure and contents thereof, except those temporary buildings, plant tools and equipment which the contractor is bound to insure under clause 1.9. Clause 1.7, as I read it, completes the apportionment of risk, by placing on the contractor the risk in respect of third party claims and obliging it to insure against such claims to the extent therein specified. I therefore accept [Counsel for defendants'] submission that, on its true construction, clause 1.7 is no more than a provision by way of indemnity (in the narrow sense) against third party claims, coupled with an obligation to insure against such claims. If it is to succeed on this appeal, the contractor nevertheless still has to surmount another major hurdle, which I will discuss in the next section of this judgment. Before doing so, however, I should draw attention to one other point for the sake of clarity.

Having reached the conclusion (contrary to my own) that the wording of clause 1.7 prima facie imposed a primary contractual liability on the contractor for the damage in question, the judge proceeded carefully to consider whether this prima facie contractual liability was displaced by the provisions of clause 2.1. For this purpose, he rightly gave close consideration to the effect of the decision of this court in *James Archdale & Co Ltd* v. *Comservices Ltd* [1954] 1 All ER 210, [1954] 1 WLR 459, the decision of Sellers J in *Buckinghamshire CC* v. *Y.J. Lovell & Son Ltd* [1956] JPL 196 and *Scottish Special Housing Association* v. *Wimpey Construction (UK) Ltd* [1986] 2 All ER 957, [1986] 1 WLR 995. The relevant contract in each of those cases contained a clause which expressly imposed on the defendant contractor liability of a primary nature in respect of certain matters, but was followed by a clause which was asserted by the contractor to place the risk in respect of the particular case of fire on the plaintiffs. Each of these three cases necessitated a close examination and comparison of the wording of the two particular clauses in question. In the first and third of them the contention of the contractor, whose negligence had caused the fire, succeeded. In the second it failed. In my judgment, however, problems similar to those which arose for decision in those three cases do not arise in the present case, where clause 1.7 imposes no express primary liability on the contractor in respect of anything (other than a liability to insure). Further reference to these three decisions therefore will not, in my judgment, assist the resolution of the present appeal . . .

That this case turns substantially on the non-standard provisions of the contract is suggested by the consideration of the insurance provisions of IFC 84 in* Scottish Newcastle plc *v.* G. D. Construction (St Albans) Ltd *(2003) 86 Con LR 1, CA; [2003] EWCA Civ 16, CA.

AIKENS J:

Analysis

15. As I have already stated, in my view it is important to note the nature of the breaches of contract and duty that the employer pleads against the contractor. They are all based on duties to take care. The allegation is that the contractor, or the sub-contractor for whose negligence the contractor is said to be liable under the terms of clause 6.1.2, failed to take care.

16. With that in mind it is necessary to consider the scope of clause 6.1.2 of the contract. It was accepted by both counsel that although cases concerning other wordings may contain general principles that this court must follow, there are no cases which have considered precisely this wording. Therefore I will consider the proper construction of the clauses first and then I shall consider the cases as may be necessary.

Construction of Clause 6.1.2

17. The relevant wording of clause 6.1.2 can be divided into two groups of phrases. The first group contains phrases defining the obligations of the contractor. These phrases provide: (i) that the contractor shall be liable for and shall indemnify the employer against (ii) any liability, loss or claim (iii) in respect of any damage whatsoever to any property, real and personal (iv) in so far as such damage arises out of or in the course of or by reason of the carrying out of the works (v) and to the extent that this damage is due to any negligence, breach of statutory duty, omission or default (vi) by the contractor, his servants or agents or (vii) any person employed or engaged by the contractor.

18. There are several points to note about this group of phrases. The first is that the clause is, in my view, defining the scope and limits of the liability of the contractor to the employer for certain types of default. Secondly the type of defaults which are covered by this clause are all based on negligence of one form or another. Hence the phrase 'to the extent that [loss, injury or damage] is due to any negligence, breach of statutory duty, omission or default [of the contractor etc.]'. In my view the words 'omission or default' are intended to embrace failures by the contractor to fulfil contractual obligations to take care.

19. Thirdly, in agreement with the submission of [Counsel for the respondent], this first group of phrases enlarges the classes of person for whose acts or omissions the contractor is liable. That is the effect of the wording 'servants, agents or of any person employed upon or engaged upon or in connection with the Works or any part thereof' and so forth to the end of that sentence. Fourthly, again in agreement with [Counsel for the respondent's] submissions, this group of phrases widens the class of loss or damage for which the contractor is liable to the employer and for which the contractor has to indemnify the employer.

20. In short, the first group of phrases lays down the general scope of the contractor's liability and obligation to indemnify the employer as a result of all types of failure to take care, whether contractual or non-contractual.

21. As part of defining the scope of the contractor's liability and obligation to indemnify, there are qualifying words in the second group of phrases in the last sentence of clause 6.1.2. This sentence limits the contractor's liability and its obligation to indemnify the employer in several ways. First, this sentence cuts down the property embraced by the liability and obligation to indemnify. Thus 'The Works' and other matters are excluded from the ambit of clause 6.1.2 by virtue of the terms of clause 6.1.3.

22. Next, this sentence deals with a case when (as under this contract) clause 6.3C1 applies. In such cases the 'liability and indemnity . . . excludes loss or damage to any property required to be insured [under clause 6.3C.1] caused by a Specified Peril'. In my view this means that the general scope of the 'liability and indemnity' embraced by the first part of the clause is cut down, when clause 6.3C1 applies. From the general scope of clause 6.1.2 is excluded any loss or damage to any property required to be insured under clause 6.3C1, where the loss or damage complained of has been caused by a specified peril. Therefore it is necessary to identify three things in order to work out the precise extent of the exclusion of the contractor's liability and obligation to indemnify the employer when clause 6.3C1 applies. The first is: what type of property has to be insured under clause 6.3C1. The second is the nature of the specified perils to be covered by the insurance. The third is whether the loss or damage complained of was caused by one of those specified perils.

Clause 6.3C1; 'Joint Names Policy' and 'Specified Perils'

23. When this clause applies, it places the employer under a contractual obligation to take out and maintain a joint names policy (as defined) in respect of property defined as 'existing structures'. A joint names policy has to include both the employer and the contractor as an

insured under the policy. It must also provide that the insurers can have no right of recourse against any person named as an assured. Clause 6.3C1 also provides that if there is a claim under the 'Joint Names Policy' then the contractor must authorise the insurer to pay all the insurance proceeds to the employer.

24. The 'Specified Perils' include 'fire'. As a matter of fact a fire can, of course, be caused by accident, inadvertence, negligence and a deliberate act of the assured or of third parties. So can some of the other perils identified as specified perils, such as 'bursting of water tanks'. But yet other specified perils, such as 'earthquake' or 'lightning', can have no element of human act or omission in their cause or creation.

The link between clause 6.1.2; clause 6.3C1 and the 'Specified Perils' in cases when clause 6.3C1 applies

25. In my view clauses 6.1.2; 6.3C1 and the definition of 'Specified Perils' and 'Joint Names Policy' are intended together to define the whole scope of the liability of the contractor to the employers for negligent acts and defaults when clause 6.3C1 applies. None of the clauses can be looked at in isolation. So the extent of the exclusion of the liability of the contractor and its obligation to indemnify the employer must depend on what the parties intended should be insured under the joint names policy. That depends in turn on what the parties intended should be included within the definitions of the 'Specified Perils' that are identified.

What is meant by 'Fire' in the 'Specified Perils' clause?

26. The clause listing the 'Specified Perils' identifies the particular perils that are to be covered by an insurance policy that has to be taken out by the employer. To my mind the parties must have intended that the words or phrases identified as specified perils be given the meaning that is normally given to them when they are used to identify a peril covered by an insurance policy. If the parties had intended otherwise, then I think that they would have said so. For nearly two hundred years when the word 'fire' has been used in an insurance policy to describe one of the perils covered by the policy, the meaning of the word 'fire' has been clear. Unless qualified by other words or a warranty in the policy, the peril 'fire' covers loss proximately caused by a fire, whether the fire was started by accident, was caused by the negligence of the assured or any third party or was caused by the deliberate act of a third party. (See e.g. *Busk* v. *Royal Exchange Assurance Co* (1818) 2 B & Ald 73; *Shaw* v. *Robberds* (1837) 6 Ad & El 75; *Mark Rowlands Ltd* v. *Berni Inns Ltd* [1985] 3 All ER 473 at 484, [1986] QB 211 at 232, per Kerr LJ and [1985] 3 All ER 473 at 486, [1986] QB 211 at 232 at 234, per Glidewell LJ.) If 'fire' is an insured peril in the policy, then a loss that is proximately caused by 'fire' is covered by the policy. It is irrelevant that the fire was itself caused by negligence or even the deliberate act of a third party. But, in the absence of express words in the policy, the parties would not have intended to cover losses by fire when that fire was caused by the deliberate act of the insured itself.

27. Under clause 6.3C1 of the contract, the employer had a contractual obligation to take out and maintain a joint names policy in respect of the existing structures which would pay for the full cost of reinstatement, repair or replacement of loss or damage due to the specified peril (amongst others) of fire. If the employer had fulfilled that obligation, then in my view that policy would have paid on a loss of the existing structure of the public house which was caused by a fire that was the result of the negligence of the contractor's sub-contractors. Moreover, if the employer had fulfilled its contractual obligation under clause 6.3.1 of IFC 84, the insurance policy would have contained a clause that stated that the insurer had no right to use the name of the employer to sue (by subrogation) the contractor, who would also be named as an assured under the policy. The effect of this 'no recourse' provision in clause 6.3.1

would have been to prevent the insurers from using the name of the employer to sue the co-assured contractor for damages on account of the negligence of its sub-contractors. (See *Co-operative Retail Services Ltd* v. *Taylor Young Partnership Ltd* (*Carillion Construction Ltd*, Part 20 defendants) (2002) 82 ConLR 1 at [65], [2002] 1 WLR 1419, per Lord Hope of Craighead. Lords Bingham of Cornhill, Mackay of Clashfern and Steyn agreed with Lord Hope on this point.) I discuss this case more fully below. The point to note here is that the contractors had to take out an 'all risks' policy against loss or damage in relation to a new construction site. Both the contractor and employer were named as joint assureds. Lord Hope concluded that the very existence of such a policy in joint names would prevent it being contended that the contractor was liable to the employer for damage resulting from a fire caused by the negligence of the contractor. The present case is even stronger, because of the existence of the 'no right of recourse' provision in clause 6.3.2 of the contract.

28. In this case the employer failed to take out the policy it should have done. But that cannot detract from the conclusion that I reach on the proper construction of these clauses. This is that the whole scheme of clauses 6.1.2; 6.3.2; 6.3C1, the provisions defining a 'Joint Names Policy' and 'Specified Perils' was to divide and allocate the risk of loss and damage to different types of property to the employer and the contractor. In my view, by the wording of those clauses the parties allocated to the employer the risk of loss and damage to existing structures by a fire which was caused by the negligence of a sub-contractor. The employer's losses were to be covered by the joint names insurance policy that the employer was contractually bound to take out and maintain. If the employer had fulfilled its obligations, then it would have obtained the insurance proceeds because, under clause 6.3C1, the contractor was obliged to authorise the insurer to pay the insurance proceeds of a claim to the employer.

JCT 98 clause 22C does not impose on the employer a duty to insure in respect of consequential loss such as loss of profits.

Kruger Tissue (Industrial) Ltd v Frank Galliers Ltd

QUEEN'S BENCH DIVISION

(1998) 57 Con LR 1

JUDGE JOHN HICKS QC:

Clause 22C

23. The issue here is whether an obligation to insure the existing structures and their contents 'for the full cost of reinstatement, repair or replacement of loss or damage' includes an obligation to cover loss of profit from and increased cost of working in a business carried on in those structures. On the natural and ordinary meaning of words the answer is in my view plainly 'no'. The expression 'cost of reinstatement, repair or replacement' is wholly apt in relation to buildings, fixtures, fittings and goods, but wholly inapt in relation to economic loss.

24. Is there anything in the documentary context or the factual setting to displace or confirm the natural meaning? [Counsel for defendant's] arguments for a wider construction which would include economic loss and thus produce the former result rested, I believe, on the implicit but fallacious assumption that the sole, or at least primary, purpose of clause 22C.1 is to serve as a qualification to clause 20.2, and that it must be interpreted as if it were a kind of proviso to the latter. On that basis it might well be right to entertain and give some weight to general policy considerations such as whether there should be a wider or narrower exception

to the contractor's liability for negligence and whether it is desirable to distinguish in that context between different kinds of damage.

25. But although clause 22C.1 does indeed serve by reference to qualify liability under clause 20.2 its primary purpose is to deal not with liability but with insurance. It must be construed in its own right and not merely as an adjunct to clause 20.2. Viewed in that light its primary function is simply to ensure, in conjunction with the rest of clause 22C, that there will be a fund available out of which damage to the existing structures of the works can be made good. That will no doubt generally be in the interests of the employer, who may indeed for the same reason extend the cover further, but no contractual obligation is needed for that reason; the contractual obligation, with the resulting expense to the employer, is imposed in the interests of the contractor and there is no policy reason why its meaning should be strained to extend more widely than to cover physical loss and damage.

26. Of equal, if not greater, significance is the undeniable consideration that the construction of clause 22C.1 must be the same for whatever purpose it is being invoked. In particular, since it imposes a free-standing and binding obligation, that obligation can be enforced by the contractor at any time, without reference to clause 20.2. The construction contended for by Galliers therefore entails that if the employer insures only for material loss, or although covered also for consequential loss adds the contractor's name only to the material loss section of the policy, the contractor can at any stage sue for breach of contract or can himself, under clause 22C.3, take out a joint names policy covering the employer's consequential loss and charge the premium to the employer by adding it to the contract price. I do not find that a plausible intention to attribute to the parties to such a contract in the absence of plain words having that effect, let alone in the teeth of plain words which have the reverse effect.

The insurance provisions in the contract between employer and contractor may have the effect not only of allocating risk as between employer and contractor but as between employer and sub-contractor (WHTSO ***v.*** Haden Young ***(1987) 37 BLR 135,*** Norwich City Council ***v.*** Harvey ***(1988) 45 BLR 14; compare*** National Trust ***v.*** Haden Young ***(1994) 41 Con LR 112). In deciding whether the contract has this effect it is necessary to consider all the provisions of the contract.***

British Telecommunications plc v. James Thomson & Sons (Engineers) Ltd

HOUSE OF LORDS

(1998) 61 Con LR 1

BT invited tenders for works of repair and refurbishment to a telephone switching station in Glasgow. MDW Ltd were the successful tenderers and the defendants were the domestic sub-contractors to MDW in respect of certain steelwork. The contract was on JCT 1980 terms. During the course of the works a fire broke out which was assumed for the purposes of the proceedings to be due to the defendants' negligence. The defendants argued that it was not fair, just and reasonable to impose a duty of care on them. This argument succeeded in both the outer and inner houses of the Court of Session, but was rejected by the House of Lords.

LORD MACKAY OF CLASHFERN: In order to examine this argument further it is necessary to set out the relevant terms of the main contract. Before doing so it is important to notice that the contractual provisions in the present case are different from those under consideration in *Wimpey Construction UK Ltd* v. *Scottish Special Housing Association* [1986] 9 Con LR 19, [1986] 1 WLR 995, and *Norwich City Council* v. *Harvey* [1989] 1 All ER 1180, [1989] 1 WLR 828. Amendment no. 2 above referred to, altered substantially provisions in the earlier contract relating to insurance and therefore while these cases are illustrations of the way in which such a matter as that presently in issue may be handled they cannot provide a direct answer in the differing contractual position of the present case . . .

Adopting this as their basis counsel for Thomson argue that having regard to the wording of the relevant provisions of the main contract the risk of fire being caused to existing structures while the works were being carried out was assumed, from the outset, by BT. In argument in the courts below counsel for BT accepted that this was the result of clauses 20.2 and 22C.1 as between BT and the main contractor. However, in the light of further consideration and in particular of the reported decision of the official referee in *Kruger Tissue (Industrial) Ltd* (formerly *Industrial Cleaning Papers Ltd*) v. *Frank Galliers Ltd* (1998) 57 Con LR 1, in your Lordships' House, he did not repeat this concession.

It is true, as counsel for Thomson argue, that BT were obliged to obtain insurance cover in respect of the existing structures, together with the contents thereof, owned by them, for the full cost of reinstatement, repair or replacement of loss or damage due to *inter alia* fire. The contractual arrangements therefore envisaged in the event of fire, BT would be indemnified by its insurers for the full cost of reinstatement irrespective of whether the loss or damage arose due to an act of God or an act or omission on the part of the main contractor or a sub-contractor such as Thomson . . .

It follows in my opinion that the terms of the provision for insurance of existing structures in respect of specified perils, while they provide for the recognition of the nominated sub-contractor as an insured under the policy or that such nominated sub-contractor shall have the benefit of a waiver of any right of subrogation which the insurer may have against him, provide no such protection for any domestic sub-contractor.

It is true, as was pointed out by the Lord Ordinary and the majority of the Second Division, that the absence of a protection against the right of subrogation does not of itself establish such a right but in considering whether the terms of the insurance policy which required to be taken out under the main contract are such as to make it unjust, unfair or unreasonable that Thomson should have a duty of care to BT, it is in my opinion necessary to take full account of all the provisions of the main contract with regard to the requirement for insurance and the terms on which such a policy should be taken out.

It is true also that in so far as the existence of the obligation on the employer to take out insurance against the specified perils in respect of existing structures relieves the main contractor from responsibility that he otherwise would have had for the negligence of sub-contractors under clause 20.2, the risk of Thomson's actings causing such loss as part of the responsibility of the main contractor will be covered. However, in considering the nature of the risk undertaken by the insurer the fact that the insurer will have a right of subrogation against a domestic sub-contractor such as Thomson will legitimately affect the question of premium. I conclude therefore that any element of double insurance which may be involved in giving effect to BT's argument is not a sustainable commercial objection to the success of that argument since practical considerations of premium will be affected by the right which the insurer has under the contract in particular, his right of recourse against Thomson if Thomson has a duty of care toward BT . . .

The question is whether or not it is fair, just and reasonable to impose a duty of care and in considering that question if the terms of a contract are to be taken into account it must be right to take account of all the terms of the contract that are relevant to the question. In my opinion it is of crucial significance in the present case that a distinction is made between nominated sub-contractors on the one hand and domestic sub-contractors on the other in the terms of the insurance policy to be provided by BT under the contract. In my view the contractual provisions reinforce rather than negative the existence of a duty of care toward BT by Thomson in the circumstances of the present case. Accordingly, in my opinion, this appeal succeeds and the case should be remitted to the Court of Session for a proof before answer.

In practice, the insurance needs of the parties will also be determined by the express contractual terms as to the transfer of risks. For instance, it is commonly provided that the works are at the contractor's risk until practical completion. The contractor therefore needs insurance cover against the possibility of the works being damaged by causes which are not his fault. The clause discussed in the following case has since been amended.

A.E. Farr Ltd v. The Admiralty

QUEEN'S BENCH DIVISION

[1953] 2 All ER 512

The plaintiffs contracted to construct a destroyer jetty at a naval base. Clause 26(2) of the contract provided:

> The works and all materials and things whatsoever including such as may have been provided by the Authority on the site in connection with and for the purposes of the contract shall stand at the risk and be in the sole charge of the contractor, and the contractor shall be responsible for, and with all possible speed make good, any loss or damage thereto arising from any cause whatsoever other than the accepted risks . . .

After the plaintiffs had done a substantial part of the work, the jetty was damaged by a destroyer (owned by the employer) which collided with it. The plaintiffs claimed the cost of repairing the jetty.

Held: Because of clause 26(2) the plaintiffs were liable for the repairs to the jetty.

> PARKER J: However one approaches this matter, one always gets back to what is the true construction of condition 26(2). It is argued forcibly by counsel for the plaintiffs that, if a party desires to be free of liability for negligence, he must say so in express words. That is undoubtedly true, and at first sight, when one looks at the facts of this case, it does seem as if very clear words were needed to make contractors liable for damage caused by the negligent navigation of an Admiralty ship which damages part of the works they have constructed, but it all comes back to the meaning of the words in that clause, and, in particular, to what is meant by 'any cause whatsoever' . . . It seems to me that the words in the present case are about as wide as they can be and that I must read 'any cause whatsoever' as if it included and expressly said 'including damage caused by the negligent navigation by an Admiralty servant of a ship'. If one reads the words in that way, there is no possibility of the plaintiffs recovering the sum

expended under condition 7(2) of the contract, because, by condition 26(2), they are responsible for the damage and have to make it good, and, therefore, the expense which they have incurred in so doing is not an expense beyond that provided for in, or reasonably contemplated by, the contract.

It is, however, said that there is an implied term in the contract that the contractor should have uninterrupted and undisturbed possession of the site, and that the building owner must not come, either negligently or deliberately, and knock down the works, and, accordingly, 'any cause whatsoever' must be read subject to that implied term. It seems to me that that is the wrong approach to the matter. Once one realises that condition 26(2) must be given the wide construction which, I think, it must be given, there is no room for any implied term, however it is worded, which, in effect, destroys the construction ordinarily given to those words. It was faintly argued by counsel for the plaintiffs that, even if he was wrong, it might be a possible view that the provision under condition 26(2) that the contractor was to make good was merely machinery, and that, having made good, it would be possible for him to obtain the sum that he has expended against the Admiralty for breach of contract or in negligence. It seems to me, however, that condition 26(2) is plain beyond words that, not merely as a matter of machinery has he got to carry out repairs, but that he is wholly responsible for the loss or damage. In my view, therefore, this claim, however hard it may be on the contractors, must fail.

An insurance policy taken out by virtue of the express insurance requirements of the contract may itself present problems of interpretation.

Manufacturers' Mutual Insurance Ltd v. Queensland Government Railways

HIGH COURT OF AUSTRALIA

(1968) 42 ALJR 181

The insurance company issued a policy of insurance to the respondents covering all loss or damage arising in connection with the respondents' contract to supply and erect a railway bridge except 'loss or damage arising from faulty design'. The bridge was to replace one built in 1897 and swept away by a flood in March 1956. In February 1958, three concrete piers erected in the bed of the river as part of the supports for the bridge were brought down by a flood higher than any previously recorded. The arbitrator found that there was no negligence in the design of the piers, but that with the benefit of hindsight and further research it was possible to say that inadequate allowance had been made for the transverse forces which operate on piers in a stream.

Held: 'Faulty design' did not imply any element of negligence, and that the loss was due to 'faulty design'.

BARWICK CJ, MCTIERNAN, KITTO AND MENZIES JJ: We have come to the conclusion that upon a proper construction of the relevant exclusion, the loss which occurred did arise from faulty design. Let it be accepted, as the arbitrator found, that the piers, as designed, failed to withstand the water force to which they were subjected because they were designed in accordance with engineering knowledge and practice which was deficient, rather than because the designer failed to take advantage of such professional knowledge as there was, nevertheless the loss was due to 'faulty design' and the arbitrator has done no more than explain how it happened that the design was faulty. We think it was an error to confine faulty design to 'the personal failure or non-compliance with standards which would be expected of designing engineers' on the part of the designing engineers responsible for the piers. To design something that won't work simply because at the time of its designing insufficient is known about the problems involved and their solution to achieve a successful outcome is a common enough instance of faulty design. The distinction which is relevant is that between 'faulty', i.e., defective, design and design free from defect. We have not found sufficient ground for reading the exclusion in this policy as not covering loss from faulty design when, as here, the piers fell because their design was defective although, according to the finding, not negligently so. The exclusion is not against loss from 'negligent designing'; it is against loss from 'faulty design', and the latter is more comprehensive than the former.

The performance of the contract is often secured by a performance bond, which is often in turn secured by a further guarantee. It is again a question of construction in what precise circumstances the bond and guarantee may be payable.

General Surety and Guarantee Co Ltd v. Francis Parker Ltd

QUEEN'S BENCH DIVISION

(1977) 6 BLR 18

DONALDSON J: These preliminary issues concern a performance guarantee and counter-guarantee in the building trade. The issue, which I am told is of general importance, is whether the terms of the counter-guarantee are such that the defendants are liable to make payment under it before the plaintiffs are in their turn liable to pay under the guarantee.

In September 1972 the Corporation of the City of Liverpool entered into a contract under which Francis Parker (Contracts) Ltd agreed to build 177 houses at a price of £1,162,326. The plaintiffs, who are a well-known insurance company, gave a performance guarantee to the Corporation limited to 10 per cent of the contract price of £116,232.60p. At the same time the plaintiffs took a guarantee from the defendants, who are the parent company of Francis Parker (Contracts) Ltd.

Disputes arose between the Corporation and the contractors, and on 25 April 1974, the Corporation gave notice purporting to determine the employment of the contractors. On 1 May 1974, the contractors in their turn purported to accept that notice as a wrongful repudiation of the contract. Whichever is right – and this still has to be determined – the employment of the contractors came to an end at that time and the Corporation thereafter contracted with different builders for the completion of the works.

On 22 August 1974, the City Solicitor wrote to the plaintiffs stating that the building contract had been terminated because the contractors had failed to proceed regularly and diligently with the works. The letter concluded:

> Your Company are sureties in respect of Phase 11 of this project and I write to give you notice that the City Council will eventually be submitting a claim under the Bond and will let you know in due course the amount involved.

This letter was not a demand for payment, but rather notice that 'eventually' such a demand would be made. This fact created a problem for the plaintiffs. On no view of the true construction of the counter-guarantee could the defendants be under any liability before the Corporation had made a claim under the principal guarantee. But 'eventually' could be a very long time. The Corporation and the contractors were likely to be locked in mortal arbitral combat, in which rounds were likely to be measured in months, if not years, rather than minutes. When the Corporation 'eventually' made a claim on the plaintiffs and the plaintiffs were in a position to make a claim under the counter-guarantee, the defendants' financial status might be very different. This is no reflection upon the defendants. It is a fact of economic life. Meanwhile the plaintiffs wanted to be able to improve the security afforded to them by the counter-guarantee by obtaining a cash payment thereunder.

The plaintiffs thereupon set about the delicate task of stimulating the Corporation to put forward a claim, without giving the slightest indication that it would be met. In a letter dated 29 November 1974, the plaintiffs referred to their wish to prepare 'our year-end claims reserves', and continued:

> In these circumstances, would it now be advisable to let us have a formal claim for payment of the Bond amount, which could be accompanied, if you so desire, by a statement that you do not propose to take action unless and until you see fit to do so.

The Corporation duly obliged by a letter dated 9 December 1974, in which the City Solicitor wrote:

> I refer to your letter of 27 November 1974, and now write to make a formal claim for payment of the Bond amount in respect of the above contract. I look forward to receiving your cheque in the sum of £116,232 in settlement at your earliest convenience.

I rather doubt whether the City Solicitor thought that the plaintiffs' 'earliest convenience' would occur in the foreseeable future, but that is by the way.

The plaintiffs in their turn wrote to the defendants on 13 December 1974, as follows:

> I refer to the above Bond and regret to advise you that we have now received a formal claim for payment of the Bond from Liverpool Corporation as per the attached copy letter, dated 9 December 1974. We herewith make formal demand for payment of this amount from yourselves in accordance with Clause 1 of the counter indemnity given by yourselves to us, a copy of which is attached.

I now turn to the wording of the guarantee. That given by the plaintiffs to the Corporation was in these terms:

> General Surety & Guarantee Co Ltd of 76, Cross Street, Manchester hereby guarantee to the Corporation the punctual, true and faithful performance and observance by the contractor of the covenant on his part contained in the said Agreement, and undertake to be responsible to the Corporation, their successors or assigns as sureties for the contractor for the payment by him of all sums of money, losses, damages, costs, charges and expenses that may become due or payable to the Corporation, their successors or assigns by or from the contractor by reason or in consequence of the default of the contractor in the performance or observance of his said covenant, but so nevertheless that the total amount to be demanded

> or recovered by the Corporation, their successors or assigns of or from us as sureties shall not exceed 10 per centum of the total prices payable under the said Agreement.

The counter-guarantee in which the defendants were described as 'the Guarantor' and the plaintiffs as 'General Surety' was, so far as is material, in these terms:

> (1) The Guarantor hereby undertakes that the contractor will duly perform and observe all its obligations under the contracts and that if General Surety is called upon to make any payment or payments under any of the said Guarantee Bonds the Guarantor will forthwith on demand pay to General Surety the full amount of such payment or payments.
> (2) The undertaking herein contained is irrevocable.
> (3) General Surety shall not be required before enforcing the undertaking herein contained to take any step to recover from any other person whether a principal debtor or surety any sum due to General Surety or be required to enforce any of General Surety's rights in any order of priority and no time given to any other person whether a principal debtor or surety or any variation or release of any obligation of any other person or release of any other security shall in any way lessen or affect the liability of the Guarantor hereunder.
> (4) General Surety shall be under no obligation to obtain the prior consent of the Guarantor or give notice to the Guarantor before or after entering into a Guarantee Bond in respect of the obligations of the contractor.

It seems to me to be clear – and [counsel] for the plaintiffs did not seek to persuade me otherwise – that the plaintiffs are only liable to the Corporation to the extent that it can be proved that the contractor has made default and that sums of money, losses, damages, costs, charges and/or expenses have become due and payable by the contractor to the Corporation by reason or in consequence of that default. Whether or not the day will come when this can be proved, I know not. But it has not yet come and the liability of the plaintiffs to the Corporation has yet to arise. Indeed, whilst no criticism can be made of the Corporation's notice of intention to lodge a claim, the claim itself was clearly premature.

But what of the counter-guarantee? [Counsel] submits that the first rule of construction is that documents are to be taken to mean what they say, unless what they say is so repugnant to good sense that some other meaning must have been intended. In this case the defendants agreed that if the plaintiffs were called upon to make any payment under the guarantee bonds, the defendants would forthwith on demand pay them the full amount of such payment. The plaintiffs have been called upon to pay the sum of £116,232.60 and the defendants are accordingly liable to make a similar payment.

I accept that effect must be given to the counter-guarantee in accordance with the meaning of the words used by the parties. But what is that meaning? The words 'if the General Surety is called upon to make any payment or payments under any of the said Guarantee bonds' show clearly that actual payment by the plaintiffs is not a condition precedent to liability on the part of the defendants. But do they go further than this? The concept of being called upon to make a payment is ambiguous. It may mean that A is liable and ought to pay B. Or it may mean merely that a demand has been made that A pay B regardless of whether or not the demand is well-founded. I think that the former is the true meaning of the words in this particular contract.

My reasons for reaching this conclusion are as follows:

(i) If this is, as it says, a contract of guarantee, the background must be one in which there is a principal debtor, namely the plaintiffs as General Surety, and a secondary debtor, namely the defendants. On the plaintiffs' construction, there need be no principal debtor.

(ii) If this is a contract of indemnity, and the argument proceeded upon the basis that it was, the background is one of a legal relationship in which liability at law arises only when the promisee (the plaintiffs) has been damnified by payment and in equity when the promisee has incurred a liability. On the plaintiffs' construction, it is unnecessary that the plaintiffs should either have been damnified or be liable to be damnified in the future.

(iii) If the counter-guarantee is a hybrid, which is probably the case, these considerations overlap.

(iv) Whatever its category, a contract of guarantee or indemnity is to be strictly construed in favour of the promisor.

(v) The promissory words in clause 1 of the counter-guarantee are ancillary to the primary undertaking that the contractor will duly perform and observe all its obligations under the contracts. If the plaintiffs are right, these words are otiose, since liability can arise whether or not the contractor is in default. So too are many of the provisions of clause (3) which exonerate the plaintiffs from taking steps which they could only take if the contractor was in default.

(vi) The promissory words require there to be a demand 'under . . . the guarantee bonds'. A premature demand may purport to be made under the guarantee bond, but is not in law so made.

(vii) If the plaintiffs are right, the defendants are liable to pay them the sum £116,232.60p, although it may eventually emerge that nothing is due from the contractors to the Corporation. I do not doubt that in such an event the money would be repayable, but it is not so certain that it would be repayable with interest. I find this situation very odd.

(viii) There is a simple and well-known way of achieving the result which the plaintiffs say that they have achieved. This is to provide that, as between the plaintiffs and the defendants, a demand by the Corporation shall be conclusive evidence of the plaintiffs' liability to the Corporation – see Bache & Co (London) Ltd v. Banque Vernes et Commerciale de Paris SA (1973). This course has not been adopted.

The argument before me has proceeded on the basis that either (a) there is an immediate liability on the part of the defendants to pay the plaintiffs the sum of £16,232.60 or (b) liability to make this payment is dependent upon the happening of any one of the events specified in the Order for the trial of the preliminary issues. No other position has been contended for. I therefore determine these issues by holding that:

There can be no liability on the part of the defendants to pay the plaintiffs under the terms of the agreement made between the parties unless and until:

(a) the architect appointed under the contract between Liverpool Corporation and the defendants' subsidiary has properly certified that the Corporation has suffered loss and damage in excess of the sum which would otherwise have been due to the subsidiary under the said contract and/or

(b) any such certificate has been opened up, reviewed and if necessary revised by an arbitrator pursuant to the said contract and/or

(c) liability to the Corporation has been admitted by the subsidiary in the sum of £116,232.60 or greater sum and/or

(d) the Corporation has successfully sued or arbitrated with the subsidiary to judgment or to an award not subject to case stated for the aforesaid or a greater sum.

. . . I have not gone into the details of this as to whether these are all conditions which trigger the guarantee, because that was not argued in front of me. I have simply taken it as a block.

Trafalgar House Construction (Regions) Ltd v. General Surety and Guarantee Co Ltd

HOUSE OF LORDS

(1995) 44 Con LR 104

The plaintiff entered into a contract with Maidstone Borough Council for construction of a new leisure complex and engaged KD Chambers Ltd as groundworks sub-contractors. The plaintiffs required Chambers to provide a bond for 10 per cent of the sub-contract value and Chambers duly provided a bond executed by themselves and the defendants. Chambers went into receivership and the plaintiff applied for summary judgment for the whole amount of the bond.

Held: The bond was a guarantee and therefore the defendant was entitled to rely on defences which would have been available to Chambers.

LORD JAUNCEY OF TULLICHETTLE:

The guarantee issue

In reaching the conclusion that the bond was not a guarantee in the ordinary sense of a 'see to it' obligation Saville LJ had regard to *Moschi* v. *Lep Air Services Ltd* [1972] 2 All ER 393, [1973] AC 331 wherein the guarantor personally guaranteed the performance by the principal debtor of its obligation to make regular instalment payments. There is, however, nothing in that case which suggests that a guarantee can only take the form of a 'see to it' obligation and Lord Reid went no further than to say that 'the authorities appear to recognise that at least most contracts of guarantee are of this nature' (see [1972] 2 All ER 393 at 399, [1973] AC 331 at 345).

There is however other material far more germane to this issue than *Moschi* v. *Lep Air Services Ltd*. Bonds in similar form have existed for more than 150 years and have been treated by the parties thereto and by the courts as guarantees. Indeed the current standard ICE Conditions of Contract contain a specimen bond in terms identical to those in the Chambers bond. In the first place the bond itself contains indications that it was intended to be a guarantee. The appellants are described as 'the surety'. There is a provision to the effect that no alteration in the terms of the sub-contract should release the surety from liability. In the absence of such provision a surety would normally be released from his obligation by any subsequent material alteration to the contractual provisions agreed between the contractor and sub-contractor . . .

My Lords, I have no doubt that the Court of Appeal were in error in concluding that the bond was not a guarantee but was akin to an on demand bond. No distinction can, in my view, properly be drawn between the effect of this bond minus the second part of the condition and the bond considered by Lord Atkin in the *Workington* case and other bonds using this or similar wording which have for many years been generally treated as guarantees (*Hudson's Building and Engineering Contracts* (11th edn, 1995) vol. 2, pp. 1499–1500, para 17–007). Thus in a second action arising out of the bond in the *Workington* case, *Workington Harbour and Dock Board* v. *Trade Indemnity Co Ltd (No. 2)* [1938] 2 All ER 101 at 105 Lord Atkin said:

> My Lords, both actions [that is the first and second actions] were brought on the money bond. It is well established that in such an action the plaintiff has to establish damages occasioned by the breach or breaches of the conditions, and, if he succeeds, he recovers judgment on the whole amount of the bond, but can only issue execution for the amount of the damages proved.

This dictum makes it clear beyond doubt that proof of damage and not mere assertion thereof is required before liability under such a bond arises.

I have therefore no hesitation in concluding that the Chambers bond without the second part of the condition would amount to a guarantee and that the appellants would be entitled to raise all questions of sums due and cross-claims which would have been available to Chambers in an action against them for damages . . .

There is no doubt that in a contract of guarantee parties may, if so minded, exclude any one or more of the normal incidents of suretyship. However, if they choose to do so, clear and unambiguous language must be used to displace the normal legal consequences of the contract – language such as was used in *Hyundai Shipbuilding & Heavy Industries Co Ltd* v. *Pournaras* [1978] 2 Lloyds Rep 502 where the letter of guarantee provided:

> The [defendant] hereby irrevocably and unconditionally guarantees the payment in accordance with the terms of the contract of all sums due or to become due by the buyer to you under the contract and in case the buyer is in default of any such payment the [defendant] will forthwith make the payment in default on behalf of the buyer . . .

This was construed as enabling the shipowner to recover from the guarantors of the buyers the amount due irrespective of the position between yard and buyers (see Roskill LJ at 508). The words relied upon by the respondents however do not clearly displace those legal consequences. Indeed the use of the word 'damages' is far more consistent with the compensation arrived at after taking into account all sums due to or by Chambers and the appellants. If the parties had intended to produce the result contended for by Mr Beloff it would have been a simple matter to use a form of words such as 'the additional expenditure incurred'. Instead they have used words which if anything point away from such a result.

I therefore conclude that the second part of the condition in no way alters the effect to be given to the remainder of the bond from which it follows that claims by Chambers of unpaid sums and set-off must be taken into account in determining the extent of the appellants' obligation . . .

Hudson's Building and Engineering Contracts (11th edn) para 17–007 refers to the almost universal practice of commercial bondsmen in clothing a very simple language in the jargon of an eighteenth century English bond. Like Lord Atkin and Hunter JA, I find great difficulty in understanding the desire of commercial men to embody so simple an obligation in a document which is quite unnecessarily lengthy, which obfuscates its true purpose and which is likely to give rise to unnecessary arguments and litigation as to its meaning.

Building contracts commonly provide in express terms that one of the parties shall indemnify the other against certain losses. Such a provision should not be read so as to impose on one party a duty to indemnify the other against the other's own negligence unless clear words to that effect are used. Indemnity clauses are strictly construed by the courts.

Walters v. Whessoe Ltd and Shell Refining Co Ltd

COURT OF APPEAL

(1960) 6 BLR 23

The plaintiff's husband was employed by Whessoe in operations on a site belonging to Shell. He was killed in an accident. Winn J held that both Whessoe and Shell were negligent, and apportioned liability 20 per cent to Whessoe and 80 per cent to Shell. Shell claimed to be entitled to indemnity from Whessoe by virtue of clause 3 of the contract between them, which provided:

> Injury and damage. The contractors shall indemnify and hold Shell their servants and agents free and harmless against all claims arising out of the operations being undertaken by the contractor in pursuance of this contract or order or incidental thereto in respect of (a) personal injury, including death and industrial disease, sustained by any employee of the contractor or of a sub-contractor, (b) loss or damage to the property and personal injury, including death, to the person of any third party and (c) loss or damage to the property, equipment or tools of the contractor, a sub-contractor or any of their employees.

Held: Shell's claim must be rejected.

> DEVLIN LJ: It is now well established that if a person obtains an indemnity against the consequences of certain acts, the indemnity is not to be construed so as to include the consequences of his own negligence unless those consequences are covered either expressly or by necessary implication. They are covered by necessary implication if there is no other subject-matter upon which the indemnity could operate. Like most rules of construction, this one depends upon the presumed intention of the parties. It is thought to be so unlikely that one man would agree to indemnify another man for the consequence of that other's own negligence that he is presumed not to intend to do so unless it is done by express words or by necessary implications. The reason for the rule is illustrated by one of the earliest cases in which it was enforced – *Phillips* v. *Clark* (1857). In that case the question was whether a shipowner who had inserted in a bill of lading the stipulation, 'not accountable for leakage or breakage', was exempted from responsibility for a loss by leakage and breakage if it arose from his own negligence. In the course of the argument Cockburn CJ said: 'I think it can hardly be permitted to him to contend that he inserted the clause for the purpose of protecting himself against negligence . . . Can we in a court of justice put so absurd a construction upon language that is susceptible of another and a more rational construction.'
>
> The law therefore presumes that a man will not readily be granted an indemnity against a loss caused by his own negligence. Such a loss is due to his own fault. No similar presumption can be made if he is made responsible for the negligence of others over whose acts he has no control. A responsibility of this sort may arise in law in the case of an employer and his workmen or an occupier and persons whom he invites to the premises. Similarly, a man may be responsible without negligence on his part for breach of statutory duty committed either by himself or by his servants or by someone for whom he was vicariously liable. In none of these cases is there any negligence on his part and there is therefore no reason to presume an intention to exclude them from the indemnity.

The indemnity which we have to construe does not contain any express words covering negligence. Its terms are wide enough to cover breaches of statutory duty and acts of independent contractors which may impose a liability on Shell without any fault on the part of themselves or their servants. It is not therefore necessary, in order to give the clause subject-matter, to construe it as covering negligence and so the indemnity does not cover the consequences of a negligent act committed by Shell's servants.

In two cases the court considered clause 14 of the then current RIBA contract (1939 edition) which provided:

(a) Injury to persons. The contractor shall be solely liable for and shall indemnify the employer in respect of any liability, loss, claim or proceedings whatsoever arising under any statute or at common law in respect of personal injury to or the death of any person whomsoever arising out of or in the course of or caused by the execution of the works, unless due to any act or neglect of the employer or of any person for whom the employer is responsible.

(b) Injury to property. Except for such loss or damage by fire as is at the risk of the employer under clause 15(b)(B) of these conditions the contractor shall be liable for and shall indemnify the employer against any loss, liability claim or proceedings in respect of any injury or damage whatsoever to any property real or personal in so far as such injury or damage arises out of or in the course of or by reason of the execution of the works, and provided always that the same is due to any negligence, omission or default of the contractor, his servants or agents or of any sub-contractor.

A.M.F. International Ltd v. Magnet Bowling Ltd

QUEEN'S BENCH DIVISION

[1968] 2 All ER 789

MOCATTA J, having read clause 14, said: It is hard to imagine anyone being proud of the draftsmanship of this condition. Nevertheless, counsel for Magnet, whilst being freely critical, claimed that sub-paragraph (b) clearly covered him. Magnet have been held liable in respect of damage to A.M.F.'s property. Such damage arose out of or in the course of or by reason of the execution of 'the works' and was due to negligence, omission or default of Trenthams. The claim was prima facie, at any rate, unanswerable.

Counsel for Trenthams advanced a number of arguments to the contrary. First 'the works' in condition 14(b) could not include the work undertaken by A.M.F. under the permission granted by condition 23. This may be so, but the provisions in items 1/5/C, 1/6/E and 2/1/B of the bill of quantities, to which I have already referred were in my judgment part of 'the works'. It is clear on the facts as I have found them that had these provisions been complied with water should not have entered the building. For these reasons I reject this argument. Secondly, counsel for Trenthams said that condition 10 prevented Magnet placing any reliance on any of these items in the bill of quantities so as to entitle them to indemnity under condition 14(b). Condition 10 reads:

> The quality and quantity of the work included in the contract sum shall be deemed to be that which is set out in the bills of quantities mentioned in clause 2 of these conditions

which bills unless otherwise expressly stated shall be deemed to have been prepared in accordance with the principles of the standard method of measurement of building works last before issued by the Royal Institution of Chartered Surveyors and the National Federation of Building Trades Employers, but save as aforesaid nothing contained in the said bills of quantities shall override, modify or affect in any way whatsoever the application or interpretation of that which is contained in these conditions.

I accept the answer of counsel for Magnet to this point. The items in question constituted part of the quantity of the work included in the contract sum within the opening words of that condition. There is therefore no conflict, as counsel for Trenthams argued, between the last part of condition 10 and condition 14(b) to prevent Magnet from relying on the latter as they do.

Thirdly, counsel for Trenthams raised a point of law of general importance and some difficulty. Relying on my anticipated (and now actual) finding that the damage to A.M.F.'s property was partly caused by the negligence of Magnet, he submitted that on authority Magnet were unable to recover anything under the indemnity clause. He relied on *Rutter* v. *Palmer* (1922), *Joseph Travers & Sons Ltd* v. *Cooper* (1915), and similar well-known cases following on them, as establishing the general principle that an exception clause does not protect against negligence unless it expressly so provides on the true construction of the words used, or unless no content can be given to the words used unless they protect against negligence on the principle laid down in *Alderslade* v. *Hendon Laundry Ltd* (1945). These points are nowhere more clearly stated than in *Canada Steamship Lines Ltd* v. *Regem* (1952) in the well-known passage, which it is unnecessary for me to read. Condition 14(b) could be given content to protect Magnet against liability to third parties arising through no negligence on their part. Counsel for Trenthams instanced *Hughes* v. *Percival* (1883) as an example. In that case the building owner was held liable in tort for damage caused to his neighbour's house by bad work by his builder, notwithstanding that he had employed a competent architect and builder. Another example relied on was *Balfour* v. *Barty-King* (1957). On this point I think counsel for Trenthams was undoubtedly right. He further argued that these principles applied as much to claims by persons suing on indemnity clauses as to defendants seeking to escape liability under exception clauses. This undoubtedly seems to have been accepted by the Privy Council in *Canada Steamship Lines Ltd* v. *Regem* (1952). It seems to have been assumed, at least as a possibility, in *A.E. Farr Ltd* v. *The Admiralty* (1953) and *Westcott* v. *J.H. Jenner (Plasterers) Ltd and Bovis Ltd* (1962), though in both those cases the person relying on the indemnity succeeded by reason of the language used being held to have been sufficiently clear to have covered his own negligence.

The most relevant and powerful authority relied on by counsel for Trenthams was *Walters* v. *Whessoe Ltd and Shell Refining Co Ltd* (1960) . . .

Counsel for Magnet sought to distinguish the present case in a variety of ways. First of all he submitted that the words of condition 14(b) were wide enough to include Magnet's negligence and breach of contract. He pointed out that the word 'solely', present in sub-paragraph (a), is absent from sub-paragraph (b). As against this counsel for Trenthams suggested that the effect must be the same by reason of the words 'in so far as' in sub-paragraph (b). Further, counsel for Magnet urged that condition 14 was a sweeping-up clause and that sub-paragraph (b) significantly lacked the words at the end of sub-paragraph (a) 'unless due to any act or neglect of the employer', etc. I have already said that the phraseology of the condition as a whole leaves much to be desired; I do not think any real assistance is to be derived from comparing the language of the two sub-paragraphs. If a party relying on an indemnity is to escape the effect of the principle laid down in the *Whessoe* case the language he relies on must be clear. In my judgment the language used in 14(b) is not clear enough to take Magnet out of the principle.

Counsel for Magnet then argued that I must distinguish between *causa causans* and *causa sine qua non*, that in the *Whessoe* case Whessoe's negligence was only a *causa sine qua non* and that the *causa causans* or real or effective cause was the negligence of Shell. Here he equated Magnet with Whessoe and Trenthams with Shell. Moreover, the passages which I have read from the judgments of Sellers and Devlin LJJ, in dealing with the principle, did not mention the partial responsibility in negligence of Whessoe. Unless one interpreted the reasoning of the Court of Appeal in this way, counsel submitted there was an insoluble logical conflict between the *Whessoe* case and a line of authorities holding that a plaintiff can recover as damages for breach of contract the damages, flowing therefrom that he has had to pay a third party for liability in tort. He referred to the well-known line of cases starting with *Mowbray* v. *Merryweather* (1895), continuing with *Scott* v. *Foley Aikman & Co* (1899) and *The Kate* (1935), and most recently exemplified by *Sims* v. *Foster Wheeler Ltd* (1966). In each of these cases the plaintiff succeeded in contract because his liability in tort to a third party was due to a breach of duty owed to that third party and not to the defendant. He further relied for his interpretation of the *Whessoe* case on what was said by Devlin J, in *Compania Naviera Maropan S.A.* v. *Bowaters Lloyd Pulp and Paper Mills Ltd* (1955).

I am unable to see how I can distinguish the reasoning and decision in the *Whessoe* case on the lines suggested. What fraction of responsibility greater than twenty per cent is required before the other party's negligence ceases to be the effective cause? The question has only to be asked to be seen to be unanswerable. If it can be answered, I can see no grounds for holding sixty per cent responsibility to be the effective cause and forty per cent only a sine qua non. Further I think that there is a real distinction in principle between the indemnity cases and the line of contract cases cited. In the latter the defendant is under a contractual obligation to provide reasonably safe tackle, or a berth or scaffolding, etc. or warrants that what he has provided is reasonably safe to use. In the indemnity cases the party giving the indemnity merely undertakes to pay the other party money in certain circumstances. If, as a matter of construction, the event that has occurred falls outside those circumstances, no liability to indemnify arises. For these reasons and considering myself bound by the *Whessoe* case, even if not strictly bound by the *Canada Steamship* case, I reject Magnet's claim to be indemnified under condition 14(b). In doing so, however, I base my decision on my earlier finding of Magnet's liability to A.M.F. in tort and express no view whether, had I only held them liable in contract, the claim under the indemnity clause would have failed.

City of Manchester v. Fram Gerrard Ltd

QUEEN'S BENCH DIVISION

(1974) 6 BLR 70

KERR J: I must turn to clause 14(b) on which ultimately everything depends. It does not appear ever to have been considered judicially except by Mocatta J in *AMF International Ltd* v. *Magnet Bowling Ltd* (1968), and he then said: 'It is hard to imagine anyone being proud of the draftsmanship of this condition'. One important issue to be decided on the short-route argument in relation to [the sub-sub-contractors] in this case, i.e. whether or not the clause covers persons in [their] position and sub-sub-contractors generally, is unclear and certainly lends support to this view.

I deal first with three general matters which are of some relevance to the plaintiffs' contentions both in relation to the long-route and short-route arguments. First, clause 14(b) deals with injury to property, whereas clause 14(a) deals with injury to persons. I have not set out the latter because [counsel] rightly did not press any argument to the effect that one gets any real assistance on the construction of sub-clause (b) from the wording of sub-clause (a) and I note that Mocatta J took the same view in the AMF case. Secondly, I approach the construction of this clause on the basis that, being an indemnity clause, it should be construed strictly, analogously to an exemption clause; at any rate to the extent to which the defendants are thereby sought to be held liable for defaults etc. by persons other than themselves over whom they have no control. This is the position here, since it is not suggested by the plaintiffs that there was any negligence, omission or default on the part of the defendants in any respect whatever. This general approach to the construction of the clause is in my view established by authority. It was the approach adopted by the Privy Council in the leading case of *Canada Steamship Lines Ltd* v. *Regem* (1952), where the Board was dealing both with an exemptions clause and an indemnity clause, and approached the construction of both in the same way. This approach was followed by Mocatta J in the *AMF* case and in the decision of the Court of Appeal in *Walters* v. *Whessoe Ltd* (1960) there cited. More recently there is further direct authority that this is the correct approach to indemnity clauses in *Gillespie Brothers & Co Ltd* v. *Roy Bowles Transport Ltd* (1973), where the *Canada Steamship Lines* case was again followed, at any rate by the majority of the court, in the context of an indemnity clause, and where Buckley LJ expressly equated the approach to the construction of indemnity clauses with that to exemption clauses. I therefore proceed on the basis that if and in so far as the wording of clause 14(b) does not clearly have the effect of rendering the defendants liable for the defaults etc. of other persons, then the plaintiffs cannot successfully rely upon it.

. . . The plaintiffs' final, and in my view most difficult, submission was that the words 'or of any sub-contractor' are wide enough to include sub-sub-contractors. They contend that [K] were sub-sub-contractors of the defendants. It could not, of course, avail the plaintiffs merely to rely on the fact that [K] were sub-contractors of Smith's, since they are not suing Smith's and since there is no indemnity clause in force between the plaintiffs and Smith's. [Counsel] said – in my view rightly so far as it goes – that any one working on a site otherwise than under some contract of employment is in general parlance often referred to as a 'sub-contractor'. But I have come to the conclusion that such a wide construction cannot properly be given to the indemnity clause in this contract. There are a number of reasons for this. First, as conceded by [counsel], the contract nowhere uses the words 'sub-contractor' in a context which requires it to be read as including sub-sub-contractors. It especially provides in clause 13(a) for the subletting of part of the works by the main contractor with the written consent of the architect, and in clause 21 for the appointment of nominated sub-contractors by the main contractor. Both these clauses therefore refer only to sub-contractors of the main contractor, and the contract nowhere deals, either expressly or by necessary implication, with further generations of sub-contractors. I therefore see no sufficient reason for attributing any wider meaning to the words 'sub-contractor' in the indemnity clause. Secondly, if it had been intended that the indemnity clause should have the wide meaning for which the plaintiffs contend, then it would have been easy to say so. Instead of referring to 'any sub-contractor' it would have been easy to say something like 'anyone executing any part of the works' or 'anyone working on the site'. Further, I do not think that anything is added to the plaintiffs' submissions, or provides any answer to the foregoing points, by relying on the main contractors' obligations in relation to insurance, to which I have already referred. [Counsel] said that the employer would wish to be covered by insurance, and therefore also by indemnity, in relation to anyone working on the site and not merely in relation to the first generation of sub-contractors. But in my view this is little more than an argument *ad misericordiam*. If this had been intended then it was incumbent upon the employer to cover it by clear language. The other side of the coin in relation to this argument is that it would lead to hardship on the main contractor. He would no doubt wish at least to have the opportunity – as he expressly has in relation to nominated sub-contractors – to cover his liability under the indemnity clause by sub-contracts of indemnity. But in practice this would be extremely difficult, to say the least, in relation to every contractor on the site with whom the main contractor has no contractual nexus, particularly when it is well known that many individual workmen nowadays require to be employed as independent contractors and not under contracts of employment.

I have accordingly come to the conclusion that the words 'or of any sub-contractor' in clause 14(b) are not wide enough to cover further generations of sub-contractors beyond those who are sub-contractors to the main contractor.

It may be possible in some circumstances to infer that one party has impliedly undertaken to indemnify the other.

Hadley v. Droitwich Construction Co Ltd

COURT OF APPEAL

[1967] 3 All ER 911

The plaintiff was injured when a small mobile crane collapsed. The crane had been hired by its owner to a builder and both owner and builder were sued by the plaintiff. The trial judge held that both defendants were at fault, the owner because he had negligently handed the crane over with an incorrect adjustment of the clearance between its rear roller and the bottom of its base and the builder because he had negligently failed to have the machine serviced properly and to provide a competent driver for it. The builder claimed that he was entitled to be indemnified by the owner against his liability to the plaintiff.

Held: The builder was not entitled to an indemnity as the owner, in a letter confirming the oral agreement, had written 'we thank you for your promise to put a competent man in charge of this machine who will operate it carefully and carry out the servicing properly'.

HARMAN LJ: [The learned judge goes on, however, to acquit the employers of all liability.] I cannot follow him. He can only make that finding, as it seems to me, if what had passed was that the owners gave an assurance, on which the employers were entitled to rely, that the crane was all right as handed over and need not be further examined. So far from that being the position, the owners, although they did not send their book of words, so to speak, or any specific information about the alterations which they had made to the machine, did write a letter to say they were glad that the employers agreed to put an experienced man on this job and to see that the machine was properly serviced. Any warranty that the owners may have impliedly given as to fitness of the machine for its task was undoubtedly, as I see it, qualified by those two conditions. Neither condition was performed: the warranty disappears, as the conditions were broken. So far from being entitled to put the whole liability on the owners, I think that the employers cannot shift it from their shoulders in so far as it was due to their own breach of duty towards their employee.

Mowbray v. *Merryweather* (1895) was quite a different case. There was there an assurance, on which the plaintiff (who had paid the workman) was entitled to rely, that everything was in order, and he was absolved from his duty vis-à-vis the second defendant to examine the chain in question. Here, owing to the presence of the letter, the boot is on the other leg, so it seems to me.

In a building contract there are nearly always several parties (contractors, sub-contractors, sub-sub-contractors, suppliers, etc.) involved. It is common for insurance policies to be taken out in the names of some or all of these parties. The legal effect of this was considered in:

Petrofina (UK) Ltd v. Magnaload Ltd

QUEEN'S BENCH DIVISION

[1983] 3 All ER 35

An accident occurred at an oil refinery at Killingholme where a major extension was being built. The refinery was owned by Lindsey Oil, the third plaintiffs, who operated it for the benefit of the first and second plaintiffs. The works were financed by the fourth plaintiff Omnium Leasing, a consortium of companies. The main contractors were Foster Wheeler, who had sub-contracted heavy lifting operations to Greenham (Plant Hire). Certain specialist lifting equipment was provided by the first defendants, Magnaload, and the second defendants, Mammoet Stoof. The plaintiffs alleged that the accident was due to the negligence of the first and second defendants. The proceedings involved a preliminary issue as to whether the first and second defendants were covered by a contractor's 'all risks' policy taken out by Foster Wheeler.

LLOYD J: Counsel for the defendants told me that it is a question of great importance for the construction industry generally; for policies such as the one I have to consider are in common use. I am sure he is right. The answer in the present case turns on the construction of this particular policy, and on general principles of law. The policy is a 'contractors' all risks insurance policy'. By section 1 it provides:

> The insurers will indemnify the insured against loss of or damage to the insured property whilst at the contract site from any cause not hereinafter excluded occurring during the period of insurance.

'The insured' are defined in the schedule as –

> Omnium Leasing (Owner) and/or Lindsey Oil Refinery and/or Foster Wheeler Ltd. and/or Contractors and/or Sub-contractors.

'The insured property' is defined as –

> Item No. 1. The works and temporary works erected . . . in performance of the insured contract and the materials . . . for use in connection therewith belonging to the insured or for which they are responsible brought on to the contract site for the purpose of the said contract . . .
>
> Item No. 2. Constructional plant comprising plant and equipment . . . if and insofar as not otherwise insured belonging to the insured and for which they are responsible brought on to the contract site for the purpose of the insured contract.

'The insured contract' is defined as 'Construction erection and testing of an extension to the Lindsey Oil Refinery at South Humberside'.

By an indorsement dated 23 April 1978 the policy was extended to include –

> The erection operation and subsequent dismantlement of the following items of plant which are being used on the Greenham (Plant Hire) Ltd sub-contract for the erection of eight large vessels: (1) . . . £600,000; (2) . . . £250,000; (3) Hydrajack system £2,150,000

giving a total of £3,000,000. The minimum premium was increased by £9,000.

The policy also provides in section 3, third party liability cover in standard terms, whereby the insurers agree to indemnify the insured –

> Against all sums for which the insured shall become legally liable to pay as damages consequent upon (a) Accidental bodily injury to or illness or disease of any person (b) Accidental loss of or damage to property occurring . . . as a result of and solely due to the performance of the insured contract happening on or in the immediate vicinity of the contract site.

There is an exception to the third party liability cover which excludes property forming the subject of the insured contract.

The first point taken by the insurers is that the defendants are not 'sub-contractors' within the meaning of the policy, and are therefore not insured at all. If they are right about that, then none of the other points arise. At an early stage of the argument counsel for the plaintiffs very properly accepted, though he never formally conceded, that the first defendants, Magnaload, must be sub-contractors; but he maintained the argument in relation to the second defendants, Mammoet.

The facts are that in the initial stages of the negotiations it was contemplated that Mommoet would enter into a contract with Greenhams for carrying out the specialist lifting operation with their hydrajack. At the end of June there was a change of direction. At a meeting which took place between Magnaload and Mammoet on 27 June 1978 it was agreed that, for administrative reasons, the contract would be in the name of Magnaload, not Mammoet. Mammoet would, however, remain responsible for the operation itself. The reasons for this change do not matter. The sub-contract price had already been agreed between Greenhams and Mammoet at £185,000. It was agreed that Mammoet would reimburse Magnaload out of that sum for any services which Magnaload performed or any costs which they incurred. On 13 July 1978 there was a meeting between Magnaload, Mammoet and Greenhams at which Greenhams agreed that the contract would be with Magnaload provided they received '100 per cent support' from Mammoet. On 24 July Greenhams sent a telex to Magnaload as follows:

> Please find the draft of the purchase order for Foster Wheeler job at Immingham. Could we have your comments as soon as possible. 'Order on Magnaload. For supply of lifting equipment and services in order to carry out the erection of heavy vessels at the Lindsey Oil Refinery expansion project, Killingholme, South Humberside. All as more fully described in the attached scope of work and conditions of order. For the lump sum price of £185,000.'

Then, under the heading scope of work and conditions of order, I will quote two paragraphs:

> Erection of vessels at Lindsey Oil Refinery expansion project, Killingholme, South Humberside. Main contractor: Foster Wheeler Ltd. Heavy lifting sub-contractor: Greenham Plant Hire Ltd. Specialist lifting contractor: Magnaload Ltd. . . . General conditions of contract in line with those which Greenham Plant Hire have accepted from Foster Wheeler are understood to apply to this order.

That telex is signed by Greenham Plant Hire.

The same day a copy of that telex was sent on by Magnaload to Mammoet. Mammoet came back with certain comments, which were sent on to Greenhams. According to W, who gave evidence at the hearing, and who had taken an active part in the negotiations as commercial director of Magnaload, all three parties reached agreement on all points of substance.

The term 'sub-contractor' is defined in the head contract as meaning –

> Any person to whom the preparation of any design the supply of any plant or the execution of any part of the works is sub-contracted, irrespective of whether the contractor is in direct contract with such person.

There is a similar definition in the sub-contract between Foster Wheeler and Greenhams.

On the facts which I have set out, there was clearly a contract between Greenhams and Magnaload, as, indeed, counsel for the plaintiffs accepted. Strictly speaking Magnaload were not sub-contractors but sub-subcontractors. But I would hold that they were nevertheless sub-contractors within the definition contained in the contracts and within the ordinary meaning of the word 'sub-contractors' as contained in the policy. To my mind the word 'sub-contractors' in the context of the policy must include sub-sub-contractors as well as sub-contractors.

The next question relates to the property insured. Counsel for the defendants submits that a 'contractors' all risk' policy in this form is an insurance on property, namely the works and temporary works belonging to the insured, or for which the insured are responsible. Counsel for the plaintiffs accepts that it is an insurance on property, but submits that the property insured must be read distributively, in other words, each insured is only insured in respect of his own property, or property for which he is responsible. I do not accept that construction. It seems to me that on the ordinary meaning of the words which I have quoted, each of the named insured, including all the sub-contractors, are insured in respect of the whole of the contract works. There are no words of severance, if I may use that term in this connection, to require me to hold that each of the named insured is only insured in respect of his own property. Nor is there any business necessity to imply words of severance. On the contrary, as I shall mention later, business convenience, if not business necessity, would require me to reach the opposite conclusion. I would hold, as a matter of construction, that each of the named insured is insured in respect of the entire contract works, including property belonging to any other of the insured, or for which any other of the insured were responsible.

But then comes the question: what is the nature of the interest insured? Counsel for the plaintiffs submits that the only possible insurable interest which one insured could have in property belonging to another, is in respect of his potential liability for loss of or damage to that property. Accordingly section 1 of the present policy is, he submits, a composite insurance. It is an insurance on property so far as it relates to property belonging to any particular insured; it is a liability insurance so far as it relates to all other property comprised in 'contract works'.

Counsel for the defendants, on the other hand, submits that section 1 of the policy is a policy on property, pure and simple.

In my judgment the submission of counsel for the defendants is correct. A very similar question arose in *Hepburn* v. *A. Tomlinson (Hauliers) Ltd* (1966). In that case a firm of carriers took out a policy of insurance of goods in transit, on a form known as 'Form J'. A quantity of cigarettes were stolen while the goods were still in transit, by reason of the negligence of the owners, Imperial Tobacco Co. It was common ground that the carriers were not liable to the owners. On the carriers making a claim under the policy, for the benefit of the owners, it was argued that the policy was a liability policy, and since the carriers were not liable to the owners, they could not recover from the insurers. The argument was rejected by Roskill J, and his judgment was upheld by the Court of Appeal and the House of Lords.

It appears that underwriters had always regarded an insurance on 'Form J' as being a liability policy, and not a policy on goods. But Lord Reid said that the language of the policy showed conclusively that the policy was a policy on goods. I can find no relevant distinction between the language of the present policy and the language in *Hepburn* v. *Tomlinson*. Section 1 of the present policy covers 'all risks of loss or damage to the insured property'. It excludes the cost of replacing or rectifying insured property which is defective in material or workmanship, and so on. There is nothing in section 1 of the policy which is appropriate to an insurance against liability, whereas everything is appropriate to a policy on property. Section 2 of the policy covers the contractual liability of the contractors to maintain the contract works during the maintenance period. Section 3 covers third party liability. The very fact that third party liability is covered in a separate section, which expressly excludes liability for damage to property forming part of the contract works, shows to my mind that section 1 of the policy is an insurance on property, and not an insurance against liability.

There is a passage in Jessel MR's judgment in *North British and Mercantile Insurance Co* v. *London, Liverpool and Globe Insurance Co* (1877) which points in the other direction. For he refers to a wharfinger's insurance on goods held in trust, or for which they are responsible, as being in terms an insurance against liability. But that dictum (for it was no more) cannot now stand against the decision of the House of Lords in *Hepburn* v. *Tomlinson*.

That brings me to the central question in the case. In *Hepburn* v. *Tomlinson* it was held, indeed it was conceded, that if the policy was an insurance on goods, then the carriers could, as bailees, insure for their full value, holding the proceeds in trust for the owners. In the present case the defendants could not be regarded as being in any sense bailees of the property insured under the policy. Does that make any difference? Can the defendants recover the full value of the property insured, even though they are not bailees? It is here that one leaves the construction of the policy, and enters, hesitatingly, the realm of legal principle.

What are the reasons why it has been held ever since *Waters* v. *Monarch Fire and Life Assurance Co* (1856) that a bailee is entitled to insure and recover the full value of goods bailed? Do those reasons apply in the case of a sub-contractor?

One reason is historical: the bailee could always sue a wrongdoer in trover. If his possessory interest in the goods was sufficient to enable him to recover the full value of the goods in trover, why should he not be able to insure that interest?

Another reason was that, as bailee, he was 'responsible' for the goods. Responsibility is here used in a different sense from legal liability. A bailee might by contract exclude his legal liability for loss of or damage to the goods in particular circumstances, e.g. by fire. But he would still be 'responsible' for the goods in a more general sense, sufficient, at any rate, to entitle him to insure the full value.

It is clear that neither of these reasons apply in the case of a sub-contractor.

But there is a third reason which is frequently mentioned in connection with a bailee's right to insure the full value of the goods. From a commercial point of view it was always regarded as highly convenient. Thus in *Waters* v. *Monarch Fire and Life Assurance* (1856) itself Lord Campbell CJ said:

> What is meant in these policies by the words 'goods in trust'? I think that means goods with which the assured were entrusted; not goods held in trust in the strict technical sense . . . but goods with which they were entrusted in the ordinary sense of the word. They were so entrusted with the goods deposited on their wharfs: I cannot doubt the policy was intended to protect such goods: and it would be very inconvenient if wharfingers could not protect such goods by a floating policy.

In the case of a building or engineering contract, where numerous different sub-contractors may be engaged, there can be no doubt about the convenience from everybody's point of view, including, I would think, the insurers, of allowing the head contractor to take out a single policy covering the whole risk, that is to say covering all contractors and sub-contractors in respect of loss of or damage to the entire contract works. Otherwise each sub-contractor would be compelled to take out his own separate policy. This would mean, at the very least, extra paperwork; at worst it could lead to overlapping claims and cross claims in the event of an accident. Furthermore, as W pointed out in the course of his evidence, the cost of insuring his liability might, in the case of a small sub-contractor, be uneconomic. The premium might be out of all proportion to the value of the sub-contract. If the sub-contractor had to insure his liability in respect of the entire works, he might well have to decline the contract.

For all these reasons I would hold that a head contractor ought to be able to insure the entire contract works in his own name and the name of all his sub-contractors, just like a bailee or mortgagee, and that a sub-contractor ought to be able to recover the whole of the loss insured, holding the excess over his own interest in trust for the others.

If that is the result which convenience dictates, is there anything which makes it illegal for a sub-contractor to insure the entire contract works in his own name? This was a question which was much discussed in the early cases on bailment. But it was never illegal at common law for a bailee to insure goods in excess of his interest. As for statute, the Marine Insurance Acts obviously do not apply. It is true that the Life Assurance Act 1774 by section 3 prohibited an insured from recovering more than his interest on the happening of an insured event. But policies on goods were specifically excluded by section 4 of the Act. Accordingly it was held that neither at common law nor by statute was there anything to prevent the bailee from insuring in excess of his interest.

What about a sub-contractor? 'Goods' in section 4 of the 1774 Act has always been given a wide interpretation, and would clearly cover contract works until they became part of the realty. Whether the works would remain 'goods' thereafter, and if so for how long, may be more difficult. But it was not suggested that the insurers would have any defence here under the Life Assurance Act, so I say no more about it. I would hold that the position of a sub-contractor in relation to contract works as a whole is sufficiently similar to that of a bailee in relation to goods bailed to enable me to hold, by analogy, that he is entitled to insure the entire contract works, and in the event of a loss to recover the full value of those works in his own name.

For the reasons which I have mentioned, I would hold, both on principle and on the authority of the *Commonwealth Construction* case (1976), that a sub-contractor who is engaged on contract works may insure the entire contract works as well as his own property, and that the defendants in this case were each so insured.

That brings me to the next question: does the fact that the defendants are fully insured under the present policy defeat the insurer's right of subrogation? In the *Commonwealth Construction* case, and in the American cases there referred to, it was assumed that it followed automatically that the insurers could have no right of subrogation. In the Commonwealth Construction case it was described as being a 'basic principle'. In one of the American cases it was said that the rule was too well established to require citation. In none of the cases is there any discussion as to the reason for the rule, except for a brief reference in the *Commonwealth Construction* case (1976) to *Simpson & Co* v. *Thompson Burrell* (1877) as follows:

> The starting point of that submission is the basic principle that subrogation cannot be obtained against the insured himself. The classic example is, of course, to be found in *Simpson & Co* v. *Thompson Burrell* (1877). In the case of true joint insurance there is, of course, no problem; the interests of the joint insured are so inseparably connected that the several insureds are to be considered as one with the obvious result that subrogation is impossible. In the case of several insurance, if the different interests are pervasive and if each relates to the entire property, albeit from different angles, again there is no question that the several insureds must be regarded as one and that no subrogation is possible.

Chapter 7
The Employer's Obligations

Some of the employer's obligations are imposed by the express terms of the building contract; others are imposed by the general law.

Possession of site

Time and extent of possession

It is an implied term in every building contract that the employer will give possession of the site to the contractor in sufficient time to enable him to complete his obligations by the contractual date.

Freemen & Son v. Hensler

COURT OF APPEAL

(1900) 64 JP 260

The plaintiff contracted with the defendant to demolish old houses and to erect new ones, the work to be completed within 6 months. The contract was dated 4 July. Under its terms all the brickwork for the new houses was to be built up simultaneously, no part being raised more than 5 feet higher than the remainder. At the defendant's request, the plaintiff agreed to a fortnight's delay from 4 July, but it was not until some weeks after the expiry of that time that possession of part of the site was given. It was nearly five months before the plaintiff got possession of the whole site. The plaintiff claimed damages for breach of contract.

Held: It was an implied term of the contract that the contractor would be given possession of the site immediately. The agreement between the parties had waived that obligation, and substituted a reasonable time. As possession had not been given within a reasonable time, the plaintiff was entitled to damages for the loss which he had sustained by reason of the delay.

Powell-Smith and Furmston's Building Contract Casebook, Fifth Edition. Michael Furmston.

> COLLINS LJ: The contract clearly involves that the building owner shall be in a position to hand over the whole site to the builder immediately upon the making of the contract. There is an implied undertaking on the part of the building owner, who has contracted for the buildings to be placed by the [contractor] on his land, that he will hand over the land for the purpose of allowing the [contractor] to do that which he has bound himself to do.

The general principle is that in a project for new works, the contractor is entitled to exclusive possession of the entire site, but the degree of possession to be given depends on the circumstances of the case.

London Borough of Hounslow v. Twickenham Garden Developments Ltd

CHANCERY DIVISION
(1970) 7 BLR 81

The facts of this case are not material for present purposes, save that it arose out of a dispute between the parties to a contract in JCT 63 terms. That contract (and JCT 80) impose an express duty on the employer to give possession of the site to the contractor. It is couched in the following terms:

> On the Date of Possession possession of the site shall be given to the Contractor who shall thereupon begin the Works . . . : JCT 80, clause 23.1.

'The site' is nowhere specifically defined, and its extent is a question of fact to be determined in light of all the circumstances.

> MEGARRY J: The contract necessarily requires the building owner to give the contractor such possession, occupation or use as is necessary to enable him to perform the contract, but whether in any given case the contractor in law has possession must, I think, depend at least as much on what is done as what the contract provides . . .

The employer must give the contractor a sufficient degree of possession to permit the execution of the work unimpeded by others.

The Queen in Right of Canada v. Walter Cabott Construction Ltd

CANADIAN FEDERAL COURT OF APPEAL
(1975) 21 BLR 42

Cabott contracted with the Crown for the erection of a hatchery building. The contract was one of six contracts for the project as a whole. The work required under two of the later contracts interfered with Cabott's work because they encroached on the site. One of the later contracts was awarded to a third party, but Cabott successfully tendered for another to mitigate the effects of that contract on its work. Cabott claimed damages for breach of implied terms relating to possession.

Held: Cabott's claim succeeded. The Crown was in breach of contract by denying Cabott part of the site of the work. A clause of the contract providing that there were no implied terms was ineffective because it is fundamental to a construction contract that working space should be provided unimpeded by others.

URIE J: The learned trial judge was correct in finding that the appellant was in breach of its contract with the respondent by denying the respondent a portion of the site of the work which she was obliged to furnish to permit compliance with the contract.

The second breach found by the learned trial judge was that the appellant failed to observe an implied term that the respondent would have a sufficient degree of uninterrupted and exclusive possession of the site to permit it to carry out its work unimpeded and in the manner of its choice.

The appellant relied on general condition 6 for demonstrating the alleged error of the learned trial judge. General condition 6 reads as follows:

> 6. No implied obligation of any kind by or on behalf of Her Majesty shall arise from anything in the contract, and the express covenants and agreements herein contained and made by Her Majesty are and shall be the only covenants and agreements upon which any rights against Her Majesty are to be founded; and, without limiting the generality of the foregoing, the contract supersedes all communications, negotiations and agreements, either written or oral, relating to the work and made prior to the date of the contract.

In my opinion, it has no application in the case at bar because it is fundamental to a building contract that work space be provided unimpeded by others. The proposition of law is succinctly put by the learned author of *Hudson's Building and Engineering Contracts*, 10th edn (1970), at p.318, as follows:

> Since a sufficient degree of possession of the site is clearly a necessary precondition of the contractor's performance of his obligations, there must be an implied term that the site will be handed over to the contractor within a reasonable time of signing the contract (see, e.g. *Roberts* v. *Bury Commissioners* (1870)) and, in most cases it is submitted, a sufficient degree of uninterrupted and exclusive possession to permit the contractor to carry out his work unimpeded and in the manner of his choice. This must particularly be so when a date for completion is specified in the contract documents.

This statement of the law was adopted by Spence J, in the *Penvidic* case (1975). The learned trial judge was, therefore, in my opinion, clearly right when he found the second breach of contract.

The employer is not in breach of his obligation to give sufficient possession of the site to the contractor if he is wrongfully excluded from the site by a third person for whom the employer is not responsible in law and over whom he has no control.

Porter v. Tottenham Urban District Council

COURT OF APPEAL

[1915] 1 KB 1041

The plaintiff contracted to build a school for the council upon land belonging to them. The contract provided that he should be entitled to enter on the site immediately, and that he should complete the work by a specified date. The only access to the site was from a road, and as the soil of the council's land and of the adjoining road was soft, the contract provided that the plaintiff lay a temporary sleeper roadway from the road to the site for access, and subsequently provide a permanent pathway. The plaintiff began work but was forced to abandon it because of a threatened injunction from an adjoining owner, who claimed that the road was his property. The third party's claims were held to be unfounded. The plaintiff then resumed and completed the works. He claimed damages against the council in respect of the delay caused by the third party's action.

Held: The plaintiff's claim must fail. There was no implied warranty by the council against wrongful interference by third parties with the free access to the site.

The position is different where the employer has it in his power to have the situation rectified, if necessary by legal action.

The Rapid Building Group Ltd v. Ealing Family Housing Association Ltd

COURT OF APPEAL

(1985) 1 Con LR 1

The plaintiffs contracted with the defendants on JCT 63 terms to construct 101 dwellings in five blocks at Stoneville Road, Chiswick, London W4. Clause 21 of the contract and the Appendix entry provided that possession of the site should be given to the contractor on 23 June 1980. At the time when, by clause 21, the defendants promised to give the contractor possession, they were unable to do so because the north-east corner of the site was occupied by squatters. Eviction proceedings were taken by the defendants and the squatters were evicted, but it was at least 19 days before the site was cleared of squatters so as to enable the contractors to occupy the whole of the site.

Held: The defendants were in breach of clause 21 and the breach caused appreciable delay.

> STEPHENSON LJ: The learned judge found that the defendants were in breach of clause 21; he further found that the breach caused appreciable delay on the part of the plaintiffs. In my judgment, both those findings must stand. The first was plainly right; [clause 21 provides that on the date stated in the appendix] possession 'shall be given' to the contractor, and it seems to me unarguably that there was a clear breach of that term by the failure of the defendants, for whatever reason, to remove these squatters until an appreciable time after they had promised to give the plaintiffs possession of the site . . . I would hold with the judge that there was a clear breach of the express term and that there is nothing inconsistent in that finding with . . . *Porter* v. *Tottenham Urban District Council* (1915), a case in which, in rather similar circumstances, but where there was no clause 21, this court held that there was no implied warranty that the site should be completely clear; nor is it in any way inconsistent with the case of *LRE Engineering Services Ltd* v. *Otto Simon Carves Ltd* (1981), where Robert Goff J, as he then was, had to construe the contractual words 'affording access' – again something that is quite irrelevant to the construction of clause 21 in this case.

Site conditions

Under the general law there is no warranty by the employer that the site is fit for the works or that the contractor will be able to construct the building on the site to the employer's design.

Appleby v. Myers

COURT OF EXCHEQUER CHAMBER

(1867) LR 2 CP 651

The plaintiff contracted to supply and erect machinery on the defendant's premises, payment to be on completion. Part of the work had been completed when a fire accidentally broke out on the defendant's premises, destroying both the premises and the works.

Held: Since neither party was at fault, both parties were excused further performance, but on the true interpretation of the contract the plaintiff was not entitled to recover any payment until the whole work had been completed.

> BLACKBURN J: The whole question depends on the true construction of the contract between the two parties. We agree with the court below in thinking that it sufficiently appears that the work which the plaintiff agreed to perform could not be performed unless the defendant's premises continued in a fit state to enable the plaintiff to perform the work . . . [and] that if by any default on the part of the defendants these premises were rendered unfit to receive the work, the plaintiff would have had the option to sue the defendants for this default, or to treat the contract as rescinded, and sue on a *quantum meruit*. But we do not agree with them in thinking that there was an absolute promise or warranty by the defendant that the premises should at all events continue so fit.

Thorn v. London Corporation

HOUSE OF LORDS

(1876) 1 App. Cas. 120

Thorn contracted with the corporation to demolish an existing bridge and to construct a new one, in accordance with plans and a specification prepared by the corporation's engineer. The plans and specification proved defective in important respects and, as a result, Thorn lost money and the bridge had to be built in a different way. The descriptions given were said to be believed to be correct but were not guaranteed. Thorn brought an action claiming reimbursement for his loss of time and labour and alleged that there was an implied warranty that the bridge could be built inexpensively in accordance with the plans and specification.

Held: His claim must fail as no such warranty could be implied. Thorn's obligation to complete the works covered substantial deviations from what had been planned.

Note: Lord Cairns LC also remarked that 'if the additional or varied work is so peculiar, so unexpected, and so different from what any person reckoned or calculated on, it may not be within the contract at all . . .'.

In the absence of any specific guarantee or definite representations by the employer as to the nature of the soil in which the works are to be executed, the contractor is not entitled to abandon the contract on discovering the nature of the soil.

Bottoms v. Lord Mayor of York

COURT OF EXCHEQUER

(1892) 2 HBC (4th edition) 208

Bottoms contracted to execute sewerage works for the defendants, intending to use poling boards for the excavations. The soil turned out to be unsuitable and necessitated extra works, which the engineer refused to authorise as a variation. Bottoms thereupon abandoned the works and sued for the value of the work done. Neither party had sunk boreholes, but before signing the contract the defendants had been advised that Bottom's price was such that he was bound to lose money in the type of soil conditions expected.

Held: The plaintiff's claim must fail. He was not entitled to abandon the contract on discovering the nature of the soil or because the engineer declined to authorise extra payment.

Where the contract incorporates the rules of a method of measurement which makes provision for extra payment in respect of unexpected ground conditions, the contractor will be entitled to recover.

C. Bryant & Son Ltd v. Birmingham Hospital Saturday Fund

KING'S BENCH DIVISION

[1938] 1 All ER 503

The contractor agreed to erect a convalescent home, the contract being in an earlier version of what is now the JCT form, clause 11 of which provided that 'the quality and quantity of the work included in the contract sum shall be deemed to be that which is set out in the bills of quantities, which bills, unless otherwise stated shall be deemed to have been prepared in accordance with the standard method of measurement last before issued by the Chartered Surveyors' Institution'. The Standard Method of Measurement required that, where practicable, the nature of the soil should be described and attention drawn to any existing trial holes, and that excavation in rock should be given separately. The bills referred the contractor to the drawings, a block plan and the site to satisfy himself as to the local conditions and the full extent and nature of the operations, but contained no separate item for the excavation of rock. The architect knew of the existence of rock on the site, but this was neither shown on any of the plans nor referred to in the Bills.

Held: The contractor was entitled to treat the excavation in rock as an extra, and to be paid the extra cost of excavation, plus a fair profit.

Note: The position would appear to be the same as regards contracts in JCT 63, JCT 80 and JCT 98 terms. JCT 98, clause 2.2.2.1, requires that the bills are to have been prepared in accordance with SMM 7, unless otherwise specifically stated, and claims for site conditions may be based on the principle laid down in this case. See now JCT 2005, clause 2.13.

The terms of the tender documents may be such as to give rise to a collateral warranty or implied term that the ground conditions are as described.

Bacal Construction (Midlands) Ltd v. Northampton Development Corporation

COURT OF APPEAL

(1975) 8 BLR 88

As part of their tender, Bacal submitted sub-structure designs and detailed priced bills of quantities for six selected blocks of dwellings and ancillary works in selected foundation conditions. These formed part of the contract documents. The foundation designs had been prepared on the assumption that the soil conditions were as indicated on the relevant borehole data provided by NDC, and the designs were adequate on that assumption. The NDC tender documents stated that the site was a mixture of Northampton shire sand and upper lias clay. As work proceeded, tufa was discovered in several areas of the site. This necessitated re-design of the foundations in those areas and the carrying out of additional work. Bacal claimed that there had been a breach of an implied term or warranty by NDC in that the ground conditions should accord with the hypotheses upon which they had been instructed to design the foundations.

Held: On the proper interpretation of the contract, there was an implied term or warranty as claimed by Bacal, and they were entitled to damages for its breach.

> BUCKLEY LJ: Bacal have submitted that there are strong commercial reasons for implying such a term or warranty in the contract as they have suggested. First, before designing the foundations for any building it is essential to know the nature of the soil conditions. Secondly, where the contract is for a comprehensive development of the kind here in question, the contractor must know the soil conditions at the site of each projected block in order to be able to plan his timetable and to estimate his requirements for materials. These are matters which relate directly to the contract price. Thirdly, if the work is interrupted or delayed by unforeseen complications, the contractor is unlikely to be able to complete his contract in time. Clause 22 of the contract requires the contractor to pay or allow liquidated damages in such an event. The corporation have in fact retained a very substantial sum by way of liquidated damages to which they claim to be entitled because Bacal did not complete the contract within 65 weeks. Clause 23 of the contract makes provision for extension of the contract time in certain circumstances, but the only provision of that clause which might possibly apply to the present

case is subparagraph (c), which permits an extension of time if by reason of the architects' instructions issued under clause 1(2), clause 11(1) or clause 21(2) the completion of the works has been delayed beyond the due completion date. The delay resulting from the discovery of tufa was not occasioned by any instructions issued by the architect, and clause 1(2) cannot, in my opinion, apply. Clause 11(1) can only apply where the architect has issued instructions requiring a variation or he has sanctioned a variation made by the contractor otherwise than pursuant to such an instruction. That sub-clause cannot apply in the present case if upon the true construction of clause 1(4) we are not concerned with any variation. Clause 21(2) only applies if the architect has issued an instruction postponing any work to be executed under the contract, which is not the present case. Accordingly, in my opinion, sub-paragraph (e) of clause 23 can have no application in the present case. The learned judge accepted Bacal's argument and held that the corporation was liable on an implied term of warranty. I agree with him . . .

Note: The contract discussed was substantially in JCT 63 form.

In an appropriate case the contractor may have a claim against the employer for misrepresentations about site and allied conditions made during pre-contractual negotiations.

Morrison-Knudsen International Co. Inc. v. Commonwealth of Australia

HIGH COURT OF AUSTRALIA

(1972) 13 BLR 114

This was an action by a contractor against the employer for negligence. The contractor claimed that basic information supplied by the employer at pretender stage 'as to the soil and its contents at the site of the proposed work was false, inaccurate and misleading . . . the clays at the site, contrary to that information, contained large quantities of cobbles'. The issue was a preliminary one, the court holding that the effect of the documents at issue could not finally be determined until all the relevant facts were established.

BARWICK CJ: The basic information in the site information document appears to have been the result of much highly technical effort on the part of [the employer]. It was information which the [contractors] had neither the time nor the opportunity to obtain for themselves. It might even be doubted whether they could be expected to obtain it by their own efforts as a potential or actual tenderer. But it was indispensable information if a judgment were to be formed as to the extent of the work to be done . . .

A factual misrepresentation made during pre-contractual negotiations by one party and relied on by the other may give rise to liability under the Misrepresentation Act 1967. In an appropriate case such liability could attach to inaccurate information about ground conditions.

Howard Marine and Dredging Co. Ltd v. A. Ogden & Sons (Excavations) Ltd

COURT OF APPEAL

(1977) 9 BLR 34

Ogden wished to hire barges for use in connection with excavation work In the course of negotiations, Howard Marine's manager misrepresented the barges' deadweight. He stated that the payload of each barge was 1,600 tonnes; it was only 1,055. The misstatement was based on his recollection of a figure given in Lloyd's Register, which was incorrect. He could have ascertained the correct figures from Howard Marine's shipping documents.

Held: Howard Marine was liable in damages for the misrepresentation under section 2(1) of the Misrepresentation Act 1967.

BRIDGE LJ: The first question then is whether [Howard] would be liable in damages in respect of [their manager's] misrepresentation if it had been made fraudulently, that is to say, if he had known that it was untrue. An affirmative answer to that question is inescapable. The judge found in terms that what [the manager] said about the capacity of the barges was said with the object of getting the hire contract for the respondents, in other words with the intention that it should be acted on. This was clearly right. Equally clearly the misrepresentation was in fact acted on by [Ogden]. It follows, therefore, on the plain language of the statute that, although there was no allegation of fraud, [the manager] must be liable unless they proved that [Howard] had reasonable ground to believe what he said about the barges' capacity.

It is unfortunate that the learned judge never directed his mind to the question whether [the manager] had any reasonable ground for his belief. The question he asked himself, in considering liability under the Misrepresentation Act, 1967, was whether the innocent misrepresentation was negligent. He concluded that if [the manager] had given the inaccurate information in the course of the April telephone conversations he would have been negligent to do so but that in the circumstances obtaining at the Otley interview in July there was no negligence. I take it that he meant by this that on the earlier occasions the circumstances were such that he would have been under a duty to check the accuracy of his information, but on the later occasions he was exempt from any such duty. I appreciate the basis of this distinction, but it seems to me, with respect, quite irrelevant to any question of liability under the statute. If the representee proves a misrepresentation which, if fraudulent, would have sounded in damages, the onus passes immediately to the representor to prove that he had reasonable ground to believe the facts represented. In other words the liability of the representor does not depend upon his being under a duty of care the extent of which may vary according to the circumstances in which the representation is made. In the course of negotiations leading to a contract the statute imposes an absolute obligation not to state facts which the representor cannot prove he had reasonable ground to believe.

Section 3 of the Misrepresentation Act 1967 as amended by the Unfair Contract Terms Act 1977 provides that clauses which seek to limit or exclude liability for misrepresentation are valid only if reasonable.

The following case has been taken to be authority for the proposition that at common law it is not possible to exclude liability for fraudulent misrepresentation. This rule was confirmed by the House of Lords in *HIH Casualty and General Insurance* v. *Chase Manhattan Bank* [2003] 1 All ER (Comm) 349, but the House left open the question whether by clear words it was possible to exclude liability for the fraud of an agent.

S. Pearson & Son Ltd v. Dublin Corporation

HOUSE OF LORDS
[1907] AC 351

The contract between the parties provided that 'the contractor is to satisfy himself as to . . . all . . . things so far as they may have any connection with the works in the contract, and to obtain his own information on all matters which can in any way influence his tender'. Plans prepared by the corporation's engineers showed a wall as existing when in fact it did not exist, and the contractor tendered on a lower price on this basis. The statement in the plans was found to be fraudulent. The contractors discovered the absence of the wall, which they had proposed to use in their works, after they had begun work. They had to revise their plans as a result and incurred considerable extra cost.

Held: The contractors were entitled to succeed.

> LORD LOREBURN LC: It seems clear that no one can escape liability for his own fraudulent statements by inserting in a contract a clause that the other party shall not rely upon them . . . [The corporation say] that, though a principal is liable for the fraudulent representation of his agent, yet that rule only applies where the representation has in fact been made by the agent. I cannot accept that contention. The principal and the agent are one, and it does not signify which of them made the incriminated statement or which of them possessed the guilty knowledge.

Co-operation

It is an implied term of every building contract that the employer will do all that is reasonably necessary on his part to bring about completion of the contract.

'If A employs B for reward to do a piece of work for him which requires outlay and effort on B's part . . . generally speaking, where B is employed to do a piece of work which requires A's co-operation . . . it is implied that the necessary co-operation will be forthcoming': *per* Viscount Simon LC in *Luxor (Eastbourne) Ltd* v. *Cooper* [1941] 1 All ER 33.

'In general . . . a term is necessarily implied in any contract, the other terms of which do not repel the implication, that neither party shall prevent the other from performing it': *per* Lord Asquith in *Cory Ltd* v. *City of London Corporation* [1951] 2 All ER 85.

It is an implied term of a contract in JCT terms that the employer will not hinder or prevent the contractor from carrying out its obligations in accordance with the terms of the contract and from executing the work in a regular and orderly manner. This implied term extends to those things which the architect must do to enable the contractor to carry out the work and the employer is liable for any breach of this duty on the part of the architect.

London Borough of Merton v. Stanley Hugh Leach Ltd

CHANCERY DIVISION

(1985) 32 BLR 51

In 1972 the contractor agreed with Merton to construct 287 dwellings. The contract was on JCT 63 terms. The parties were in dispute over delayed completion, the contractor contending that the delay was almost entirely due to lack of diligence and care and lack of co-operation by Merton's architect and alleged that Merton were in breach of implied terms of the contract, *inter alia*:

(1) That Merton would not hinder or prevent the contractor from carrying out its obligations in accordance with the terms of the contract and from executing the works in a regular and orderly manner; and

(2) that Merton would take all steps reasonably necessary to enable the contractor to discharge its obligations and to execute the works in a regular and orderly manner.

Held: Terms were to be implied as alleged.

VINELOTT J: Before turning to the detailed sub-paragraphs it will be convenient to make two general observations. First, the cases in which the courts have considered whether a term should be implied into a contract cover a very wide variety of situations. The only feature common to them all is that the court is asked to add to or deduce from a contract something which may fairly be said to be a part but not an explicit part of the rights and obligations conferred or imposed by the contract.

The wide variety of circumstances in which the court is invited to take that step was emphasised by Lord Wilberforce in *Liverpool City Council* v. *Irwin* (1977). The main issue in that case was whether there should be read into tenancies of flats in a tower block an obligation on the part of the landlord to keep the means of access to the flats in reasonable repair.

Lord Wilberforce, having observed that the contracts of tenancy were partly but not wholly in writing that the landlords reserved the common parts including lifts and stairs and that the block comprised a number of flats and maisonettes, the respondent being on the ninth floor, the occupiers of which, though they would have to use the common parts, were not under any express obligation to maintain them, said:

> To say that the construction of a complete contract out of these elements involves a process of 'implication' may be correct; it would be so if implication means the supplying of what is not expressed. But there are varieties of implication which the courts think fit to make and they do not necessarily involve the same process. Where there is, on the face of it, a complete, bilateral contract, the courts are sometimes willing to add terms to it, as implied

> terms: this is very common in mercantile contracts where there is an established usage: in that case the courts are spelling out what both parties know and would, if asked, unhesitatingly agree to be part of the bargain. In other cases, where there is an apparently complete bargain, the courts are willing to add a term on the ground that without in the contract will not work – this is the case, if not of *The Moorcock* (1889) itself on its facts, at least the doctrine of *The Moorcock* as usually applied. This is, as was pointed out by the majority of the Court of Appeal, a strict test – though the degree of strictness seems to vary with the current legal trend – and I think that they were right not to accept it as applicable here. There is a third variety of implication, that which I think Lord Denning MR favours, or at least did favour in this case, and that is the implication of reasonable terms. But though I agree with many of his instances, which in fact fall under one or other of the preceding heads, I cannot go so far as to endorse his principle; indeed it seems to me, with respect, to extend a long, and undesirable, way beyond sound authority.
>
> The present case, in my opinion, represents a fourth category, or I would rather say a fourth shade on a continuous spectrum. The court here is simply concerned to establish what the contract is, the parties not having themselves fully stated the terms. In this sense the court is searching for what must be implied.

Lord Cross drew the same distinction when he said:

> When it implies a term in a contract the court is sometimes laying down a general rule that in all contracts of a certain type – sale of goods, master and servant, landlord and tenant and so on – some provision is to be implied unless the parties have expressly excluded it. In deciding whether or not to lay down such a *prima facie* rule the court will naturally ask itself whether in the general run of such cases the term in question would be one which it would be reasonable to insert. Sometimes, however, there is no question of laying down any *prima facie* rule applicable to all cases of a defined type but what the court is being in effect asked to do is to rectify a particular – often a very detailed – contract by inserting in it a term which the parties have not expressed. Here it is not enough for the court to say that the suggested term is a reasonable one the presence of which would make the contract a better or fairer one; it must be able to say that the insertion of the term is necessary to give – as it is put – 'business efficacy' to the contract and that if its absence had been pointed out at the time both parties – assuming them to have been reasonable persons – would have agreed without hesitation to its insertion. The distinction between the two types of case was pointed out by Viscount Simonds and Lord Tucker in their speeches in *Lister* v. *Romford Ice and Cold Storage Co Ltd* (1957), but I think that Lord Denning MR in proceeding – albeit with some trepidation – to 'kill off' MacKinnon LJ's 'officious bystander' (*Shirlaw* v. *Southern Foundries* (1926) Ltd (1939)) must have overlooked it.

He went on to reject an argument that an obligation on the part of a landlord to keep the common parts in repair could be implied by the application of the officious bystander test but later agreed with the majority that it was implicit in the relationship between the landlord and the tenants in the circumstances of that case that the landlords were to use reasonable care to keep the common parts in reasonable repair and efficiency. Lord Edmund-Davies similarly rejected the argument that some such obligation should be implied to give business efficacy to the contract or from the application of the 'officious bystander test' but agreed that such an obligation was placed on the landlords 'by the general law, as a legal incident of this kind of contract, which the landlords must be assumed to know as well as anyone else'.

It is important to bear in mind the different nature of these processes of 'supplying what is not expressed' because (as the decision in *Irwin* v. *Liverpool County Council* demonstrates) limitations appropriate to a process of implication at one end of the spectrum may not be applicable to the process of implication at the other end of the spectrum. At one end of the spectrum the court imports into a contract apparently complete a term which on examination of the surrounding circumstances can be seen to be one which was the obvious but unexpressed intention of the parties or (if it is a different test) one which must of necessity be implied to give business efficacy to the contract. Similarly the court may import a term because it is customary in the trade (though one of the parties may have been unaware of the custom) or by virtue of some statutory provision (though there neither of the parties may have been actually aware of it). In a case where it is sought to import a term on the ground that it was an obvious but unexpressed part of the agreement between the parties or to give business efficacy to the contract it is for the party who seeks to import that term to establish the grounds of the implication and also to show that the term can be formulated with reasonable precision. At the other end of the spectrum the process of implication is one of spelling out or deducing what is implicit in the contract in the sense of being part of the legal context appropriate to contracts generally or to contracts of the particular type under consideration and so inherent in the legal relationship between the parties created by the contract. In such cases it may not be possible to formulate that which it is sought to imply with any degree of precision. This is stressed by Lord Simonds in *Lister* v. *Romford Ice and Cold Storage Co Ltd* (1957) where the question was whether an employer was obliged by a contract of employment to ensure that the driver of one of his vehicles was protected by insurance against third party claims. He said:

> For the real question becomes, not what terms can be implied in a contract between two individuals who are assumed to be making a bargain in regard to a particular transaction or course of business; we have to take a wider view, for we are concerned with a general question, which, if not correctly described as a question of status, yet can only be answered by considering the relation in which the drivers of motor-vehicles and their employers generally stand to each other. Just as the duty of care, rightly regarded as a contractual obligation, is imposed on the servant, or the duty not to disclose confidential information (see *Robb* v. *Green* (1985)), or the duty not to betray secret processes (see *Amber Side and Chemical Co* v. *Menzel* (1913)), just as a duty is imposed on the master not to require his servant to do any illegal act, just so the question must be asked and answered whether in the world in which we live today it is a necessary condition of the relation of master and man that the master should, to use a broad colloquialism, look after the whole matter of insurance. If I were to try to apply the familiar tests where the question is whether a term should be implied in a particular contract in order to give it what is called business efficacy, I should lose myself in the attempt to formulate it with the necessary precision. The necessarily vague evidence given by the parties and the fact that the action is brought without the assent of the employers shows at least ex post facto how they regarded the position. But this is not conclusive for, as I have said, the solution of the problem does not rest on the implication of a term in a particular contract of service but upon more general considerations.

Lord Wilberforce, after referring to this passage, said:

> We see an echo of this in the present case, when the majority in the Court of Appeal, considering a 'business efficacy term', i.e. a '*Moorcock*' term . . . found themselves faced with five

> alternative terms and therefore rejected all of them. But that is not, in my opinion, the end, or indeed the object, of the search.

As will be seen, the terms which it is sought to imply in the instant case fall within what Lord Wilberforce described as his fourth category.

The second general observation relates to the position of the architect. The case for Merton, which is central to its contentions on this and other issues, is that the architect is not, under the contract, the servant or agent of the building owner or employer. It is said that the architect is introduced into the conditions solely to hold the ring between the building owner and the contractor. The contractor when he enters into the contract is free to price and plan the work as he thinks fit. He is free in the course of carrying out the work to alter his original plan to meet unexpected difficulties. The contract recognises that his planning and pricing may be affected by circumstances beyond his control and for which he cannot fairly be held to be responsible. Moreover the building owner may require the works contracted for to be varied or additional work to be carried out. The contract provides an elaborate and, it is said, exhaustive machinery to meet these divergencies and to ensure that the contractor is properly compensated. At the same time this machinery is designed to protect the building owner both from excessive delay and from unreasonable escalation of costs. The machinery embodied in the contract . . . is designed to achieve a fair balance between the conflicting interests of contractor and building owner. Central to that machinery is the architect. He is introduced to hold the balance between a contractor and a building owner. In fulfilling that role he does not act as a servant or agent of either of them, neither therefore is responsible for any act or failure to act on his part. The contract provides the contractor with express remedies if the architect fails to carry out some administrative duty (for instance to supply instructions or drawings requested by the contractor in due time). Insofar as the contract confers discretion on the architect, both the contractor and the building owner have the right to refer his decisions to arbitration. Viewed in that way, it is said, the architect acts independently from both contractor and building owner.

I do not think this is a possible view. It is to my mind clear that under the standard conditions the architect acts as the servant or agent of the building owner in supplying the contractor with the necessary drawings, instructions, levels and the like and in supervising the progress of the work and in ensuring that it is properly carried out. He will of course normally though not invariably have been responsible for the design of the work. There are very few occasions when a building owner himself is required to act directly without the intervention of the architect . . . To the extent that the architect performs these duties the building owner contracts with the contractor that the architect will perform them with reasonable diligence and with reasonable skill and care. The contract also confers on the architect discretionary powers which he must exercise with due regard to the interests of the contractor and the building owner. The building owner does not undertake that the architect will exercise his discretionary powers reasonably; he undertakes that although the architect may be engaged or employed by him he will leave him free to exercise his discretions fairly and without improper interference by him.

A contractor must be prepared to be willing to accept a contract which confers these discretionary powers on the servant or agent of the building owner firstly because the 'architect/supervising officer' is normally a qualified architect (and if he is not he will be a responsible employee with experience of building works) and in either case will be under a duty to act fairly between the parties; secondly because (as was made clear by the Court of Appeal in *Northern Regional Health Authority* v. *Crouch* (1984) his decision is subject to review by an arbitrator appointed under clause 35. As Sir John Donaldson MR pointed out:

> (Clause 35) goes far further than merely entitling [the arbitrator] to treat the [arbitrator's] certificate, opinions, decisions, requirements and notices as inconclusive in determining the rights of the parties. It enables, and in appropriate cases requires, him to vary them and so create new rights, obligations and liabilities in the parties. This is not a power which is normally possessed by any court and again it has a strong element of personal judgment by an individual nominated in accordance with the agreement of the parties.

As I have said, to the extent that the architect exercises these discretions his duty is to act fairly; 'The building owner and the contractor make their contract on the understanding that in all such matters the architect will act in a fair and unbiased manner and it must therefore be implicit in the owner's contract with the architect that he shall not only exercise due care and skill but also reach such decisions fairly, holding the balance between his client and the contractor' (see *Sutcliffe* v. *Thackrah* (1974) *per* Lord Reid).

That the architect under these (and similar) standard conditions acts as the servant or agent of the building owner has been recognised in many cases and is so stated in the leading text books. The contention now advanced by [Merton] so far as I am aware has never before been advanced. [Merton] submitted that the position of the architect must be re-considered in the light of the decision of the Court of Appeal in *Crouch*. I can see nothing in the judgments in *Crouch* which is in any way inconsistent with the accepted view of the role of the architect. It is now clear that insofar as the architect exercises discretionary powers and the exercise of his discretion can be reviewed by the arbitrator, the arbitrator stands in the shoes of the architect and does not exercise a purely arbitral role. To that extent the arbitrator (as [Merton] expressed it) is part of the machinery of the contract. But to the extent that the architect acts as the agent or servant of the building owner in discharging obligations imposed on the building owner by the contract his acts are not subject to review by the arbitrator – though they may found a claim for damages for breach of contract the extent of which will fall to be determined by the arbitrator.

I turn now to the terms which Leach seeks to imply. Implied Terms (i) and (ii) are as follows:

(i) [Merton] would not hinder or prevent Leach from carrying out their obligations in accordance with the terms of the contract [and from executing the works in a regular and orderly manner].

(ii) [Merton] would take all steps reasonably necessary to enable [Leach] so to discharge their obligations [and to execute the works in a regular and orderly manner].

The arbitrator held that these terms ought to be implied.

In my judgment, the arbitrator was clearly right as regards the first of these terms. Vaughan Williams LJ observed in *Barque Quilpué Ltd* v. *Brown* (1904):

> There is an implied contract by each party that he will not do anything to prevent the other party from performing a contract or to delay him in performing it. I agree that generally such a term is by law imported into every contract.

The implication of such a term seems to me to fall clearly within Lord Wilberforce's fourth category. The implied undertaking not to do anything to hinder the other party from performing his part of the contract may, of course, be qualified by a term express or to be implied from the contract and the surrounding circumstances. But the general duty remains save so far as qualified. It is difficult to conceive of a case in which this duty could be wholly excluded. 'Parties are free to agree to whatever exclusions or modifications of all kinds of obligations as they please within the limits that the agreement must retain the legal characteristics of a contract' (see *Photo Production Ltd* v. *Securicor Ltd* (1980) *per* Lord Diplock) (my emphasis). What is less clear is whether on the facts pleaded there has been any breach of the general duty modified as it must be in the way I have indicated. However that is a matter which the arbitrator will have to consider in the second stage.

As regards the second of these two terms it is well settled that the courts will imply a duty to do whatever is necessary in order to enable a contract to be carried out. The principle was expressed in a well-known passage in the speech of Lord Blackburn in *Mackay* v. *Dick* (1881) where he said:

> Where in a written contract it appears that both parties have agreed that something should be done which cannot effectively be done unless both concur in doing it, the construction of the contract is that each agrees to do all that is necessary to be done on his part for the carrying out of that thing though there may be no express words to that effect.

However, the courts have not gone beyond the implication of a duty to co-operate whenever it is reasonably necessary to enable the other party to perform his obligations under a contract. The requirement of 'good faith' in systems derived from Roman law has not been imported into English law. The limitations of the principle stated in *Mackay* v. *Dick* were stressed by Devlin J (as he then was) in *Mona Oil Equipment Co* v. *Rhodesia Railways* (1949) where he said:

> I can think of no term that can properly be implied other than one based on the necessity for co-operation. It is, no doubt, true that every business contract depends for its smooth working on co-operation, but in the ordinary business contract, and apart, of course, from express terms, the law can enforce co-operation only in a limited degree – to the extent that it is necessary to make the contract workable. For any higher degree of co-operation the parties must rely on the desire that both of them usually have that the business should get done.

However it is, I think, clear that the implementation of a building contract embodying the JCT general conditions does require close co-operation between the contractor and the architect. The arbitrator . . . gives illustrations of situations where the contractor must rely upon the co-operation of the architect if the work is to be completed expeditiously and efficiently. Again it is recognised in all the leading text books that a contract incorporating the JCT standard terms falls within the category of contracts in which a requirement that a building owner will 'do all that is necessary to bring about completion of the contract' will be implied (see the passage from *Keating on Building Contracts* cited by the arbitrator). In *Holland Hannen & Cubitts* v. *WHTSO* (1981) it was conceded that under the contract there in issue (which incorporated similar standard terms) 'the building owner would do all things necessary to enable the contractor to carry out the work' and Judge Newey clearly thought that that concession was rightly made. For the reasons I have given I think that this implied undertaking by the building owner extends to those things which the architect must do to enable the contractor to carry out the work and that the building owner is liable for any breach of this duty on the part of the architect.

Implied Term (iii). This was stated in the following terms:

> The Architect would provide [Leach] full, correct and co-ordinated information concerning the works.

The arbitrator held that this term should be implied subject to the deletion of the words 'full' and 'co-ordinated' which he held to be unnecessary. Leach do not appeal from this part of his decision. The term as modified seems to me a particular application of the more general term in (ii); the case for Leach is that construed in the light of that implied term in the contract in so far as it imposes an obligation on the architect to supply the contractor with drawings and other information in the course of the work: it imposes an obligation to supply the contractor with accurate drawings and information.

The case for Merton is that clause 1(2) of the conditions expressly provides that if the contractor finds any discrepancy or divergence between the contract drawings and contract bills he must give notice to the architect and that the architect must then give an instruction. Then 'As and when from time to time may be necessary the Architect Supervising Officer, without charge to the Contractor, shall furnish him with two copies of such drawings or details as are reasonably necessary either to explain and amplify the Contract Drawings or to enable the Contractor to carry out and complete the Works in accordance with these Conditions' (clause 3(4)). The contract, it is said, thus contains a complete code which governs the provision of drawings and other information required by the contractor and in conjunction with clause 23(f) provides a remedy if the drawings or other information provided prove to be incorrect and to require correction or amplification by the architect.

I have found this point one of some difficulty but I am not persuaded that the arbitrator was wrong to imply this term. Clause 3(4) imposes on the architect an obligation to furnish the contractor with drawings and details as and when necessary. I agree with the arbitrator that it must have been in the contemplation of the parties that the architect would act with reasonable diligence and would use reasonable care and skill in providing this information. The contract does not impose a duty on a contractor to check the drawings to see if there are discrepancies or divergencies and discrepancies and divergencies may come to light at a time when it is too late for the contractor to call for an instruction or for further drawings and details under clause 3(4).

[Counsel for Merton] criticised the observation by the arbitrator that:

> The parties having agreed that there should be remedies open to the contractor if he were involved in delay or expense because of discrepancies in the contract drawings, would as reasonable people agree that there should also be a remedy if discrepancies appeared in an explanatory non-contract drawing. Knowing that they have provided in express terms for the architect to communicate full information to the contractor, they would readily agree that such information should be correct.

He submitted that at this point the arbitrator overlooked the principle recognised by him that a term will not be implied merely upon the ground that it is reasonable. I think that all the arbitrator was saying in this paragraph is that the fact that a contractor has a remedy when discrepancies or divergencies are discovered in good time would not be regarded by a reasonable man familiar with the operation of the contract as absolving the building owner from liability for failure by the architect to use reasonable skill and care in supplying drawings and other information in a case where that machinery does not afford the contractor with an adequate remedy. The general implication (which is an implication of law founded on the consideration of the provisions of the contract as a whole) is not therefore excluded by clause 3(4).

Implied Terms (iv) and (v). These were stated in the following terms:

> (iv) The architect would issue to [Leach] in due time such instructions, drawings, details and other information as [Leach] needed to carry out and complete the works in a regular

and orderly manner and in accordance with their programme and/or in accordance with their obligations under the contract.

(v) If any variations were to be made the architect would issue variation instructions, together with such drawings, details and information as were needed by such time as would enable the claimants to carry out and complete the work in a regular and orderly manner and in accordance with their programme and/or in accordance with their obligations under the Contract.

The arbitrator rejected these implied terms on the ground that they are repetitive of the express provisions in the contract (in particular as regards implied term (iv) by clause 3(4) of the conditions). Leach do not appeal from this part of his decision.

Implied Term (vi) is as follows:

> The Architect will administer the Contract in an efficient and proper manner and in accordance with the practice and procedure normally followed by architects administering a substantial building contract, in particular [Leach] will contend that it was normal practice and procedure:
>
> (a) That the issue of every drawing in the course of construction should be accurately recorded by means of architect's instruction sheets or some other similar procedure.
>
> (b) Every drawing issued by the architect should have a unique reference.
>
> (c) Whenever a drawing is issued with any amendments a note of the amendment should be made in the note box and the fact of the amendment should be indicated by the reference.

The arbitrator held that a term should be implied to the effect that the architect would administer the contract in an efficient and proper manner and that the efficient and proper administration of the contract required that drawings be identified in a manner set out in sub-paragraphs (b) and (c).

The first part of this implied term (that the architect would administer the contract in an efficient and proper manner) seems to me no more than a restatement or particular application of Merton's contractual undertaking that there would at all times be a person answering the definition of 'the Architect' in clause 3 of the articles of agreement and that the architect would be a reasonably competent person and would use that degree of diligence, skill and care in carrying out the duties assigned to him under the contract that could reasonably be expected of an architect appointed to that position. As the arbitrator observed, it is clearly impossible to define accurately and comprehensively the various acts of an architect which together make up efficient and proper administration.

As regards sub-paragraph (b) the arbitrator held that:

> I accept the basic contention that the issue of each drawing used for the erection of a building should be accurately recorded. Such a requirement is necessary for business efficacy as, if there were no such record the only result could be of uncertainty in the minds of the architect and contractor leading to the possibility of endless disputes. The basic contention is therefore an example of efficient administration by the architect.
>
> Sub-paragraph (b) is much more precise in its wording and reflects not only efficient and proper administration of the contract but also the normal practice followed by architects. Even in small jobs, I have not come across any examples of architect's or consultant's drawings which did not bear a unique reference number or cipher. The scope of the reference, ranging from simple to complicated, has undoubtedly varied but a precise, unique

> reference other than a description of the contents or subject-matter of the drawing is invariably present. The normal practice is for an identification panel to be incorporated in the drawing in a standard position, usually in the bottom right-hand corner of the drawing.

And then he sets out the information normally contained in that panel. He continued:

> Therefore I have no hesitation in accepting such a procedure as an example of proper and efficient administration by the architect.

His conclusions on sub-paragraph (c) were that:

> Sub-paragraph (c) states an objective which is undoubtedly highly desirable. It is well accepted in the building industry that certain drawings are amended on more than one occasion and that it is neither possible nor practicable to withdraw all earlier copies. Thereby the risk arises of two apparently similar drawings containing different details and this can give rise to confusion, delay and expense. To minimise this risk, draftsmen have developed practices of adding further information to the identification panel on the drawing to draw attention to the amendment. But the method adopted can vary from the detailed practice of recording the date, brief description of the amendment(s) made and the addition to the drawing number of a further cipher such as the letter 'A' for the first amendment; 'B' for the second amendment and so on to the mere addition of ciphers 'A', 'B' etc. or '/1', '/2' etc. to the original drawing number. I therefore consider that the procedure described is over-precise in requiring a note of the amendments but I accept that the fact of the amendment should be indicated by an addition to the reference number or cipher.

I feel considerable doubt whether a finding in these terms is appropriate to Stage 1 of the arbitration or one which the court can be asked to review on an application for the determination of preliminary points of law. The question what steps the architect was bound to take to discharge his duty to administer the contract in an efficient manner seems to me a matter appropriate to Stage II and one on which the arbitrator will then be entitled to rely on his expert knowledge. In this application I propose only to say I can see no ground upon which I would be justified in expressing any contrary view.

See also *J. & J. Fee Ltd* v. *The Express Lift Co Ltd*, p. 40.

Instructions

An employer's failure to issue plans, drawings and other information necessary for the execution of the works and at the proper time is a breach of contract by the employer, through the agency of the architect, unless the express terms of the contract state otherwise. In the absence of an express term, the employer's obligation is to issue instructions within a reasonable time. What is a reasonable time depends on the circumstances of the case and on the express terms of the contract. Under JCT terms the obligation is to provide drawings and details 'as and when from time to time may be necessary'. The implication is that the information will be issued from time to time as the contract progresses at a time which is reasonable in all the circumstances.

Neodox Ltd v. Borough of Swinton & Pendlebury

QUEEN'S BENCH DIVISION

(1958) 5 BLR 34

The dispute arose under a civil engineering form of contract, but it is submitted that the same principle is applicable to JCT terms.

> DIPLOCK J: It is clear . . . that to give business efficacy to the contract, details and instructions necessary for the execution of the works must be given by the engineer from time to time in the course of the contract and must be given in a reasonable time. In giving such instructions, the engineer is acting as agent for his principals, the corporation, and if he fails to give such instructions within a reasonable time, the corporation are liable in damages for breach of contract.
>
> What is a reasonable time does not depend solely upon the convenience and financial interests of the [contractors]. No doubt it is to their interest to have every detail cut and dried on the day the contract is signed, but the contract does not contemplate that. It contemplates further details and instructions being provided, and the engineer is to have a time to provide them which is reasonable having regard to the point of view of him and his staff and the point of view of the corporation, as well as the point of view of the contractors.
>
> In determining what is a reasonable time as respects any particular details and instructions, factors which must obviously be borne in mind are such matters as the order in which the engineer has determined the works shall be carried out . . . , whether requests for particular details or instructions have been made by the contractors, whether the instructions relate to a variation of the contract which the engineer is entitled to make from time to time during the execution of the contract, or whether they relate to part of the original works, and also the time, including any extension of time, within which the contractors are contractually bound to complete the works.
>
> In mentioning these matters, I want to make it perfectly clear that they are not intended to be exhaustive, or anything like it. What is a reasonable time is a question of fact having regard to all the circumstances of the case . . .

Where a contractor submits a programme showing completion earlier than the contractual date of completion stated in the appendix no term will be implied that the employer by himself, his servants or agents, should so perform the contract as to enable the contractor to carry out the works in accordance with the programme and to complete the works on the earlier programmed completion date.

Under JCT 98 terms the contractor is entitled to complete the works on a date earlier than the date of completion stated in the appendix. He is further entitled to carry out the works in such a way as to enable him to achieve an earlier completion date whether or not the works were programmed.

Glenlion Construction Ltd v. The Guinness Trust

QUEEN'S BENCH DIVISION

(1987) 11 Con LR 126

The parties entered into a with quantities contract in JCT form for a residential development at Bromley, Kent. Clause 3.13.4 of the bills contained the following provision:

Progress chart
Provide within 1 week from the date of possession, a programme chart of the whole of the works, including the works of nominated sub-contractors and suppliers and contractors and others employed direct including public utility companies and showing a completion date no later than the date for completion. The chart to be a bar chart in an approved form. Forward 2 copies to the architect, 1 copy to the quantity surveyor and keep up to date. Modify or redraft.

Disputes arose between the parties and these were referred to arbitration. The arbitrator found against Glenlion on a number of issues. Glenlion appealed against the arbitrator's interim award under the Arbitration Act 1979, section 1(2). For the purposes of the appeal, it was assumed that Glenlion had provided a programme in accordance with the requirements of clause 3.13.4 of the bills.

Held: The full wording of clause 3.13.3 was a contract provision. On the true construction of the contract if and in so far as the contractor's programme showed a completion date before the date for completion stated in the appendix, the contractor was entitled to carry out the works in accordance with the programme and to complete the works by the programmed completion date. The contractor was entitled to so complete whether or not he produced a programme with an earlier date and whether or not he was contractually bound to produce a programme. If and in so far as the programme showed a completion date before the date for completion specified in the appendix there was no implied term that the employer, by himself, his servants or agents, should so perform the contract as to enable the contractor to carry out the works in accordance with the programme and to complete the works by the programmed completion date.

JUDGE JAMES FOX-ANDREWS QC: [In] the light of the wording of condition 21 it is self evident that Glenlion were entitled to complete before the date for completion. And the contractor was entitled to complete on an earlier date whether or not he produced a programme with an earlier date and whether or not he was contractually bound to produce a programme. It would follow that if he was entitled to complete before the date for completion he was entitled to carry out the works in such a way as to enable him to achieve the earlier completion date whether or not the works were programmed.

It is not suggested by Glenlion that they were both entitled and obliged to finish by the earlier completion date. If there is such an implied term it imposed an obligation on the Trust but none on Glenlion. It is unclear how the variation provisions would have applied. Condition 23 operates, if at all, in relation to the Date for Completion stated in the Appendix. A fair and reasonable extension of time for completion of the works beyond the date for completion stated in the appendix might be an unfair and unreasonable extension from an earlier date.

It is not immediately apparent why it is reasonable or equitable that a unilateral absolute obligation should be placed on an employer.

As long ago as 1970 Mr Duncan Wallace, editor of *Hudson's Building and Engineering Contracts*, 10th edition, wrote at p.603:

> In regard to claims based on delay, litigious contractors frequently supply to architects or engineers at an early stage in the work highly optimistic programmes showing completion

> a considerable time ahead of the contract date. These documents are then used (a) to justify allegations that the information or possession has been supplied late by the architect or engineer and (b) to increase the alleged period of delay, or to make a delay claim possible where the contract completion date has not in the event been extended.

In *Wells* v. *Army & Navy Co-operative Society* (1902) Wright J at first instance had said:

> The plaintiffs must within reasonable limits be allowed to decide for themselves at what time they are to be supplied with detail.

Mr Keating in his supplement – albeit after the date of this contract said:

> Further, it is sometimes said relying upon the dictum of Wright J . . . that the prime consideration is the contractor's own decision as to the time when he is to be supplied with instructions. Again, it is thought that this is too wide.
>
> Such decision is a factor but not decisive. The contractor cannot unilaterally determine what is a reasonable time. Thus a contractor does not prove a claim for delay in instructions merely by establishing non-completion with requests for instructions or a schedule of dates for instructions he has served upon the architect. But agreement by the architect with such a schedule, or even, acquiescence may, it is submitted, be relevant evidence as to the question of what is reasonable. Sometimes contractors at the commencement of or early in the course of a contract prepare and submit to the architect a programme of works showing completion at a date materially before the contract date. The architect approves or accepts without comment such programme. It is then argued that the contractor has a claim for damages for failure by the architects to issue instructions at times necessary to comply with the programme. While every case must depend upon the particular express terms and circumstances, it is thought that, upon the facts set out, the contractors' argument is bad; and that is the case even though the contractor is required to complete "on or before" the contract date. (See the JCT form 1963 and 1980 edition.) There is no authority on this point.

It appears to me that the fact that the programme is required to be provided by the contractor does not in itself make the position different.

By 10 July 1981 *Hudson's Building Contracts*, 10th edition, had gone through some five impressions. The relevance of Mr Keating's comments in his supplement written after the date of the contract merely suggest that whether or not *Hudson* was in all respects accurate the views expressed in *Hudson* were not out of line with a generally held view.

It appears a foregone conclusion that Glenlion cannot establish that such a term was so obvious that it went without saying. The contract as drawn is efficacious in the sense that it produces what appears to me to be the desired effect. The unilateral imposition of a different completion date would result in the whole balance of the contract being lost.

The position would be no different if the obligation imposed on the Trust was instead of being absolute, a requirement that the Trust should act reasonably.

Payment

Most modern building contracts provide for payment by instalments as the work proceeds, normally against an architect's certificate. Failure to make payment when due does not normally amount to a repudiation of the contract, nor entitle the contractor to terminate the contract. The contractor's remedy is to sue on the certificate because when the certificate has been issued and falls due, a debt arises from the employer to the contractor, e.g., JCT 80, clause 30.1.1. For further discussion of certificates, see Chapter 10.

The express terms of the contract may confer additional rights on the contractor, e.g. JCT 80, clause 28.1.1., which makes the employer's failure to pay on any certificate a ground on which the contractor may determine his employment under the contract. See now JCT 2005, clause 8.9.1.

Lump sum contracts

A lump sum contract is one to complete the whole work for a lump sum

In an ordinary lump sum contract, the contractor is entitled to payment if he achieves 'substantial completion' and he is then entitled to the contract price less a sum by way of set-off or counterclaim for any defects.

H. Dakin & Co. Ltd v. Lee

COURT OF APPEAL

[1916] 1 KB 566

Dakin contracted to carry out alteration works to a house in accordance with a specification for a lump sum of £264. The work was not done strictly in accordance with the specification. The employer refused any payment.

Held: Dakin was entitled to recover £264, subject to a deduction of the amount necessary to make the work correspond with that contracted to be done. The defects and omissions amounted only to a negligent performance of the contract, and not to its abandonment or a failure to complete.

Hoenig v. Isaacs

COURT OF APPEAL

[1952] 2 All ER 176

The plaintiff was employed by the defendant to decorate and furnish a flat for the sum of £750, the terms of payment being 'net cash, as the work proceeds; and balance on completion'. The defendant paid £400 by instalments, but refused to pay the balance of £350 on the ground that some of the design and workmanship was defective.

Held: There had been a substantial performance of the contract and the defendant was liable for £750, less the cost of remedying the defects, which was assessed at £56.

DENNING LJ: This case raises the familiar question: Was entire performance a condition precedent to payment? That depends on the true construction of the contract. In this case the contract was made over a period of time and was partly oral and partly in writing, but I agree with the official referee that the essential terms were set down in the letter of 25 April 1950. It describes the work which was to be done and concludes with these words:

> The foregoing, complete, for the sum of £750 net. Terms of payment are net cash, as the work proceeds; and balance on completion.

The question of law that was debated before us was whether the plaintiff was entitled in this action to sue for the £350 balance of the contract price as he had done. The defendant said that he was only entitled to sue on a *quantum meruit*. The defendant was anxious to insist on a *quantum meruit*, because he said that the contract price was unreasonably high. He wished, therefore, to reject that price altogether and simply to pay a reasonable price for all the work that was done. This would obviously mean an inquiry into the value of every item, including all the many items which were in compliance with the contract as well as the three which fell short of it. That is what the defendant wanted. The plaintiff resisted this course and refused to claim on a *quantum meruit*. He said that he was entitled to the balance of £350 less a deduction for the defects.

In determining this issue the first question is whether, on the true construction of the contract, entire performance was a condition precedent to payment. It was a lump sum contract, but that does not mean that entire performance was a condition precedent to payment. When a contract provides for a specific sum to be paid on completion of specified work, the courts lean against a construction of the contract which would deprive the contractor of any payment at all simply because there are some defects or omissions. The promise to complete the work is, therefore, construed as a term of the contract, but not as a condition. It is not every breach of that term which absolves the employer from his promise to pay the price, but only a breach which goes to the root of the contract, such as an abandonment of the work when it is only half done. Unless the breach does go to the root of the matter, the employer cannot resist payment of the price. He must pay it and bring a cross-claim for the defects and omissions, or, alternatively, set them up in diminution of the price. The measure is the amount which the work is worth less by reason of the defects and omissions, and is usually calculated by the cost of making them good . . . It is, of course, always open to the parties by express words to make entire performance a condition precedent. A familiar instance is when the contract provides for progress payments to be made as the work proceeds, but for retention money to be held until completion. Then entire performance is usually a condition precedent to payment of the retention money, but not, of course, to the progress payments. The contractor is entitled to payment pro rata as the work proceeds, less a deduction for retention money. But he is not entitled to the retention money until the work is entirely finished, without defects or omissions. In the present case the contract provided for 'net cash, as the work proceeds; and balance on completion'. If the balance could be regarded as retention money, then it might well be that the contractor ought to have done all the work correctly, without defects or omissions, in order to be entitled to the balance. But I do not think the balance should be regarded as retention money. Retention money is usually only ten per cent, or fifteen per cent, whereas this balance was more than fifty per cent. I think this contract should be regarded as an ordinary lump sum contract. It was substantially performed. The contractor is entitled, therefore, to the contract price, less a deduction for the defects.

In considering whether there is substantial performance it is relevant to take into account both the nature of the defects and the proportion between the cost of rectifying them and the contract price.

Bolton v. Mahadeva

COURT OF APPEAL

[1971] 2 All ER 1322

The plaintiff agreed to install a central heating system in the defendant's house. The contract price was a lump sum of £560. On completion, the defendant complained that the system was defective and refused payment. The defects were such that the central heating did not perform effectively. It gave off fumes and failed adequately to heat the house. The cost of remedying the defects was £174.

Held: A plaintiff was only entitled to recover on a lump-sum contract, subject to set-off in respect of defects, where he had substantially performed the contract. In considering this, it was relevant to take into account the nature of the defects and the proportion between the cost of remedying them and the contract price. As installed, the system failed to heat the house adequately. Taking into account the cost of remedying the defects, it was impossible to say that there had been substantial performance.

CAIRNS LJ: In considering whether there was substantial performance I am of opinion that it is relevant to take into account both the nature of the defects and the proportion between the cost of rectifying them and the contract price. It would be wrong to say that the contractor is only entitled to payment if the defects are so trifling as to be covered by the *de minimis* rule.

The main matters that were complained of in this case were that, when the heating system was put on, fumes were given out which made some of the living rooms (to put it at the lowest) extremely uncomfortable and inconvenient to use; secondly, that by reason of there being insufficient radiators and insufficient insulation, the heating obtained by the central heating system was far below what it should have been. There was conflicting evidence about those matters. The judge came to the conclusion that, because of a defective flue, there were fumes which affected the condition of the air in the living rooms, and he further held that the amount of heat given out was such that, on the average, the house was less warm than it should have been with the heating system on, to the extent of 10 per cent. But, while that was the average over the house as a whole, the deficiency in warmth varied very much as between one room and another. The figures that were given in evidence and, insofar as we heard, were not contradicted, were such as to indicate that in some rooms the heat was less than it should have been by something between 26 and 30 per cent.

The learned judge, having made those findings, came to the conclusion that the defects were not sufficient in degree to enable him to hold that there was not substantial performance of the contract . . .

For my part, I find it impossible to say that the judge was right in reaching the conclusion that in those circumstances the contract had been substantially performed. The contract was a contract to install a central heating system. If a central heating system when installed is such that it does not heat the house adequately and is such, further, that fumes are given out, so as to make living rooms uncomfortable, and if the putting right of those defects is not something which can be done by some slight amendment of the system, then I think that the contract is not substantially performed . . .

The actual amounts of expenditure which the judge assessed as being necessary to cure those particular defects were £40 in each case. Taking those matters into account and the other matters making up the total of £174, I have reached the conclusion that the judge was wrong in saying that his contract had been substantially completed; and, on my view of the law, it follows that the plaintiff was not entitled to recover under that contract.

I have reached that conclusion without taking into account an argument that was pressed on us in this court by the defendant to the effect that it was a term of the contract that no payment should become due until the work had been completed to such an extent and in such a manner that he could properly sign a satisfaction note to be handed to an insurance company which was guaranteeing payment of the contract price. That contention must necessarily depend on the existence of some implied term to that effect, because there is nothing expressly in the contract about it. If the defendant wanted to rely on such an implied term, I think it was necessary for him to plead it, which he did not. It does not seem that any evidence was directed at the hearing to any such implied term, and no reference to the matter is made in the notice of appeal.

Other contracts

Questions of payment under other types of contract, such as 'measure and value' contracts and 'cost plus percentage' contracts depend on the terms of the contract in question and no general principle can be laid down.

The general rules relating to payment and related matters under building contracts were summarised as in the following case.

Holland Hannen & Cubitts (Northern) Ltd. v. Welsh Health Technical Services Organisation

QUEEN'S BENCH DIVISION

(1981) 18 BLR 89

The facts of this case are not material for present purposes; it is further referred to on p. 242.

JUDGE JOHN NEWEY QC: It seems to me that the law is clear and can be shortly stated as follows:

(1) An entire contract is one in which what is described as 'complete performance' by one party is a condition precedent to the liability of the other party: *Cutter* v. *Powell* (1795), and *Munro* v. *Butt* (1858).

(2) Whether a contract is an entire one is a matter of construction; it depends upon what the parties agreed. A lump sum contract is not necessarily an entire contract. A contract providing for interim payments, for example, as work proceeds, but for retention money to be held until completion is usually entire as to the retention moneys, but not necessarily the interim payments: Denning LJ in *Hoenig* v. *Isaacs* (1952).

(3) The test of complete performance for the purposes of an entire contract is in fact 'substantial performance': *H. Dakin & Co Ltd* v. *Lee* (1916) and *Hoenig* v. *Isaacs*.

(4) What is substantial is not to be determined on a comparison of cost of work done and work omitted or done badly: *Kiely & Sons Ltd* v. *Medcraft* (1965) and *Bolton* v. *Mahadeva* (1972).

(5) If a party abandons performance of the contract, he cannot recover payment for work which he has completed: *Sumpter* v. *Hedges* (1898).

(6) If a party has done something different from that which he contracted to perform, then, however valuable his work, he cannot claim to have performed substantially: *Forman & Co Proprietary Ltd* v. *The Ship 'Liddesdale'* (1900).

(7) If a party is prevented from performing his contract by default of the other party, he is excused from performance and may recover damages: dicta by Blackburn J in *Appleby* v. *Myers* (1866); *Mackay* v. *Dick* (1880).

(8) Parties may agree that, in return for one party performing certain obligations, the other will pay him on a *quantum meruit*.

(9) A contract for payment of a *quantum meruit* may be made in the same way as any other type of contract, including conduct.

(10) A contract for a *quantum meruit* will not readily be inferred from the actions of a landowner in using something which has become physically attached to his land: *Munro* v. *Butt* (1858).

(11) There may be circumstances in which, even though a special contract has not been performed there may arise a new or substituted contract; it is a matter of evidence: *Whitaker* v. *Dunn* (1886).

Retention

Where the contract provides that retention money is trust money, as by using the phrase 'the employer's interest in any amounts so retained shall be fiduciary as a trustee', the employer must, if called upon by the contractor, set aside and hold the retention money in a separate bank account, subject to the discretion of the court.

Rayack Construction Ltd v. Lampeter Meat Co Ltd

CHANCERY DIVISION

(1979) 12 BLR 30

The plaintiffs contracted to build meat processing plants for the defendants, the contract being in JCT 63 form. There was a retention of 50 per cent to be retained during the whole of a five year defects liability period. The plaintiffs claimed that the defendants were obliged to pay the retention into a separate trust bank account.

Held: The retention clause imposed an obligation on the defendants to set aside the money retained into a separate trust account and, in the circumstances, a mandatory injunction would be granted to that effect.

Henry Boot Building Ltd v. The Croydon Hotel & Leisure Co Ltd

COURT OF APPEAL

(1986) 36 BLR 41

The parties contracted in JCT 63 terms, and disputes between them were referred to arbitration.

The plaintiffs sought a mandatory injunction ordering the defendants to pay retention money into a separate bank account. The architect had issued a clause 22 certificate entitling the defendants to liquidated damages in excess of the retention money.

The trial judge held that the effect of the certificates issued under clause 22 was to discharge the defendants from any obligation under clause 30(4)(a) to appropriate and set aside a separate trust fund and that the relationship between the defendants and the plaintiffs was at most that of debtor and creditor and that he could not order security to be given for payment of a disputed debt. The plaintiffs appealed.

Held: Although the employer was under an obligation under clause 30(4)(a) to appropriate and set aside amounts retained as a separate trust fund and that obligation could be enforced by the grant of a mandatory injunction, no injunction could be granted at a time when the employer was entitled to deduct a greater amount of liquidated or ascertained damages because there was no subsisting obligation to appropriate and set aside.

NOURSE LJ: The question depends upon the true construction and effect of clauses 22 and 30(4)(a) of the Conditions. The convenient course is to start with clause 22. It is clear that if the architect issues a certificate under that clause the employer may deduct the stated amount of the liquidated and ascertained damages from the sums retained by him by virtue of clause 30(3) since those sums are 'monies . . . to become due to the Contractor under this Contract'. Turning to clause 30(4)(a), in *Rayack Construction Ltd* v. *Lampeter Meat Co Ltd* (1979) Vinelott J decided – in my view correctly and the contrary has not been argued – that that provision imposes on the employer an obligation to appropriate and set aside the amounts retained by virtue of clause 30(3) as a separate trust fund and that the court can enforce that obligation by the grant of a mandatory injunction. But it seems to me to be clear that if, as here, the court is asked to grant an injunction at a time when the employer is entitled to deduct a greater amount of liquidated and ascertained damages, no injunction can be granted for the simple reason that there is no subsisting obligation to appropriate and set aside. In *Rayack* no certificate had been issued under clause 22. The non-existence of any such obligation on the facts of the present case is confirmed by that part of clause 30(4)(a) which subjects the retentions to the right of the employer to have recourse thereto, although it seems to me that provision can only have direct effect in a case where a separate fund has first been constituted.

[Counsel for the plaintiff] has argued that once it has been accepted that there is a serious question to be tried as to the validity of the architect's certificates the case falls to be decided according to the balance of convenience, which he says is markedly in favour of the plaintiff. I was at one time attracted by that argument, but on reflection I find that it would involve a departure from the agreement to which the parties have committed themselves. In my view they are bound by certificates issued under clause 22 unless and until they are set aside. As to that there is no serious question to be tried.

I am therefore of the opinion that there is no subsisting obligation on the defendant to appropriate and set aside a separate trust fund under clause 30(4)(a) or, in other words, that the defendant cannot at present be joined with the plaintiff in the relationship of trustee and beneficiary. At the most their relationship is, as the learned judge has held, that of debtor and creditor. It necessarily follows that no injunction can be granted, it being the invariable practice of the court not to order security for the payment of a disputed debt: see e.g. *Lister* v. *Stubbs* (1890).

Wates Construction (London) Ltd v. Franthom Property Ltd

COURT OF APPEAL

(1991) 53 BLR 23

The plaintiffs were contractors to the defendants for the erection of a hotel. The contract was on JCT 1980 terms but clause 30.5.3 had been deleted. Practical completion took place on 11 August 1989. Soon afterwards the plaintiffs discovered that the retention had not been paid into a separate bank account. They requested the defendants to do so and the defendants refused.

Held: The defendants were required to open a separate bank account despite the deletion of clause 30.5.3 because clause 30.5.1 created a clear trust in favour of the contractor and nominated sub-contractors.

BELDAM LJ: The second argument put forward on behalf of the employer was that he was only bound to appropriate a sum to the fund and was not bound to set it aside in the sense of placing it in a separately identifiable account. It was argued that the statement in the interim certificates which had been issued was a sufficient appropriation of the fund for that purpose. As the corollary of this argument Mr Ullstein said that it was open to the employer to use the fund for the purposes of his own business as working capital and that it would be quite wrong to require him to open up a separate account. The answer to this contention seems to me to lie in the first duty of the trustee, which is to safeguard the fund in the interests of the beneficiaries. It would be, in my judgment, a breach of trust for any trustee to use the trust fund in his own business, and there can be found a clear instance of that in the case of *Re Davies* [1902] 2 Ch 314 where the trustee who, from the best of motives used the trust fund for the purpose of providing the beneficiary with an increased income and to reduce the overdraft of a business in which he had an interest, conceded that that was not a proper investment or use to make of a trust fund. So I reject that contention . . .

The final argument advanced by Mr Ullstein for the employer is more powerful. It is based on the failure of the judge to consider the effect of the deleted clause 30.5.3 to which I have referred. It is argued that the fact that clause 30.5.3 was deleted by the parties is indicative of a common intention on their part that the employer should be under no such obligation as the contractor now contends the contract provides . . .

It may well be that exceptionally, in the case of a standard form of contract contained in a printed document, the fact that a particular deleted provision may assist in the resolution of an ambiguity in another part of the agreement and so justifies looking at the deleted part. But I have no doubt that the general rule is that stated by Viscount Sumner, Lord Hatherley and Lord O'Hagan. In any event, even if there may be exceptional cases, in my view, they are of no assistance in construing the agreement in this case. Firstly, it seems to me that there is no ambiguity about the part of the agreement which remains. The words of clause 30.5.1 under which the trust is created are quite clear. Secondly, the fact of deletion in the present case is of no assistance because the parties in agreeing to the deletion of clause 30.5.3 may well have had different reasons for doing so and it is not possible to draw from the deletion of that clause a settled intention of the parties common to each of them that the ordinary incidence of the duties of trustee clearly created by clause 30.5.1 were to be modified or indeed removed. It may have been thought by one of the parties to have been unnecessary to have included clause 30.5.3. It may have been that one of them thought that the employer should have been liable to account for any interest to the contractor if the retention fund was placed in a separate account. But there may be various reasons, which it is not possible to set out in full, why the clause was deleted and it is quite impossible to draw any clear inference from the fact of deletion. I therefore would reject an argument based upon the fact of deletion and can see no ambiguity upon which reference to that deleted clause could assist.

I therefore return to the principle which underlies the contractor's application in this case. It seems to me that the only way in which the interest of the beneficiaries (the contractor and sub-contractors) in the retention fund could be safeguarded and preserved is if that fund is placed in a separate account and is not used for the purposes of the employer's business or hazarded in any other way. The provisions in the contract between the parties in this case seem to me to be indistinguishable in their effect from the provision in the case of *Rayack Construction Ltd* v. *Lampeter Meat Co Ltd* [1979] 12 BLR 34. That was a decision at first instance of Vinelott J in which he had to construe a clause the effect of which was, as I have said, in my view indistinguishable. The plaintiff contractors in that action were claiming a declaration that they were entitled to be paid forthwith. There was a mandatory injunction ordering the employers to pay the retention fund into a separate bank account and to apply the retention monies in accordance with the trust which was specified in a condition of the contract between the parties in that case. In particular, I would refer to the reasons given by Vinelott J, with which I agree.

Quantum meruit

Where the contractor does work under a contract and no price has been agreed he is entitled to be paid a reasonable sum for work and materials. There may also be a claim on a quantum meruit where there is a contract for specified work and, at the employer's request, the contractor does work outside the contract. Whether or not such a claim may be made is a question of interpreting the contract.

Sir Lindsay Parkinson & Co Ltd v. Commissioners of Works

COURT OF APPEAL

(1950) 1 All ER 208

Contractors agreed to erect an ordnance factory for £3,505,834 by 30 January 1939. Under the contract the employers were entitled to require the contractors to execute additional work on a measured basis, and the Bills of Quantities stated that its value would be in the order of £500,000. Owing to delay in receipt of information from the employers, the contractors became entitled to extensions of time. The employers were anxious that work be completed by 30 January 1939, and so the parties entered into a deed of variation under which it was agreed that the contractors should adopt uneconomic working methods to achieve completion by the due date, at an estimated additional cost of £1 million, and that the contractors would be allowed a net profit of at least £150,000 but not more than £300,000 on the cost of the works, i.e. a net profit of 3 per cent and 6 per cent respectively on the total contract price of £5 million. As a result of extras ordered by the employer, the works executed eventually amounted to in excess of £6½ million.

Held: On the true interpretation of the deed of variation, it was not within the contemplation of the parties that there would be such large increases in the amount of work to be executed. The parties contemplated work to the approximate value of £5 million, and the maximum profit of £300,000 was agreed on that basis. Since the additional work had been performed by the contractors at the employer's request, the contractors were entitled to be paid a reasonable sum in respect of it.

Serck Controls Ltd v. Drake & Scull Engineering

QUEEN'S BENCH DIVISION

(2000) 73 Con LR 100

Between June 1995 and January 1997 the claimant (Serck) carried out design and installation work for the defendant (DS) on a control system at a site near Preston. The work was part of the construction of a replacement research facility for British Nuclear Fuels. If all the contracts originally envisaged had been made, DS would have been the mechanical and electrical sub-contractor and Serck its sub-sub-contractor. No formal contract between DS and Serck was ever made.

There was a trial of a preliminary issue as to whether there was a contract between Serck and DS. Mr Recorder Marrin QC held that there was not.

It was accepted that Serck were entitled to be paid a reasonable sum for the work they had done. The present proceedings were concerned with the question whether a reasonable sum should be measured by the value of the work to DS or what would be a reasonable remuneration to Serck for doing it.

JUDGE JOHN HICKS QC:

34. A *quantum meruit* claim may, however, arise in a wide variety of circumstances, across a spectrum which ranges at one end from an express contract to do work at an unquantified price, which expressly or by implication must then be a reasonable one, to work (at the other extreme) done by an uninvited intruder which nevertheless confers on the recipient a benefit which for some reason, such as estoppel or acquiescence, it is unjust for him to retain without making restitution to the provider. It is clear from the passage in *Hudson* immediately preceding the one cited by [counsel for the defendant] that the words quoted above do not relate to the former category of case. Moreover, in one of the very passages from the authorities relied on by [counsel for the defendant], Robert Goff J in *BP Exploration (Libya) Ltd* v. *Hunt*

(No 2) [1979] 1 WLR 783 at 805 describes the basic measure of recovery in restitution as being 'the reasonable value of the plaintiff's performance – in a case of services, a *quantum meruit* or reasonable remuneration . . .' (my emphasis.)

35. At the first end of the spectrum described in the first sentence of the last paragraph the measure should clearly be the reasonable remuneration of the claimant; at the other it should be the value to the defendant. In between there is a borderline, the position of which may be debatable. It is, however, unnecessary and inappropriate to conduct that debate in the abstract here if, as I believe to be the case, it is clear on which side of the line the present facts lie.

36. The letter of intent of 1 December 1994, after the instruction to proceed quoted in para 16 above, continued:

> In the event that we are unable to agree satisfactory terms and conditions in respect of the overall package, we would undertake to reimburse you with all reasonable costs incurred, provided that any failure/default can reasonably be construed as being on our part.

It was, rightly, not contended by DS that in this case the proviso operated to exempt it from the obligation thus undertaken.

37. In my judgment that plainly created a contractual relationship, albeit not the formal 'contract' of which the existence was in dispute between the parties and was the subject of the preliminary issue. There was, moreover, an express term as to remuneration: 'all reasonable costs incurred'. It is true that 'costs' is a very odd word, implying the exclusion of any profit element and perhaps of overheads, but it is one very commonly used in such contexts and there was no suggestion that it had any such exclusionary effect here.

38. The letter of 27 April 1995 containing what the Scott Schedule treats as the 'primary instruction' under which most of the remaining work was executed read simply: '. . . you are instructed to proceed with all activities to achieve the contract programme'. There was, of course, no 'contract' in the sense intended and, as Mr Marrin found, no agreed programme, but the instruction stood and was acted upon. If, which I do not believe to be the case, there could be any doubt that in the context that imported the same obligation to 'reimburse . . . all reasonable costs incurred' as was expressed in the letter of 1 December 1994, that doubt was resolved by the letter of 18 May 1995 (item 100 in the Scott Schedule), which was an 'official instruction' to extend the letter of intent –

> to cover the necessary supply, delivery, off load and installation, as may be required to complete the full controls package relative to the [BNFL Springfield site].

39. In those circumstances I have no doubt that the sum due to Serck should be assessed by reference to what would be reasonable remuneration for executing the work. . . .

55. What emerges from the authorities, in my view, is that distinctions need to be drawn. If the value is being assessed on a 'costs plus' basis, for example from time sheets and hourly rates for labour, then deductions should be made for time spent in repairing or repeating defective work, and for inefficient working or (as is one of the allegations here) excessive tea-breaks and the like. If the value is being assessed by reference to quantities the claimant stands to gain nothing from such activities or inactivities and, if attributable to the claimant or his sub-contractors, they are irrelevant to the basic valuation; extra time and expense enter into the picture at this stage only if relied upon by the claimant as arising without fault on his part, as discussed in paras 47 to 51 above. If such a claimant makes a claim based on extra time or expense which was in truth his own fault he should fail, but that is simply an issue of fact; Serck says that it has excluded such elements from its valuations.

56. A second distinction is that between defects made good during the course of the work, which are covered by the discussion in the last paragraph, and those remaining at completion. There should clearly be a deduction for the latter, if pleaded and proved, whatever the mode of valuation, simply because the work as handed over is thereby worth less, but no such plea is advanced here.

57. The third distinction is between what I have called the 'basic valuation', which is the subject of the last two paragraphs, and matters which, even if expressed in terms of a 'reduction' or 'diminution' of the valuation, are in essence 'cross-claims', in the words of Bingham LJ in the *Crown House Engineering Ltd* case. They are in essence cross-claims because what the defendant seeks is in truth compensation for loss or expense suffered or liabilities incurred by reason of the claimant's conduct. The examples given in the above extracts are 'tardy performance' and 'unsatisfactory performance', but there may be others. It is, as I understand it, only to this last category that the extracts from the *Crown House Engineering Ltd* case and the last sentence of that from *Lachhani*'s case above apply.

58. If that is the nature of such claims they must depend upon breach of some duty by the claimant, so the first question is as to the nature and extent of the duties owed in the absence of express terms, when carrying out such work, and in particular duties as to progress and co-operation with other trades, for no breach of any other duty seems to be at all relevant here. There is clearly no duty to adhere to any particular contractual programme, for there is no contract, and indeed in the present case it was precisely inability to agree upon a programme which was one of the reasons for failure to enter into a contract.

59. In fact no duty of any relevant kind, nor any breach of such a duty, is pleaded or relied upon by DS (except in relation to the separate point dealt with in para 84 below), nor do I understand from [counsel for the defendant's] closing submissions on the law that any cross-claim of the kind now under discussion is in the end pursued; his contention is simply that DS should not be 'penalised' by inefficient working or inadequate supervision on Serck's part, or be required to pay for hours spent in rectifying defective work. These are all matters within the scope of the principles discussed in paras 55 and 56 above and do not involve the issue left open in the Crown House Engineering Ltd case. I do not therefore have to decide that issue.

The same principle may apply where work is done pursuant to a letter of intent: see British Steel Corporation ***v.*** Cleveland Bridge & Engineering Co Ltd (1981), pp. 3–6.

Set-off

Where money is due to a contractor under a certificate the employer will often refuse to pay or to pay in full on the grounds that he is entitled to damages, for example, for delay or defective work. Similar arguments are often used by contractors to justify non-payment of sub-contractors.

At one time it was thought that there were special rules about set-off in building contracts. This was based on a line of decisions of the Court of Appeal of which *Dawnays Ltd* v. *F.G. Minter* is the best known, but this view was rejected by the House of Lords in *Gilbert-Ash (Northern) Ltd* v. *Modern Engineering (Bristol) Ltd* which affirmed the basic principle that clear words are necessary to justify the argument that a contract has taken away a party's normal remedies.

Dawnays Ltd v. F.G. Minter Ltd

COURT OF APPEAL
(1971) 1 BLR 16

Clause 13 of the 'green' form of nominated sub-contract provided:

> The contractor shall notwithstanding anything in this sub-contract be entitled to deduct from or set off against any money due from him to the sub-contractor (including any retention money) any sum or sums which the sub-contractor is liable to pay to the contractor under this sub-contract.

There was also a comprehensive arbitration clause. Main contractors refused to pay over to the sub-contractors moneys due to them in an interim certificate on the grounds that they were entitled to damages for delay caused by the plaintiffs.

Held: On the true construction of the contract, the only sums deductible under clause 13 are liquidated and ascertained sums which are established or admitted to be due.

Gilbert-Ash (Northern) Ltd v. Modern Engineering (Bristol) Ltd

HOUSE OF LORDS
(1973) 1 BLR 73

The sub-contract in this case was not in standard form, the set-off clause providing:

> If the sub-contractor fails to comply with any of the conditions of this sub-contract, the contractor reserves the right to suspend or withhold payment of any monies due or becoming due to the sub-contractor. The contractor also reserves the right to deduct from any payments certified as due to the sub-contractor and/or otherwise to recover the amount of any bona-fide contra accounts and/or other claims which he, the contractor, may have against the sub-contractor in connection with this or any other contract.

Held: On the true construction of the contract, the main contractors were entitled to set-off claims in respect of alleged breaches of contract by the sub-contractors and that the common law and equitable right of set-off was not excluded by the terms of the contract.

> VISCOUNT DILHORNE: My lords, the question raised in this appeal, one of great importance to the building trade, has been considered by the Court of Appeal in a number of cases in recent years. This is the first occasion on which it has come to this House. Shortly stated, it is whether, when under a contract in the RIBA form an architect has given an interim certificate, the employer is bound to pay to the contractor the amount certified without deduction or set-off save as permitted by the contract; and where the architect has told the contractor the amount forming part of the total certified attributable to work executed by a sub-contractor, the con-

tractor is bound to pay that amount to the sub-contractor without deduction save as permitted by the sub-contract.

In his judgment in this case Lord Denning MR put the matter thus:

> When the main contractor has received the sums due to the sub-contractor – as certified or contained in the architect's certificate – the main contractor must pay those sums to the sub-contractor. He cannot hold them up so as to satisfy his cross-claims. Those must be dealt with separately in appropriate proceedings for the purpose. This is in accord with the needs of business. There must be a 'cash flow' in the building trade. It is the very lifeblood of the enterprise. The sub-contractor has to expend money on steel work and labour. He is out of pocket. He probably has an overdraft at the bank. He cannot go on unless he is paid for what he does as he does it. The main contractor is in a like position. He has to pay his men and buy his materials. He has to pay the sub-contractors. He has to have cash from the employers, otherwise he will not be able to carry on. So, once the architect gives his certificates, they must be honoured all down the line. The employer must pay the main contractor, the main contractor must pay the sub-contractor, and so forth. Cross-claims must be settled later.

Counsel for the respondents sought to support this passage with one qualification. He agreed that deductions might be made from the amount certified where authorised by the contract and that deductions authorised by the sub-contract might be made by the contractor from the sum which he had received from the employer attributable to work done by the sub-contractor and that the contractor was then only obliged to pay the balance.

Counsel for the appellants challenged this. He contended that neither the contract nor the sub-contract excluded the right of the employer to rely on his common law and equitable rights of counterclaim and set-off when sued by the contractor for the amounts certified and the right of the contractor to do so when sued by a sub-contractor for the amount contained in the certificate attributable to his work.

Which view is right must in every case depend on the terms of the contract or sub-contract in question. Changes are made from time to time in the form of the RIBA contracts and sub-contracts may take very different forms. Prior to 1971 and the decision of the Court of Appeal in *Dawnays Ltd* v. *F.G. Minter Ltd* (1971) it appears to have been generally thought that an employer was entitled to set off or counterclaim against the amount certified. In *Halsbury's Laws of England* it is said:

> the contractor is entitled to immediate payment thereof [the amount certified] subject to the terms of the contract and any right of the employer to any set-off or counterclaim (for example, for liquidated damages).

Indeed, no scintilla of authority prior to the decision in the *Dawnays'* case can be found for the proposition that the amount certified by the architect in an interim certificate as the value of work done and consequently payable to the contractor is in a special position in that the employer cannot lawfully counterclaim and set off against the liability amounts due to him from the contractor . . . A great deal has been said in *Dawnays'* case and the cases which followed it, as well as in this case, as to the importance of a 'cash flow' in the building industry. I cannot think that the building industry is unique in this respect. It is, of course, true that the contract makes provision for payments as the work proceeds, but, it is to be observed, a fact to which I feel insufficient attention has been paid, that the contractor is only entitled to be paid for work properly executed. He is not entitled to be paid on interim certificates for work which is defective. The architect should only value work executed properly, that is to say, to his reasonable satisfaction (clause 1); and no interim certificate is of itself conclusive evidence that the work was in accordance with contract (clause 30(8)).

Under clause 30(2) it is for the architect to deduct the retention money and the instalments previously paid and the balance is the amount to which the contractor is by clause 30(1) entitled to payment.

When one has regard to other provisions of the contract this does not mean payment without deduction. For instance, if the contractor has failed to comply with an instruction and the employer has employed others to do so, all costs incurred in doing so are recoverable from the contractor by the employer as a debt 'or may be deducted by him from any monies due or to become due to the contractor under this contract' (clause 2(1)).

Similarly if the contractor does not comply with the insurance requirements of the contract the employer may himself insure 'and may deduct a sum equivalent to the amount paid in respect of premiums from any monies due or to become due to the contractor' (clause 19) (see also clause 20(a)). The employer may also deduct under this contract £400 per week as liquidated and ascertained damages for non-completion by the due date (clause 22).

The existence of these provisions in the contract lends no support to the contention that the amount certified is of a special sacrosanct character which must be paid without deduction. If these deductions can be made, as they clearly can, from the amount certified, the amount which clause 30(1) says the contractor is entitled to be paid, why should it be inferred that the contract impliedly, for there is nothing express, excludes reliance by the employer on his common law and equitable rights to counterclaim and set-off for the amount certified. I see no ground for any such inference.

In practice standard forms (particularly of sub-contracts) often contain clauses designed to limit the right of set-off, particularly by requiring the giving of prior notice of alleged rights of set-off.

There are at least three separate doctrines:

(1) Legal set-off. This is where there are ascertained liquidated sums due in each direction *B. Hargreaves Ltd* v. *Action 2000 Ltd* (below).

(2) Abatement. This is an application in a building contract of the principles stated in relation to sale of goods in *Mondel* v. *Steel* (1841) 8 M&W 837, that is that the buyer can deduct the value of a defect from the price and pay the balance. In *Acsim (Southern) Ltd* v. *Dancon Danish Contracting and Developing Co Ltd* (below) the Court of Appeal held that a contractual provision limiting set-off did not apply to abatement.

Abatement can only operate to diminish or extinguish the price paid or payable under the contract. This amount cannot exceed the price *C.A. Duquemin Ltd* v. *Slater* (below).

Abatement can only be invoked where the other party's defective performance leads to a diminution in value: *Mellowes Archital Ltd* v. *Bell Products Ltd* (below).

(3) Equitable set-off. See *C.M. Pillings & Co Ltd* v. *Kent Investments* and *Rosehaugh Stanhope* v. *Redpath Dorman Long Ltd* (below).

B. Hargreaves Ltd v. Action 2000 Ltd

COURT OF APPEAL

(1992) 36 Con LR 75

BALCOMBE LJ: Before I turn to the facts upon which he seeks to rely, I should state first what I understand to be the law on this subject. The modern law of common law set-off at law derives from *Stooke* v. *Taylor* [1880] 5 QBD 569. I need not deal with the facts of that case. I refer to two passages from the judgment of Cockburn CJ, which have since been referred to time and again with approval. Cockburn CJ said (at 575):

> Set-off and counter-claim are both the creation of statute, the common law now admitting of the action of a plaintiff against a defendant being met by an independent claim of the defendant against the plaintiff, but leaving the defendant to his cross-action. The effect of these two modes of proceedings must, therefore, be sought in the statutes by which they were introduced, and in their results; and when these are looked at, it will be seen how essentially these two forms of procedure differ. By the statutes of set-off this plea is available only where the claims on both sides are in respect of liquidated debts, or money demands which can be readily and without difficulty ascertained. The plea can only be used in the way of defence to the plaintiff's action, as a shield, not as a sword.

Then he said (at 576):

> In the case of set-off the claim being for liquidated damages, its existence and its amount must be taken to be known to the plaintiff, who should have given credit for it in his action against the defendant.

In the well known case of *Hanak* v. *Green* [1958] 2 All ER 141 at 149–150, [1958] 2 QB 9 at 23 the first of the passages from Cockburn CJ's judgment which I have read was cited with approval by Morris LJ. More recently in this court, in *Axel Johnson Petroleum AB* v. *M.G. Mineral Group AG* [1992] 2 All ER 163 at 166, [1992] 1 WLR 270 at 272 Leggatt LJ said, and he is referring, as is clear from the opening words, to set-off at common law as opposed to equitable set-off:

> For set-off to be available at law the claim and cross-claim must be mutual, but they need not be connected. They need not be debts strictly so called, but may sound in damages. The question is, in the language of Tindal CJ in *Morley* v. *Inglis* [1837] 4 Bing NC 58 at 71, 132 ER 711 at 716, 'whether the demand is capable of being liquidated, or ascertained with precision at the time of pleading'. A claim under guarantee, therefore, cannot be set off until the amount of the liability in respect of which the guarantee was given has been established. As Lord Ellenborough CJ said in *Crawford* v. *Stirling* [1802] 4 Esp 207 at 209, 170 ER 693 at 694:
>
>> To make the sum admissible as a set-off, the sum must be settled in monies numbered. . . .

Leggatt LJ then said ([1992] 2 All ER 163 at 167, [1992] 1 WLR 270 at 274):

> By the time of *Stooke* v. *Taylor* [1880] 5 QBD 569 at 575 the principle relevant for present purposes could be stated by Cockburn CJ in a sentence: 'By the statutes of set-off this plea is available only where the claims on both sides are in respect of liquidated debts, or money

demands which can be readily and without difficulty ascertained.' That principle was echoed more recently by Lord Denning MR in *Henriksens Rederi A/S* v. *PHZ Rolimpex* [1973] 3 All ER 589 at 593, [1974] QB 233 at 246, when he said, referring to the statutes of set-off and citing *Bullen and Leake's Precedents of Pleadings* (3rd edn, 1868) p. 679: 'These apply only "where the claims on both sides are liquidated debts or money demands which can be ascertained with certainty at the time of pleading" . . .'.

From these figures it seems to me clear that in each of the cases on which Mr Sheridan seeks to rely – and I have used those two cases as examples only and the same point can be made in relation to the others – the sum claimed by way of set-off depends upon Mr Austin's valuation of the work done, his estimation of what had been omitted, his estimation of whether certain work had been done inadequately and the value he attributed to that inadequate work (which is the way Mr Sheridan prefers it to be put rather than what I think is the normal way, which is the cost of making good the defects). All this makes it clear to me that it is impossible to bring these claims within the way in which it was put by Cockburn CJ in *Stooke* v. *Taylor*, 'money demands which can readily and without difficulty be ascertained'. These claims can only be ascertained by litigation or arbitration, and to say therefore that they were readily ascertainable at the relevant date – and it is common ground that the relevant date was the date upon which notice of the appointment of a receiver was given – seems to me to be quite impossible and indeed, as I think Mr Sheridan might concede, would suggest that the courts have been labouring for the last 150 years, if not longer, under a misapprehension as to the nature of set-off at law.

Acsim (Southern) Ltd v. Dancon Danish Contracting and Development Co Ltd

COURT OF APPEAL

(1989) 19 Con LR 1

RALPH GIBSON LJ: The same is true, in my judgment, of this case. The 'rights of the parties in respect of set-off' were in this case agreed to be those set out in clause 15 and 'no other rights whatsoever shall be implied as terms of this sub-contract relating to set-off': see clause 15(4). I am not sure what the businessmen who use this form understand by the phrase 'rights . . . in respect of set-off'. It seems to me, however, that 'set-off', in the context of this contract, means 'set-off' in our law as defined by the decisions of the court. Mr Reese QC felt unable to argue otherwise. If parties wish to subject to the requirements of a clause like this clause 15, including conditions precedent as to prior notice, all grounds for contesting or reducing the sum claimed by a sub-contractor in respect of an interim payment, they can, without difficulty, find words apt for that purpose. The words of clause 15 as they stand do not, in my judgment, affect the right of a contractor to defend a claim for an interim payment by showing that the sum claimed includes sums to which the sub-contractor is not entitled under the terms of the contract or to defend by showing that, by reason of sub-contractor's breaches of contract, the value of the work is less than the sum claimed under the ordinary right of defence established in *Mondel* v. *Steel* (1841) 8 M&W 858, 151 ER 1288.

It is to be emphasised, so far as this case is concerned, that the defence which Dancon wish to put forward, and have substantiated, at least to the extent of showing it to be arguable, by Mr Hamilton's evidence, is concerned not only with showing that by breach of contract the work is worth less but also with showing that, as to part of the claim, the work had not been done at all, and/or that the sum claimed in respect of it was not included within the sum promised to be paid. Such a defence does not necessarily raise a breach of contract and may consist merely of asserting that the sum claimed has not been earned. There is nothing whatever in clause 15, in my view, to support the contention that the contractor has agreed to comply with the provisions there set out as to notice in order to be entitled to dispute an invoice on such grounds. Clause 15 is dealing with the right of set-off 'against any money otherwise due under this sub-contract'. If, for example, the claim is in respect of work not done, or if the claim is calculated on the basis that the price has been increased by variations which have not been agreed, or on the basis that the price includes 25% of the original cost of an item which has been wholly omitted, then to the relevant extent the sum claimed is not 'money otherwise due'.

C.A. Duquemin Ltd v. Slater

QUEEN'S BENCH DIVISION

(1993) 35 Con LR 147

JUDGE JOHN NEWEY QC: My judgment and the order gave the arbitrator power to 'abate' the contractors' claim, which meant that he should deduct the difference between the value of the work and materials supplied at the date when they were supplied and their value if they had not been defective, but had been in accordance with the contract. The result could have been to reduce the contractors' claim to nil, but could not have produced what might be described as 'negative value'.

Instead, however, the arbitrator purported to deduct the estimated cost of putting right the defects from the contractors' claim and so produced what he himself called 'excess payments' and correctly described as 'indistinguishable from an award of damages'. I do not think that he was right to rely on *Lintest Builders Ltd* v. *Roberts* (1978) 10 BLR 120, a case in which the employer was in a position to set off or counterclaim.

In my view the arbitrator has acted without jurisdiction. I do not think that I need consider whether by doing so he has misconducted himself; the proceedings must in any event have been void.

Mellowes Archital Ltd v. Bell Products Ltd

COURT OF APPEAL

(1997) 58 Con LR 22

BUXTON LJ: I respectfully consider that that statement accurately sets out the limits of the rule of abatement in *Mondel* v. *Steel* (1841) 8 M&W 858, (1835–42) All ER Rep 511 as demonstrated by the earlier authority that I have ventured to cite. That rule does not apply to claims based on delay. I do not see any oddity or inconvenience in that. The difference between abatement and set-off is only of significance in very particular situations, namely special issues of limitation, such as was the case in *The Brede* and the *Aries Tanker* case; or where, as in our case, a contractual limitation on remedies confines itself to 'set-off'.

The historical limits on abatement do not, in the normal case, in any way affect the ultimate rights of a party, any more than in our case they in law affect the ultimate rights of

Bell; because, as Lord Denning MR says in the passage that I have just ventured to cite from *The Brede*, those rights will in any event normally be able to be asserted by set-off. And I would say further that although the *Aries Tanker* case was concerned with the special issue of application of abatement of freight, and therefore does not directly affect the issue before us, the speeches of their Lordships in that case at least strongly suggest that we are not free to depart, and should not depart, from the established limits and ambit of the defence of abatement.

I hold, therefore, the defence asserted in this case is not available to Bell, and judgment should be entered against them.

HOBHOUSE LJ: It is therefore clear that, for a party to be able to rely upon the common law right to abate the price which he pays for goods supplied or work done, he must be able to assert that the breach of contract has directly affected and reduced the actual value of the goods or work – 'the thing itself'. In other words any other loss or damage, if it is to be relied upon by way of answer to a claim for the price, has to arise from the principle of equitable set-off. In most contractual relationships there would be no need to draw a distinction between the two types of defence. But under DOM/1 it is necessary to do so.

Whilst it may be possible to conceive of a case in which delay has affected the value of the thing itself, the normal effect of breaches of the obligation of timeous performance will be to cause losses to the other contracting party which are consequential upon the breach and therefore can only be relied upon, if at all, under the principle of equitable set-off. In the present case the factual situation is clear. The plaintiff sub-contractor's claim to be paid the price is based upon the valuation of the goods supplied and work done. Indeed the valuation was the defendant main contractor's own valuation, and is undisputed. The case of the defendants is that the plaintiffs' delays caused them serious losses through the prolongation of the head contract, the disruption of their own contractual works and those of other sub-contractors, the need to accelerate other work, and the reduced contribution to their own overhead expenses. Thus, the defendants' case is based upon financial losses which they say they have suffered as a consequence of the plaintiffs' breaches of their obligations of timeous performance. Subject to the terms of clause 23, those losses can be relied upon to support an equitable set-off but they cannot justify the legal defence of abatement of the price.

Payment of the amount of an interim certificate is not a condition precedent to arbitration under contracts in JCT 98 and similar forms. If the employer raises a bona fide arguable contention that a payment certificate may have been overvalued he is entitled to have that issue arbitrated and the contractor is not entitled to summary judgment on the certificate.

C.M. Pillings & Co Ltd v. Kent Investments Ltd

COURT OF APPEAL

(1986) 4 Con LR 1

The defendants let a contract to the plaintiffs under the JCT Fixed Fee Form of Prime Cost Contract (1967 edition, revised 1976) for the extension of a large house, the provision of an indoor swimming pool and the addition of a new master bedroom suite. The estimated prime cost written into the contract was £300,500. As work progressed, a large number of variations were ordered by the architect and delays to completion occurred. The estimated final cost was in the order of £630,000.

A certificate of practical completion was issued by the architects on 14 September 1984. On 5 October 1984 the architects issued an interim certificate for £110,000. This was subsequently revised to £101,529. The defendants disputed the sum certified and wished to challenge the accuracy of the certificate. They refused to honour the certificate, and when the plaintiffs issued a writ and sought summary judgment for £101,529, the defendants asked for an order that the action should be stayed under section 4 of the Arbitration Act 1950 and referred to arbitration.

Held: (1) There was just sufficient evidence to entitle the employers to raise a bona fide arguable contention about the accuracy of the interim certificate.

(2) There is a right to raise cross-claims in any action for payment under a certificate unless the terms of the contract expressly exclude that right. On its true construction, the arbitration clause made it clear that such cross-claims could be raised by its specific reference to disputes as to 'whether or not a certificate has been improperly withheld or is not in accordance with this agreement'.

KERR LJ: The decision of the House of Lords in *Modern Engineering (Bristol) Ltd* v. *Gilbert-Ash (Northern) Ltd* (1973) made it clear that, unless the contract on its true construction had the effect that the contractor was entitled to immediate payment, cross-claims which would be admissible as equitable set-offs could be raised by the employer, or head contractor as the case may be, as a reason for withholding amounts certified as being payable by him. In the ultimate analysis the House of Lords was concerned with a particular clause in the sub-contract which, on its true construction, as I think some of their Lordships put it, not only did not exclude the right to raise cross-claims against certified amounts, but made it quite clear, albeit unnecessarily as a matter of law, that such cross-claims could be raised.

In the present case the cross-claim in question – the challenge to the accuracy of the certificate – cannot be questioned as an admissible cross-claim; indeed it is expressly referred to in [the arbitration clause]. The only question is therefore whether the contract on its true construction somehow provides that such cross-claim cannot be asserted without the contractor having first been paid under the interim certificate. As I have already said, I cannot find any provision which has that effect. That appears to have been the conclusion of the majority of their Lordships in *Gilbert-Ash*, to the extent to which they referred to the head contract in that case, which appears to have been indistinguishable in principle from the contract which we have here, although it was not a prime cost plus fee contract. The numbering of the contract was clearly different, but it is not in dispute between counsel, who are very experienced in these matters, that the substance of the provisions there referred to was substantially the same as in the present case. Indeed, one can identify a number of identical provisions referred to in those speeches of their Lordships which I have also mentioned earlier in this judgment to which they draw attention in connection with the present issue. The upshot of those speeches is that, in order to exclude the right to assert cross-claims admissible as equitable set-offs, and thus to exclude rights of deduction or non-payment of certified amounts based on such cross-claims, it is necessary to find some clear and express provision in the contract which has that effect. There was none in the head contract in *Gilbert-Ash*, and there is none here.

The learned editor of *Hudson's Building Contracts* (10th edn supplement, paragraphs 492–494) takes the view, with which I respectfully agree, that standard contracts such as the present contain no sufficient provision which has that effect. We were also referred to the decision of Tudor Evans J at Leeds Crown Court, in *Vickers* v. *Lynn & Jones* (1981), where he reached precisely the same conclusion.

On the other side of the line, this court recently had before it a contract which expressly excluded those consequences. I refer to *Tubeworkers Ltd* v. *Tilbury Construction Ltd* (1985). It was concerned with the FASS Form of sub-contract. There were certain provisions there which unmistakably provided that the *Gilbert-Ash* situation, if I may so put it, should be altered by express terms. That standard form was clearly revised for that purpose. There is nothing of that kind in the present contract sufficiently strong to displace reliance upon cross-claims admissible by way of set-off, as upheld in *Gilbert-Ash*.

The Arbitration Act 1996 renders the discussion of a stay in the above case obsolete but the discussion of set-off is still useful.

Rosehaugh Stanhope v. Redpath Dorman Long Ltd

COURT OF APPEAL

(1990) 26 Con LR 80

This was a non-standard management contract for one of the phases of the Broadgate development in the City of London. The plaintiffs were the employers; Bovis/Schat joint venture were the 'construction manager'; there was no main contractor and the defendants were trade contractors for the supply and erection of structural steel.

Clause 19(3) of the contract provided:

> If the Trade Contractor is in breach of any of his obligations under sub-clause (1) of this Condition, he shall, without prejudice to and pending the final ascertainment or agreement between the parties as to the amount of the loss or damage suffered or which may be suffered by the Client in consequence thereof, forthwith pay or allow to the client such sum as the Construction Manager shall *bona fide* estimate as the amount of such loss or damage, such estimate to be binding and conclusive upon the Trade Contractor until such final ascertainment or agreement. If a Payment Certificate is issued after the date of the Construction Manager's estimate any such amount may be deducted from the amount which would otherwise be stated as due in such a certificate. Such loss or damage may include, inter alia, any additional professional fees, interest charges or loss of rental income suffered or incurred by the Client and any loss and/or expense suffered or incurred by others engaged by the Client including the Construction Manager for which the Client is or may be liable under his contracts with such others.

BINGHAM LJ: The question is this: are the defendants in breach of any of their obligations under clause 19(1) if they do not complete within the period stated in the programme, as extended by the construction manager? I have not found this an easy question and I see great force in the plaintiffs' submissions which the judge accepted. But I have with some hesitation reached a contrary view. The defendants can become subject to no obligation under clause

19(3) unless they are in breach of an obligation under para (1). If a breach is admitted or proved, or if the defendants can show no arguable grounds for denying a breach, para (3) may be operated. But para (1) acknowledges that the period for completion stated in the programme is subject to such fair and reasonable extensions of time as the construction manager may grant under clause 20. I think this assumes that the construction manager has granted fair and reasonable extensions of time. On occasion this may be admitted. Sometimes it may be proved. More often it may be incontestable or virtually so. But I do not think the two paragraphs, read together, envisage that the defendants may be in breach for purposes of para (3) when there is a live and arguable issue whether the construction manager has made fair and reasonable extensions to the programmed completion date which, if made, would exonerate the defendants. Paragraph (3) provides that the construction manager's bona fide estimate shall be binding and conclusive until final ascertainment (presumably under para (5) or by the court), but there is no corresponding provision with regard to breach, and it could not in my view be argued that his ruling on liability under the last sentence of para (1) or under para (4) is binding. I incline to this construction the more readily since I cannot believe the parties intended one of them to be subject to a potentially crippling obligation upon a contingency. In any event, I consider these provisions to be ambiguous and so adopt the construction less favourable to the plaintiffs whose document it is.

The possibility of using set-off as a defence has in practice been greatly restricted by the provisions of the Housing Grants, Construction and Regeneration Act 1996. Sections 110 and 111 of the Act provide:

Dates for payment

110. – (1) Every construction contract shall –

(a) provide an adequate mechanism for determining what payments become due under the contract, and when, and

(b) provide for a final date for payment in relation to any sum which becomes due.

The parties are free to agree how long the period is to be between the date on which a sum becomes due and the final date for payment.

(2) Every construction contract shall provide for the giving of notice by a party not later than five days after the date on which a payment becomes due from him under the contract, or would have become due if –

(a) the other party had carried out his obligations under the contract, and

(b) no set-off or abatement was permitted by reference to any sum claimed to be due under one or more other contracts, specifying the amount (if any) of the payment made or proposed to be made, and the basis on which that amount was calculated.

(3) If or to the extent that a contract does not contain such provision as is mentioned in subsection (1) or (2), the relevant provisions of the Scheme for Construction Contracts apply.

Notice of intention to withhold payment

111. – (1) A party to a construction contract may not withhold payment after the final date for payment of a sum due under the contract unless he has given an effective notice of intention to withhold payment.

The notice mentioned in section 110(2) may suffice as a notice of intention to withhold payment if it complies with the requirements of this section.

(2) To be effective such a notice must specify –
(a) the amount proposed to be withheld and the ground for withholding payment, or
(b) if there is more than one ground, each ground and the amount attributable to it, and must be given not later than the prescribed period before the final date for payment.
(3) The parties are free to agree what that prescribed period is to be.
In the absence of such agreement, the period shall be that provided by the Scheme for Construction Contracts.
(4) Where an effective notice of intention to withhold payment is given, but on the matter being referred to adjudication it is decided that the whole or part of the amount should be paid, the decision shall be construed as requiring payment not later than –
(a) seven days from the date of the decision, or
(b) the date which apart from the notice would have been the final date for payment, whichever is the later.

Miscellaneous

Nomination of sub-contractors and specialists

Where the contract provides that the employer may nominate sub-contractors, a term may be implied that the specialist will be nominated in time for the contractor to incorporate the work into his programme. A term may also be implied that the nomination will be of a specialist willing to enter into a sub-contract which is consistent with the terms of the main contract.

Leslie v. Metropolitan Asylums District Managers

COURT OF APPEAL
(1901) 1 LGR 862

ROMER LJ: The [main contractors] clearly could not be compelled to enter into a [sub-]contract which did not adequately protect them . . . If the terms of those [sub-]contracts prove insufficient properly to protect the [main contractors], that is their fault or misfortune and they cannot hold [the employers] liable in any way.

Objection to nominees

Where the contract gives the main contractor the right to object to a proposed nominee, as does JCT 98, clause 35.5.1 (JCT 80, clause 35.4.1, JCT 63, clause 27(a)) the contractor's remedy is to exercise that right of objection in either of the foregoing cases, and if he does not he cannot rely on any such implied term.

Percy Bilton Ltd v. Greater London Council

COURT OF APPEAL
(1981) 17 BLR 9

For the facts see pp. 328–9, 425–7 where extracts from the speeches in the House of Lords are given. The following statement is unaffected by the decision in the House of Lords.

STEPHENSON LJ: [The plaintiff argued] that at the time of the application for re-nomination, the new sub-contractor's date for completion was later than the plaintiff's date for completion and that, since this would make it impossible for the plaintiff both to accept the new sub-contractor and to comply with the provision in their own contract as to the time for completion, therefore the time provision must go completely . . .

I do not accept that argument. The contractor, faced with a sub-contract with such a provision as to completion, would be entitled to refuse to accept that sub-contractor under clause 27 . . .

Displacement of implied terms

Any implied term will be displaced to the extent that there is an express term dealing with the same matter.

Form of sub-contracts

Where the main contract contemplates the use of nominated sub-contractors, there is no implied term that a particular form of sub-contract should be used, but the express terms of a particular form of contract, such as JCT 98, may raise this implication or make the use of a particular form mandatory.

James Longley & Co Ltd v. Reigate & Banstead Borough Council

COURT OF APPEAL

(1982) 22 BLR 31

The facts of this case, which was decided under JCT 63 terms, are not relevant for present purposes.

WALLER LJ: [The employers] sought to support their case by referring to the sub-contract which is in a standard form and in which [they] submitted it was an implied term between the building owner and the main contractor that it should be used. In our opinion this case must be decided on the main contract alone, and no reference should be made to the sub-contract . . . While the sub-contract refers to the main contract, there is nothing in the main contract which dictates the form of the sub-contract.

Effect of failure

If a nominated sub-contractor or supplier fails during the continuance of the works, the employer is under a duty to re-nominate, and to pay the actual cost of the substituted sub-contractor's work. See: North-West Metropolitan Regional Hospital Board v. T.A. Bickerton & Son Ltd (1970), p. 421.

Default in supply of goods

Where goods are to be provided by or on behalf of the employer, a term may be implied that they will be supplied in time to enable the contractor to carry out the works expeditiously and in accordance with his planned progress, and the employer will be liable for his supplier's default.

Thomas Bates & Son Ltd v. Thurrock Borough Council

COURT OF APPEAL
22 October 1975

Thurrock employed Crux Limited as management consultants in connection with a housing estate which they were developing. Crux were not paid fees, but it was agreed that only Crux components would be used in the houses to be erected, and that they were to be paid a price for the components so used. Thurrock contracted with Bates in JCT 63 terms for the construction of the houses. Crux components were detailed in the specification as 'to be supplied by Crux' and were not p.c. items. Crux were described as 'appointed suppliers'. The contract overran by 25 weeks and Bates claimed damages against Thurrock, attributing the delay and disruption to them through the shortcomings of Crux as their agents. They said that Crux failed to supply the components in due time and that the components were of inferior quality. Bates had no contract with Crux. The council settled the claim for £65,000 and claimed this amount from Crux. The contract between Thurrock and Bates provided that Crux's components 'shall be incorporated in the works and be delivered to the site in accordance with a programme to be agreed between the architect, Bates and Crux'.

Held: Crux were liable. Having agreed to supply the components it was right to imply that they would live up to their programme so that the contract works could be kept up to date. As Thurrock had undertaken to supply the components they were liable to Bates on the basis of an implied term that the components would be supplied in time to avoid disruption and delay. Thurrock could pass that obligation over to Crux.

> LORD DENNING MR: [It is clear from the main contract] that Thurrock were undertaking this obligation that Crux items should be incorporated in the works. Thurrock were undertaking that the obligation would be fulfilled. It is Thurrock's obligation and Thurrock's undertaking. The fact that payment was to be made by Bates to Crux is neither here nor there . . . It is plain that Bates can sue Thurrock because Thurrock promised that the items shall be supplied. If A promises B that C will supply goods to B in a reasonable time and C does not supply them or supplies them late, then B can sue A for damages for C's default. Equally, Thurrock can pass over the obligation to Crux.

Note: Before the official referee Bates alleged four terms were to be implied in their contract with Thurrock:

(1) That the architect and/or his agents Crux would furnish Bates with all necessary information in due time to enable Bates to carry out and complete the works expeditiously and in accordance with the contract.
(2) That Thurrock by their agents Crux would supply Crux components in accordance with the programme envisaged by the contract.
(3) That neither Thurrock, the architect, nor Crux would hinder or prevent Bates from carrying out and completing the works expeditiously.
(4) That Crux components when supplied would be of good quality and fit for their intended purpose.

Thurrock admitted (1), (3) and (4); (2) was held to be an implied term.

> JUDGE WILLIAM STABB QC: Commonsense seems to me to dictate that the scheme was designed to provide that Crux were to supply drawings and materials and that Thurrock were to be responsible for seeing that the contractor used those drawings and took those materials. In so doing Thurrock were . . . assuming responsibility to the contractor for the Crux drawings and for the materials which they, Thurrock, were requiring the contractor to use and take, and, accordingly, would be liable to the contractor in the first instance for default or defect in the drawings or materials with a write-over against Crux who supplied them.

Chapter 8
Ownership and Vesting of Materials

Materials and goods

A building contract is one for work and materials. Materials belonging to the contractor and brought on site remain his property, but when they are incorporated in the works they become the property of the employer, unless the contract provides otherwise.

Appleby v. Myers

COURT OF EXCHEQUER CHAMBER

(1867) LR 2 CP 651

The facts are set out on p. 137.

> BLACKBURN J: Materials worked by one into the property of another become part of that property. This is equally true whether it be fixed or moveable property. Bricks built into a wall become part of the house; thread stitched into a coat which is under repair, or planks and nails and pitch worked into a ship under repair, become part of the coat or ship, and, therefore, generally, and in the absence of something to show a contrary intention, the bricklayer, or tailor, or shipwright, is to be paid for the work and materials he has done and provided . . .

The express terms of the contract may alter the position as between employer and contractor, but terms in the contract between a supplier and the contractor cannot affect the employer's position since there is no contractual relationship between the supplier and the employer. As soon as the on-site goods and materials are incorporated in the works, they become the property of the employer, even if supplied to the contractor under a contract containing a retention of title clause.

Reynolds v. Ashby

HOUSE OF LORDS

[1904] AC 406

The appellant sold machines to a factory tenant on hire-purchase terms. The machines were fixed to concrete beds on the factory floor by means of nuts and bolts, and could

Powell-Smith and Furmston's Building Contract Casebook, Fifth Edition. Michael Furmston.

be removed without damaging the building. The tenant mortgaged the factory, together with its fixtures and fittings, to Ashby, who entered into possession. The tenant defaulted under the hire-purchase agreement, and the appellant claimed the machines or their value from Ashby.

Held: The machines were fixtures and part of the building and Ashby was entitled to retain them as against the appellant.

> LORD LINDLEY: I pass now to consider whether, in this case, the fact that the mortgagor Holdway had not acquired the ownership of the machines by paying for them entitles the appellant to recover their value from the respondents. The title to chattels may clearly be lost by being affixed to real property by a person who is not the owner of the chattels . . . Holdway agreed to buy the machines; but the appellant knew what he wanted them for, and that, before paying for them, he intended to put them up in his factory and to use them. The appellant knew that the factory was mortgaged, and he ran the risk of the machines being claimed as fixtures. In effect, Holdway was authorised by the appellant to convert the chattels into fixtures, subject to the right of the appellant to enter and retake them if he did not pay them. But, apart from this knowledge and authority, the result would be the same, although not so obvious. After the machines were fixed, and before the appellant claimed them, the second mortgagee took possession; the appellant's right to enter and remove the machines, resting as it did on his contract with Holdway, ceased to be exercisable. The appellant has no greater rights against the respondents than he had against the second mortgagee, whose rights the respondents have acquired. The machines had ceased to be chattels belonging to the appellant; they were not chattels wrongfully detained by the respondents. The machines had become fixtures which the appellant was not entitled to remove from the possession of the mortgagees.

Hobson v. Gorringe

COURT OF APPEAL
[1897] 1 Ch 182

Hobson sold K an 11 horse power gas engine on hire purchase terms. The contract provided that if K defaulted on the instalment payments, Hobson was at liberty to remove and repossess the engine, K having no claim for any money paid. It was also provided that the contract should not operate as a contract of sale, but only as an agreement for the hire of the engine, unless and until K had paid all the instalments.

K installed the engine in his saw-mill. It was fixed by screws and bolts to a concrete base laid in the soil. K defaulted on the instalments and mortgaged his saw-mill and fixtures to a third party, Gorringe, who took possession of the premises together with the engine.

Held: On the true interpretation of the agreement, coupled with the annexation of the engine to the soil, the engine became a fixture, subject as between Hobson and K to Hobson's right to unfix it and retake possession if the hirer defaulted. This right imposed no legal obligation on the third party and, therefore, the owner's remedy was by action against the hirer personally. The action against Gorringe failed.

In re Morrison, James & Taylor, Ltd

COURT OF APPEAL

[1914] 1 Ch 50

By an agreement made between the applicants and the company, the applicants agreed to erect and install a sprinkler system in the company's premises. The sprinkler was to be paid for by instalments, on terms that the system would remain the property of the applicants until the entire price was paid. Subsequently, the company issued debentures and thereafter a receiver was appointed on behalf of the debenture holders. The applicants were not paid the price of the system and claimed to be entitled to enter the company's premises and remove it.

Held: The applicants were entitled to remove the installation, notwithstanding the appointment of the receiver.

The position is different in the case of unfixed goods and materials, whether on or off site. In that case, even if the employer has paid for the goods and materials, unless the contractor is their owner he cannot pass title to the employer. A vesting clause in the building contract is only effective to transfer property if the contractor has good title himself.

Dawber Williamson Roofing Ltd v. Humberside County Council

QUEEN'S BENCH DIVISION

(1979) 14 BLR 70

The council contracted in JCT 63 form with a main contractor for the erection of a school. The main contractor sub-contracted the roofing work to the plaintiffs on the 'blue' form of sub-contract which provided by clause 1(1) that 'The sub-contractor shall be deemed to have knowledge of all the provisions of the main contract . . . '. The 'blue' form did not expressly provide when property in unfixed goods and materials was to pass to the main contractor, although clause 14(1) of JCT 63 provided that, subject to certain conditions, where the value of any unfixed goods or materials 'has been included in an interim certificate under which the contractor has received payment, such materials and goods shall become the property of the employer'. The plaintiffs delivered 16 tons of slates to the site, and their value was included in an interim certificate issued under the main contract. The main contractor did not pay the plaintiffs and subsequently became insolvent. The council refused to return the slates, maintaining that they were their property. The plaintiffs sued for the return of the slates and damages, or the value of the slates.

Held: The plaintiffs were entitled to succeed as title in the slates had not passed to the main contractor. Clause 1(1) of the sub-contract did not have the effect of making main contract clause 14(1) a term of the sub-contract.

MAIS J: The plaintiffs submit that their contract with the main contractor was a 'supply and fix' contract. They were not selling the slates to the main contractor. However, there was no privity of contract between the plaintiffs and the defendants, and my attention was drawn to *Hudson's Building Contracts*, 10th edition, at page 742, and the following pages and authorities.

The defendants drew attention to clause 1(1) of the sub-contract between the plaintiffs and the main contractor. By this clause the plaintiffs are deemed to have notice of all the provisions of the main contract. Whilst this is so, that does not, in my judgment, mean that there is any privity of contract between the plaintiffs and the defendants. [Counsel for the plaintiffs], however, submitted that material brought on to the site on a 'supply of material and fix' contract remained the property of the sub-contractor until it was fixed. He referred me to *Tripp* v. *Armitage* (1839) . . . I would refer also to the case of *Seath* v. *Moore* (1886), where Lord Watson said:

> Materials provided by the builder and portions of the fabric, whether wholly or partially finished, although intended to be used in the execution of the contract, cannot be regarded as appropriated to the contract or as 'sold' unless they have been affixed to or in a reasonable sense made part of the corpus . . .

and unless there be any good reason to the contrary, this appears to me to conclude the case against the defendants.

The defendants submit, however, and contend that they are entitled to the property in the slates, and they rely, first of all, on clause 14(1) of the main contract, whereby:

> any unfixed materials and goods delivered to, and placed on, or into the works, shall not be removed except for use upon the works, unless consent in handwriting be given, and when the value of the goods has been included in any interim certificate under which the contractor has received payment, such material and goods shall become the defendants' property.

In my judgment, this presupposes there is privity between the defendants and the sub-contractor, which there is not in the present instance, or the main contractor has good title to the material and goods. If the title has passed to the main contractor from the sub-contractor, then this clause has force.

Secondly, as I indicated earlier, the defendants at one time relied upon section 25(2) of the Sale of Goods Act 1893, but [counsel for the defendants] rightly says that he can no longer rely upon that clause. In my judgment, there being no privity of contract between the plaintiffs and the defendants, the defendants are not entitled to claim property in the slates by reason of any clause in a contract they had with the main contractor.

Note: JCT 80 and JCT Nominated Sub-contracts NSC/4 and 4a were amended in an attempt to overcome this decision. See now JCT 98, clause 16.1.

Where goods are supplied to the main contractor under a sub-contract, and the main contractor becomes insolvent before paying, the sub-contractor or supplier has no right to claim against the employer.

Pritchett & Gold and Electrical Power Storage Co v. Currie

COURT OF APPEAL

[1916] 2 Ch 515

Currie contracted with the second defendant for the supply of an electrical installation, which included a storage battery. The second defendant entered into a sub-contract with the plaintiffs for the supply and erection of the battery. The battery was delivered, and shortly afterwards the second defendant became insolvent. The plaintiffs claimed the price of the battery from Currie.

Held: The claim must fail.

Note: Although at first sight the sub-contract appears to be one for 'work and materials' it was in effect a contract for the sale of goods.

PICKFORD LJ: The plaintiffs sued the defendant, Mrs Currie, for the value of materials which they had supplied to the defendant company for the purpose of their being erected by the defendant company on Mrs Currie's premises as an electric battery as part of a much larger contract for the whole electrical installation, amounting to nearly £1,850. The master made an order that without prejudice to any questions between the plaintiffs and the defendant company Mrs Currie should lodge in court a certain sum of money, a portion of the balance of the contract price alleged to be payable by her to the defendant company, and it was ordered that on such lodgment all proceedings should be stayed against Mrs Currie and that the plaintiffs should within fourteen days deliver a statement of their claim to the defendant company who were within fourteen days thereafter to deliver their defence and be at liberty therein to set up any counter-claim. The defendant company had been added on the application of Mrs Currie as defendants. The result of that was that a certain sum of money payable by Mrs Currie in respect of the main contract was paid into court, and the plaintiffs were to deliver a claim by which they were to establish their right to that sum of money as against the defendant company . . .

To establish that right, counsel for the plaintiffs, concedes, quite rightly, that he must establish that he has a lien upon or a right to follow that particular sum. He also concedes that, unless such a right is given to him by the authority of *Bellamy* v. *Davey* (1891), he cannot make out such a right because it is the assertion by a sub-contractor of a lien on the whole of the money payable by the principal debtor to the head contractor under the contract. He admits that, unless that is given to him by the authority of that case, it is contrary to the general rules of law. I do not think that it is in the least necessary for me to consider whether *Bellamy* v. *Davey* is right or is wrong. I think that it is enough to say that, assuming that the decision in that case was absolutely right, it is no authority for the proposal for which the plaintiffs contend in the present case. That was a case in which the contractor and the sub-contractor were concerned only with the article that was supplied by the sub-contractor. No doubt the contractor had charged a larger price to the building owner than the sub-contractor charged to him. But that and that only was the subject of the contract. It may be – I do not say whether it is or is not so – that in that case the sub-contractor may have a lien upon the amount payable to the contractor for the amount due to him, and that only the surplus went to the contractor, but it seems to me that that does not in any way establish that where there is a sum of money payable to a contractor, not in respect of one article only, but in respect of the whole contract which includes many things other than the subject-matter of the sub-contract, the sub-contractor has a lien upon a large sum which is not connected with him except indirectly for the payment of his debt. It seems to me that *Bellamy* v. *Davey* is no authority for such a proposition as that. The plaintiffs have to establish their right as against the defendant company. The defendant company are entitled to have this money paid to them by Mrs Currie unless the plaintiffs have established that they have a lien upon it. The plaintiffs have not established any such lien. Therefore, it seems to me that they have not established that they are entitled to the money as against the defendant company or anybody else. I think that the action fails on that ground.

Vesting clauses

The majority of building contracts contain a clause vesting in the employer the property in unfixed materials on site. Such a clause sometimes extends to the contractor's plant during the currency of the contract. JCT 98, clause 16, is a typical vesting clause. Whether a vesting clause is effective to transfer property to the employer without qualification depends on its terms. See now JCT 2005, clauses 2.24 and 2.25.

Bennett & White (Calgary) Ltd v. Municipal District of Sugar City No. 5

PRIVY COUNCIL

[1951] AC 786

A construction contract between the appellants and the Canadian government provided that 'all machinery, tools, plant, materials . . . and things whatsoever provided by the contractors . . . for the works . . . shall from the time of their being so provided become and until the final completion of the said work shall be the property of His Majesty for the purposes of the said work'. The appellants provided plant and equipment, including motor vehicles, on site. The respondent local authority assessed all the plant to tax, and the appellants contended that His Majesty was the owner of the property and that they were not liable.

Held: The Crown was at all material times the 'owner' of the plant, materials and equipment in question, but in fact there was a re-delivery of possession to the contractors for the purpose of carrying out the works and, therefore, they were 'in legal possession' of the property for the purposes of the taxing statute.

Where a vesting clause provides for certification of the amount payable to the contractor upon delivery of materials, property passes when the certificate is issued.

Banbury & Cheltenham Railway Company v. Daniel

CHANCERY DIVISION

(1884) 54 LJ Ch 265

A contract between the parties for the construction of a railway provided that the engineer should certify, on a monthly basis, the amount payable to the contractor in respect of 'the value of materials delivered', and that such certificates were payable by the employer seven days after presentation.

Held: The property in the materials delivered passed to the company upon their being certified for by the engineer.

Vesting on insolvency

Some building contracts provide that, in the event of the contractor's insolvency, the employer shall be entitled to seize materials and plant on site and vest them in himself for the purpose of completing the unfinished work. Normally, such a clause will be good against the contractor's liquidator (or trustee in bankruptcy if the contractor is not a limited company). Such a provision will be invalid against other third parties, such as the owner of plant which is hired to the contractor.

See further Chapter 18.

Chapter 9
Acceptance and Defects

The building owner is not prevented from complaining of defects in the works by the fact that he has paid the price in full, even where he has so paid to settle an action by the builder for the price.

Davis v. Hedges

COURT OF QUEEN'S BENCH

(1871) LR 6 QB 687

HANNEN J: The plaint was for damages for the non-performance and improper performance of certain works which the plaintiff had employed the defendant to execute.

The defence set up was, that the defendant had sued the plaintiff for the price of the work now alleged to have been improperly done, and that the plaintiff had settled by paying the whole amount then sued for; and as the plaintiff might have given the non-performance and defective performance now complained of in evidence, in reduction of damages, the plaintiff was precluded from bringing a cross action for that which he might have availed himself of as a defence to the former suit.

The county court judge decided in favour of the defendant, holding that all the damages now sued for might have been used in reduction of damages in the former action, and that for anything which might have been so used no cross action could be maintained.

We are of opinion that the decision of the learned judge was erroneous . . .

It appears, from . . . *Mondel* v. *Steel* (1841), that the present practice of allowing the defence of the inferiority of the thing done to that contracted for to be applied in reduction of damages was introduced (on the same principle as the statutes of set-off were passed) for the benefit of defendants. It would greatly diminish the benefit, and in some cases altogether neutralise it, if the defendant was not allowed an option in the matter. The hypothesis is, that the plaintiff suing for the price is in default. The conditions on which he can bring his action are usually simple and immediate. The warranted chattel has been delivered, or the work contracted for has been done, and the right to bring an action for the price, unless there is some stipulation to the contrary, arises. On the other hand, the extent to which the breach of warranty or breach of contract may afford a defence is usually uncertain; it may take some time to ascertain to what amount the value of the article or work is diminished by the plaintiff's default. It is unreasonable, therefore, that he should be able to fix the time at which the money value of his default shall be ascertained. In many cases the extent to which the value of works may be diminished by defect in their execution may be altogether incapable of discovery until some time after the day of payment has arrived. Surely the right to redress for the diminution of value, when discovered, ought not to depend on the accident whether the contracting party in the wrong had or had not issued a writ for the price.

Another inconvenience which would result from holding that the inferiority of the thing done to that contracted for must, if an action be commenced, be used by way of defence, is, that instead of furthering the object for which this defence was permitted, namely, the prevention of circuity of action, it would in many cases tend to complicate and increase litigation. The cases are, perhaps, rare in which the consequences of defective performance of work are limited to the depreciation of the value of the work done; they usually involve consequential damage by reason of the necessity of repairing the defective work; and for this the case of *Mondel* v. *Steel* decides a separate action must be brought. Parke B., there says: 'All the plaintiff could by law be allowed in diminution of damages on the former trial was a deduction from the agreed price, according to the difference at the time of delivery between the ship as she was, and what she ought to have been according to the contract; but all claim for damages beyond that on account of the subsequent necessity for more extensive repairs could not have been allowed in the former action, and may now be recovered.' So that two litigations instead of one would be frequently necessary, with the additional inconveniences that it would be very difficult to discriminate in each action between that which ought to be allowed in diminution of price, and that which should be compensated for as special damage.

Where one party fails substantially to perform the contract, the other party may be entitled to terminate it, but is not obliged to do so. In some circumstances, the conduct of that other party may amount to an acceptance of the work done, which will give rise to an obligation to pay for it.

Tannenbaum and Downsview Meadows Ltd v. Wright-Winston Ltd

ONTARIO COURT OF APPEAL

(1965) 49 DLR (2d) 386

The defendant was the builder of a housing project and was required under the terms of his mortgage to construct a twelve inch trunk sewer from the project to a sewage treatment plant some 3,000 ft away. The approximate cost was $65,000 ($38,000 for the sewer and the balance for a pumping station). The plaintiffs owned land nearby and agreed to build a 36in trunk sewer and pumping station to serve both properties in return for the defendant contributing $55,000 to the cost. The plaintiff built the sewer at a cost of $110,000 but, in breach of contract, did not build the pumping station, apparently because of a new plan under which the local authority was to extend the trunk sewer to a new and larger treatment plant. The defendant decided that it was too expensive to build a pumping station as a temporary measure and instead built a 'pumping main' at a cost of about $18,000 and cut into the sewer constructed by the plaintiffs.

Held: Although the plaintiffs' failure to build the pumping station was a breach of contract of a kind which entitled the defendants to terminate the contract, they had elected not to and by using the sewer they became bound to pay the agreed price, less the cost of building the 'pumping main'.

MCLENNAN JA: I turn now to the main ground of appeal which is that the plaintiffs contracted to build the works for a lump sum, that is, it was an entire contract which the plaintiffs abandoned for their own purposes and are not entitled to claim under it on the principle stated in *Sumpter* v. *Hedges* (1898) and *Bradley* v. *Horner* (1957). A subsidiary proposition is that there is no evidence upon which to fix any amount on a *quantum meruit*. With that subsidiary proposition I agree and the plaintiffs' case must stand or fall on a claim under the contract.

In *Sumpter* v. *Hedges*, the plaintiff, a builder, agreed to erect buildings on the defendant's land for a lump sum. After he had done part of the work he abandoned the work for lack of funds. The defendant completed the buildings. In *Bradley* v. *Horner*, the plaintiff, a painter, agreed to paint the interior of a house for a lump sum but abandoned the work before it was half finished. In both cases the actions were dismissed because of the rule that where there is a contract to do work for a lump sum the price cannot be recovered until the work is completed. Other cases such as *Dakin & Co. Ltd* v. *Lee* (1916), and *Hoenig* v. *Isaacs* (1952), illustrate the modification of the rule. In cases where there is substantial performance of the contract, a defendant who gets the benefit of substantial performance is liable on the contract for the agreed price less the cost of making good the omissions.

There was no substantial performance in this case. The construction of a pumping station was a term of the contract, performance of which was essential to its completion because the whole purpose of the contract would be defeated if the pumping station were not erected. In *Sumpter* v. *Hedges* and cases of a like nature, where there was no substantial performance, the argument was advanced that because the defendant took possession and received the benefit of the plaintiff's work he ought to pay for it on a *quantum meruit* but as the judges in *Sumpter* v. *Hedges* pointed out the defendant was always in possession and had no option as to whether he would take the benefit of the work or not.

There is, of course, a vital distinction between the facts in *Sumpter* v. *Hedges* and in *Bradley* v. *Horner* and this case. The works which the plaintiffs agreed to construct were not on the defendant's land and the contract was in the nature of a joint venture in that the completed works were for the mutual use and advantage of both parties and the defendant accepted the advantage and used the works. In *Sumpter* v. *Hedges*, A.L. Smith LJ, in relation to the facts of the case said, 'the plaintiff having abandoned the contract, what was the owner to do? He cannot keep buildings in an unfinished state on his lands'. The answer to the same question put about the defendant in this case is I think quite a different one. The works were not being constructed on the defendant's land, and the defendant did not have to accept the benefit of the works constructed by the plaintiffs but could have constructed its own sewer on the easements it held when it became aware of the plaintiffs' breach.

Counsel for the defendant laid great stress on the conclusion of the learned trial judge that the plaintiffs had abandoned the contract, as if abandonment discharged the contract and put an end to the obligations on both sides. It may well be that the inference made by the learned trial judge that the plaintiffs abandoned the contract is warranted but whether it was an abandonment or a refusal or neglect to perform an essential term is largely, I think, a question of semantics. I would prefer to put it this way. Because the plaintiffs did not perform an essential term of the contract the defendant had a right to terminate it and put an end to its obligation but until it did the contract was not discharged nor was the defendant's obligation discharged.

The authorities to which I refer for this proposition are recent decisions of the House of Lords: *Heyman* v. *Darwins Ltd* (1942), and *White and Carter (Councils) Ltd* v. *McGregor* (1962).

The rule laid down in those cases is that where one party repudiates a contract by making it clear to the other party that he will not perform his obligations the other party has a choice. He may accept the repudiation and, if he chooses, sue for damages for breach or he may disregard the repudiation and the contract remains in full force. Repudiation by one party does not terminate the contract. Termination requires repudiation by one party and acceptance of the repudiation by the other.

The effect of that rule on this case is this. When the defendant became aware that the plaintiffs were not going to build a pumping station it could have notified the plaintiffs that it was treating their breach as discharging the contract and its obligation to pay, constructed its own sewer and pumping station and brought an action for damages for the difference between its cost and the contract price of $55,000. It will be recalled that the evidence disclosed the approximate cost to the defendant of the 12in sewer and pumping station was $64,000. The defendant did not adopt this course. It took the plaintiffs over 8 weeks to construct the sewer – from 16 July to 13 September. The defendant knew of the plaintiffs' breach at the end of the first 2 weeks of that period. It did nothing while the plaintiffs were completing the most costly item and after that part of the works had been finished by the plaintiffs the defendant completed the works, used what the plaintiffs had constructed and accepted the benefits thereof. I am therefore of the opinion that the learned trial judge was right in holding that the defendant was liable on the contract. This result is consistent with the dicta of Denning LJ in *Hoenig* v. *Isaacs* (1952). While that was a case of substantial performance of a contract to decorate and furnish a flat Denning LJ held that, even if entire performance was a condition precedent to payment, by using the defective furniture the defendant waived the condition and must pay the contract price subject to a deduction to remedy the breach.

Chapter 10
Approval and Certificates

General

Unless the terms of the contract specify a set form of certificate, all that is required is that 'it must clearly appear that the document relied upon is the physical effect of a certifying process'. One should, therefore, have some regard to the factors of 'form', 'substance' and 'intent' of which Devlin J spoke in . . . *Minster Trust Ltd* v. *Traps Tractors Ltd* (1954) . . . The document should be 'the expression in a definite form of the exercise of the . . . opinion . . . of the . . . architect . . . in relation to some matter provided for by the terms of the contract': per Edmund Davies LJ in *Token Construction Co Ltd* v. *Charlton Estates Ltd* (1973).

Cantrell v. Wright and Fuller Ltd

QUEEN'S BENCH DIVISION

(2003) 91 Con LR 97; [2003] EWHC 1545 (TCC)

The parties entered into a contract on 1 April 1997 for the construction of phase 1 of a new extension to a nursing home in Suffolk on JCT 80 terms. The claimants were the owners of the nursing home and the defendants the contractors. The contract works achieved practical completion on 23 February 1998, the defects liability period expired on 23 August 1998 and a document relied on by the defendant as a final certificate was issued on 29 March 1999.

The claimant alleged defects in the works and there was a reference to arbitration. The arbitrator decided as a preliminary issue that the certificate was a final certificate and the claimants appealed.

JUDGE ANTHONY THORNTON QC:

87. Thus, the certificate must be one which clearly expresses the relevant opinion of the architect in a form that shows that the opinion is that of the architect, is the one which the contract calls for and which addresses and only addresses the matters called for.

Powell-Smith and Furmston's Building Contract Casebook, Fifth Edition. Michael Furmston.

88. If the certificate is, on its face, clear and unambiguous, there will be little or no need to consider extraneous contemporary material in order to be satisfied that it fulfils the form, substance and intent test. However, where, as here, the certificate is itself ambiguous, recourse may be made to any covering letter or other contemporaneous document which was produced with or as part of the certifying process so long as that additional document is properly to be regarded as being issued as part of the certificate.

89. If it is contended that the certificate was not the product of the architect and therefore not an expression of his own intent, additional evidence will be both admissible and required. A particular example of this situation would be where, as in this contract, it is alleged by the employer that the necessary measurements and valuations have been wholly delegated to a subordinate or sub-contracted to a quantity surveyor, who had no authority under the contract to undertake an independent role in valuations, to such an extent that the resulting certificate was in no sense the expression of opinion of the certifier but was that of someone to whom he had abnegated all effective responsibility. If such a delegation had occurred, the resulting final certificate would not be, in intent, that required by the conditions.

90. A question can arise, as in this contract, as to whether the certificate is in form invalid or whether it remains valid despite containing an obvious error or departure from the contractual requirements for such a certificate. What, in other words, is the dividing line between substantial errors or inaccuracies that invalidate from inconsequential errors that do not? This question was considered in two decisions, one cited to the arbitrator and one raised by me with the parties during the course of argument. These were the *Hosier & Dickinson* case already referred to and raised by me during the course of the argument and *Emson Contractors Ltd* v. *Protea Estates Ltd* (1987) 13 ConLR 41, cited to the arbitrator.

91. In the *Hosier & Dickinson* case, the final certificate was initially issued with the employer named as Transloyd Ltd because of an erroneous belief that that company was to be treated as the successors of the employer. That certificate was invalid and a subsequently issued certificate in the name of Kaye was treated as the final certificate. In the *Emson Contractors* case, the certificate referred to the contractors as Emson Construction Ltd whereas the contract was in fact with Emson Contractors Ltd. Moreover, the certificate recited that the date of the contract was 16 October 1984 whereas it had in fact been 23 October 1984. Judge Fox-Andrews found that these errors were immaterial and did not, nor would reasonably have, misled anyone. The validity of that final certificate was, in consequence, upheld.

92. In summary, therefore, an error or a departure from the contractual requirements in a certificate will only invalidate the certificate if its nature or effect is such that it is no longer clearly and unambiguously the required certificate in form, substance or intent or if, applying an objective standard, the error does not mislead or does not have the potential of misleading either of the parties to whom it is addressed as to its form, substance or intent.

93. Overall, the construction of the certificate in question and any other document to be read with it is subject to the same rules of construction as any commercial contract. Thus, that construction exercise is to be undertaken against the factual background surrounding its issue known to both parties. This is because both parties have agreed that they will be bound by, and abide by the terms of, the final certificate issued by a third party, namely the architect. This construction exercise is to be undertaken as part of a consideration of the meaning and effect of the certificate as a whole, of any parts of it and of any potential error or omission that it might contain. . . .

95. The JCT 80 conditions constitute a detailed and carefully crafted set of interrelated clauses that define a series of interlocking events and periods of time which involve the close collaboration and co-operation of the parties, the architect and the party undertaking the functions of the quantity surveyor which are intended to culminate in the issuing of the final

certificate. Following practical completion, these events include the identification and making good of defects and any associated valuation of an abatement to take account of defects; the defects liability period and the certificate of making good defects; the operation of the provisions concerned with the final determination of the contractor's extension of time entitlement; the finalisation of the completion date and the operation of the machinery concerned with the payment of liquidated damages by the contractor; the accounting procedures, including the provision of documents, that lead to the finalisation of the adjusted contract sum and which involve the contractor and each nominated sub-contractor; the procedures concerned with the final payment to each nominated sub-contractor; the final ascertainment of the amount of value added tax to be paid by the employer in relation to the adjusted contract sum; and, finally, the preparation and issuing of the final certificate itself.

96. If either party disputes the content of the final certificate or the make-up of the adjusted contract sum, particularly with regard to any valuation, to any allowance for defects, to any decision it records as to the quality of workmanship or to extensions of time, there is a defined window of time immediately following the issue of the final certificate in which either party may commence arbitration or other proceedings and, if this is not done, the final certificate is conclusive evidence in relation to the four categories of dispute provided for in clause 30.9.1.

Role of the architect or certifier

97. Each of the events provided for by the conditions involve, in varying degrees, the employer, contractor, architect, quantity surveyor and each nominated sub-contractor and each event is linked to a preceding and succeeding event. The conditions contain a significant number of time limits and time requirements that relate to the period of time within which, or the date by which, particular events are to be performed that are expressed in different but invariably mandatory terms.

98. The majority of the events that I have referred to involve the architect, as certifier, in the performance of administrative functions which usually include his having to issue a certificate or other decision recording his conclusion as to a particular state of affairs or as to a valuation he has performed. Some of these decisions involve the quantity surveyor but, in this contract, the functions of the quantity surveyor were to be performed by the architect.

99. In undertaking these functions, the architect does not act as the agent of the employer but, since he is engaged by the employer, he has a contractual obligation to act fairly, impartially and in accordance with the powers given to him by the conditions. The employer may not interfere in the timing of the issue of any certificate but is not himself in breach of contract if a particular certificate is not issued or is erroneous unless he is directly responsible for that failure. However, if and when it comes to his notice that the architect has failed to comply with his administrative obligations, by for example failing to issue a certificate required by the contract, the employer has an implied duty to instruct the architect to perform that function in so far as it remains within the power of the architect to perform it and the employer is in breach of the contract with the contractor to the extent that he does not intervene to arrange for the correct or a correcting step to be taken by the architect. The architect's powers in relation to these administrative steps are derived exclusively from the conditions of contract, he has no power to act in any other way than as defined by the conditions and once the last step, the issuing of the final certificate, has been taken, his authority to act and his role under the contract cease. He then becomes, in traditional language, *functus officio*.

Most modern building contracts contain provisions that the work is to be executed to the satisfaction or reasonable satisfaction of the architect. When so certifying, the architect must act independently as between employer and contractor. Whether or not the work is to the 'reasonable satisfaction' of the architect is to be judged objectively.

Minster Trust Ltd v. Traps Tractors Ltd

QUEEN'S BENCH DIVISION

[1954] 3 All ER 136

This case involved a contract for the sale of re-conditioned machines, under which the sellers were to provide a 'Hunt Engineering Certificate that [the machines] have been fully reconditioned to [Hunt's] satisfaction'. The following abstract from the judgment deals with certifiers generally.

DEVLIN J: There is no general rule of law prohibiting the influencing of certifiers. Apart from fraud, a duty not to influence can only be imposed by an implication arising from contract. Such an implication must, in accordance with settled principles, be both reasonable and necessary, and the contract must be examined to see what it yields in this respect. There is, after all, nothing to prevent a party from requiring that work shall be done to his own satisfaction. He might then choose to act on the recommendation of an agent. If an agent is named in the contract, it may be plain that he is to function only as the alter ego of his master, and then his master can tell him what to do: see *Ranger* v. *Great Western Railway Co* (1854) per Lord Cranworth LC and per Lord Brougham. Whether it be the act of the master or the servant, there may be a question (again depending on the implication to be drawn from the contract) whether the dissatisfaction must be reasonable, or whether it can be capricious or unreasonable so long as it is conceived in good faith. The tendency in modern cases seems to be to require the dissatisfaction to be reasonable.

What has to be ascertained in each case is whether the agent is or is not intended to function independently of the principal. The mere use of the word 'certificate' is not decisive. Satisfaction does not necessarily alter its character because it is expressed in the form of a certificate. The main test appears to be whether the certificate is intended to embody a decision that is final and binding on the parties. If it is, then it is in effect an award, and it has the attributes of its arbitral character. It cannot be attacked on the ground that it is unreasonable, as the opinion of a party or the certificate of one who is merely an agent probably can. On the other hand, it must be made independently, for independence is the essence of the arbitral function. If two parties agree to appoint an arbitrator between them, it would be, I think, implied in the contract in order to give it business efficacy (it is not now in practice necessary to consider the point because of the control over the arbitrator which the Arbitration Acts give to the court) that neither side would seek to interfere with his independence. If a party to a contract is permitted to appoint his agent to act as arbitrator in respect of certain matters under the contract, a similar term must be implied; but it is modified by the fact that a man who has to act as arbitrator in respect of some matters, and as servant or agent in respect of others, cannot remain as detached as a pure arbitrator should be.

There is another distinction between certifiers. If work under a contract has to be completed to the satisfaction of a certifier, it may mean that his duty is merely to see that the requirements of the contract are met, or it may mean that he is entitled to impose a standard of his own. It may be that his standard is that to which the parties submit and that it constitutes the only provision in the contract about quality; or it may be that his standard is an added protection, so that performance under the contract must satisfy both the contract requirements and the certifier . . . So far as I am aware, there is no case in which a certifier who has to certify only according to his own standards has been held to be an arbitrator. It is not, however, possible under this head to draw a hard and fast line between the two categories. There must always be some room for the discretion of the certifier, however tightly the contract may be drawn. Even if he has to certify only whether an express term has been complied with, there must be room for the exercise of his judgment. *Cammell Laird & Co. Ltd* v. *Manganese Bronze & Brass Co Ltd* (1934), is a case in which it was held that the satisfaction of a third party operated only within the requirements of the contract; that is to say, it was not permissible for the third party to put on the seller obligations as to quality outside the contract. Lord Wright said:

> It is true that the clause cannot be intended to deal with matters outside the scope of the respondents' obligations under the contract, or to enlarge these obligations except in so far as these are enlarged because there is super-added the judgment of a third party.

Allied to this, there is a distinction between a certificate that is given only for the purposes of a particular contract and one that, although it may be called for by a particular contract, is not particularly related to that contract. The certifier may not then be an agent of either party. That is so in the case of a Lloyd's certificate or of a certificate by a public analyst, or others that may be obtainable on a fee. It has now become quite common to include among documents which have to be tendered against payment a certificate of inspection or of quality. Certificates of this sort are addressed to all the world or to all who may be concerned. If the phrase is not used as more than a label, they might be called certificates *in rem* as compared with certificates *in personam*, which deal only with particular contracts and are addressed only to particular parties. The former carry the same meaning to all who read them. The latter may have to be interpreted in the light of particular contractual requirements or of information known only to the addressees. A certificate *in rem* certifies a standard of quality extraneous to the contract. It may be the certifier's own standard or it may be taken from some public or independent source. A certificate *in personam* may be based on the certifier's own standards or on standards prescribed by the contract. It may be a certificate of quality or a certificate that a contract has been carried out.

Finally, there is a distinction between a certifier whose function has been completed before the contract is entered into and one who has a function to perform under the contract. In the latter case there is room for an implied undertaking that he will not be improperly influenced. In the former case there can be no room for an undertaking that relates to the future. If anything is to be implied, it must be a warranty about the past. There may be grounds for implying a warranty that the certifier has not been improperly influenced, but, so far as I know, the point has never been decided. All the leading cases on the topic, such as *Hickman & Co* v. *Roberts* (1913) and *Panamena Europea Navigacion (Compania Limitada)* v. *Frederick Leyland & Co Ltd* (1947) deal with certifiers whose duties arose out of the contract. It seems clear that there cannot be the same necessity for implying a warranty in respect of the past as there is for implying an undertaking for the future. If a building owner makes the obligation of payment conditional on his architect's certificate of quality, he must not, for example, instruct his architect not to be content with less than three coats of paint, for he has impliedly undertaken that he will leave his architect free to judge independently, either by reference to the contract requirements or to his own standards of quality, whichever the case may be, how many coats of paint are required. But if he sells a completed house and agrees to supply with it his architect's certificate of quality, does he thereby warrant that he has not told his architect to be content with two coats of paint? It may be argued that there is no necessity for an implied warranty of that sort, and that the buyer is sufficiently protected by the fact that the seller would, in his own interests, have been concerned to get the best quality he could.

Panamena Europea Navigacion Compania Limitada v. Frederick Leyland & Co Ltd

HOUSE OF LORDS

[1947] AC 428

Leyland had contracted to repair Panamena's ship, payment to be made after the issue of a certificate by Panamena's surveyor 'that the work has been satisfactorily carried out': clause 7. The surveyor contended that these words meant that his function was not confined to certifying the actual quality of the works done, but also covered the amount and value of the materials and labour used. He refused to grant a certificate unless he was supplied with information about these points, to which he considered himself entitled. Leyland had done everything under the contract which they were bound to do. Panamena supported their surveyor's view.

Held: The contention of the surveyor was wrong and, in the circumstances, Leyland were entitled to recover the amount due to them without the surveyor's certificate.

Note: The undermentioned extract is taken from the judgment of the Court of Appeal: see (1943) 76 Lloyds LR 113.

SCOTT LJ: The problem calls for answers to two questions: (1) What is the scope of the function assigned to the certifying third party by the agreement of both the contracting parties? (2) What, if any, is the contractual undertaking, express or implied, of the party whose servant, or agent, or nominee such third party expert is? . . . As I have already indicated, the submission of the appellants, made both here and below, is that the function of the surveyor, as defined by the contract was not merely to consider the quality of the work carried out by the respondents and then to express his satisfaction or dissatisfaction therewith, but to investigate the value of that work in relation to the amounts charged by the respondents . . . The respondents, on the other hand, contend that the appellants' surveyor was concerned only with quality . . . The proper interpretation of the sentence in clause 7 which defines the surveyor's function assists materially in answering my second question, namely, is there any and what implied undertaking by the shipowner towards the repairer that the surveyor shall keep within the scope of the function assigned to him by the contract? That question of construction seems to me a fundamental one in the present contract; and I cannot help thinking that it is one which would have helped greatly towards the decision of many cases of building and similar contracts which have been before the courts, and have been decided without conscious or express reference to it, as an aid to adjudication . . .

The first thing seems to me to be to define the legal relationship between [the surveyor] and the appellants, for it is primarily on the right solution of this problem of construction

> that the legal consequences of the absence of [the surveyor's] certificate in the light of the actual facts of the case must depend. To shut out common complications which do not arise in the present case, it is important to note first that the surveyor appointed by the shipowners under the present contract, [the surveyor] (and he was obviously so appointed with the approval of the repairers), is not either an arbitrator, or a quasi arbitrator (a phrase which has no clear meaning and is best avoided), but only an expert whose opinion in regard to the quality of the work done, the contractor has to satisfy, as a condition precedent to the owners' undertaking to pay; secondly, that there is no arbitration clause to which the validity of the surveyor's action or inaction can be submitted by the aggrieved party; and thirdly, that he is put in the position of a person whose duty it is to act with scrupulous independence of both parties.

Where the contract provides for the issue of certificates by the architect, it is an implied term that the employer will do nothing to prevent the independent exercise of the architect's powers as certifier. If he does so, the need for a certificate is dispensed with and the contractor can sue without a certificate.

Hickman & Co v. Roberts

HOUSE OF LORDS

[1913] AC 229

The employer instructed the architect not to issue a certificate until he had received the contractor's account for extras. The architect wrote to the contractor saying: 'Had you better not call to see my clients, because in face of their instructions to me I cannot issue a certificate whatever my own private opinion in the matter'.

Held: The employer had improperly influenced the architect. The contractor was entitled to payment even in the absence of a certificate.

It is a further implied term of the contract that the employer will do all that he reasonably can to ensure that the architect properly performs his duties as certifier.

Perini Corporation v. Commonwealth of Australia

SUPREME COURT OF NEW SOUTH WALES

(1969) 12 BLR 82

A building contract contained a provision that the Director of Works, a government employee, could extend the time for completion 'if he thinks the cause sufficient . . . for such period as he shall think adequate'. The contractors applied to the Director of Works for extensions of time, but many of these applications were refused because of departmental policy.

Held: As certifier, the Director of Works had duties imposed on him under the contract. He had a discretion as to whether or not to grant an extension of time, and was entitled to rely on other people to supply him with information on which to exercise his discretion. He was bound to give his decision within a reasonable time and was not entitled to defer it. Departmental policy could properly be taken into account, but he should not regard it as controlling him. Terms were implied into the contract that the employer would not interfere with the Director's duties as certifier and that the employer would ensure that the Director did his duty as certifier.

MACFARLAN J: I must . . . express my opinion upon an important argument submitted by counsel for the defendant. This argument was that, not only was there not any authority justifying the implication of a term in an agreement of this type but that in fact the authorities were to the contrary. The argument was principally founded upon *Hickman & Co* v. *Roberts* (1913), and *Panamena Europea Navigation Compania Limitada* v. *Frederick Leyland & Co Ltd* (1947). It was argued that these authorities decided that if there had been a wrongful, in the sense of unauthorised, exercise of the powers by a certifier with the knowledge of the employer of the certifier, the employer being the other party to the contract pursuant to which the certifier was appointed, the only right of the contractor was that he was entitled to disregard the provisions of the agreement with respect to time and either to sue for the price or resist a claim for liquidated damages by way of penalty: *Dixon* v. *South Australian Railways Commissioner* (1923). While it is, in my opinion, the law that a contractor is entitled to disregard the provisions of the agreement with respect to time . . . it does not follow, nor has it been decided that if the contractor has otherwise suffered damage he is not entitled to sue upon an implied term . . . I was referred to the decision of the Court of Appeal in *Panamena* . . . In that case the certifier adopted a wrong understanding of his functions and his employer, which was the other contracting party, adopted and encouraged the understanding of the certifier. In the result the certifier did not issue his certificate and the claim of the plaintiff was for the contracted price, or in the alternative, for damages in the same amount. The claim of the plaintiff was based in the alternative upon the implication of a term in the agreement that the defendant, building owner, would not encourage or influence the certifier wrongly to withhold the issue of his certificate. The defence was that a certificate had not been issued and that in its absence and in accordance with the provisions of the agreement the plaintiff was not entitled to recover . . . Scott, LJ, said that the problem for decision called for answers to two questions which he stated as follows:

> (1) What is the scope of the function assigned to the certifying third party by the agreement of both the contracting parties?
> (2) What, if any is the contractual undertaking, express or implied, of the party whose servant, or agent, or nominee such third party expert is?

The learned Lord Justice also said:

> Let us consider the position. The repairers were entitled to rely on the surveyor doing what the contract said he was to do, that is, keeping to matters of quality only. It seems to me plain if the shipowners had known that he was departing from his proper function under the contract, it would have been their duty to stop him and tell him what the function was for which the contract provided . . . It obviously was not the contractual duty of the repair-

> ers to bring him to book. It is equally obvious that they would count on his carrying out his proper function. In those circumstances I think the Court ought to imply an undertaking by the owners that in the event of its becoming known to them that their surveyor was departing from the function which both parties had agreed he was to perform, they would call him to book, and tell him what his real function was. This seems to me an implication exactly on the lines of all the authorities on implied terms from *The Moorcock* (1889) down to *Luxor (Eastbourne) Ltd* v. *Cooper* (1941).

At the same page, Scott LJ said:

> In the result, I am of opinion that they were under a contractual duty to keep their surveyor straight on the scope of what I metaphorically called his 'jurisdiction' by which I do not mean that he was in any sense an arbitrator, but only that as an expert entrusted with the duty of impartiality within a certain sphere he had to form his opinion with judicial independence within that sphere. It follows from my premises that in failing to inform Dr Telfer that he was going outside and away from the limits of his function, they broke their implied undertaking.

Luxmoore LJ agreed with the judgment of Scott LJ and also with that of Goddard LJ. Goddard LJ as I read his judgment . . . agreed with the construction of the contract stated by Scott LJ, and did not propose to add to it. Accordingly it is apparent that all members of the Court held the same opinion on the construction of the contract. However, Goddard, LJ then dealt with another question, namely, whether the defendant was entitled to rely on the absence of a certificate and he held, applying the principles of *Hickman & Co* v. *Roberts* (1913) that the defendant was not so entitled. Both these grounds were stated as grounds for dismissing the appeal in the Court of Appeal. It is significant, I think, that although the claim of the plaintiff was stated in the alternative to be for damages for breach of the implied term the damages claimed were the same amount as was claimed firstly for the contract price. It was accordingly sufficient in order that the case might be decided in favour of the plaintiff. In that sense *Panamena Europea Navigation Compania Limitada* v. *Frederick Leyland & Co Ltd* (1947) is different from the present case where the plaintiff, having been paid its contract price claims other damages for breach of the implied term which all members of the Court of Appeal in *Panamena* held to have existed.

When the case came to the House of Lords . . . the appeal was dismissed and it is true . . . that the House of Lords did not deal in any way with the plaintiff's claim based upon the existence of an implied term . . . The defendant submitted that this was because the House of Lords was of the opinion that there was not any implied term in the contract and he relied upon the approval given by Lord Thankerton, with whose speech the other noble and learned Lords concurred, to the judgment of Goddard LJ, in the Court of Appeal. However, in my opinion the argument of the plaintiff is correct about the true extent of the decision of the House of Lords in *Panamena*. [From] the report it appears that specific questions were submitted to their Lordships in the course of argument and that the argument of counsel and the subsequent decision of the House was confined to those points. The fourth point only is material and that is stated in the following words:

> Are the respondents entitled to recover the amount claimed in the action – and not merely such amount as may be held to be a reasonable price – without producing a certificate of the appellant's surveyor in pursuance of clause 7 of the contract?

The House of Lords answered this question in the affirmative and dismissed the appeal. In my opinion in so treating the appeal the House of Lords directed its attention to the alternative, albeit the first alternative ground, upon which the plaintiffs in the action rested their claim. The noble and learned Lords expressed their opinions on that point and in my opinion it is correct to say they approved the judgment of Goddard LJ . . .

I find some comfort in the conclusion I have reached about *Panamena* from the circumstance that Lord Devlin was counsel in *Panamena*, both at the trial and in the two appeals. That learned judge referred to *Panamena* in his judgment in *Minster Trust Ltd* v. *Traps Tractors Ltd* (1954), and it would be surprising in the circumstance if, having referred to *Panamena*, without making any such criticism as a judge of first instance may be entitled to make of a decision of the House of Lords, he should then have added in the course of his judgment the passages which I have cited about the implication of a term in such a contract. In my opinion Lord Devlin must have held the same opinion of the effect of the *Panamena* judgments as I have expressed.

I will now consider the affirmative aspect of the term which . . . the plaintiff argued must be implied in this agreement. Fundamentally he argued it is essential to consider this aspect against the background that the Director of Works was at all times a servant of the [government]; that the position he occupied as certifier was part of the machinery set up by the agreement of the [government] and the plaintiff for certifying applications for extension of time. As such a servant of the [government] the Director of Works was obliged by law to obey all lawful orders of [his employer] . . . If, for example, the [government] were to say to the Director: 'You are ordered to act in a particular manner', and if that order constituted a breach of the contractual mandate conferred on him by clause 35, the Director would none the less be still obliged to act in compliance with the [government] order. Against this background I will assume that the certifier, with the knowledge of both parties to the agreement, acted in breach of his obligation. Could the [government] in those circumstances say: 'I will not do anything to insure that he carries out these duties'? In my opinion an application of the test stated in *The Moorcock* (1889), would require the answer that the [government] was contractually bound to order him to carry out his duties. In my opinion the plaintiff and the defendant, being the parties bound by his agreement, are bound to do all co-operative acts necessary to bring about the contractual result. In the case of the defendant this is an obligation to require the Director to act in accordance with his mandate if the defendant is aware that he is proposing to act beyond it. In *Mackay* v. *Dick* (1881) Lord Blackburn said:

> I think I may safely say as a general rule that where in a written contract it appears that both parties have agreed that something shall be done which cannot effectually be done unless both concur in doing it, the construction of the contract is that each agrees to do all that is necessary to be done on his part for the carrying out of that thing, though there may be no express words to that effect.

I am accordingly of the opinion that a term must be implied in the present agreement binding the defendant to insure that the Director of Works, its servant, performs his duties under clause 35 in accordance with this mandate.

The liability of an employer for delay by the architect in issuing interim certificates depends on the knowledge of the employer.

Hong Huat Development Co (Pte) Ltd v. Hiap Hong & Co Pte Ltd

SINGAPORE COURT OF APPEAL
(2000) 82 Con LR 89

This case involved an appeal from an arbitrator who had held that the employer was liable for late certification by the architect. The court held that such liability depended on the employer's knowledge.

CHAO HOCK TIN JA: We entirely agree with the learned judge that the employer could be liable for the default of the architect in issuing the interim certificates but only if the employer was aware of such default. This proposition of law is amply supported by authorities: *Frederick Leyland & Co Ltd* v. *Cia Panamena Europea Navigacion Ltda* (1943) 76 Ll L Rep 113 and *Perini Corp* v. *Commonwealth of Australia* (1969) 12 BLR 82. The prerequisite of knowledge is essential before an employer is required to act to ensure that his architect complies with the terms of the building agreement between the employer and the contractor. The standard works on the subject have also reiterated this principle, e.g. *Hudson's Building and Engineering Contracts*, *Emden's Construction Law* and *Keating on Building Contracts*.

The rationale for this rule is obviously to take into account the special position of the architect in a building contract, even though he is employed and paid by the employer. When an architect acts as a certifier under a building contract for interim payments, he is to act fairly and impartially between the parties, and the employer is not to interfere with the architect's exercise of this function. The arbitrator, in fact, recognised this point in his award. Such functions of the architect, among others, must be exercised by him independently as an expert. It would be unreasonable to expect a lay employer to warrant the performance of the architect in respect of such functions without establishing that the employer knew the architect had gone wrong: see *Lubenham Fidelities and Investment Co* v. *South Pembrokeshire DC* (1986) 6 ConLR 85 at 99.

From the reasoning of the arbitrator which we have cited above, it is clear that the arbitrator has laid down the duties of an employer in too wide a term when he said that 'the employer is thus liable for any breach of this (certifying) duty on the part of the architect'. The issue is not simply whether there could be implied a term under which the appellants as employers could conceivably be liable for the architect's default, but more specifically, what is the nature and/or extent of that implied obligation.

The respondents do not in substance dispute the foregoing, for they accept that the implied duty was premised on knowledge. However, they submitted that the appellants had failed in their duty, as they knew that no interim certificates had been issued and they did nothing to ensure that the architect issued the certificates on time. The arbitrator must have taken this into account, as had the learned judge (who referred specifically to the need for knowledge on the part of the employer in para 14 of his grounds of judgment).

With respect, we think this argument ignores the clear basis on which the arbitrator made his findings. The implied term as formulated by the arbitrator would render the employer liable for any of the architect's breaches. Thus, the sentences in the award read:

> I find it hard to give due consideration to the fact that the act of repeated late issuing by the architect, and late honouring of interim payment certificates by the respondents could possibly have been acquiesced to by the claimants, since such late payment pattern was one that appeared to have been rather consistent. There was clearly no evidence to support the respondents' seemingly inexcusable repeated delays in honouring the interim payment

> certificates. It is therefore not open to the respondents to deny that they are in breach of clause 30(1) of the building contract.

The reference to the consistent pattern of late payments on the part of the appellants was to counter the argument that the respondents had acquiesced in the late payment, not the appellants' knowledge of the wrongful delay or decision on the part of the arbitrator in the issuance of the interim certificates. The arbitrator had not made any finding that the appellants knew that the architect had wrongfully delayed the issue of interim certificates or that the architect had wrongfully certified.

Indeed, the mere fact that there was a consistent pattern of late interim certificates does not necessarily imply that the appellants must have known that the architect was in default of clause 30(1). Although the arbitrator has found that the architect was in fact in breach of clause 30(1), it is a separate question whether the appellants were aware of this. There is correspondence to suggest that it appeared to the appellants that the delay in issuing interim certificates was attributable to the respondents' consistently improper submission of claims, resulting in certification delays by the quantity surveyor, or at least because there were disputes between the architect, quantity surveyor and respondents over the sums to be certified.

In many cases the architect's certificate will be the subject of an arbitration and sometimes the arbitrator will decide that the architect's figures are wrong (in practice the material presented to the arbitrator will be different from that which was before the architect. By the time the arbitration takes place the parties will usually have researched the matter thoroughly and decided how to present it to the best advantage).

Under ICE Conditions, Fifth, Sixth and Seventh Editions, the contractor is entitled to interest where the engineer fails to certify amounts due. For this purpose, however, it is not sufficient to show simply that the arbitrator considers the sum due to be greater than the sum certified by the engineer.

Secretary of State for Transport v. Birse-Firr Joint Venture

QUEEN'S BENCH DIVISION

(1993) 35 Con LR 8

HOBHOUSE J: It is convenient to start with a consideration of the scheme of clause 60 and the contract of which it forms part. It is well recognised that progress payments are an integral and important part of construction contracts. They have been described as the 'life blood' of the contractor. If the contractor is deprived of the proper progress payments he may be placed in financial difficulties and incur substantial financing costs. However, it has also been pointed out that the position is not all one way and that the employer too may be relying upon finance in order to perform the contract and that overpayments by the employer may equally cause him loss. It is relevant to the construction of clause 60(6) that it merely deals with an obligation to pay interest by the employer to the contractor and does not apply the other way round. It is not a provision for the adjustment in later certificates of sums earlier certified together with an interest allowance either way; nor does it as such apply to any decision of the engineer under clause 66.

This leads on to the next point, which was expressly recognised by the arbitrator. Certification may be a complex exercise involving an exercise of judgment and an investigation and assessment of potentially complex and voluminous material. An assessment by an engineer of the appropriate interim payment may have a margin of error either way. It may be subsequently established that it was too generous to the contractor just as it may subsequently be established that the contractor was entitled to more. Further the sum certified may be made up from a large number of constituent figures, some of which may likewise be assessed favourably to one party or the other. It may be that a contractor can say that under a certain heading he did not have certified as high a figure as can later be seen to be appropriate but that under another heading he has to accept that the figure certified can be shown to have been an over-certification. At the interim stage it cannot always be a wholly exact exercise. It must include an element of assessment or judgment. Its purpose is not to produce a final determination of the remuneration to which the contractor is entitled but is to provide a fair system of monthly progress payments to be made to the contractor. The strict 28-day period in sub-clause (2) can be contrasted with the three months allowed in sub-clause (3) 'after receipt of this final account and of all further information reasonably required for its verification'.

The engineer is appointed by the employer and is in many respects his agent. But that is not the engineer's sole function in a contractual scheme such as that contained in the ICE Conditions. This has been spelt out in relation to the parallel situation of the architect under the JCT Conditions . . .

A distinction clearly emerges from this case between the issue of a certificate which, bona fide, assesses the value of the work done at a lower figure than that claimed by the contractor and a certificate which, because it adopts some mistaken principle or some error of law, presumably in relation to the correct understanding of the contract between the parties, produces an under-certification. In the latter case there can be said to have been a withholding of a certificate. This case also confirms that under clause 66, an aggrieved contractor can proceed to arbitration straight away without waiting till the completion of the work and obtain whatever award he is entitled to at that time . . .

The opinion which the engineer is required to form and express in his certificate is a contractual opinion. It must be a bona fide opinion arrived at in accordance with a proper discharge of his professional functions under the contract. In subclause (3) there is an express reference to 'the amount which in his opinion is finally due under the contract'. It is implicit in sub-clause (2) that the sum certified is that which in his opinion he considers to be due under the contract as an interim payment for that month. If it therefore should be the case that the engineer's opinion is based upon a wrong view of the contract then it can be said that he has failed to issue a certificate in accordance with the provisions of the contract. This was the situation in *A.E. Farr Ltd* v. *Ministry of Transport*. Therefore, leaving on one side all questions of bad faith or improper motive, and none is suggested in the present case, a contractor who is asserting that there has been a failure to certify must demonstrate some misapplication or misunderstanding of the contract by the engineer. For example, it certainly does not suffice that the contractor should merely point to a later certification by the engineer of a sum which had been earlier claimed but not then certified . . .

In my judgment the correct construction of the language used in clause 60 having regard to the overall scheme of the contract and the general commercial circumstances which surround any contract of this kind is that the words 'failure by the engineer to certify' refer to, and refer only to, some failure of the engineer which can be identified as the failure by the engineer to respect and give effect to the provisions of the contract. Those words do not refer to an under-certification which does not involve any contractual error or misconduct of the engineer. It follows that the approach of the contractors in the present case cannot on any view be supported. The letter of 23rd April 1992 was written on the basis of a simple comparison of sums which had been claimed at some earlier date but not then certified and were subsequently certified or otherwise allowed by the engineer. Such an exercise does not, without more, demonstrate any failure by the engineer to fulfil his contractual obligations. The scheme of the contract, including clauses 52, 60 and 66, recognises that prior to the final certificate there will be an on-going assessment of the works and their value and the sums to which the contractors are entitled. The contractors have not so far attempted to prove or even, so far as I can tell, allege that the engineer in some way disregarded his contractual obligations. The award accepted the indiscriminate approach of the contractors and accordingly, quite apart from the other points that have been raised, cannot stand in its present form . . .

However, I must respectfully disagree with the statement of that extremely experienced official referee that:

> if the arbitrator revises his certificate so as to increase its amount it follows that the engineer has failed to certify the right amount.

To take the simplest example, it may be that different evidence is available to the arbitrator to that which was available to the engineer and that the engineer on the material available to him arrived at the correct conclusion. If on the other hand the basis of the refusal by the engineer to certify certain sums is a construction which the engineer placed upon the contract and the arbitrator subsequently holds that the construction was wrong, then it does not follow that the engineer has failed to certify in accordance with sub-clauses (2) and (3); he has failed to perform his obligations under the contract in accordance with the terms of the contract.

Nature and effect of certificates

In considering the effect of certificates there is a fundamental distinction between interim and final certificates. Interim certificates generate a right to payment but as their name suggests they are in many respects provisional. Final certificates may however be final. How final depends on the precise wording of the contract.

East Ham Borough Council v. Bernard Sunley & Sons Ltd

HOUSE OF LORDS

[1965] 3 All ER 619

Sunley erected a school for the council under RIBA terms. Clause 24 provided:

> . . . (f) Upon expiration of the defects liability period . . . the architect shall . . . issue a final certificate and such final certificate . . . save as regards all defects and insufficiencies in the works or materials which a reasonable examination would not have disclosed, shall be conclusive evidence as to the sufficiency of the works and materials. (g) Save as aforesaid no certificate of the architect shall of itself be conclusive evidence that any works or materials to which it relates are in accordance with this contract.

The arbitration provision (clause 27) provided:

> ... the arbitrator shall have power ... to open up, review or revise any certificate ... in the same manner as if no such certificate ... had been given.

Some two years after the architect had issued his final certificate, stone panels fixed to the exterior walls fell off owing to defective fixing by Sunley and were repaired by the employer, who sought to recover the cost from Sunley. One of the matters in dispute was whether the arbitrator had power to reopen the architect's final certificate.

Held: Clause 24(f) and (g) placed the architect's final certificate in a special position for a special purpose and therefore the wide general words of clause 27 could not be read literally because so to do would deprive clause 24(f) and (g) of any effect or operation. In order to give effect to both clauses, clause 27 must be read as being subject to clause 24(f) and (g). Accordingly the final certificate of the architect was conclusive and could not be reopened by the arbitrator save in the exceptional circumstances specified.

> VISCOUNT DILHORNE: How are these two apparently conflicting provisions to be reconciled? It is unfortunate that a contract, in a form sanctioned by the Royal Institute of British Architects and presumably widely used, should contain such apparent inconsistencies.
>
> I cannot see any way of interpreting this contract so as to give full effect to the words in clause 24(f) and in clause 27.
>
> The appellants contended that the provision in clause 27 must be read as overriding and qualifying clause 24(f). Their argument in effect was that clause 24(f) must be read as if it provided that the certificate should be conclusive evidence save and except when there was an arbitration. The respondents, on the other hand, contend that the provision in clause 27 must be read as subject to clause 24(f), with the result that the arbitrator, despite the wording of clause 27, was bound to treat the certificate as conclusive evidence of sufficiency.
>
> The appellants suggested that the conclusive effect of the final certificate was to be interpreted solely as terminating the contractor's liability under clause 12 to recall to remedy defects. I do not consider this suggestion well founded. The contractor's liability to recall ended at the end of the defects liability period and the termination of that liability was in no way dependent on the issue of the final certificate.
>
> They further contended that to treat the certificate as conclusive evidence in litigation but not in an arbitration was not unreasonable as the parties might have been ready to agree to investigation of the sufficiency of the work before an expert arbitrator but not willing for that to take place in litigation. I do not myself regard it as right to base the interpretation of his contract on any such assumption as to the intention of the parties. Further, it appears to be thought in some quarters that, if special powers are given to an arbitrator, they devolve on the court should there be litigation. I do not regard the decision in *Robins* v. *Goddard* (1905) as establishing or, indeed, supporting such a proposition. In that case, as I understand it, the Court of Appeal held that the arbitration clause, which gave power to an arbitrator to open up, review and revise a certificate, showed beyond doubt that the certificates in that case were not conclusive and, the certificates not being conclusive, the court was not obliged to treat them as if they were.
>
> If, however, it be the case that in litigation the court has such powers as the contract may give to an arbitrator, then if the appellants are right, the final certificate is not to be treated as conclusive evidence of sufficiency either in litigation or in an arbitration.

The appellants sought also to rely upon the words 'Provided always' at the commencement of clause 27. They contended that these words showed that the rest of the contract was subject to clause 27. I do not read the contract this way. These words form part of the sentence entitling either party to go to arbitration and, in my view, make it clear that the whole contract was subject to arbitration. They do not form part of the final sentence of the clause and I do not think it is to be inferred from the inclusion of these words at the beginning of the first sentence in the clause that it was intended that the powers given to the arbitrator in the final sentence should override the conclusive effect of the final certificate.

If the final sentence of clause 27 had been intended to make it possible for the arbitrator to treat the final certificate as not conclusive on the very matters on which clause 24(f) said it was to be conclusive, one would have expected this to have been made clear beyond doubt. As the interim certificates were not made conclusive by clause 24(a) and (b), it is not possible to say with any certainty why it was necessary to say that the arbitrator had power to determine any matter as if no certificate had been given. Perhaps the explanation may be that these words in clause 27 were retained in the form of contract sanctioned by the RIBA lest, after the decision in *Robins* v. *Goddard*, their omission might lead to it being argued that the certificates were conclusive.

To read clause 24(f) as subject to the final sentence of clause 27 would, in my view and as the respondents contended, rob the final certificate almost entirely of its conclusive effect. To treat clause 27 as applying to all the certificates, interim and final, but not as giving power to the architect to ignore the conclusive effect of the final certificate, in my opinion does less violence to the language of the contract.

On this question I agree with the Court of Appeal. I think the answer to it is in the affirmative. In my opinion, the specific provision as to the conclusiveness of the final certificate prevails over the generality of the words in the final sentence of clause 27.

In *Windsor Rural District Council* v. *Otterway & Try Ltd* (1954) Devlin J came to the contrary conclusion when considering the same question on the same form of contract. For the reasons I have stated I am unable to agree with his conclusion.

Note: After the Sunley case, and until July 1976, JCT 63 clause 30(7) declared that the final certificate was, subject to certain provisos about the commencement of proceedings or the appointment of an arbitrator, declared to be 'conclusive evidence in any proceedings that the works have been properly carried out and completed in accordance with the terms of this contract' and subject to three exceptions set out.

P. & M. Kaye Ltd v. Hosier & Dickinson Ltd

HOUSE OF LORDS

[1972] 1 All ER 121

Contractors agreed to build a warehouse and offices for the employers on JCT 63 terms. Work on the warehouse was substantially completed in June 1967 although, with the contractors' consent, the employers had taken possession in April 1967. The architect issued interim certificates in April and July and the employers paid sums on account, leaving a balance of £14,861 unpaid. The employers complained that the warehouse floor was faulty. The contractors relaid the floor and then, on still not being paid, commenced proceedings. The employer did not seek to refer the dispute to arbitration, but sought leave to defend the proceedings by alleging that the flooring was still faulty and counterclaimed some £13,500. The action was referred to the Official Referee but then postponed. The contractor alleged that he tried to repair any further defects, but about 12 months later wrote to the architect requesting the final certficate, claiming that it was due since the works could not now be other than properly carried out.

The architect issued the final certificate showing a balance due to the contractor of £2,360. Three days after its issue the employer asked the contractor to concur in the appointment of an arbitrator. The contractor refused, alleging that it was now too late, and relying on clause 30(7) which then provided as follows:

> Unless a written request to concur in the appointment of an arbitrator shall have been given . . . by either party before the final certificate has been issued . . . the said Certificate shall be conclusive evidence in any proceedings arising out of this contract . . . that the works have been properly carried out and completed in accordance with the terms of this contract . . .

The contractors issued a second writ claiming the amount of the final certificate.

Held: The words in clause 30(7) that the final certificate shall be 'conclusive evidence in any proceedings arising out of this contract' prevented any further legal action, including the legal proceedings started long before the certificate was issued.

Note: Lord Diplock gave a strong dissenting judgment, which is quoted below. Clause 30(7) was revised by the JCT as a result of this case.

> LORD MORRIS OF BORTH-Y-GEST: Of the issues which are open for decision, the most difficult, in my opinion, is whether the words 'conclusive evidence in any proceedings arising out of this Contract' should be limited to proceedings commenced after the date of the final certificate or whether they also cover proceedings previously begun. As a matter of language it can hardly be doubted which alternative is to be preferred: to accept the former involves writing in a limitation which is not there. But there are more substantial issues involved, which relate, broadly, to the interaction between an architect's certificate and the power and duty of the courts to decide disputes. Can the parties, it may be asked, by a contractual stipulation, exclude the courts from their constitutional responsibility? If one considers the closely analogous field of arbitration, one may contrast the willingness of the courts to stay court proceedings when there has been a submission to arbitration, while still retaining ultimate control, with their unwillingness, once court proceedings have started, to allow the question before the court to be decided by subsequent arbitration – see *Doleman & Sons* v. *Ossett Corporation* (1912), not cited before the Court of Appeal but referred to by Cairns LJ in his judgment. The present case resembles, it is said, the situation in that case.
>
> My Lords, I am impressed, as was Lord Denning MR, with this argument, but on the whole I have come to a conclusion against it. To describe clause 30(7), as sought to be invoked here, as an ouster of the court's jurisdiction, seems to me to beg the question and in fact to misdescribe the effect attributed to it. The court proceedings, as can be seen from the pleadings, raise the question whether the work done and the materials used were such as should have

been done and used under the contract. An essential question must be, what standard is to be set? If the proper standard to which that work and those materials ought to conform were one to be fixed by the court (for example, by reference to what is reasonable), I could see good reasons for not allowing this matter to be decided conclusively outside the court by another person. But that is not what the contract provides. Throughout the contract, the architect is the person to pronounce on these matters. This appears right from the inception: in clause 1(1) where it is said that the works are to be carried out and completed (it may be noted that this language is repeated in clause 30(7)) to the reasonable satisfaction of the architect; and as regards materials it is for the architect to be satisfied that they are in accordance with the contract bills. This being so, what objection can there be to a clause which provides that, as regards these matters, as to which it is the architect's standard that is relevant, the architect's final certificate is to be conclusive evidence? The court has to find the facts. It has to do so in accordance with the contract. The clause provides a means – the means – of establishing the facts. The court retains ultimate control in seeing that the architect acts properly and honestly and in accordance with the contract. The method of proof, chosen by the parties, is legitimate, and by its terms binding, and I can see no reason for denying the contractual effect of the evidence to proceedings previously commenced, or, as I would prefer to put it, for reading in a limitation to subsequent proceedings. Allowing to *Doleman & Sons* v. *Ossett Corporation* its full force (and I note that four very eminent judges were equally divided even there) the present case is altogether different.

LORD DIPLOCK (dissenting): Clause 30(7) nevertheless provides that the final certificate shall be 'conclusive evidence' not only of a matter to which it does refer, i.e. the adjustment of the contract sum, but also a matter to which it does not refer at all, i.e. 'that the Works have been properly carried out and completed in accordance with the terms of this Contract'. In their natural meaning, it seems to me that these words are dealing not primarily with the activities of the contractor but with the state of the works as a result of the activities of the contractor, and are dealing with their state at the time of issue of the certificate. They mean no more than that the works are in accordance with the contract drawings and contract bills, subject to any variations authorised by the architect under clause 11, and that the workmanship and materials are of the quality required by the contract. They do not mean that at no time previously to the issue of the final certificate were there defects in the works which required remedying and had been remedied, for sub-clause (6) expressly contemplates that the final certificate must be issued notwithstanding this; and the parties should not lightly be held to have intended the final certificate to be conclusive evidence of the truth of anything that the certifier knew to be a lie.

This being, as I think, the natural meaning of the words directly under consideration, I return to see whether it requires to be qualified because of the remaining provisions of sub-clause (7). It is to be observed that the final certificate is to have no evidential value as to any of the matters referred to in the latter part of the paragraph unless –

> a written request to concur in the appointment of an arbitrator shall have been given under clause 35 of these Conditions by either party before the Final Certificate has been issued or by the Contractor within 14 days after such issue . . .

This is obviously an elliptical phrase. 'A written request to concur in the appointment of an arbitrator' forms no part of the procedure in referring a dispute to arbitration under clause

35 if the parties agree on an arbitrator themselves. It would be quite irrational to make the evidential effect of a final certificate dependent on whether or not in an arbitration pending at the time of the issue of the final certificate the arbitrator had been appointed as a result of a written notice to concur instead of by agreement without such notice. Both parties, indeed, concede that it is a necessary implication that an agreement to refer a dispute to arbitration by an agreed arbitrator must have been intended to have the same consequences on the evidential effect of the final certificate as a written notice to concur in the appointment of the arbitrator. Furthermore, the phrase says nothing expressly about the nature of the dispute which must have been referred to arbitration. Yet disputes can arise under the contract which have nothing to do with the state of the works at any stage before or after the end of the construction period or the defects liability period. For instance, there may be a dispute arising under clause 7 which deals with the right of the employer to indemnify for infringements of patents, or clause 19 which deals with the obligation of the contractor to insure. It would be quite irrational for the parties to make the evidential effect of a final certificate dependent on the existence or non-existence of a pending arbitration in a dispute which had nothing to do with any of the matters of which the final certificate was to be conclusive evidence. It seems to me to be also a necessary implication that a prior reference to arbitration was intended to deprive the final certificate of its conclusive character only in those cases where the dispute referred to arbitration was one concerning the subject matter of which the final certificate was made conclusive evidence.

My Lords, if clause 30(7) is so construed, the system of certification provided for in the contract culminating in the final certificate makes business sense. The primary obligation of the contractor is to carry out and complete the work to the reasonable satisfaction of the architect. Prima facie the standard to be attained is a subjective one set by the architect, subject only to the qualification that it must be reasonable.

In the course of the progress of the works during the construction period until practical completion the architect signifies his satisfaction by the issue of interim certificates. By issuing his certificate of practical completion he signifies his satisfaction with the state of the works at the end of the construction period; but this is subject to any latent defects which may become apparent to him during the defects liability period. If none becomes apparent, he signifies his satisfaction with the state of the works at the end of the defects liability period by refraining from issuing a schedule of defects within the next 14 days. If latent defects do become apparent during the defects liability period and are either the subject of a schedule of defects or of written instructions by the architect to make good, he signifies his satisfaction that they have been made good by issuing a certificate to that effect under clause 15(4).

But the contract also provides that the parties may dispute the reasonableness of the architect's satisfaction as to any of these matters. The only method of doing so provided by the contract is by arbitration under clause 35. It is expressly provided by clause 30(8) that none of these previous certificates of the architect shall be conclusive evidence as to the condition of the works, materials or goods to which it relates and, by clause 35(3) that any of them may be treated by the arbitrator as if it had never been issued.

My Lords, in a contract of this character, there is good business sense in agreeing on a time limit, shorter than that laid down by general statues of limitation, within which any claims must be brought either for breaches of contract generally or alternatively, for breaches of contract of a particular kind. If it was intended to do the former it would be very simple to state this in an express provision. There are many precedents to be found in business contracts, but there is no such provision in the contract. Even clause 35 itself imposes no time limit within which disputes must be referred to arbitration.

Disputes under the contract may be of two kinds: disputes, on the one hand, as to whether the architect acted reasonably in signifying his satisfaction of the way in which the works had been carried out or completed or in ordering adjustments to be made to the contract sum, and disputes, on the other hand, as to matters in respect of which the architect has no functions to perform. In a contract by which the parties have agreed to accept 'the reasonable satisfaction of the Architect' as the prima facie criterion as to whether particular obligations under the contract have been performed, there is good business sense in distinguishing breaches of these obligations from other obligations arising under the contract, and in providing a time limit after which neither party can dispute the reasonableness of any satisfaction he has expressed as to the performance of any such obligation. In my opinion, that is what is done by clause 30(6) and (7).

The date of issue of the final certificate is so fixed by clause 30(6) that the parties will have already had an opportunity of referring to arbitration under clause 35 any dispute about the reasonableness of the satisfaction signified by the architect as to the condition of the works at the time that any previous certificate was issued by him, or about the reasonableness of any decision by him as to any adjustments to the contract sum which is necessary in accordance with the terms of the contract. If either party has availed himself of this opportunity, the final certificate has no evidential effect. But if neither party has done so before the issue of the final certificate (or in the case of the contractor within 14 days thereafter), both parties are deprived of any further right to dispute that the architect's satisfaction was not reasonable or that the obligation of which performance to his reasonable satisfaction was the criterion had not been performed; or that the adjustment to the contract sum was not properly made.

This, in my judgment, is the effect of the provision in clause 30(7) which makes the final certificate conclusive evidence that the works have been properly carried out and completed in all respects in accordance with the conditions of this contract. I can see no grounds in language or in reason for limiting its evidential effect to proceedings started after the date of the certificate. It is relevant evidence only if a fact to be proved in the proceedings is that the works have been properly carried out and completed to the reasonable satisfaction of the architect. There is no principle of law which prevents parties to litigation from agreeing in advance what shall be treated as conclusive evidence of a relevant fact – at any rate if such agreement is not a mere sham to induce the court to decide a hypothetical question on a hypothesis of fact known by both parties to be false; *Kerr* v. *John Mottram Ltd* (1940).

The way in which clause 30(7) works in practice is well-illustrated by the two actions which have been consolidated and are the subject of the instant appeal. The first action was brought during the defects liability period for the part of the works of which possession had been taken under the provisions for sectional completion to be found in clause 16. Two claims were advanced by the employers. The first was as to the liability of the contractors for consequential damage resulting from latent defects which had appeared during the defects liability period in respect of which remedial measures had already been undertaken by the contractors on the instructions of the architect. Under the contract the architect had the duty of satisfying himself whether there were defects but he had no functions in respect of any claim for consequential damage. His opinion whether such a claim was justified or not was irrelevant. The second claim was for unliquidated damages representing the cost of further remedial measures [which] had already been undertaken by the contractors to make the works comply with the terms of the contract. As I have pointed out earlier, this kind of claim was premature, even if it were well-founded, since under the contract the contractor had the right, as well as the duty, to undertake any further remedial works himself in accordance with the instructions, and to the reasonable satisfaction of, the architect. By the date at which the reply and defence to counterclaim had been delivered, the final certificate had been issued under clause 30(6).

Accordingly, the architect must have been satisfied either that no further remedial measures were needed at the date of the first action to bring the works into conformity with the terms of the contract at the end of the defects liability period, or that such further measures as were needed had been satisfactorily completed. Of the fact that his satisfaction was reasonable, the final certificate was conclusive evidence. By the time of the hearing of the preliminary issue the employers had lost their opportunity to dispute this and, accordingly, of proving their claim, which had been prematurely advanced, for the cost of any remedial measures. The final certificate, however, was irrelevant to the claim for consequential damage. It could have no effect on it.

In the second action, brought after the date of the final certificate, identical claims were made in the defence and counterclaim, except that the date at which it claimed that further remedial measures were still necessary to make the works conform with the requirements of the contract was after the expiry of the period during which the contractor would have had the right, as well as the duty, to execute them himself. But the same principle applies. The final certificate is conclusive evidence that all necessary remedial works had been executed by the time of its issue. It, too, is irrelevant to the claim for consequential damage in respect of defects which had been found during the defects liability period.

Crown Estate Commissioners v. John Mowlem & Co Ltd

COURT OF APPEAL

(1994) 40 Con LR 36

In this case the Court of Appeal had to consider the effect of clause 30.9.1 of JCT 1980 which provides:

> Except as provided in clauses 30.9.2 (and save in respect of fraud) the Final Certificate shall have effect in any proceedings arising out of or in connection with this Contract (whether by arbitration under article 5 or otherwise) as
>
> 30.9.1.1 conclusive evidence that where the quality of materials or the standard of workmanship are to be to the reasonable satisfaction of the Architect the same are to such satisfaction.

(This wording follows closely that of the 1977 revision of JCT 1963. The original 1963 version was significantly more final in its wording.)

STUART-SMITH LJ: The respondents submit that the answer is to be found in clause 8.1, which they contend contemplates that the parties will stipulate contractual criteria for the standards and quality of work and materials. They may or may not do so, there being three possible options:

Case A Criteria stipulated in the contract documents, for example to British Standard Specifications.

Case B Standards and quality not stated in the contract documents in which case there is an implied term that materials will be of a reasonable quality and fit for their purpose and workmanship will be to a reasonable standard.

Case C Standards and quality expressed to be to the architect's satisfaction . . .

The effect of these provisions is that the architect has to approve the quality of materials and standard of workmanship. Where a standard is stated in the contract (case A), he must be satisfied that the works comply with the standard; where there is no express provision in the contract (case B) he must be satisfied that they have been executed to a reasonable standard. Case C, it is accepted, comes within clause 2.1.4, which imposes an objective standard and does not permit an arbitrary or subjective view of the architect to prevail . . .

In my judgment, the wider construction is to be preferred. I agree with Mr Seymour that the word 'approval' in clause 2.1.4 supports the construction that this clause is concerned with compliance and not merely setting standards. Furthermore, I think it would be strange that the contract should specify an objective standard of reasonableness only for the setting of standards for materials and workmanship, and not the far more important question of compliance. In this context, Mr Marrin has only been able to identify one instance where the standard is arguably set to the reasonable satisfaction of the architect . . .

At best Mr Marrin could only suggest very few examples where this type of provision is still found in modern contracts. Moreover, while it is understandable that the architect may approve a sample of materials, I find it very difficult to see how he can set in advance a standard of workmanship to his satisfaction. The contractor would have no way of knowing in advance to what standard he had to carry out the works. Finally, it seems to me that if the narrow construction is correct, clause 2.1.4 would contain more explicit words, such as '. . . is expressed to be a matter for the opinion of the architect'.

Note: This decision caused considerable surprise in the construction industry. It was rapidly followed by a substantial amendment of JCT 1980 to make the final certificate less final. See now JCT 98, clause 30.9.

There was much discussion between 1984 and 1998 of the extent to which interim certificates were free from challenge (other than by an arbitration started in good time where the contract provides for arbitration). These doubts derived largely from the decision of the Court of Appeal in Northern Regional Health Authority v. Derek Crouch Construction Co Ltd. [1984] 2 All ER 175. This decision has now been overruled.

Beaufort Development (NI) Ltd v. Gilbert Ash NI Ltd

HOUSE OF LORDS

(1998) 59 Con LR 66

LORD HOFFMANN: My Lords, the question before your Lordships is whether an arbitrator appointed to decide a dispute arising under a building contract in the JCT standard form has a power to review decisions and certificates of the architect which is not available to a court. The English Court of Appeal so held in *Northern Regional Health Authority* v. *Derek Crouch Construction Co Ltd* [1984] 2 All ER 175, [1984] QB 644, but your Lordships are invited to say that they were wrong.

The clause which is said to give the arbitrator these exceptional powers is 41.4, of which the relevant parts are as follows:

. . . the Arbitrator shall, without prejudice to the generality of his powers, have power to rectify the contract so that it accurately reflects the true agreement made by the Employer

> and the Contractor, to direct such measurements and/or valuations as may in his opinion be desirable in order to determine the rights of the parties and to ascertain and award any sum which ought to have been the subject of or included in any certificate and to open up, review and revise any certificate, opinion, decision . . . requirement or notice and to determine all matters in dispute which shall be submitted to him in the same manner as if no such certificate, opinion, decision, requirement or notice had been given.

The words particularly relied upon are those which confer a power to 'open up, review and revise any certificate, opinion, decision . . . requirement or notice' and determine matters in dispute as if they had not been given. The Court of Appeal in the Crouch case said that these were special powers conferred exclusively upon the arbitrator. Browne-Wilkinson LJ ([1984] 2 All ER 175 at 186–187, [1984] QB 644 at 667) said that in an action 'questioning the validity of an architect's certificate or opinion', the jurisdiction of the court would be limited to deciding whether or not the certificate or opinion was invalid for bad faith or excess of power. It could not revise the certificate on the ground that the court thought it was wrong. A clause such as 41.4, on the other hand, gave the arbitrator 'power not only to enforce the contractual obligations but to modify them'. Donaldson MR also said that the arbitrator could vary the certificates to create new rights and obligations which would not otherwise arise from the contract. Dunn LJ said that one could not imply a term that if the dispute was litigated instead of arbitrated, the court should have a similar power.

My Lords, I have no doubt that it is open to the parties to enter into an agreement of the kind described by the Court of Appeal in the *Crouch* case. I put aside the purely theoretical question of whether it is right to speak of the architect or arbitrator having power to modify the contractual obligations of the parties. I find this a strange concept. The powers of the architect or arbitrator, whatever they may be, are conferred by the contract. It seems to me more accurate to say that the parties have agreed that their contractual obligations are to be whatever the architect or arbitrator interprets them to be. In such a case, the opinion of the court or anyone else as to what the contract requires is simply irrelevant. To enforce such an interpretation of the contract would be something different from what the parties had agreed. Provisions of this kind are common in contracts for the sale of property at a valuation or goods which comply with a specified description. The contract may say that the value of the property or the question of whether the goods comply with the description shall be determined by a named person as an expert. In such a case, the agreement is to sell at what the expert considers to be the value or to buy goods which the expert considers to be in accordance with the description. The court's view on these questions is irrelevant.

It is less usual, though certainly theoretically possible, to add a second tier to arrangements of this kind, and to provide that a party who is dissatisfied with the view of one expert shall be entitled to call for the opinion of another, which shall be final and binding. From the point of view of the court, the final outcome is no different from that in the case of a single expert. The contractual obligations of the parties depend upon the opinion of the one expert or the other and not upon its own view of the matter.

It is this two-tier arrangement which the Court of Appeal in the *Crouch* case considered that the JCT contract had created; what Donaldson MR afterwards called, cryptically but vividly, an 'internal arbitration' (see *Benstrete Construction Ltd* v. *Hill* (1987) 38 BLR 115 at 118). It is internal in the sense that it does not adjudicate upon the rights and duties of the parties but is part of the machinery for determining what they are. The court appears to have considered that in the absence of a second-tier power of the arbitrator to open up, review and revise the architect's certificates, they would (if given in good faith and within the ambit of the relevant contractual provisions) be binding upon the parties. So the critical question is whether, upon the true construction of the contract, such certificates are binding. Unless they are, there is no need for any special second-tier arrangement. They will be open to review by any tribunal called upon to determine the rights of the parties, whether arbitral or judicial.

The judgments of the Court of Appeal contain no very detailed analysis of the provisions of the contract which are said to confer upon the architect this power to issue binding certificates. Although none of the judges say so expressly, there is an implied suggestion that one can infer such a power from the very fact that the arbitrator is given a power to 'open up, review and revise'. This is the argument from redundancy; the parties are presumed not to say anything unnecessarily and unless the decisions of the architect were binding, there would be no need to confer upon the arbitrator an express power to open up, review and revise them. The later judgment of Donaldson MR in *Benstrete Construction Ltd* v. *Hill* (1987) 38 BLR 115, in which he distinguished *Crouch* on the ground that the contract in the latter case did not have a similar arbitration clause, tends to support the view that he had adopted this form of reasoning.

I think, my Lords, that the argument from redundancy is seldom an entirely secure one. The fact is that even in legal documents (or, some might say, especially in legal documents) people often use superfluous words. Sometimes the draftsmanship is clumsy; more often the cause is a lawyer's desire to be certain that every conceivable point has been covered. One has only to read the covenants in a traditional lease to realise that draftsmen lack inhibition about using too many words. I have no wish to add to the anthology of adverse comments on the drafting of the JCT standard form contract. In the case of a contract which had been periodically renegotiated, amended and added to over many years, it is unreasonable to expect that there will be no redundancies or loose ends. It is therefore necessary to make a careful examination of the contract as a whole in order to discover whether upon its true construction it does confer binding power upon the decisions of the architect or whether there is some other explanation for the 'open up, review and revise' power in clause 41.4. It is also important to have regard to the course of earlier judicial authority and practice on the construction of similar contracts. The evolution of standard forms is often the result of interaction between the draftsmen and the courts and the efforts of the draftsman cannot be properly understood without reference to the meaning which the judges have given to the language used by his predecessors.

The substantive provisions of the agreement state the principal obligations of the parties in clear and objective terms. The contractor is obliged by clause 2.1 of the conditions to 'carry out and complete the Works in accordance with the Contract Documents, using materials and workmanship of the quality and standards therein specified . . . '. In this particular contract, the preliminary articles defined the 'Works' as the construction of a nine-storey office block as described in contract documents. Clause 8.1.3 provides that all work is to be carried out in a proper and workmanlike manner and by clause 23.1.1 the contractor is to proceed 'regularly and diligently' with the works and complete them on or before the completion date. The contract specified 14 January 1995 as the completion date and said that the contract price was to be £1,700,000.

This framework of carefully defined contractual obligation is not easily reconcilable with a broad discretion, said to be conferred in the first instance upon the architect and subject to review by an arbitrator, to vary or modify the rights of the parties or to have them conclusively determined by the judgment of one or the other. The parties have agreed that a particular building is to be constructed out of specified materials in a workmanlike manner and that the work should proceed regularly and diligently to completion by a specified date. No doubt within this framework there is room for judgment about what amounts to proper workmanship and diligent progress. But one would not ordinarily describe the exercise of such judgment as a power to modify the contractual rights. These are questions which require the application of objective standards and with which the courts are routinely familiar.

The contract provides for the issue by the architect of certificates or statements in writing as to his opinion on various matters. For present purposes, these documents may be treated as similar. As Devlin J said in *Minster Trust Ltd* v. *Traps Tractors Ltd* [1954] 3 All ER 136 at 145: 'The mere use of the word "Certificate" is not decisive'. In the absence of express words, the parties are highly unlikely to have intended that some of these statements of opinion should be binding and others not. I shall give a few examples. Clause 30 provides for the issue of interim certificates of the value of the work for which the contractor is from time to time entitled to payment. Clause 30.1.1 provides that the contractor is entitled to payment within 14 days after the issue of the certificate. Clause 25, which deals with extension of time, lists a number of 'relevant events' such as *force majeure* or failure to provide instructions or information which the parties accept as capable of delaying completion beyond the completion date without breach of the contractor's primary obligation to proceed diligently with the works. By clause 25.3, if the architect is of opinion that a relevant event is likely to delay completion beyond the completion date, he must give the contractor an appropriate extension of time by a written notice fixing a new completion date. Clause 26 deals with claims for loss and expense caused by deferment of giving possession of the site or various matters such as provision of information for which the employer or architect is responsible. Here again, the architect is required to state his opinion that loss or expense has been caused, or is likely to be caused, by one of the specified matters, whereupon the amount is ascertained by the quantity surveyor and added to the contract price. Finally, clause 30.8 provides for the issue of a final certificate stating the balance due from employer to contractor or vice versa.

Clause 30.9 expressly makes the final certificate conclusive evidence as to various matters. But there is no other express provision which says that any certificate or expression of opinion is to be binding upon the parties in the same way as the determination of an expert. Clause 30.10, immediately after the provisions dealing with the final certificate, says:

> Save as aforesaid no certificate of the Architect shall of itself be conclusive evidence that any works, materials or goods to which it relates are in accordance with this Contract.

This clause has itself been the subject of refined arguments of the *inclusio unius, exclusio alterius* variety. The clause refers to certificates, therefore it must have been intended that other statements of opinion by the architect should be conclusive. The clause refers only to works, materials and goods being in accordance with the contract, therefore it must have been intended that certificates as to other matters such as extensions of time should be conclusive. In a contract such as this, such arguments are just as dangerous as the argument from redundancy, of which they are in truth merely a variety. If arguments of this kind are to be pursued, what seems to me much more compelling is that the contract contains express and elaborate terms which provide for conclusiveness as to various matters for one certificate and one only, namely the final certificate.

If the certificates are not conclusive, what purpose do they serve? If one considers the practicalities of the construction of a building or other works, it seems to me that parties could reasonably have intended that they should have what might be called a provisional validity. Construction contracts may involve substantial work and expenditure over a lengthy period. It is important to have machinery by which the rights and duties of the parties at any given moment can be at least provisionally determined with some precision. This machinery is provided by architect's certificates. If they are not challenged as inconsistent with the contractual terms which the parties have agreed, they will determine such matters as when interim payments are due or completion must take place. This is something which the parties need to know. No doubt in most cases there will be no challenge.

On the other hand, to make the certificate conclusive could easily cause injustice. It may have been given when the knowledge of the architect about the state of the work or the effect of external causes was incomplete. Furthermore, the architect is the agent of the employer. He is a professional man but can hardly be called independent. One would not readily assume that the contractor would submit himself to be bound by his decisions, subject only to a challenge on the grounds of bad faith or excess of power. It must be said that there are instances in the nineteenth century and the early part of this one in which contracts were construed as doing precisely this. There are also contracts which provided that in case of dispute, the architect was to be arbitrator. But the notion of what amounted to a conflict of interest was not then as well understood as it is now. And of course the inclusion of such clauses is a matter for negotiation between the parties or, in a standard form, the two sides of the industry, so that what is acceptable will to some extent depend upon the bargaining strength of one side or the other. At all events, I think that today one should require very clear words before construing a contract as giving an architect such powers.

The language and practical background of the JCT contract does not therefore suggest that any certificates other than the final certificate were intended to have conclusive effect. I return, therefore, to clause 41.4, from which the Court of Appeal in the *Crouch* case drew the opposite conclusion. It is worth noticing in passing that, in addition to the power to 'open up, review and revise', it also confers express powers to rectify the contract and to direct measurements and valuations. It seems plain that the reason for the inclusion of these powers is to 'determine the rights of the parties' which would be possessed by a court. If the power to 'open up, review and revise' was intended to be peculiar to the arbitrator, it would at any rate be different in its purpose from the other powers.

Nevertheless, it seems to me that cases since *Crouch* show that the decision has caused such uncertainty and even injustice that its dicta should be disapproved. I refer in particular to the recent decision of the Court of Appeal in *Balfour Beatty Civil Engineering Ltd* v. *Docklands Light Rly Ltd* (1996) 49 Con LR 1. It was a claim for extension of time and loss and expense under the ICE Conditions of Contract, which had been amended, first, by substituting the employer's representative for the engineer and, secondly, by deleting the arbitration clause. The contract provided for the employer to certify extensions of time and loss and expense claims. But there was no provision that they were to be binding or conclusive. Nevertheless, the court held that there was no power to 'open up, review and revise' them such as an arbitrator might have had if there was an arbitration clause in the usual form and that, as a matter of construction, 'the contractor's entitlement was to depend on the employer's judgment': see 49 Con LR 1 at 10 per Bingham MR. Your Lordships will remember that in *Benstrete Construction Ltd* v. *Hill* (1987) 38 BLR 115 at 118 Donaldson MR appeared to be saying that the Crouch construction of the certification clauses, as conclusive in litigation was based upon the fact that the contractors were held, even in the absence of an arbitration clause or any express language as to the certificates being conclusive, to have subjected themselves to the judgment of the employer. It is true, as Bingham MR remarked (at 57): 'It is not for the court to decide whether the contractor made a good bargain or a bad one; it can only give fair effect to what the parties agreed.' On the other hand, in deciding exactly what the parties did agree, it seems to me that in the absence of express language, one should not assume so uncommercial a bargain. I do not think that the *Balfour Beatty* case would have been decided as it was if not for the shadow of the Crouch decision that certificates, even if not declared to be conclusive, can be questioned only for bad faith or excess of power. I do not think that anyone in the industry can be said to have acted in reliance on the Crouch case and I would therefore overrule it. I must acknowledge the assistance which I have had in reaching this conclusion from the writings of Mr I.N. Duncan Wallace QC.

Henry Boot Construction Ltd v Alstom Combined Cycles Ltd (No. 2)

COURT OF APPEAL

(2005) 101 Con LR 52, [2005] EWCA Civ 814

The appellant (Boot) was the main contractor for civil engineering works in connection with the construction of a power station at Connah's Quay. The respondent (Alstom) was the employer. The contract, which was not under seal, was basically on ICE conditions (6th edn).

Boot started work on 11 April 1994 and achieved substantial completion on 28 May 1996.The defects correction certificate was issued on 15 August 2000. Boot submitted its final account in stages, the final part being submitted on 29 June 2001. The total sum claimed in the final account was £102.08m. The engineer issued the final certificate on 9 October 2002 in sum of £44.43m.

In March 2003 both Boot and Alstom served notices of dispute calling on the engineer to make decisions under clause 66 of the contract. On 23 May 2003 the engineer gave a decision that the value of the final account was £44.38m and that a sum of at least £2.9m was overdue for payment. He made no decision on limitation issues raised by Alstom.

An arbitration was commenced on 27 May 2003 and his Honour Judge Humphrey Lloyd QC was appointed as judge arbitrator. The limitation defence was tried as a preliminary issue. It was held that all or nearly all of Boot's claim was statute-barred because time began to run when the work was done, which was more than six years before the commencement of the arbitration. Boot appealed on the ground that a right to payment did not exist until a certificate had been issued.

DYSON LJ:

In my judgment on the true construction of this contract, certificates are a condition precedent to Boot's entitlement to payment under clause 50(2) and (4), and they are not merely evidence of the engineer's opinion. By 'condition precedent' I mean that the right to payment arises when a certificate is issued or ought to be issued, and not earlier. It does not, however, follow from the fact that a certificate is a condition precedent that the absence of a certificate is a bar to the right to payment. This is because the decision of the engineer in relation to certification is not conclusive of the rights of the parties, unless they have clearly so provided. If the engineer's decision is not binding, it can be reviewed by an arbitrator (if there is an arbitration clause which permits such a review) or by the court. If the arbitrator or the court decides that the engineer ought to have issued a certificate which he refused to issue, or to have included a larger sum in a certificate which he did issue, they can, and ordinarily will, hold that the contractor is entitled to payment as if such certificate had been issued and award or give judgment for the appropriate sum. (see further at [40]–[45] below). It is convenient to make such an award or to enter such a monetary judgment in order to avoid the risk of further proceedings in the event that the employer does not pay. For the reasons that follow, I consider that the right to payment arises when a certificate is issued or ought to be issued, and not when the work is done (although the doing of the work is itself a condition precedent to the right to a certificate).

For the purposes of the first issue, I shall concentrate on interim payments. Clause 60(1) requires Boot to submit at monthly intervals a statement showing the estimated value of the permanent works executed up to the end of that month as well as the other matters referred to in cl 60(1)(b)–(d). I shall examine cl 60(1) in more detail when I deal with the second issue. Clause 60(2) requires the engineer within 28 days of the delivery of the monthly statement to certify and Alstom to pay the amount which in the opinion of the engineer 'on the basis of the monthly statement' is due on account of sub-cl 1(a)–(d). Thus what the engineer has to certify and Alstom to pay is not the true final value of the work in fact done and materials in fact supplied etc, but what in the opinion of the engineer is due on the basis of the monthly statement. If Boot omits an item of work from the statement, even if the work has been done, the engineer is not obliged to include its value in the certificate. Moreover, the engineer is required to certify and Alstom to pay within 28 days of the delivery of the statement. It follows that the certificate may be issued at least 28 days after the end of the month to which the statement relates, and, if Boot takes some time to deliver the statement, possibly even later than that. In my view, it is clear that Boot's entitlement to interim payment does not arise until the engineer issues his certificate. The machinery for interim payment is provided exclusively by cl 60(1) and (2), and it is inconsistent with the proposition that Boot's cause of action in respect of interim payments arises any earlier than the date on which an interim certificate is issued, or (where a certificate is not issued) the date on which a certificate ought to be issued. I do not see how it is possible to construe this contract as meaning that the right to interim payments arises brick-by-brick or day by day or is in any other way unrelated to certificates.

As Mr Furst points out, it was not suggested in the *Beaufort Developments* case (still less did the House hold) that the certification process was irrelevant to the contractor's rights and merely evidence of the true content of those rights. On the contrary this passage in the speech of Lord Hoffmann suggests that certification was an essential element in determining the parties' rights and obligations even if, due to its provisional status, a certificate could subsequently be challenged. It is a *non sequitur* to reason that, because interim certificates are not conclusive and may subsequently be reviewed, therefore they are not an essential part of the contractual machinery for payment. To assert that a certificate (a) is a condition precedent to a right of action, and (b) is not conclusive, since it can be reviewed by an arbitrator or court, is not to advance two inconsistent propositions. It is, therefore, not surprising that their Lordships did not say anything about the question of what constitutes the cause of action or when it arises. Nor is it surprising that decisions such as *Dunlop & Ranken, Lubenham Fidelities and Costain* were not mentioned. They were not relevant to the issue that the House of Lords had to decide.

The question that arises in relation to the second issue is whether a claim to have a sum included in a final certificate in respect of work or events for which a sum should have been included in an interim certificate is a fresh cause of action, or merely a repeat of the earlier cause of action. If the causes of action are the same, and the claim for inclusion of the sum in the final certificate is made more than six years after the cause of action first arose, then it is statute-barred. In that event a separate question that arises is whether the engineer is obliged to include in the final certificate a sum in respect of that cause of action even though it is statute-barred. That question is the subject of the third issue.

In my view, the cause of action in respect of an engineer's failure to include a sum in an interim certificate is not the same as the cause of action in respect of the failure to include a sum in the final certificate, even if the two sums happen to be the same. This is because interim certificates are no more than provisional estimates of the sum to which Boot is entitled by way of instalment payments. The engineer is required to certify the amount which on the basis of Boot's monthly statement in his opinion is due 'on account of clause 60(1)(a) and (d)'. Clause 60(1)(a) requires the monthly statement to show the 'estimated contract value of the Permanent Works'. Clause 60(1Xd) requires the statement 'to show the estimated amounts to which the Contractor considers himself entitled in connection with all other matters . . .' (i.e. other claims). The engineer is, therefore, required to decide, on the basis of estimates, what in his opinion is due as a payment 'on account'. The contract does not state what supporting documentation *if any* Boot must put forward in support of its estimates. Boot is not even obliged to claim the full estimated values or amounts, because the whole of cl 60(1) is subject to the qualification: 'unless in the opinion of the Contractor such values and amounts together will not justify the issue of an interim certificate'. These words may do no more than reflect the language of cl 60(3). More importantly, the engineer is not required to carry out a detailed and accurate valuation each month. His obligation is no more than to certify the amount which in his opinion is due *on account* of cl 60(1)(a) and (d). This is of particular relevance in relation to claims for extra expense pursuant to various clauses of the contract, where the calculation of Boot's true entitlement can be very complicated and time-consuming, and where it is often impossible to form an accurate view at the interim stage of the extra expense that will finally be payable.

In my view, the nature of the exercise required by the contract to be performed by Boot and the engineer at the interim stage is so different from that required at the final account stage that it cannot be said that a failure by the engineer to perform his obligations in accordance with cl 60(2) can start time running in relation to a cause of action based on the failure by the engineer to perform his obligations in accordance with cl 60(4). It seems to me that the submissions of Mr ter Haar fail to take account of the fundamental differences between what the engineer is required to do at the interim stage and what he is required to do at the final stage. The position would be otherwise if this were, say, a contract for a fixed sum of £1m whose only payment provision was that the price was to be paid by ten equal monthly instalments of £100,000. In such a case, the right to claim the first instalment would accrue at the end of the first month, and the right to sue for that instalment would become statute-barred six years after the end of the first month. This is because the right to payment of £100,000 at the end of the first month would be the cause of action in respect of the work done in that month. The right to payment of £100,000 for that work would be unaffected by any other provisions of the contract. It would not be reviewed or recalculated at the end of the contract. But this kind of contractual arrangement is quite different from that provided by the contract in this case, where there is a single contract price and one set of elaborate provisions for dealing with instalment payments on account and a very different set of elaborate provisions for ascertaining the contract price finally payable in the light of the work actually done and the events that occur during the carrying out of the works.

There is no requirement that the architect should observe the rules of natural justice before issuing a certificate, unless the express terms of the contract so require.

London Borough of Hounslow v. Twickenham Garden Developments Ltd

CHANCERY DIVISION

(1970) 7 BLR 81

For the facts, see p. 383.

MEGARRY J: It seems to me that an architect under a building contract has to discharge a great number of functions, both great and small, which call for the exercise of his skilled professional judgment. He must throughout retain his independence in exercising that judgment; but provided he does this, I do not think that, unless the contract so provides, he need go further and observe the rules of natural justice, giving due notice of all complaints and affording both parties a hearing. His position as an expert and the wide range of matters that he has to decide point against any such requirement; and an attempt to divide the trivial from the important, with natural justice applying only to the latter, would be of almost insuperable difficulty. It is the position of independence and skill that affords the parties the proper safeguards, and not the imposition of rules requiring something in the nature of a hearing. For the rules of natural justice to apply, there must . . . be something in the nature of a judicial situation; and this is not the case.

My view is fortified by two decisions, not cited in argument, which seem to me to be adverse to . . . the contractor's submissions. First, there is *Panamena Europea Navigacion (Compania Limitada)* v. *Frederick Leyland & Co Ltd* (1947). In this, Lord Thankerton, with whom the other Lords of Appeal agreed, indicated the true basis of the decision in *Hickman & Co* v. *Roberts* (1913), the limits of which, he said, had been misapprehended. The decision was based on the fact that the architect:

> . . . had referred to the owners for instructions and had accepted and acted on their instructions in reference to the matters submitted to him as arbitrator, regardless of his own opinion.

The *Panamena* case concerned a contract for the repair of a ship under the supervision of the shipowner's surveyor, and it is clear from Lord Thankerton's speech that the House of Lords did not regard certain discussions between the shipowner and the surveyor, and certain expressions of opinion by the surveyor about what he would later have to certify, as being improper conduct.

Secondly, there is the New Zealand decision in *A.C. Hatrick (NZ) Ltd* v. *Nelson Carlton Construction Co Ltd* (1964). That concerned a certificate given by an engineer under a building contract, certifying that in his opinion the contractor had 'failed to make such progress with the works' as the engineer deemed sufficient to ensure their completion within the period specified in the contract. In a long and careful reserved judgment, Richmond J considered . . . the applicability of the rules of natural justice. The observations of Channell J in *Page* v. *Llandaff & Dinas Powis Rural District Council* (1901) to the effect that if the architect hears one side, he must also hear the other, were duly cited, but in the end the judge rejected the submission. The architect must be impartial and independent; but unless the contract so requires, expressly or by implication (and Richmond J stressed the importance of the terms of the contract), the architect is not required to hold anything in the nature of a hearing.

Interference or obstruction with the issue of a certificate means actual intermeddling rather than mere negligence or omission.

R.B. Burden Ltd v. Swansea Corporation

HOUSE OF LORDS

[1957] 3 All ER 243

The facts of this case are set out at p. 372.

LORD TUCKER: I think, without attempting an exhaustive enumeration of the acts of the employer which can amount to obstruction or interference, that the clause is designed to meet such conduct of the employer as refusing to allow the architect to go on to the site for the purpose of giving his certificate, or directing the architect as to the amount for which he is to give his certificate or as to the decision which he should arrive at on some matter within the sphere of his independent duty. I do not think that negligence or errors or omissions by someone who, at the request, or with the consent, of the architect is appointed to assist him in arriving at the correct figure to insert in his certificate can amount to interference. Interference, to my mind, connotes intermeddling with something which is not one's business, rather than acting negligently in the performance of some duty properly undertaken. Nor do I think that the conduct found against [the quantity surveyor] obstructed the issue of a certificate; it may have resulted in the issue of a certificate for a smaller sum than that which was due, but that is, in my view, a matter for arbitration under clause 27 and not a ground for repudiation under clause 20. I think the use by the learned official referee of the words 'the sort of certificate' to which the appellants were entitled shows the difference between the interpretation which he put on these words and that which I have endeavoured to suggest as correct.

A certificate which is good on its face and within the authority of the architect is 'in accordance with these conditions' for the purposes of JCT 63, clause 35(2).

Killby & Gayford Ltd v. Selincourt Ltd

COURT OF APPEAL

(1973) 3 BLR 104

The plaintiff contractors sought payment of some £15,000 shown as due to them on an interim certificate. The defendants disputed the architect's authority to issue the certificate and, while admitting that extra work had been ordered and carried out, also disputed its amount.

Held: Since the certificate had been properly issued under the terms of the contract, it must be honoured.

LORD DENNING MR: In the first place, the sixth interim certificate was given in accordance with the conditions of the contract. It must be honoured. Condition 30(1) says that 'the contractor shall be entitled to payment within 14 days from the issue of that certificate'.

In the second place, there is no suggestion that the work was done badly, or that there were any defects or delays. So there is nothing in the nature of a set-off or counterclaim . . .

In the third place, Selincourt Ltd have produced no evidence to show that the variations or extras were not properly ordered. The architect was their agent. They should ask him about it. They cannot call upon the contractors to produce the written orders. The contractors can rely on the certificate without more ado.

If there were any ground for saying that the certificate was 'not in accordance with these conditions', that is a matter which under condition 35(2) could be the subject of an interim arbitration before the works were completed . . . So long as a certificate is good on the face of it and is within the authority given by the contract, then it is in accordance with the conditions. It must be honoured. I do not think it is open to the employers of the contractors to challenge an interim certificate by saying that it is too much, or too little, or includes this or omits that, or that the extras were not sanctioned in writing. Such matters must be left till after the practical completion of the work. Then they will all be considered in the final arbitration at the end of the contract. In my opinion there is no ground whatever for saying that this interim certificate is not in accordance with the conditions of the contract. It should be honoured.

Unless and until the final certificate is issued to the employer as required by clause 3(8) of JCT 63 it has no effect and is not binding on the parties.

The mere signature by the architect of a certificate does not amount to its issue under clause 3(8).

By issuing an interim certificate for payment the architect is not thereby expressing a final view of the state of the works. He is entitled to take a fresh view each time he issues an interim certificate and his opinion does not become conclusive until he issues the final certificate.

London Borough of Camden v. Thomas McInerney & Sons Ltd

QUEEN'S BENCH DIVISION

(1986) 9 Con LR 99

The defendants constructed 219 dwellings for the plaintiffs, including a 20 storey block of flats clad with bricks. The contract was in JCT 63 Standard Form. Although the plaintiffs' director of architecture was the designated architect under the contract, the day-to-day supervision was done by 'job architects' in the plaintiffs' employ. Work began in 1969 and practical completion was achieved in stages between February 1972 and December 1973. The final account was audited in July 1976, and taking into account interim payments which had already been received by the defendants, there was a balance due to them of £3,653.37. It was intended that an interim certificate should be issued under clause 30(1) in the sum of £653.37, and a final certificate for £3,000. The relevant council committee authorised the expenditure necessary to settle the audited final account, and a technical assistant duly prepared the certificates, in a book of printed certificates. They were in fact never sent because, following a memorandum to the job architect asking for confirmation 'that there are no latent defects, or other reasons why the final certificate should not be issued' his inspection revealed cracks in the bricks on the 20 storey block, and in February 1983 the plaintiffs commenced proceedings against the defendants claiming damages for breach of contract and negligence, alleging bad workmanship in relation to the bricks. The defendants disputed the claim and counterclaimed alleging negligence by the plaintiffs' servants.

Shortly before the trial, the plaintiffs' junior counsel came across a folder which contained, amongst other things, the printed book of certificates which included the interim certificate for £653.37 and the final certificate for £3,000 which had been prepared by the technical assistant. They were signed in the name of the director of architecture by the plaintiffs' chief quantity surveyor, who had authority to sign. The existence of this book of certificates and other documents were immediately disclosed to the defendants. In light of the discovery of the signed final certificate an application was made for the trial of a preliminary issue and the trial date was vacated.

Held: (1) The final certificate prepared by the technical assistant and signed in the name of the Director of Architecture was not a valid final certificate issued to the employer by the architect under the contract and was not therefore conclusive evidence under clause 30(7) that 'the works had been properly carried out and completed in accordance with the terms of this contract'. The court was not therefore precluded by the Court of Appeal decision in *Derek Crouch Construction Co Ltd* v. *Northern Regional Health Authority* (1984) from going behind the certificate, which was a nullity. Something more was needed for a certificate to be issued as required by clause 3(8) than the mere signature of the architect upon it, whether he was employed under a contract of service or as an independent agent.

(2) The signing of the final certificate did not constitute giving the opinion under clause 30(7) that the works had been 'properly carried out and completed in accordance with the terms of this contract'. Since it was not issued, it was of no effect, and the certificates, opinions and decisions referred to in clause 35(3) and in the judgments in *Crouch* were only those which had a binding effect on the parties. The final certificate only comes to life as a document which is legally enforceable as a certificate, opinion or decision of the architect when he issues it as required by clause 3(8).

(3) The unissued interim certificate was likewise of no force or effect. As to the earlier interim certificates, their issue did not mean that the architect had thereby given an opinion that the work done up to the date of the certificate was properly done. The architect was entitled to take a fresh view as to the state of the works each time he issued an interim certificate and his opinion did not become conclusive until he issued his final certificate. The fact that the architect had issued certificates that defects had been remedied in no way affected the position. The interim certificates were not binding on the plaintiffs as to the quality of the defendants' workmanship under the contract.

JUDGE ESYR LEWIS QC: The purpose of the requirement of clause 3(8) that the final certificate and interim certificate no. 56 should be issued to the employer was to put the plaintiffs on notice that payments were due to the defendants under the contract. The plaintiffs were obliged to honour the interim certificates within 21 days of their issue and to pay any balance due to the defendants under the final certificate within 14 days of its issue (see clauses 30(1) and 30(6)). They could only do this if the treasurer's department was aware that the obligation had arisen by means of the issue of a certificate. The only way in which an architect in the service of the plaintiffs could sensibly carry out his obligation under clause 3(8) was by seeing that the certificates signed by him were sent promptly to the plaintiffs' treasury department so that payment could be duly made. I bear in mind in this context the special position of an architect when he is obliged to give his opinion by way of a certificate which gives rise to obligations and rights in the contracting parties. It is long established that the architect must act fairly and impartially when carrying out this duty. This applies, in my judgment, whether or not he is in the employment of the building owner under a contract of service. To act fairly and impartially in this context he must stand apart from his employer. It seems to me that it would be strange in these circumstances if he could keep to himself decisions which were binding upon his employer and which required his employer to take action by way of payment. [The quantity surveyor] did not send a copy of the certificates to the treasurer's department so that the plaintiffs could make payment to the defendants, as their own procedure required. He did not even detach the certificates from the book in which they were kept. He did not send a duplicate copy to the defendants as required by clause 3(8). In my view, he never put the certificates into circulation. It may well be, as [was] suggested, that he signed the certificates by mistake and, having discovered his error, refrained from issuing them. Whatever the reason, in my judgment the final certificate and interim certificate no. 56 were never issued and the circumstances indicate that [the quantity surveyor] never intended that they should be. This answers the question raised by the first preliminary issue. It is unnecessary, therefore, for me to decide whether, as argued by [counsel for the plaintiffs, the quantity surveyor] had no authority to sign and issue the final certificate and interim certificate no. 56 after June 1978 in the absence of the approval of the job architect.

If the final certificate was not issued to the employer as required by clause 3(8) of the conditions, as I have found, it seems to me that it has no effect at all. It is not binding on the parties. As I have said, it may have been signed in error by [the quantity surveyor] who, upon discovering his error refrained from issuing it. In any event, there is no reason that I can see why an architect should not change his mind after signing a certificate before issuing it and then deciding that he will not issue it. The mere act of signing a document which appears to be a certificate cannot, in my view, have any binding or conclusive effect when the contract under which it is issued requires it to be brought to the attention of the employer by the issue of it to him. I consider that the mere fact that the final certificate and interim certificate no. 56 were signed has no significance.

I do not consider that by writing his signature on a certificate the architect in the context of this contract is irrevocably committed to the opinion which the certificate purports to give. In my view, the final certificate only comes to life as a document which is legally enforceable as a certificate, opinion or decision of the architect if he issues it as required by clause 3(8). Before that, in my view, the final certificate was never more than a piece of paper with a signature on it.

The second point taken by [counsel for the defendants] relates to the interim certificates including interim certificate no. 56. He argues that each interim certificate carries with it the opinion of the architect that the work done by the contractor up to the date of the issue of an interim certificate was properly done on the grounds that the architect was obliged under clause 30(2) to state in the interim certificate the total value of the work properly executed by the contractor. For the same reasons that I have given in relation to the final certificate, I consider that interim certificate no. 56 has no force or effect. As to the earlier interim certificates, which were unquestionably issued and acted upon, I conclude that there is a very short answer. The primary purpose of the interim certificates in this kind of contract is to ensure that the contractor will receive regular stage payments as his work progresses. In this context, it is expressly provided by clause 30(8) of the conditions that, except for a final certificate which is to be 'conclusive evidence in any proceedings arising out of this contract . . . that the works have been properly carried out and completed in accordance with the terms of this contract' provided that a written request to concur in the appointment of an arbitrator in accordance with clause 30(7) has not been made, 'no Certificate of the Architect/Supervising Officer shall of itself be conclusive evidence that any works, material or goods to which it relates are in accordance with this Contract'. This means that each time an interim certificate is issued the architect is only expressing a provisional view that the works have been properly executed up until the date when he makes his certificate. No doubt, as Judge Sir William Stabb held in *Sutcliffe* v. *Chippendale & Edmondson* (1971), the architect is under a duty only to certify payment for work properly done when he issues an interim certificate. This does not mean that the architect has expressed a final view as to the state of the works in relation to the contractual obligations of the contractor. He is plainly free to take a fresh view each time he issues an interim certificate and his opinion about the quality of the works cannot become conclusive in my view until he issues his final certificate. I do not think that the fact that the architect, as in this case, has issued certificates that defects required by him to be remedied by the contractor have been remedied affects this position at all. If defects emerge before the architect issues his final certificate, he is entitled to and, indeed, obliged, in order to carry out his duty to his employer, to withhold that certificate. It follows, in my view, that the interim certificates are not binding upon the plaintiffs as to the quality of the defendants' workmanship under the provisions of the contract. I can therefore see no reason at all why they cannot pursue their claim for damages for breach of contract on the grounds of bad workmanship. Their claim does not, in my judgment, involve opening up or reviewing the interim certificates issued under the contract. There is in reality no opinion, decision or certificate of the architect to open up in the context of this case. The observations I have quoted from the judgments in *Crouch* relate in my judgment to certificates, opinions and decisions of the architect which have a binding effect and are subject to review only by an arbitrator appointed under the provisions of the contract. For these reasons, I consider that the answer to the question raised by the second preliminary issue is 'no'. In these circumstances, the question raised in the third preliminary issue does not arise.

Under JCT terms, once the architect has issued the final certificate he is* functus officio *unless notice of arbitration has been given.

H. Fairweather Ltd v. Asden Securities Ltd

QUEEN'S BENCH DIVISION

(1979) 12 BLR 40

The contract between the parties was in JCT 63 form, the date for completion being 5 March 1974. Thirty-one weeks' extensions of time were granted by the architect, giving a revised date for completion of 10 October 1975. The date of practical completion was certified to be 1 December 1975. The architect issued a final certificate under clause 30(6) on 28 July 1977. On 26 March 1978 the architect issued a certificate under clause 22 to the effect that the works ought reasonably to have been completed by 10 October 1975. The defendants withheld £13,300 from sums otherwise certified in favour of the contractors, relying on the certificate of 26 March 1978.

Held: The contractors were entitled to recover £13,300, together with interest. The balance certified to be due in the final certificate was the amount due, and the subsequent purported certificate was invalid.

JUDGE WILLIAM STABB QC: If the architect were permitted to issue a certificate under clause 22 after the final certificate, an air of uncertainty would be introduced in the sense that neither party would know when the account could be regarded as finalised. I think that once the architect has issued his final certificate, and provided no notice of arbitration is given by either party within the permitted time, and provided that none of the three specific exceptions apply, the architect should be regarded as *functus officio*, and he is precluded thereby from issuing thereafter any valid certificate under clause 22 of the conditions of the contract.

There is no express or implied term of JCT contracts which gives the quantity surveyor power or authority to decide liability. He has power under the express terms of the contract to decide* quantum *only.

John Laing Construction Ltd v. County & District Properties Ltd

QUEEN'S BENCH DIVISION
(1982) 23 BLR 1

Laing was main contractor under a JCT 63 (July 1969 revision) form for the construction of a shopping development at Chatham. The work began in 1971 and ended in 1976. Disputes arose between the parties about various matters, including the amount of fluctuations payable, under clause 31(d).

Held: The quantity surveyor has no power to agree liability as to an amount payable to or allowable by the contractor.

WEBSTER J: [The provisions of JCT 63] together with the implication that [the quantity surveyor] will probably have prepared the bills of quantity referred to in the recital to the articles of agreement, lead me to the conclusion that, save for the fact that if he prepares the bills he will have specified the quality of the works, his functions and his authority under the contract are confined to measuring and quantifying and that the contract gives him authority, at least in certain instances, to decide *quantum* but that it does not in any instance give him authority to determine any liability, or liability to make any payment or allowance . . .

Recovery without certificate

Where the contractor can show that the architect (or other certifier) was disqualified from issuing a valid certificate, he can sue for moneys otherwise due to him.

Page v. Llandaff and Dinas Powis Rural District Council

QUEEN'S BENCH DIVISION
(1901) 2 HBC 316

Page was employed by the council to construct certain works. The contract provided that 'the decision of the surveyor with respect to the value, amount, state and condition of any part of the works (etc.) shall be final and without appeal'. The surveyor duly gave his final certificate and Page alleged that it 'was not honestly made or given in the exercise of, or in reliance upon, his own judgment, but was made and given by reason of the interference of, and in obedience to, the directions and orders of the council.' The council had, in fact, instructed their surveyor not to apply day-work rates for part of the works but that he was to use estimated quantities and apply a measured rate.

Held: This amounted to interference with the surveyor's functions. Accordingly, the final certificate was not conclusive and binding on the contractor.

Chapter 11
Variations

Most building contracts empower the architect to issue 'Variation Orders' in certain (often widely defined) circumstances. However, it is clear that the work which forms part of the contract cannot be taken away from the contractor and given to others, even where the variation clause is couched in wide terms.

Carr v. J.A. Berriman Pty Ltd

HIGH COURT OF AUSTRALIA
(1953) 27 ALJR 273

The defendant contracted to erect a factory for the plaintiff. The contract provided for the free issue of steel by the plaintiff and its fabrication by the defendant to the architect's instructions. The architect subsequently wrote to the defendant, stating that the plaintiff had awarded the contract for the fabrication of the structural steel to another company.

Held: Even if the architect's letter were treated as a variation order, it was outside the powers conferred by the variation clause and therefore amounted to a breach of contract entitling the defendant to terminate the contract.

FULLAGAR J: The relevant part of clause 1 of the conditions . . . is contained in the words: 'The Architect may in his absolute discretion and from time to time issue . . . written instructions or written directions . . . in regard to the . . . omission . . . of any work . . . The Builder shall forthwith comply with all Architect's Instructions.' Clause 1 is part of a printed form, and the powers conferred upon the architect extend to the giving of directions on a great variety of matters in addition to the 'omission of any work.' The clause is a common and useful clause, the obvious purpose of which – so far as it is relevant to the present case – is to enable the architect to direct additions to, or substitutions in, or omissions from, the building as planned, which may turn out, in his opinion, to be desirable in the course of the performance of the contract. The words quoted from it would authorize the architect (doubtless within certain limits, which were discussed in *R.* v. *Peto* (1826) to direct that particular items of work included in the plans and specifications shall not be carried out. But they do not, in my opinion, authorise him to say that particular items so included shall be carried out not by the builder with whom the contract is made but by some other builder or contractor. The words used do not, in their natural meaning, extend so far, and a power in the architect to hand over at will any part of the contract to another contractor would be a most unreasonable power, which very clear words would be required to confer.

Commissioner for Main Roads v. Reed & Stuart Pty Ltd

HIGH COURT OF AUSTRALIA

(1974) 12 BLR 55

Under a contract for road-works, contractors were to move and spread top-soil as part of the works. The specification provided that if there was insufficient top-soil on site, 'the Engineer may direct the Contractor in writing to obtain top-soil' elsewhere, payment to be made at a prescribed rate. Instead of invoking this provision the engineer instructed a third party to bring more top-soil onto the site.

Held: This was a breach of contract on the employer's part. Although the contract entitled the employer to omit work from the contract, it did not entitle him to take away part of the contract work so that it might be given to another contractor.

STEPHEN J: The interaction of two factors, the shortage of on-site top-soil and the underestimation of the total amount of material required to carry out all necessary top-soiling, necessarily produced, in the case of a contract such as the present, an acute conflict between the interests of the respective parties to it. The underestimation meant that the contractor had, for the same lump sum reward, to perform considerably more top-soiling than was contemplated when tendering; but if, because of the shortage of on-site top-soil, it could invoke the fourth paragraph of clause B3.03 the entire financial position would alter, it would then be paid at what was apparently a profitable rate, for all imported top-soil needed to make good that shortage and this regardless of the initial underestimation.

When the shortfall of top-soil manifested itself the contractor sought to, but the commissioner refused to, invoke these provisions of the contract which were designed to deal with that eventuality; instead the commissioner adopted a quite different course. The commissioner's Engineer decided that, rather than incur the rate of £3 per cubic yard, he would instead, by the exercise of what he regarded as powers available to him under the contract, arrange for the work of importing top-soil onto the site to be done by a third party, no doubt at cheaper rates.

Were he legally entitled to do so it would, I think, run counter to a concept basic to the contract, namely that the contractor, as successful tenderer, should have the opportunity of performing the whole of the contract work. By the contract the contractor had covenanted that for the bulk sum of almost £5 million it would perform the works and supply all the materials shown in the other contract documents. That this included the placing of all top-soil called for by the contract drawings is clear from those drawings, from the definition of 'Works' in the general conditions of contract and from the concluding words of clause A1.22 of the specification, which expressly includes in the contract work the placing of top-soil as shown in the contract drawings. Consistently with this the second paragraph of clause B3.03 provides that top-soil shall be placed by the contractor 'on batters, medians and at other locations shown on the Drawings or as determined by the Engineer . . . '. Then the third paragraph of clause B3.03 gives meaning to the figure of 49,700 cubic yards of top-soil appearing in the schedule to the tender by describing it as the estimated quantity of top-soil required in place within the limits of the completed Works. Finally, to meet the contingency of a shortfall in on-site top-soil, there appears the fourth paragraph of clause B3.03, which I have already set out in full.

The engineer sought to rely upon two provisions of the contract to attain his purpose. He regarded the fourth paragraph of clause B3.03 as conferring upon him, or rather, in his words, upon 'the Department', an option and initially told the contractor that in exercise of that option separate arrangements would be made for the supply of the shortfall of top-soil. He at the same time told the contractor that the Department would 'relieve your company of the responsibility of undertaking any further part of the work required under clause B3.03 of the specification' and that a deduction from the lump sum price would be made, at the rate of 15s per cubic yard, in respect of the difference between the quantity of on-site top-soil already spread and the quantity, as estimated at time of tender, of 49,700 cubic yards; he relied upon clause 18 of the general conditions of contract for this action.

Clause 18 is a common enough provision to be found in engineering contracts and permits of the omission from time to time by the proprietor of portions of the contract works. What it clearly enough does not permit is the taking away of portions of the contract work from the contractor so that the proprietor may have it performed by some other contractor: *Carr* v. *J.A. Berriman Pty Ltd* (1953). Yet this was what the engineer sought to do in the present case in relation to spreading of top-soil.

The commissioner was, in my view, in breach of his contractual obligations under the contract. First there was the failure of the engineer to direct the contractor to obtain additional top-soil from outside the site once it was decided that the contract work of spreading top-soil to the extent shown in the contract drawings should proceed despite the shortfall of on-site top-soil. Secondly, there was the closely allied act of taking away from the contractor the balance of top-soil placement work, using imported top-soil, and, in consequence, the deduction from the lump sum price of an amount calculated by reference to the uncompleted portion of the originally estimated cubic yardage of top-soil required to be placed on site.

The shortfall having become apparent and it having been decided that all the originally contemplated top-soil spreading should nevertheless be carried out, the engineer, by failing to give a direction under the fourth paragraph of clause B3.03, rendered it impossible for the contractor to perform its contractual obligations; without such a direction it was confined to the use of on-site top-soil of which there was insufficient. The resultant situation is not dissimilar to that which arose on the facts in *Carr* v. *J.A. Berriman Pty Ltd*, where there had been a failure to give possession of the building site to the contractor, thus constituting a breach of contract, and see *Freeman & Son* v. *Hensler* (1900), where the members of the Court of Appeal, in separate judgments, each regarded as a term necessarily to be implied into a building contract the giving of possession of the site to the contractor. As Lord Atkin said, in a quite different context, in *Southern Foundries (1926) Ltd* v. *Shirlaw* (1940), there is 'a positive rule of the law of contract that conduct of either promiser or promisee which can be said to amount to himself "of his own motion" bringing about the impossibility of performance is in itself a breach'.

It is often difficult to decide whether a departure from what the parties expected to be done under the contract is a variation. It must depend principally on careful construction of the contract, since it may be that the contractor has agreed to do all that is necessary to finish the job.

Sharpe v. San Paulo Brazilian Railway Co

COURT OF APPEAL IN CHANCERY

(1873) 8 Ch App 597

Contractors agreed to construct a railway. The contract provided that the company should not, under any circumstances, be liable to pay to the contractors a greater sum than £1,745,000, and the contractors bound themselves to execute and provide not only all the work and materials specified, but also such other works and materials as in the judgment of the employer's engineer were necessarily or reasonably implied in and by, or inferred from, the specification, plans and sections. The contract also provided that the engineer's certificate of the ultimate balance of the account should be 'final and conclusive'.

After completion of the works, the contractors claimed to be entitled to payment in respect of extra works executed under the engineer's direction, alleging that they had done so upon the faith of his promise that they would be paid.

Held: (1) The contractors were bound by the contract to complete the work for a specified sum.

(2) The engineer had no power or authority to alter the terms of the contract and that his promise to pay, if made, was unenforceable.

(3) This was so even if the amount of work was understated in the specification.

(4) In the absence of fraud, the engineer's certificate was conclusive between the parties.

SIR WILLIAM JAMES JW: The first contract was that the line should be completed for a fixed sum. But the plaintiffs say they are, upon several heads, entitled to a great deal more than that sum. The first head is that the earthworks were insufficiently calculated, that the engineer had made out that the earthworks were two million and odd cubic yards, whereas they turned out to be four million and odd cubic yards. But that is precisely the thing which they took the chance of. They were to judge for themselves. There was no fraud; it is not alleged that [the engineer] had wilfully made miscalculations for the purpose of deceiving them; and if so, that would be the personal fraud of [the engineer] himself. But he made the calculations apparently to the best of his ability, and calculated that the earthworks would be of a certain amount. The plaintiffs says it is quite clear that this was a miscalculation. But that was a thing the contractors ought to have looked at for themselves. If they did not rely on [the engineer's] experience and skill as an engineer, they ought to have looked at the consequences and made out their own calculations. I think probably they did look at it, and satisfy themselves to some extent. I do not know whether they calculated it over again, or whether they did not take 5s. a cubic yard, which seems a large sum for earthworks, for the purpose of covering them in all possible contingencies in that respect. But that is one of the things which, in my mind, was clearly intended to be governed by the contract, the company virtually saying, 'Whether the earthwork is more or whether it is less, that is the sum we are to pay'.

Then there was a considerable item as to the inclines . . . but every statement in the bill, it seems to me, puts the plaintiffs completely out of Court as to that. The bill says that the original specification was not sufficient to make a complete railway, and that it became obvious that something more would be required to be done in order to make the line. But their business, and what they had contracted to do for a lump sum, was to make the line from terminus to terminus complete, and both these items seem to me to be on the face of them entirely included in the contract. They are not in any sense of the words extra works.

Then it is alleged that the engineer, finding out that this involved more expense than he had calculated upon, promised that he would make other alterations in the line, making a corresponding diminution so as to save the contractors from loss on account of that mistake. And then in the vaguest possible way it is said that all these promises of the engineer were known to and ratified by the company. I am of opinion you cannot in that way alter a contract under seal to do works for a particular sum of money. The plaintiffs cannot say that the company is to give more because the engineer found he had made a mistake and promised he would give more, and the company verbally, or in some vague way, ratified that promise.

To my mind it was a perfectly *nudum pactum*. It is a totally distinct thing from a claim to payment for actual extra works not included in the contract.

Another situation which often arises is that the contract is of a 'measure and value' kind. Here the stated quantities are often no more than working figures and the agreement is to do so much of the work as may be required at the agreed rate.

Arcos Industries Pty Ltd v. The Electricity Commission of New South Wales

NEW SOUTH WALES COURT OF APPEAL

(1973) 12 BLR 65

The plaintiffs contracted to undertake extensive building work for the defendants at a power station on a 'schedule of rates' contract. When the work was in fact done, it was discovered that the amounts of earthworks and concrete works actually to be done fell short of the estimates in the schedule by more than 10 per cent. The plaintiffs sought a declaration that this was an omission which could not be made without their approval.

Held: The plaintiffs' claim must fail.

JACOBS P: Incorporated in the contract is a large schedule of rates covering the work in all its aspects. In the schedule there are at the beginning certain general items, then items related to earth works followed by items related to drainage, cable conduits, concrete work of many kinds, steel work, metal work, demolition, brick work, roof and wall sheeting, carpentry and joinery, plastering and so on through many different subject matters of construction. In respect of all of these items there are carried across the schedule various columns expressing first the unit of measurement of quantity, then the quantity of such units, then the rate per unit and lastly a multiplication of the quantity by the rate. This last column is described as the price. The price is added up in a summary of prices at the end of the schedule and comes to $6,797,512.38. The tender addressed to the Electricity Commission stated that Arcos offered to execute the works described for the prices and at the rates stated in the documents.

It then noted that the total amount of the tender calculated in accordance with the above-mentioned prices and rates and the estimated quantities (if any) stated in such documents is $6,797,512.38.

So far, then, there can, in my opinion, be no doubt of the intention of the parties. The quantities stated in the schedule were estimated quantities only. The contractor was to be paid only for work actually done, and the price which it was to receive was to be the product of the actual quantities multiplied by the relevant rate per unit of quantity. If more were required to support this construction of the intention of the parties, it is to be found in a number of places. Particularly in the specification in its introductory section the attention of the tenderer and contractor is drawn to the likelihood of a variation between the levels and dimensions shown on the drawings provided as a basis for tendering and the actual levels and dimensions which will be required in execution of the works . . .

The clause deals with variations of the work by way of addition thereto or omission therefrom. Such a variation is described by inference as an extra and in appropriate cases where there is authority of the engineer as an authorised extra. Then, with what to me appears to be a reference back to paragraph 1 of the clause, there is a provision in the nature of a proviso that in certain cases the approval of the contractor is required. There is to be no variation by way of addition or omission exceeding ten per cent of the contract price. The clause does not at this point refer to 'variation' because it proceeds at once to deal with additions and omissions; but such additions or omissions have already in the clause been described as things which 'vary the work' and 'extras', if they so vary the work. Then it is necessary to determine a basis of valuation so that the percentage of ten per cent or fifteen per cent may be applied to it. In the case of a lump sum contract there is no difficulty. But in the case of a schedule of rates contract a base must be specified. Therefore 'contract price' is given a very special meaning in paragraph 3 because, as the parties know, there is no fixed contract price in such a contract. So, the contract price is defined in such a case as the total value of the work; not the total estimated value, but the total value. Before it is possible to determine the total value of the work it is necessary to determine what the work is and this to me is the crucial point. The reason for having a schedule of rates contract is that the extent of the work cannot at the outset be predicated so that a contract price may be firmly stated. The nature of the work is certain, but its extent is not. The work must be completed, but the extent of it depends on information not available at the contract stage. So for the special purposes of clause 11 there must be a contract price, even in the case of a schedule of rates contract. However, that price is not the price stated in the contract as the estimated value; it is to be the true value of the work of the nature agreed to be done at the schedule of rates. This work is to be defined irrespective of any extras, that is to say, of variations in the nature of the work to be done. But variations in the levels and dimensions of the work are integral to a schedule of rates contract. The 'total value of the work' can never be determined until the extent of the work has been determined. Until then there is only a total estimated value, and that is not what the contract refers to.

I am reinforced in this view by the fact that a contrary view would seem to run counter to the whole basis of a schedule of rates contract. Its basis is that only actual work will be paid for at the rates in the schedule. Quantities as a result of the uncertainty of level or dimensions cannot initially be ascertained. If it were intended that the basis clearly expressed in clause 2(a) should be substantially modified, as it would be on the suggested interpretation of clause 11, one would expect to find such a modification of the apparent contractual basis otherwise than in a clause which on any view is primarily directed towards true variations in the works, a separate subject matter.

Grinaker Construction (Transvaal) Pty Ltd v. Transvaal Provincial Administration

SUPREME COURT OF SOUTH AFRICA

(1981) 20 BLR 30

The plaintiffs contracted to build a road for the defendants on a schedule contract. The plaintiffs argued that increases or decreases of the quantities listed in the schedule of quantities amounted to a variation.

Held: This argument should be rejected.

> VILJOEN JA: I do not agree with this argument. I agree with counsel that sub-clause (4) should be construed in the light of the sub-clauses which precede it. When as construed clause 49(4) does not, in my view, include the so-called automatic increases and decreases. The basic flaw in counsel's argument is that it equates a variation of the quantity of the works as envisaged in sub-clause (1) with an increase or decrease of the quantities in the schedule of quantities. His entire further argument on the linguistic construction to be put upon clause 49 precede from this fallacious premise. In the normal course of the execution of the work in accordance with the drawings and specifications, even though the quantities involved may differ from those appearing in the schedule of quantities, there is no cause for the engineer to intervene and to make any variation of the form, quality or quantity of the works or part thereof. From the conditions of the contract quoted above it is abundantly clear that the schedule of quantities contains 'approximate' quantities only and that the final contract price is to be computed according to the actual quantities measured. Since the quantities therein specified are expressly stated to be 'approximate', the contract itself envisages that the final computation, after due completion of the works, may result in an increase or decrease in the quantities stated in the schedule of quantities.
>
> The definition of 'Schedule of Quantities' emphasises the fact that the quantities entered therein are approximate quantities of work. This is repeated frequently throughout the contract. In my view it simply means that the quantities are an estimate by the engineer and the fact that the contractor is required to satisfy himself as to the conditions likely to influence the work is an indication that the employer does not warrant that the engineer's estimates are correct or even nearly correct. It stands to reason that in a contract of this nature no accurate or nearly accurate measurements in respect of certain items of work can be made unless intensive and extensive exploratory operations are embarked upon, which may be too costly from a practical point of view.

Where a departure from the specified work has been authorised by or on behalf of the employer this may amount to a variation and entitle the contractor to payment even though the contractor is in breach of contract.

Simplex Concrete Piles Ltd v. Borough of St Pancras

QUEEN'S BENCH DIVISION

(1958) 14 BLR 80

Simplex contracted with the defendants for pile-driving work. The contract incorporated the then current RIBA conditions, which authorised the architect to issue variation instructions and added that if compliance therewith 'involves the contractor in loss or expense beyond that provided for in . . . this contract, then, unless such instructions were issued by reason of some breach of this contract by the contractor, the amount . . . shall be added to the Contract Sum'. Another clause provided that Simplex were to satisfy the engineer as to the stability and adequacy of the piles at all stages of construction and 'shall be liable to make good at his own cost any failure or inadequacy of the work . . . due to faulty design, materials or construction irrespective of any approvals that may have been given'.

As work proceeded, tests showed that the piles were unlikely to be able to carry the working load specified. Simplex suggested to the architect that they should employ a third-party to install bored piles at extra cost and wrote to him seeking his 'instructions and views as to the extra cost which will be involved'. The architect replied that the council were 'prepared to accept your proposal that piles supporting Block "A" should be of the bored type in accordance with the quotations submitted by' the third party. Simplex proceeded to use bored piles and claimed the extra cost involved on the basis that the architect's letter was a variation instruction. Simplex conceded that, but for the work sanctioned by the letter, they would have been in breach of contract.

Held: The architect's letter constituted a variation instruction and Simplex were entitled to payment.

EDMUND DAVIES J: [It was] submitted that, unless the letter amounted to a variation, the plaintiffs cannot succeed, whatever the architect may have thought about the matter. In my judgment, it is not immaterial that the architect intended to authorise a variation, thought he was authorising variation, and thought further that, as the employers were being rendered liable, he must check on the extra expense involved, all of which factors I find were present in [the architect's] mind when he wrote his letter.

On these findings, the one question that remains is: Was it a variation which, under contract, the architect was empowered to authorise? If it was, then the liability of the defendants would appear to be clear.

From clause 1 of the general conditions it is clear that the architect's instruction does not necessarily involve any variation at all. [Counsel] submitted that a variation must involve either (1) the doing of something extra which (in the absence of instruction) the contractor is not required to do, or (2) the omission of something which he would otherwise be obliged to do, or (3) a combination of both (1) and (2), but that the 30 July letter merely gave the contractor precise instructions as to the mode of doing that which he had already bound himself to do by clause 3(a) of the general conditions and preliminaries. He relied strongly upon clause 7 thereof, requiring the contractor:

> to make good at his own cost any failure or inadequacy . . . due to faulty design, materials or construction irrespective of any approvals that may have been given

and I accept that the piles driven by the plaintiffs were gravely inadequate by reason of 'faulty design' or 'construction' or a combination thereof, and I have already said that in my view the plaintiff company would have had no defence to an action for breach of contract when on 16 July they made it clear that their original design of 164 driven piles had proved in the circumstances to be an impracticable and inadequate proposition. On any view of the case, they were accordingly very fortunate in finding so amenable and co-operative an architect . . .

[Counsel] is clearly right in his submission that an architect (whatever he himself may think) has in law no power to instruct or authorise as an extra something which the contractor is already bound to perform, but is that the correct way to describe what this particular architect did? I think not. In my judgment, it is an oversimplification of the case to say that, following upon the 30 July letter, the plaintiffs did no more than they were contractually bound to do and that, accordingly, they cannot claim extra payment. If they failed to carry out the Section 'A' works, their liability was in damages, which might as events turned out be substantial or small, according as to what alternative arrangements the defendants might have been able to make. What the contract did not oblige them to do was to get the work done by subcontractors in a wholly different way at considerable extra expense.

By clause 1 of the general conditions:

> The architect may in his absolute discretion . . . issue written instructions . . . in regard to (a) The variation or modification of the design quality or quantity of the works or the addition or omission or substitution of any work.

Accordingly it is not right to say that a variation must involve an addition to or omission from the contracted works. The contractor's detailed specifications (schedule I) here are for 32ft long, 18-inch diameter Simplex Cast Insitu driven piles, and such specifications constitute or describe in part the 'design' or 'quality' of the works. I regard as immaterial the fact that such 'design' emanated from the contractor rather than from the employer, just as I do the fact that the first suggestion of bored piles also came from the contractor. I agree with [counsel for the plaintiffs], that the architect's letter of 30 July contained an instruction involving a variation in the design or quality (or both design and quality) of the works which the plaintiffs were being instructed to perform, and I have already indicated my view that he did so in circumstances in which he was accepting on the employers' behalf that they would be responsible for the extra cost involved. Such an action fell, in my judgment, within the 'absolute discretion' vested in him by clause 1 and was motivated by his great desire 'to get the job moving', as he put it, and regardless of the legal position of the plaintiffs under their contract. It was an action which led to the plaintiffs doing something different from that which they were obliged to do under their contract, and it was an action which involved the defendants in responsibility for the extra expense which it entailed.

In some cases the parties disagree as to exactly what the contract requires. The architect may insist that the contractor carries out the contract according to what he regards as its strict meaning. Supposing the court later holds that the contractor's view as to the meaning of the contract was correct, does the architect's insistence amount to a variation order? Somewhat different views have been taken on this question.

Molloy v. Liebe

PRIVY COUNCIL

(1910) 102 LT 616

A contract provided that no works beyond those included in the contract would be paid for without a written order from the employer and architect. While the works were in progress the employer required certain work to be done, which he insisted was within the contract. The contractor insisted that the work was not within the contract and must be paid for.

Held: It was open to the arbitrator, if he concluded that the works were not in fact included in the contract, to hold that the employer had impliedly promised to pay for them.

Peter Kiewit Sons' Company of Canada Ltd v. Eakins Construction Ltd

SUPREME COURT OF CANADA

(1960) 22 DLR (2d) 465

The appellant was engaged under a contract with the local authority for work on the approach road for a bridge across Vancouver Harbour. The appellant sub-contracted the piling work to the respondent. During the course of the piling work the authority's engineer insisted that many of the piles be driven to a greater depth than the respondent considered was required by the sub-contract. The respondent complained to the appellant that the work was not required under the contract but was told that he must comply with the instructions of the engineer. Both the appellant and the engineer on behalf of the authority made it clear that they would not authorise extra payment.

The respondent claimed extra payment both from the authority and the appellant. Its action against the authority was dismissed by the trial judge and this judgment was affirmed by the British Columbia Court of Appeal and no appeal was made to the Supreme Court. The action against the appellant also failed at first instance but an appeal was allowed by the British Columbia Court of Appeal. The trial judge held that the engineer was entitled to act as he did, but the Court of Appeal held that the respondent's view of the contract's meaning was correct.

Held (without deciding): that on the correct interpretation of the contract the appeal succeeded and the respondent had no claim to extra payment. The majority took the view that if the engineer's instructions were not justified, the respondent's correct course was to refuse to carry them out without a variation order. Cartwright J, dissenting, took the view that this was an unrealistic course of action to propose.

CARTWRIGHT J (dissenting): Howell, with the fullest knowledge that the respondent was taking the position that it was being called on to do work entirely outside its contract and would expect and demand to be paid for it (a position which, in my opinion, both in fact and in law it was justified in taking) persisted in ordering that work to be done. In these circumstances the law implied an obligation on the part of the appellant to pay for that work of the performance of which it has had the benefit. I find some difficulty in basing the appellant's liability on an implied contract when the evidence shows that the respondent was repeatedly pressing the appellant to agree that it would pay for the work which it was doing and which did not fall within the terms of the sub-contract, and the appellant instead of so agreeing was making only 'nebulous statements' to the effect that the respondent ought to be paid or that 'there was something coming to' the respondent. I prefer to use the terminology which has the authority of Lord Mansfield and Lord Wright . . . and to say that the appellant having received the benefits of the performance by the respondent of the work which the latter did at the insistence of the former, the law imposes upon the appellant an obligation to pay the fair value of the work performed.

It is said that the respondent (who held what turns out to be the right view as to the meaning of the sub-contract) should have had the courage of its convictions and refused to perform any work beyond that which was required by the sub-contract, and if this resulted in its being put off the job should have sued the appellant for damages. It must, however, be remembered that the sub-contract was so difficult to construe that there has been a difference of judicial opinion as to its true meaning. The appellant (who held what turns out to be a mistaken view as to the meaning of the sub-contract) threatened the respondent with what might well amount to financial ruin unless it did the additional work which the sub-contract did not obligate it to do. To say that because in such circumstances the respondent was not prepared to stop work and so risk the ruinous loss which would have fallen on it if its view of the meaning of the contract turned out to be erroneous the appellant may retain the benefit of all the additional work done by the respondent without paying for it would be to countenance an unjust enrichment of a shocking character, which, in my opinion, can and should be prevented by imposing upon the appellant the obligation to pay to which I have referred above.

JUDSON J: Had it been necessary to choose between these two views of the legal relations between the parties, I would have preferred the view of the learned trial Judge that the Eakins company was performing no more than its contractual duty. But quite apart from this, it is to me an impossible inference in this case that the parties agreed to substitute a new contract for the original one. From the very beginning, the Eakins company knew of this added term. It began to protest late in the day that the term imposed added obligations. The engineer, who had clearly defined duties under the main contract, denied any such interpretation. Nothing could be clearer. One party says that it is being told to do more than the contract calls for. The engineer insists that the work is according to contract and no more, and that what is asserted to be extra work is not extra work and will not be paid for. The main contractor tells the sub-contractor that it will have to follow the orders of the engineer and makes no promise of additional remuneration. In these circumstances the sub-contractor continued with the work. It must be working under the contract. How can this contract be abrogated and another substituted in its place? Such a procedure must depend upon consent, express or implied and such consent is entirely lacking in this case. Whatever Eakins recovers in this case is under the terms of the original sub-contract and the provisions of the main contract relating to extras. The engineer expressly refused to order as an extra what has been referred to throughout this case as 'overdriving'. The work was not done as an extra and there can be no recovery for it on that basis. When this position became clear, and it became clear before any work was done, the remedy of the Eakins company was to refuse further performance except on its own interpretation of the contract and, if this performance was rejected, to elect to treat the contract as repudiated and to sue for damages. In the absence of a clause in the contract enabling it to leave the matter in abeyance for later determination, it cannot go on with performance of the contract according to the other party's interpretation and then impose a liability on a different contract. Having elected to perform in these circumstances, its recovery for this performance must be in accordance with the terms of the contract.

Blue Circle Industries Plc v. Holland Dredging Company (UK) Ltd

COURT OF APPEAL

(1987) 37 BLR 40

Blue Circle invited Holland to tender for dredging works in Lough Larne. The contract conditions were said to be substantially ICE Conditions Fifth Edition. A special condition 72 said that the deposit of the dredged material would be the subject of negotiations with local authorities. In due course Blue Circle responded in a way which was treated as an unqualified acceptance.

There were parallel discussions between Blue Circle, Holland and various public bodies and as a result it was agreed that the dredged material should be used to form a kidney shaped island in the lough, appropriate for use as a bird sanctuary. In due course Holland quoted for this work and Blue Circle accepted by letter saying 'an official works order will follow in due course'.

It was held that the construction of the island was wholly outside the scope of the original contract and was a new contract and not a variation.

PURCHAS LJ: In considering whether a particular turn of events comes within clause 51 of the General Conditions as a variation, as Mr Joseph correctly submitted, the question must be posed.

Could the employer have ordered the work required by it against wishes of the contractor as a variation under clause 51? If the answer is 'No', then the agreement under which such work is carried out cannot constitute a variation but must be a separate agreement. The original dredging contract provided that the spoil from excavating the channel should be deposited in 'areas within Lough Larne to be allocated . . . upon approval by the local authorities'. In the event, as a result of local pressures and the attitude of the licensing authorities, this term of the contract was impossible to fulfil legally. The only alternatives were dumping at sea or the creation of an artificial bund with the formation of an island. Either of these two solutions was wholly outside the scope of the original dredging contract and therefore, had Holland not been willing, they could not, in my judgment, have been obliged to accept the work as a variation. This is supported by the acceptance of the tender by the order form dated 14 August 1978 in the simple terms already recited, whilst collateral negotiations were in hand for the solution to the problem of discharging the dredged material and the creation of an island. In contrast, however, in the case of the island creation contract the official order form contains in considerable detail aspects of the contract under confirmation. In my judgment, Mr Joseph's submission that the island contract is separate from the dredging contract is correct.

Mr Thornton submitted that in reality if not in form the effect of the negotiations leading to the tender and acceptance of the island creation contract was a variation of the original dredging contract. Having referred to clause 51 he submitted that there had been an oral agreement by negotiations to accept a variation which was subsequently confirmed in writing in accordance with the general conditions. Attractive as this submission was I find myself unable to accept it in the reality of the context of the parties' conduct in reaching the agreement for the island creation scheme and the formal acceptance of the dredging tender in the course of the negotiations for the island creation agreement.

Another case in which the parties differed as to whether the work was an extra was* C. Bryant & Son Ltd *v.* Birmingham Hospital Saturday Fund *(1937), on p. 138. The decisive factor there was that the bills of quantities incorporated the Standard Method of Measurement which required excavation in rock to be given separately.

Even where it is clear that the work is an extra, it may be necessary to construe the contract to discover the appropriate basis of payment.

Sir Lindsay Parkinson & Co Ltd v. Commissioners of Works

COURT OF APPEAL

[1950] 1 All ER 208

ASQUITH LJ: The question in this appeal is whether the original contract between the parties, as varied by the deed of August 18, 1937, is to be read as producing this extraordinary result, namely, first, that in no circumstances may the contractor's profit exceed £300,000, but, secondly, that nevertheless, after he has earned that profit, the Commissioners are entitled to require him to perform further additions and variations till the end of time, without paying him an extra penny in respect of them. For brevity, I will call such variations and additions 'extras'. I propose to consider, in its bearing on this problem, first the construction of the original contract and, secondly, the effect of the superimposition on it of the deed of August, 1937.

Under general condition 33 of the original contract, read literally, the commissioners were entitled to require the contractor to carry out 'extras' without limit of quantum or time. They could require him to erect a series of buildings on the site, pull them down, and put them up again, with or without alterations, as often, and for as long, as they chose. When the original contract alone was concerned, this was not such an unreasonable stipulation as on its face it appears, since two factors would in practice militate against the abuse by the commissioners of their strict legal rights. First, under that contract they would have to pay for all extras under either sub-clause (a) or sub-clause (b) of condition 35 and, secondly, pay for them out of public funds. The contractor might feel some legitimate confidence that, whatever their strict rights, the commissioners would not in practice choose, or perhaps in a democratic country be allowed, to waste the taxpayers' money wantonly. A further consideration suggests some doubt whether such a course of conduct would even be within their strict legal rights. There is a note to item 57 of the first bill of quantities. (The bills of quantities are among the documents in which the original contract was embodied.) The note in question, which my Lord has read, but which I will repeat, reads as follows: 'It is probable that further work to the value of approximately £500,000 will be ordered on a measured basis under the terms of the contract.' So far as the original contract is concerned, it is not suggested that this note was per se a term of that contract, in the strict sense that the commissioners were undertaking to order not less or not more than about £500,000 by value of 'extras'. The very words 'it is probable' rule out any such construction, for they necessarily imply that, in less probable events, the commissioners may order less (or more) than this quantum of extras, a possibility which would not exist if they were not entitled to do so. It was, however, suggested that the presence of this note is indirectly relevant as supporting an implied limitation of the literal scope of condition 33, a limitation to extras on some such scale as £500,000, and at all events precludes so literal a reading of condition 33 as would entitle the commissioners to require extras to be carried out to the tune of, say, £100,000,000 or indeed £2,000,000. It is not necessary to decide this point. If it were, I should for myself feel serious doubt whether any such limitation could validly be implied.

When, however, the original contract is read subject to the overriding effect (where the two are inconsistent) of the deed of August, 1937, the case for reading condition 33 as subject to an implied restriction on its literal tenor is much stronger. Two new considerations here intervene, the first derived from the terms of the deed itself; the second derived from the circumstances which surrounded and led up to its execution. To deal with these in turn, first, if the original contract plus the deed are read without any implied limitation on their literal meaning, the result, as indicated above, is that after £300,000 profit has been earned by the contractor, he can be compelled to labour like the Danaids without reward or limit, on any further 'extras' which the commissioners may elect to exact from him, 'till the last syllable of recorded time'. The restraining practical factor that such 'extras' will have to be paid for, and paid for out of the taxpayers' funds, no longer operates, because of the proviso to clause 4 of the deed imposing a maximum profit. Only the most compelling language would induce a court to construe the combined instruments as placing one party so completely at the mercy of the other. Where the language of the contract is capable of a literal and wide, but also of a less literal and a more restricted, meaning, all relevant circumstances can be taken into account in deciding whether the literal or a more limited meaning should be ascribed to it.

The reasoning in Parkinson ***was applied in the following case.***

Cana Construction Co Ltd v. The Queen

SUPREME COURT OF CANADA

(1973) 21 BLR 12

The appellants were main contractors for the construction of a new postal terminal at Edmonton, Alberta. The contract provided for the main contractor to enter into a sub-contract for the installation of the mail handling equipment with a sub-contractor to be selected by the Crown. The contractual documents required the tenderers not to include supply and installation of the mail handling equipment but to include in the tender figure for overheads, supervision and profit on this installation. The invitation to tender indicated that the estimated cost of the installation was $1,150,000. In fact the contractor was instructed to enter into a sub-contract with the selected installer at a price of $2,078,543. The contractor requested a variation order for $2,171,398 being the sub-contract tender plus 10% of the difference between the sub-contract tender and $1,150,000 to cover overheads and profits. The Crown refused to pay any increased overheads and profits.

Held: In the circumstances the contractor was entitled to a payment for overheads and profits calculated on the real sub-contract price and not on his tender figure, which was based on the estimate.

SPENCE J: Applying this same method of interpretation [i.e., that of Asquith LJ in *Parkinson*] to the present case, I have come to the conclusion that when the parties entered into the original contract then it was on the basis that the appellants would be required to enter into a sub-contract for the installation of the mechanical handling equipment at a cost of about

$1,150,000 and that therefore their bid should include overhead, supervision and profit for that amount. As it turned out, the appellant was required to enter into a sub-contract for an amount of over $2,000,000 and therefore the overhead, supervision and profit on that larger amount was not within the contemplation of either party. None of the authorities cited by the respondent during the argument have convinced me that I should not approach the interpretation of this business contract in a fashion which I regard as business-like reaching the result which I have reached.

In some cases, the architect's failure to issue variation orders may constitute breach of contract by the employer.

Holland-Hannen and Cubitts (Northern) Ltd v. Welsh Health Technical Services Organisation

QUEEN'S BENCH DIVISION

(1981) 18 BLR 80

The plaintiffs were the main contractors for the erection of a hospital at Rhyl under JCT 63 (1969 revision). Crittalls, who were nominated sub-contractors for the supply and installation of the window assemblies, began work in September 1973. By January 1974 it was clear that the window assemblies were failing to keep rainwater out. In due course the judge held that the major reasons for the leaks were design defects although there were also defects in workmanship.

Crittalls proposed remedial works which were discussed at site meetings. (Crittalls accepted that they would be liable either under the sub-contract, or under the direct warranty which they had given to the employer.) The architects took the position that the defects were due to bad workmanship and that Cubitts should put forward Crittalls' remedial proposals as their own. (Evidence at the trial showed that at another Welsh hospital being built at the same time with the same architect and same window sub-contractor, a different main contractor had acceded to this course and a variation order had been issued.) Cubitts took the position that the defects were due to faulty design and that it was impossible to complete the contract without a variation of the design.

Held: (inter alia): By their failure to issue a variation order, the architects had made it impossible for Cubitts and Crittalls to complete the contract works.

Note: In the judgment, the architects are referred to as 'PTP'.

JUDGE JOHN NEWEY QC: In my view, PTP, by their failure to issue a variation instruction, made it impossible for Crittalls to complete their sub-contract, and made it impossible for Cubitts to complete the main contract so far as the windows are concerned.

From then onwards, PTP kept up a public face of disapproval of Crittalls proceeding with installation of windows to a revised design, shown by the seeming clause 6(4) notices while from 6 May 1974 they permitted Cubitts to proceed with finishes, which were wholly dependent on the efficacy of the remedial measures, and repeatedly offered to issue a variation instruction if only Cubitts would recommend the changes in design. What can only be described as the hypocrisy of PTP's position is demonstrated by what happened at Gurnos, where Douglas, more accommodating than Cubitts, gave their support to Crittalls' identical proposals and a variation instruction followed.

It is quite clear from PTP's internal memoranda and from Mr Keegan's remarks at the inspection on 27 September 1978, that PTP were satisfied that Crittalls had by the 1974 and later changes in the windows, overcome both defects in design and workmanship, although they quite properly wished to preserve WHTSO's right to complain if any further defects in workmanship manifested themselves. The promise made by Mr Pickup at the meeting of 8 September 1977 that, subject to water penetration being eliminated, he would instruct PTP to release money to Crittalls indicates that he, representing the mind of WHTSO themselves, was of the same view.

I think that PTP's failure to issue a variation instruction in 1974 rendered Cubitts' and Crittalls' obligation to supply windows pursuant to the entire contracts into which they had entered impossible of performance.

Crittalls went on to provide windows 'at their own risk', meaning that, if the windows should still leak, no payment would be made to them, but in the expectation that, if the windows did not leak, they would be paid. In my view, PTP and WHTSO, by their words and conduct, agreed to that arrangement. I think that the result was a new contract between Crittalls and WHTSO, to which Cubitts were not parties, that, in consideration of Crittalls providing window assemblies for the hospital, WHTSO would pay to them a *quantum meruit*.

PTP's failure to issue a variation instruction when defects in design had become apparent, when they had come to believe that Critalls' remedies were overcoming the defects and when they had no alternative proposals of their own, may possibly be excused on the grounds that they were labouring under a mistake of law as to Cubitts' responsibilities. However, I find it impossible to believe that architects in charge of a great building project, which has been brought to a stop by an unexpected difficulty, are entitled to adopt a passive attitude, as PTP did in this case. PTP's failures were ones of omission rather than of commission, but I think that they none the less amounted to breach of contract.

The same conclusion as I have reached in regard to the issue of a variation instruction applies, I think, to the grant to Cubitts of an extension of time.

In practice many disputes arise as to how variations are to be valued. Each of the standard contracts contains provisions as to this question and of course the key question is the correct construction of the relevant clause. It is the case, however, that where, as is often the position, the contract requires the agreement of rates for particular work, these rates will often form the basis of pricing variations.

Henry Boot Construction Ltd v. Alstom Combined Cycles Ltd

COURT OF APPEAL

(2000) 69 Con LR 27

Boot entered into a contract with Alstom for civil engineering works at Connah's Quay. The contract was substantially on ICE Conditions, 6th edition. The contractual provision for the pricing of variations was clause 52(1) which provided:

> The value of all variations ordered by the Engineer in accordance with clause 51, shall be ascertained by the Engineer after consultation with the Contractor in accordance with the following principles.
> [Rule 1]
> (a) Where work is of similar character and executed under similar conditions to work priced in the Bill of Quantities it shall be valued at such rates and prices contained therein as may be applicable
> [Rule 2]
> (b) Where work is not of a similar character or is not executed under similar conditions or is ordered during the Defects Correction Periods the rates and prices in the Bill of Quantities shall be used as the basis for valuation so far as may be reasonable
> [Rule 3]
> failing which a fair valuation shall be made.
> Failing agreement between the Engineer and the Contractor as to any rate or price to be applied in the valuation of any variation, the Engineer shall determine the rate or price in accordance with the foregoing principles and he shall notify the Contractor accordingly.

(The words in square brackets above are not part of the original text but were used for ease of reference in the arguments and the judgments.)

The contract provided for the incorporation into the contract of post-tender exchanges. Boot tendered on the basis that the cooling water pipe system would be installed throughout the site at a depth of 4.45 m AOD (above datum). Alstom decided to lower the system to 3.35 m AOD. By an exchange of faxes in March 1994 Boot submitted a price of £250,880 for additional and different temporary works and this was accepted by Alstom.

Boot's fax described the figure of £250,880 as being 'for additional and different temporary works only, required in the turbine hall'. If applied to the work done in the turbine hall the figure produced a rate of £89 per square metre. There was extensive extra work outside the turbine hall and if the rate of £89 per square metre were applied to this Boot would be entitled to £2,284,128.

The arbitrator held that the exchange of faxes was incorporated into the contract. He also held that Boot had made a mistake in the calculations but one which he had no power to correct, although he went on to apply a fair valuation.

Judge Humphrey LLoyd QC allowed an appeal from the arbitrator and applied the rate of £89. The majority of the Court of Appeal (Ward LJ dissenting) agreed.

> LORD LLOYD: Although Lord Neill's argument was presented with all his usual skill, I am not persuaded. The meaning of Rule 2 does not, I think, admit of much doubt. It provides a half-way house between Rule 1 and Rule 3. Like Rule 1, Rule 2 is mandatory. It applies when the work covered by the variation order is of a different character from the work priced in the bill of quantities, or is executed under different conditions. If the differences are relatively small, the engineer is obliged to use the rates set out in the bill of quantities as the basis for his valuation, making such adjustment as may be necessary to take account of the differences. But the

differences may be very great, as, for example, where the variation order calls for the excavation of foundations in solid rock, instead of clay. In those circumstances, the engineer may take the view that it would not be 'reasonable' to base his valuation on the rates contained in the bill of quantities. He is then thrown back on Rule 3. That is the sole function of the words 'so far as may be reasonable' in Rule 2. They call for a comparison between the work covered by the variation order and the work priced in the bill of quantities. They do not enable the engineer to open up or disregard the rates on the ground that they were inserted by mistake. As Beldam LJ put it in the course of argument, it is the use of the rates in the changed circumstances brought about by the variation order that must be reasonable, not the rates themselves.

Lord Neill argued that it would be wrong to regard the rates as immutable; he gently derided the learned judge's use of the word 'sacrosanct' in that connection. But it seems to me that that is exactly what the proviso to clause 55(2) requires:

> Provided that there shall be no rectification of any errors, omissions or wrong estimates in the descriptions, rates and prices inserted by the contractor in the Bill of Quantities.

Clause 52(2) on which Lord Neill relied, provides an exception. But it is an exception which proves the rule. It applies 'if the nature or amount of the variation relative to the nature or amount of the contract work' is such as to make it unreasonable to apply the contract rates to the variation. In such a case, the engineer may fix a reasonable rate. Thus, clause 52(2) creates a limited exception to Rules 1 and 2 where the scale or nature of the variation makes it unreasonable to use the contract rates. It certainly does not justify displacing the rates themselves because they were inserted by mistake or are too high or too low or otherwise unreasonable.

The same applies to another provision on which Lord Neill relied, namely, clause 56(2). It enables the engineer to increase or decrease the rates where 'the actual quantities executed in respect of any item [is] greater or less than those stated in the Bill of Quantities'. These limited exceptions underline the basic rule that the rates themselves are not subject to correction.

Any other view would have far reaching consequences. If the engineer were free to open up the rates at the request of one party or the other because they were inserted in the bill of quantities by mistake, it would not only unsettle the basis of competitive tendering, but also create the sort of uncertainty in the administration of building contracts which should be avoided at all costs.

For the above reasons, and the reasons set out in Judge Lloyd's admirable judgment, with which I am in complete agreement, I would dismiss the appeal, and remit the award to the arbitrator in accordance with para 2 of his order. I can well understand the arbitrator's reluctance to extend the effect of the mistake in the bill of quantities. But having held correctly, as I have said, that he had no power to correct the mistake, he should have carried his reasoning through to its logical conclusion. He was bound to disregard the mistake when applying Rule 2. The language of Rule 2 does not permit of any other construction. It follows that in failing to apply Rule 2 the arbitrator erred in law.

Chapter 12
Damages for Breach of Construction Contracts

If the contractor does construction work defectively the contract may give the employer the right to require the contractor to remedy the defect. This right is usually limited to a relatively short period (often called in the contract something like the 'defects liability period') after practical completion. Usually the contract gives the contractor the right as well as the duty to do the remedial work. (It will nearly always be cheaper for the contractor to do the work than anyone else).

If there is no right to call for remedial work, either because this period has elapsed or because there was no such provision in the contract, the employer will normally have an action for damages. How are such damages to be assessed? The cost of remedial work is one measure, but it is not the only one. In some cases the court will award the difference in value between the building as it is and as it would have been if the contract had been properly performed.

Ruxley Electronics and Construction Ltd v. Forsyth

HOUSE OF LORDS

(1995) 45 Con LR 61

The defendant, Mr Forsyth, wished to have a swimming pool built at his house in Kent. He entered into contracts with the plaintiff companies, which were controlled and substantially owned by the same person, for the pool and for a building to enclose it. It was an express term of the contract for the pool that the maximum depth should be 7ft 6in. The present actions were brought by the plaintiffs for the balance of the agreed contract sums.

The maximum depth of the pool was in fact not 7 ft 6 in but 6 ft 9 in. Further, this was exactly at one end of the pool. A person diving normally from the deep end of the pool would be diving into a depth of 6 feet only. It was accepted that the failure to provide the required depth was a breach of contract. Mr Forsyth counterclaimed for breach of contract. The trial judge awarded £2,500 general damages for loss of pleasure and amenity but awarded no special damages on the ground that as built the pool was safe for diving from the side of the pool and that there was no difference in value between the pool as built and the pool as contracted for. The trial judge held that the

Powell-Smith and Furmston's Building Contract Casebook, Fifth Edition. Michael Furmston.

cost of remedial work was £21,500 but that Mr Forsyth could only recover this sum if he could show that he intended to do the remedial work and that it was reasonable to do so. He held that Mr Forsyth did not intend to do the remedial work and that it would have been unreasonable of him to do so. Mr Forsyth appealed.

The Court of Appeal held (Dillon LJ dissenting) (36 Con LR 103) that it is unreasonable of a plaintiff to claim an expensive remedy if there is some cheaper alternative which would make good his loss, but if there is no alternative course which will provide what he requires, or none which will cost less, he is entitled to the cost of repair or replacement even if that is very expensive. Mr Forsyth was therefore entitled to the cost of the remedial work.

Held: The purpose of damages for breach of contract was to compensate the plaintiffs for the loss which had been suffered as a result of the defendant's breach of contract. The central question was how to measure this loss. Where the breach of contract consisted of failure perfectly to carry out building work, loss might be measured either in relation to the cost of remedying the defective work or in relation to the effect of the defective work on the building as a whole. It would be inappropriate to use the cost of building work where it would in the circumstances, as in the present case, be unreasonable to carry out the building work.

Where, as in the present case, the trial judge had decided that, because it would be unreasonable to carry out remedial work, the difference in value was the primary measure, it could be appropriate to make an award for general damages and to compensate the building owner for the disappointment that he experienced because the building work was not as per the agreement.

> LORD MUSTILL: In my opinion there would indeed be something wrong if, on the hypothesis that cost of reinstatement and the depreciation in value were the only available measures of recovery, the rejection of the former necessarily entailed the adoption of the latter; and the court might be driven to opt for the cost of reinstatement, absurd as the consequence might often be, simply to escape from the conclusion that the promisor can please himself whether or not to comply with the wishes of the promisee which, as embodied in the contract, formed part of the consideration for the price. Having taken on the job the contractor is morally as well as legally obliged to give the employer what he stipulated to obtain, and this obligation ought not to be de-valued. In my opinion, however, the hypothesis is not correct. There are not two alternative measures of damage, at opposite poles, but only one: namely the loss truly suffered by the promisee. In some cases the loss cannot be fairly measured except by reference to the full cost of repairing the deficiency in performance. In others, and in particular those where the contract is designed to fulfil a purely commercial purpose, the loss will very often consist only of the monetary detriment brought about by the breach of contract. But these remedies are not exhaustive, for the law must cater for those occasions where the value of the promise to the promisee exceeds the financial enhancement of his position which full performance will secure. This excess, often referred to in the literature as the 'consumer surplus' (see e.g. the valuable discussion by Harris, Ogus and Phillips, 'Contract Remedies and the Consumer Surplus' (1979) 95 *LQR* 581) is usually incapable of precise valuation in terms of money, exactly because it represents a personal, subjective and non-monetary gain. Nevertheless, where it exists the law should recognise it and compensate the promise if the mis-performance takes it away. The lurid bathroom tiles, or the grotesque folly instanced in argument by my

noble and learned friend Lord Keith of Kinkel, may be so discordant with general taste that in purely economic terms the builder may be said to do the employer a favour by failing to install them. But this is too narrow and materialistic a view of the transaction. Neither the contractor nor the court has the right to substitute for the employer's individual expectation of performance a criterion derived from what ordinary people would regard as sensible. As my Lords have shown, the test of reasonableness plays a central part in determining the basis of recovery, and will indeed be decisive in a case such as the present when the cost of reinstatement would be wholly disproportionate to the non-monetary loss suffered by the employer. But it would be equally unreasonable to deny all recovery for such a loss. The amount may be small, and since it cannot be quantified directly there may be room for difference of opinion about what it should be. But in several fields the judges are well accustomed to putting figures to intangibles, and I see no reason why the imprecision of the exercise should be a barrier, if that is what fairness demands.

My Lords, once this is recognised, the puzzling and paradoxical feature of this case, that it seems to involve a contest of absurdities, simply falls away. There is no need to remedy the injustice of awarding too little by unjustly awarding far too much. The judgment of the trial judge acknowledges that the employer has suffered a true loss and expresses it in terms of money. Since there is no longer any issue about the amount of the award, as distinct from the principle, I would simply restore his judgment by allowing the appeal . . .

LORD LLOYD OF BERWICK: In building cases, the pecuniary loss is almost always measured in one of two ways: either the difference in value of the work done or the cost of reinstatement. Where the cost of reinstatement is less than the difference in value, the measure of damages will invariably be the cost of reinstatement. By claiming the difference in value the plaintiff would be failing to take reasonable steps to mitigate his loss. In many ordinary cases, too, where reinstatement presents no special problem, the cost of reinstatement will be the obvious measure of damages, even where there is little or no difference in value, or where the difference in value is hard to assess. This is why it is often said that the cost of reinstatement is the ordinary measure of damages for defective performance under a building contract.

But it is not the only measure of damages. Sometimes it is the other way round. This was first made clear in the celebrated judgment of Cardozo J giving the majority opinion in the Court of Appeals of New York in *Jacob & Youngs Inc* v. *Kent* (1921) 230 NY 239. In that case the building owner specified that the plumbing should be carried out with galvanised piping of 'Reading manufacture'. By an oversight, the builder used piping of a different manufacture. The plaintiff builder sued for the balance of his account. The defendant, as in the instance case, counterclaimed the cost of replacing the pipework even though it would have meant demolishing a substantial part of the completed structure, at great expense . . .

Cardozo J's judgment is important because it establishes two principles which I believe to be correct and which are directly relevant to the present case: first, the cost of reinstatement is not the appropriate measure of damages if the expenditure would be out of all proportion to the good to be obtained, and secondly, the appropriate measure of damages in such a case is the difference in value, even though it would result in a nominal award . . .

Intention

I fully accept that the courts are not normally concerned with what a plaintiff does with his damages. But it does not follow that intention is not relevant to reasonableness, at least in those cases where the plaintiff does not intend to reinstate. Suppose in the present case Mr Forsyth had died, and the action had been continued by his executors. Is it to be supposed

> that they would be able to recover the cost of reinstatement, even though they intended to put the property on the market without delay? . . .
>
> In the present case the judge found as a fact that Mr Forsyth's stated intention of rebuilding the pool would not persist for long after the litigation had been concluded. In these circumstances it would be mere pretence to say that the cost of rebuilding the pool is the loss which he has in fact suffered. This is the critical distinction between the present case and the example given by Staughton LJ of a man who has had his watch stolen. In the latter case, the plaintiff is entitled to recover the value of the watch because that is the true measure of his loss. He can do what he wants with the damages. But if, as the judge found, Mr Forsyth had no intention of rebuilding the pool, he has lost nothing except the difference in value, if any . . .

There have been many cases where houses have been bought in reliance on a surveyor's report which turns out to be negligent. Here the courts have consistently awarded the diminution in value rather than the cost of remedial work.

Watts and Watts v. Morrow

COURT OF APPEAL

(1991) 26 Con LR 98

In 1986, the plaintiffs, who were then married, were looking for a holiday and weekend home. They wanted a country house which would be, as far as possible, trouble-free and into which they could move without undertaking any substantial repair works. They found a farmhouse, but the price was somewhat outside their budget. They made an offer of £177,500, which was the most that they could spend. The offer was accepted, but before proceeding, they instructed the defendant to carry out a full structural survey of the property. Judge Peter Bowsher QC held that the defendant had been negligent in making his survey and report and awarded the plaintiffs damages assessed on the basis of the cost of repairs carried out (amounting to £33,961.35), together with interest thereon at 15%, as well as general damages for distress and inconvenience in the sum of £8,000. The defendant appealed, contending, firstly, that the judge was wrong in law to award damages based on the cost of repairs and that the award should have been on the basis of diminution in value or excess purchase price paid, namely £15,000; secondly, that any award for distress and inconvenience was wrong in law, or alternatively that the award of £8,000 was excessive.

Held (allowing the appeal in part):

(1) The appropriate measure of damages in the case of a negligent survey is the diminution in value of the property together with interest thereon. The task of the court was to award to the plaintiff that sum of money which would, as far as possible, put the plaintiff into as good a position as if the contract for the survey had been properly fulfilled. In this case that figure was £15,000.

(2) Although in the particular circumstances, the plaintiffs were entitled to recover general damages for mental distress caused by physical discomfort or inconvenience resulting by the breach, the award was excessive as only modest compensation is recoverable under this head. A figure of £750 was substituted

RALPH GIBSON LJ: The decision in *Philips* v. *Ward* was based upon that principle: in particular, if the contract had been properly performed the plaintiff either would not have bought, in which case he would have avoided any loss, or, after negotiation, he would have paid the reduced price. In the absence of evidence to show that any other or additional recoverable benefit would have been obtained as a result of proper performance, the price will be taken to have been reduced to the market price of the house in its true condition because it cannot be assumed that the vendor would have taken less.

The cost of doing repairs to put right defects negligently not reported may be relevant to the proof of the market price of the house in its true condition: see *Steward* v. *Rapley* [1989] 1 EGLR 159; and the cost of doing repairs and the diminution in value may be shown to be the same. If, however, the cost of repairs would exceed the diminution in value, then the ruling in *Philips* v. *Ward*, where it is applicable, prohibits recovery of the excess because it would give to the plaintiff more than his loss. It would put the plaintiff in the position of recovering damages for breach of a warranty that the condition of the house was correctly described by the surveyor and, in the ordinary case as here, no such warranty has been given . . .

It is clear, I think, that the judge was regarding the contract between these plaintiffs and the defendant as a contract in which the subject matter was to provide peace of mind or freedom from distress within the meaning of Dillon LJ's phrase in *Bliss*'s case [1987] ICR 700 at 718 cited by Purchas LJ in *Hayes* v. *James & Chales Dodd* [1990] 2 All ER 815 at 826. That, with respect, seems to me to be an impossible view of the ordinary surveyor's contract. No doubt house buyers hope to enjoy peace of mind and freedom from distress as a consequence of the proper performance by a surveyor of his contractual obligation to provide a careful report, but there was no express promise for the provision of peace of mind or freedom from distress and no such implied promise was alleged. In my view, in the case of the ordinary surveyor's contract, damages are only recoverable for distress caused by physical consequences of the breach of contract. Since the judge did not attempt to assess the award on that basis this court must reconsider the award and determine what it should be . . .

BINGHAM LJ:

Damages for distress and inconvenience

A contract-breaker is not in general liable for any distress, frustration, anxiety, displeasure, vexation, tension or aggravation which his breach of contract may cause to the innocent party. This rule is not, I think, founded on the assumption that such reactions are not foreseeable, which they surely are or may be, but on considerations of policy.

But the rule is not absolute. Where the very object of a contract is to provide pleasure, relaxation, peace of mind or freedom from molestation, damages will be awarded if the fruit of the contract is not provided or if the contrary result is procured instead. If the law did not cater for this exceptional category of case it would be defective. A contract to survey the condition of a house for a prospective purchaser does not, however, fall within this exceptional category.

In cases not falling within this exceptional category, damages are in my view recoverable for physical inconvenience and discomfort caused by the breach and mental suffering directly related to that inconvenience and discomfort. If those effects are foreseeably suffered during a period when defects are repaired I am prepared to accept that they sound in damages even though the cost of the repairs is not recoverable as such. But I also agree that awards should be restrained, and that the awards in this case far exceeded a reasonable award for the injury shown to have been suffered. I agree with the figures which Ralph Gibson LJ proposes to substitute . . .

However, it does not necessarily follow that where the survey is negligent, the claimant's only recovery is the difference in value between the valuer's figure and the correct figure.

Farley v. Skinner

HOUSE OF LORDS

(2001) 79 Con LR 1

In 1990 the plaintiff was thinking of buying a house in East Sussex. He engaged the defendant, a chartered surveyor, to inspect and report on the house. He specifically asked the defendant to report on aircraft noise. The defendant advised that he thought it unlikely 'that the property will suffer greatly from such noise'. The trial judge held that this was negligent since there was a point nearby where aircraft had to stack while waiting for a landing slot at Gatwick.

The trial judge held that the purchase price coincided with the open-market value of the house taking into account aircraft noise but he awarded the plaintiff £10,000 as damages for distress and inconvenience.

The defendant appealed. The appeal was heard first by a two-judge court which was equally divided and then by a three-judge court which allowed the appeal by a majority (Clarke LJ dissenting). The plaintiff appealed and the House of Lords unanimously restored the trial judge's decision.

LORD STEYN:

V. Recovery of non-pecuniary damages

16. The examination of the issues can now proceed from a secure foothold. In the law of obligations the rules governing the recovery of compensation necessarily distinguish between different kinds of harm. In tort the requirement of reasonable foreseeability is a sufficient touchstone of liability for causing death or physical injury: it is an inadequate tool for the disposal of claims in respect of psychiatric injury. Tort law approaches compensation for physical damage and pure economic loss differently. In contract law distinctions are made about the kind of harm which resulted from the breach of contract. The general principle is that compensation is only awarded for financial loss resulting from the breach of contract (*Livingstone* v. *Rawyards Coal Co* (1880) 5 App Cas 25 at 39 *per* Lord Blackburn). In the words of Bingham LJ in *Watts* v. *Morrow* (1991) 26 Con LR 98 at 126, [1991] 1 WLR 1421 at 1445, as a matter of legal policy 'a contract-breaker is not in general liable for any distress, frustration, anxiety, displeasure, vexation, tension or aggravation which his breach of contract may cause to the innocent party' (my emphasis). There are, however, limited exceptions to this rule. One such exception is damages for pain, suffering and loss of amenities caused to an individual by a breach of contract (see *McGregor on Damages* (16th edn, 1997) pp 56–57 para 96)). It is not material in the present case. But the two exceptions mentioned by Bingham LJ, namely where the very object of the contract is to provide pleasure (proposition (2)) and recovery for physical inconvenience caused by the breach (proposition (3)), are pertinent. The scope of these exceptions is in issue in the present case. It is, however, correct, as counsel for the surveyor submitted, that the entitlement to damages for mental distress caused by a breach of contract is not established by mere foreseeability: the right to recovery is dependent on the case falling

fairly within the principles governing the special exceptions. So far there is no real disagreement between the parties.

VI. The very object of the contract: the framework

17. I reverse the order in which the Court of Appeal considered the two issues. I do so because the issue whether the present case falls within the exceptional category governing cases where the very object of the contract is to give pleasure, and so forth, focuses directly on the terms actually agreed between the parties. It is concerned with the reasonable expectations of the parties under the specific terms of the contract. Logically, it must be considered first.

18. It is necessary to examine the case on a correct characterisation of the plaintiff's claim. Stuart-Smith LJ thought ((2000) 73 ConLR 70 at 79) that the obligation undertaken by the surveyor was 'one relatively minor aspect of the overall instructions'. What Stuart-Smith and Mummery LJJ would have decided if they had approached it on the basis that the obligation was a major or important part of the contract between the plaintiff and the surveyor is not clear. But the Court of Appeal's characterisation of the case was not correct. The plaintiff made it crystal clear to the surveyor that the impact of aircraft noise was a matter of importance to him. Unless he obtained reassuring information from the surveyor he would not have bought the property. That is the tenor of the evidence. It is also what the judge found. The case must be approached on the basis that the surveyor's obligation to investigate aircraft noise was a major or important part of the contract between him and the plaintiff. It is also important to note that, unlike in *Addis* v. *Gramophone Co Ltd* [1909] AC 488, [1908–10] All ER Rep 1, the plaintiff's claim is not for injured feelings caused by the breach of contract. Rather it is a claim for damages flowing from the surveyor's failure to investigate and report, thereby depriving the buyer of the chance of making an informed choice whether or not to buy resulting in mental distress and disappointment.

19. The broader legal context of *Watts* v. *Morrow* must be borne in mind. The exceptional category of cases where the very object of a contract is to provide pleasure, relaxation, peace of mind or freedom from molestation is not the product of Victorian contract theory but the result of evolutionary developments in case law from the 1970s. Several decided cases informed the description given by Bingham LJ of this category. The first was the decision of the sheriff court in *Diesen* v. *Samson* 1971 SLT (Sh Ct) 49. A photographer failed to turn up at a wedding, thereby leaving the couple without a photographic record of an important and happy day. The bride was awarded damages for her distress and disappointment. In the celebrated case of *Jarvis* v. *Swans Tours Ltd* [1973] 1 All ER 71, [1973] QB 233, the plaintiff recovered damages for mental distress flowing from a disastrous holiday resulting from a travel agent's negligent representations (compare also *Jackson* v. *Horizon Holidays Ltd* [1975] 3 All ER 92, [1975] 1 WLR 1468). In *Heywood* v. *Wellers (a firm)* [1976] 1 All ER 300, [1976] QB 446, the plaintiff instructed solicitors to bring proceedings to restrain a man from molesting her. The solicitors negligently failed to take appropriate action with the result that the molestation continued. The Court of Appeal allowed the plaintiff damages for mental distress and upset. While apparently not cited in *Watts* v. *Morrow, Jackson* v. *Chrysler Acceptances Ltd* [1978] RTR 474 was decided before *Watts* v. *Morrow*. In the *Chrysler Acceptances* case the claim was for damages in respect of a motor car which did not meet the implied condition of merchantability in section 14 of the Sale of Goods Act 1893. The buyer communicated to the seller that one of his reasons for buying the car was a forthcoming touring holiday in France. Problems with the car spoilt the holiday. The disappointment of a spoilt holiday was a substantial element in the award sanctioned by the Court of Appeal.

20. At their Lordships' request counsel for the plaintiff produced a memorandum based on various publications which showed the impact of the developments already described on

litigation in the county courts. Taking into account the submissions of counsel for the surveyor and making due allowance for a tendency of the court sometimes not to distinguish between the cases presently under consideration and cases of physical inconvenience and discomfort, I am satisfied that in the real life of our lower courts non-pecuniary damages are regularly awarded on the basis that the defendant's breach of contract deprived the plaintiff of the very object of the contract, viz pleasure, relaxation, and peace of mind. The cases arise in diverse contractual contexts, e.g. the supply of a wedding dress or double glazing, hire purchase transactions, landlord and tenant, building contracts, and engagements of estate agents and solicitors. The awards in such cases seem modest. For my part what happens on the ground casts no doubt on the utility of the developments since the 1970s in regard to the award of non-pecuniary damages in the exceptional categories. But the problem persists of the precise scope of the exceptional category of case involving awards of non-pecuniary damages for breach of contract where the very object of the contract was to ensure a party's pleasure, relaxation or peace of mind.

21. An important development for this branch of the law was *Ruxley Electronics and Construction Ltd* v. *Forsyth* (1995) 45 Con LR 61, [1996] AC 344. The plaintiff had specified that a swimming pool should at the deep end have a depth of 7ft 6ins. The contractor failed to comply with his contractual obligation: the actual depth at the deep end was the standard 6ft. The House found the usual 'cost of cure' measure of damages to be wholly disproportionate to the loss suffered and economically wasteful. On the other hand, the House awarded the moderate sum of £2,500 for the plaintiff's disappointment in not receiving the swimming pool he desired. It is true that for strategic reasons neither side contended for such an award. The House was, however, not inhibited by the stance of the parties. Lord Mustill and Lord Lloyd of Berwick justified the award in carefully reasoned judgments which carried the approval of four of the Law Lords. It is sufficient for present purposes to mention that for Lord Mustill ((1995) 45 Con LR 61 at 71, [1996] AC 344 at 360) the principle of *pacta sunt servanda* would be eroded if the law did not take account of the fact that the consumer often demands specifications which, although not of economic value, have value to him. This is sometimes called the 'consumer surplus': see Harris, Ogus and Philips 'Contract Remedies and the Consumer Surplus' (1979) 95 *LQR* 581. Lord Mustill rejected the idea that 'the promisor can please himself whether or not to comply with the wishes of the promisee which, as embodied in the contract, formed part of the consideration for the price'. Lord Keith of Kinkel and Lord Bridge of Harwich agreed with Lord Mustill's judgment and with Lord Lloyd's similar reasoning. Labels sometimes obscure rather than illuminate. I do not therefore set much store by the description 'consumer surplus'. But the controlling principles stated by Lord Mustill and Lord Lloyd are important. It is difficult to reconcile this decision of the House with the decision of the Court of Appeal in the present case. I will in due course return to the way in which the majority attempted to distinguish the *Ruxley Electronics* case. At this stage, however, I draw attention to the fact that the majority in the Court of Appeal ((2000) 73 ConLR 70 at 79) regarded the relevant observations of Lord Mustill and Lord Lloyd as obiter dicta. I am satisfied that the principles enunciated in the *Ruxley Electronics* case in support of the award of £2,500 for a breach of respect of the provision of a pleasurable amenity have been authoritatively established.

VII. The very object of the contract: the arguments against the plaintiff's claim

22. Counsel for the surveyor advanced three separate arguments each of which he said was sufficient to defeat the plaintiff's claim. First, he submitted that even if a major or important part of the contract was to give pleasure, relaxation and peace of mind, that was not enough. It is an indispensable requirement that the object of the entire contract must be of this type.

Secondly, he submitted that the exceptional category does not extend to a breach of a contractual duty of care, even if imposed to secure pleasure, relaxation and peace of mind. It only covers cases where the promiser guarantees achievement of such an object. Thirdly, he submitted that by not moving out of Riverside House the plaintiff forfeited any right to recover non-pecuniary damages.

23. The first argument fastened onto a narrow reading of the words 'the very object of [the] contract' as employed by Bingham LJ in *Watts* v. *Morrow* (1991) 26 Con LR 98 at 126, [1991] 1 WLR 1421 at 1445. Cases where a major or important part of the contract was to secure pleasure, relaxation and peace of mind were not under consideration in *Watts* v. *Morrow*. It is difficult to see what the principled justification for such a limitation might be. After all, in 1978 the Court of Appeal allowed such a claim in the Chrysler Acceptances case in circumstances where a spoiled holiday was only one object of the contract. Counsel was, however, assisted by the decision of the Court of Appeal in *Knott* v. *Bolton* (1995) 45 ConLR 127 which in the present case the Court of Appeal treated as binding on it. In Knott's case an architect was asked to design a wide staircase for a gallery and impressive entrance hall. He failed to do so. The plaintiff spent money in improving the staircase to some extent and he recovered the cost of the changes. The plaintiff also claimed damages for disappointment and distress in the lack of an impressive staircase. In agreement with the trial judge the Court of Appeal disallowed this part of his claim. Reliance was placed on the dicta of Bingham LJ in *Watts* v. *Morrow* (1991) 26 Con LR 98 at 126, [1991] 1 WLR 1421 at 1445.

24. Interpreting the dicta of Bingham LJ in *Watts* v. *Morrow* narrowly, the Court of Appeal in *Knott*'s case ruled that the central object of the contract was to design a house, not to provide pleasure to the occupiers of the house. It is important, however, to note that Knott's case was decided a few months before the decision of the House in the *Ruxley Electronics* case. In any event, the technicality of the reasoning in *Knott*'s case, and therefore in the Court of Appeal judgments in the present case, is apparent. It is obvious, and conceded, that if an architect is employed only to design a staircase, or a surveyor is employed only to investigate aircraft noise, the breach of such a distinct obligation may result in an award of non-pecuniary damages. Logically the same must be the case if the architect or surveyor, apart from entering into a general retainer, concludes a separate contract, separately remunerated, in respect of the design of a staircase or the investigation of aircraft noise. If this is so the distinction drawn in *Knott*'s case and in the present case is a matter of form and not substance. David Capper 'Damages for Distress and Disappointment – The Limits of *Watts* v. *Morrow*' (2000) 116 *LQR* 553 at 556 has persuasively argued:

> A ruling that intangible interests only qualify for legal protection where they are the 'very object of the contract' is tantamount to a ruling that contracts where these interests are merely important, but not the central object of the contract, are in part unenforceable. It is very difficult to see what policy objection there can be to parties to a contract agreeing that these interests are to be protected via contracts where the central object is something else. If the defendant is unwilling to accept this responsibility he or she can say so and either no contract will be made or one will be made but including a disclaimer.

There is no reason in principle or policy why the scope of recovery in the exceptional category should depend on the object of the contract as ascertained from all its constituent parts. It is sufficient if a major or important object of the contract is to give pleasure, relaxation or peace of mind. In my view *Knott*'s case was wrongly decided and should be overruled. To the extent that the majority in the Court of Appeal relied on *Knott*'s case their decision was wrong.

25. That brings me to the second issue, namely whether the plaintiff's claim is barred by reason of the fact that the surveyor undertook an obligation to exercise reasonable care and did not guarantee the achievement of a result. This was the basis upon which Hale LJ after the first hearing in the Court of Appeal thought that the claim should be disallowed. This reasoning was adopted by the second Court of Appeal and formed an essential part of the reasoning of the majority. This was the basis on which they distinguished the *Ruxley Electronics* case. Against the broad sweep of differently framed contractual undertakings, and the central purpose of contract law in promoting the observance of contractual promises, I am satisfied that this distinction ought not to prevail. It is certainly not rooted in precedent. I would not accept the suggestion that it has the pedigree of an observation of Ralph Gibson LJ in *Watts* v. *Morrow* (1991) 26 Con LR 98 at 122, [1991] 1 WLR 1421 at 1442: his emphasis appears to have been on the fact that the contract did not serve to provide peace of mind, and so forth. As far as I am aware the distinction was first articulated in the present case. In any event, I would reject it. I fully accept, of course, that contractual guarantees of performance and promises to exercise reasonable care are fundamentally different. The former may sometimes give greater protection than the latter. Proving breach of an obligation of reasonable care may be more difficult than proving breach of a guarantee. On the other hand, a party may in practice be willing to settle for the relative reassurance offered by the obligation of reasonable care undertaken by a professional man. But why should this difference between an absolute and relative contractual promise require a distinction in respect of the recovery of non-pecuniary damages? Take the example of a travel agent who is consulted by a couple who are looking for a golfing holiday in France. Why should it make a difference in respect of the recoverability of non-pecuniary damages for a spoiled holiday whether the travel agent gives a guarantee that there is a golf course very near the hotel, represents that to be the case, or negligently advises that all hotels of the particular chain of hotels are situated next to golf courses? If the nearest golf course is in fact 50 miles away a breach may be established. It may spoil the holiday of the couple. It is difficult to see why in principle only those plaintiffs who negotiate guarantees may recover non-pecuniary damages for a breach of contract. It is a singularly unattractive result that a professional man, who undertakes a specific obligation to exercise reasonable care to investigate a matter judged and communicated to be important by his customer, can in Lord Mustill's words in the *Ruxley Electronics* case (1995) 45 Con LR 61 at 72, [1996] AC 344 at 360: '. . . please himself whether or not to comply with the wishes of the promisee which, as embodied in the contract, formed part of the consideration for the price'. If that were the law it would be seriously deficient. I am satisfied that it is not the law. In my view the distinction drawn by Hale LJ and by the majority in the Court of Appeal between contractual guarantees and obligations of reasonable care is unsound.

26. The final argument was that by failing to move out the plaintiff forfeited a right to claim non-pecuniary damages. This argument was not advanced in the Court of Appeal. It will be recalled that the judge found as a fact that the plaintiff had acted reasonably in making 'the best of a bad job'. The plaintiff's decision also avoided a larger claim against the surveyor. It was never explained on what legal principle the plaintiff's decision not to move out divested him of a claim for non-pecuniary damages. Reference was made to a passage in the judgment of Bingham LJ in *Watts* v. *Morrow* (1991) 26 Con LR 98 at 126, [1991] 1 WLR 1421 at 1445. Examination showed, however, that the observation, speculative as it was, did not relate to the claim for non-pecuniary damages (see the criticism of Professor M.P. Furmston 'Damages – Diminution in Value or Cost of Repair? – Damages for Distress' (1993) 6 *JCL* 64 at 65). The third argument must also be rejected.

There have been a group of cases in which the plaintiff sought damages in respect of loss which has been suffered by someone else. At one time it was believed that a plaintiff could only recover in respect of its own loss. It is clear that this is no longer always the case but at the moment the law may be said to be in a fluid state.

St Martins Property Corporation Ltd v. Sir Robert McAlpine & Sons Ltd

HOUSE OF LORDS

(1993) 36 Con LR 1

On 19 October 1974, the first plaintiffs (St Martins Property) contracted with the defendants (Sir Robert McAlpine) for the construction of Kings Mall, Hammersmith, the contract being in JCT 63 form. Clause 17 of the contract provided:

> (1) The employer shall not without the written consent of the contractor assign this contract.
> (2) The contractor shall not without the written consent of the employer assign this contract, and shall not without the written consent of the architect (which consent shall not be unreasonably withheld to the prejudice of the contractor) sub-let any portion of the works.

Prior to making the contract, St Martins Property had entered into a development agreement with the London Borough of Hammersmith, the effect of which was that St Martins Property were in effect leaseholders of the completed development for a term of 150 years.

During the construction of the development, the group of which St Martins Property were part was reorganised for tax reasons and Kings Mall was transferred to the second plaintiffs (St Martins Investments), who were a separate legal entity.

By deed of assignment dated 25 March 1986, St Martins Property transferred their interest in Kings Mall to St Martins Investments. The deed also assigned 'the full benefit of all contracts and engagements whatsoever entered into by the assignor and existing at the date hereof for the construction of and completion of the development' and continued with an *habendum* 'To hold the same . . . unto the assignee absolutely subject to all the covenants, agreements, obligations, liabilities whatsoever of the assignor'.

Consent to the assignment was neither sought from nor given by Sir Robert McAlpine, but notice of the assignment was given to them in 1986.

Defects developed in the completed development, allegedly caused by Sir Robert McAlpine's breach of contract. St Martins Property, as managing agents of the development, paid for the remedial works and were reimbursed by St Martins Investments.

On 25 January 1989 the plaintiffs issued writs against Sir Robert McAlpine for breach of contract for the cost of the remedial works. The following preliminary issues were ordered to be tried:

1. Was the benefit of the contract dated 29 October 1974 between St Martins Property and Sir Robert McAlpine validly assigned by St Martins Property to St Martins Investments?

2. Was there an implied term of the deed of assignment dated 25 March 1976 and of the Agency Agreement dated 1976 and 1983 as pleaded in paras 7 and 7A of the amended statement of claim?
3. On the assumption that the matters pleaded in para 8 of the statement of the claim are correct then:
   (a) Do St Martins Investments have a valid claim against Sir Robert McAlpine for damages, other than nominal damages, for breach of the contract dated 29 October 1974 as pleaded in para 10 of the statement of claim?
   (b) Do St Martins Property have a valid claim against Sir Robert McAlpine for damages, other than nominal damages, for breach of the contract dated 29 October 1974 as pleaded in para 11 of the statement of claim?
   (c) Do St Martins Property have a valid claim for damages, other than nominal damages, for breach of the contract dated 29 October 1974 as pleaded in para 12 of the statement of claim?

For the purposes of the trial of the preliminary issues only it was assumed, but otherwise denied, that Sir Robert McAlpine had been in breach of contract and that any breaches of contract had occurred after the purported assignment.

Held: (1) (affirming the Court of Appeal) The purported assignment was ineffective to pass any interest to the second plaintiffs who therefore had no title to sue.

(2) (rejecting the reasoning of the Court of Appeal) The first plaintiffs had not themselves suffered any substantial loss but nevertheless the first plaintiffs were entitled to recover substantial damages because:

(1) (per Lord Browne-Wilkinson, Lord Keith, Lord Bridge and Lord Ackner concurring) it was proper to treat the parties as having entered into the contract on the footing that the first plaintiff would be entitled to enforce contractual rights for the benefit of those who suffered from defective performance but who, under the terms of the contract, could not acquire any right to hold McAlpine's liable for breach; or

(2) (per Lord Griffiths) the first plaintiffs were entitled to recover the cost of the remedial work whether or not they had a proprietary interest in the subject matter of the contract at the date of breach . . .

LORD GRIFFITHS: In my view neither of these considerations provide McAlpine's with a defence to Corporation's claim. I cannot accept that in a contract of this nature, namely for work, labour and the supply of materials, the recovery of more than nominal damages for breach of contract is dependent upon the plaintiff having a proprietary interest in the subject matter of the contract at the date of breach. In everyday life contracts for work and labour are constantly being placed by those who have no proprietary interest in the subject matter of the contract. To take a common example, the matrimonial home is owned by the wife and the couple's remaining assets are owned by the husband and he is the sole earner. The house requires a new roof and the husband places a contract with a builder to carry out the work. The husband is not acting as agent for his wife; he makes the contract as principal because only he can pay for it. The builder fails to replace the roof properly and the husband has to call in and pay another builder to complete the work. Is it to be said that the husband has suffered no damage because he does not own the property? Such a result would in my view be absurd and the answer is that the husband has suffered loss because he did not receive the bargain for which he had contracted with the first builder and the measure of damages is the cost of securing the performance of that bargain by completing the roof repairs properly by the second builder. To put this simple example closer to the facts of this appeal – at the time the husband employs the builder he owns the house but just after the builder starts work the couple are advised to divide their assets so the husband transfers the house to his wife. If the roof turns out to be defective the husband can recover from the builder the cost of putting it right and thus obtain the benefit of the bargain that the builder had promised to deliver. It was suggested in argument that the answer to the example I have given is that the husband could assign the benefit of the contract to the wife. But what if, as in this case, the builder has a clause in the contract forbidding assignment without his consent and refuses to give consent as McAlpine's have done? It is then said that neither husband nor wife can recover damages; this seems to me to be so unjust a result that the law cannot tolerate it.

The principal authority relied upon by McAlpine's in support of the proposition that the contracting party suffers no loss if they did not have a proprietary interest in the property at the time of the breach was *The Albazero, Albacruz (cargo owners)* v. *Albazero (owners)* [1976] 3 All ER 129, [1977] AC 774. The situation in that case was however wholly different from the present. *The Albazero* was not concerned with money being paid to enable the bargain, i.e. the contract of carriage, to be fulfilled. The damages sought in *The Albazero* were claimed for the loss of the cargo, and as at the date of the breach the property in the cargo was vested in another with a right to sue it is readily understandable that the law should deny to the original party to the contract a right to recover damages for loss of the cargo which had caused him no financial loss. In cases such as the present the person who places the contract has suffered financial loss because he has to spend money to give him the benefit of the bargain which the defendant had promised but failed to deliver. I therefore cannot accept that it is a condition of recovery in such cases that the plaintiff has a proprietary right in the subject matter of the contract at the date of breach.

The second ground upon which the recovery of damages is resisted is that Investments in fact reimbursed Corporation for the money they spent on the repairs. But here again in my view who actually pays for the repairs is no concern of the defendant who broke the contract. The court will of course wish to be satisfied that the repairs have been or are likely to be carried out but if they are carried out the cost of doing them must fall upon the defendant who broke his contract . . .

LORD BROWNE-WILKINSON: In contracts for the sale of goods, the purchaser is entitled to damages for delivery of defective goods assessed by reference to the difference between the contract price and the market price of the defective goods, irrespective of whether he has managed to sell on the goods to a third party without loss: *Slater* v. *Hoyle & Smith Ltd* [1920] 2 KB 11, [1918–19] All ER Rep 654; see also as to non-delivery *Williams Bros* v. *E.T. Agius Ltd* [1914] AC 510. In those cases the judgments contained no consideration of the person in whom the property in the goods was vested although it appears that some of the sub-contracts had been made prior to the breach of contract.

If the law were to be established that damages for breach of a supply contract were not quantifiable by reference to the beneficial ownership of goods or enjoyment of the services contracted for but by reference to the difference in value between that which was contracted for and that which is in fact supplied, it might also provide a satisfactory answer to the problems raised where a man contracts and pays for a supply to others, e.g. a man contracts with a restaurant for a meal for himself and his guests or with a travel company for a holiday for his family. It is apparently established that, if a defective meal or holiday is supplied, the contracting party can recover damages not only for his own bad meal or unhappy holiday but also for that of his guests or family: see *Jackson* v. *Horizon Holidays Ltd* [1975] 3 All ER 92, [1975] 1 WLR 1468 as explained in *Woodar Investment Development Ltd* v. *Wimpey Construction UK Ltd* [1980] 1 All ER 571 at 576, 585, 588, 591, [1980] 1 WLR 277 at 283–284, 293–294, 297, 300–301.

There is therefore much to be said for drawing a distinction between cases where the ownership of goods or property is relevant to prove that the plaintiff has suffered loss through the breach of a contract other than a contract to supply those goods or property and the measure of damages in a supply contract where the contractual obligation itself requires the provision of those goods or services. I am reluctant to express a concluded view on this point since it may have profound effects on commercial contracts which effects were not fully explored in argument. In my view the point merits exposure to academic consideration before it is decided by this House . . .

In my judgment the present case falls within the rationale of the exceptions to the general rule that a plaintiff can only recover damages for his own loss. The contract was for a large development of property which, to the knowledge of both Corporation and McAlpine's, was going to be occupied, and possibly purchased, by third parties and not by Corporation itself. Therefore it could be foreseen that damage caused by a breach would cause loss to a later owner and not merely to the original contracting party, Corporation. As in contracts for the carriage of goods by land, there would be no automatic vesting in the occupier or owners of the property for the time being who sustained the loss of any right of suit against McAlpine's. On the contrary, McAlpine's had specifically contracted that the rights of action under the building contract could not without McAlpine's consent be transferred to third parties who became owners or occupiers and might suffer loss. In such a case, it seems to me proper, as in the case of the carriage of goods by land, to treat the parties as having entered into the contract on the footing that Corporation would be entitled to enforce contractual rights for the benefit of those who suffered from defective performance but who, under the terms of the contract, could not acquire any right to hold McAlpine's liable for breach. It is truly a case in which the rule provides 'a remedy where no other would be available to a person sustaining loss which under a rational legal system ought to be compensated by the person who has caused it'.

Counsel for the defendant submitted that it would be wrong to distort the law in order to meet what he described as being an exceptional case. He said that this was a one-off or exceptional case since the development was sold before any breach of contract had occurred and there was an express contractual prohibition on assignment. He submitted that to give Corporation a right to substantial damages in this case would produce chaos when applied to other cases where the contractors have entered into direct warranties with the ultimate purchasers of the individual parts of a development. I am not impressed by these submissions. I am far from satisfied that this is a one-off or exceptional case. We are concerned with standard forms of building contracts which prohibit the assignment of the benefit of building contracts to the ultimate purchasers. In the prolonged period of recession in the property market which this country has experienced many developments have had to be sold off before completion, thereby producing the risk that the ownership of the property may have become divided from the right to sue on the building contract at a date before any breach occurs. As to the warranties given by contractors to subsequent purchasers, they will not, in my judgment, give rise to difficulty. If, pursuant to the terms of the original building contract, the contractors have undertaken liability to the ultimate purchasers to remedy defects appearing after they acquired the property, it is manifest the case will not fall within the rationale of *Dunlop* v. *Lambert*. If

> the ultimate purchaser is given a direct cause of action against the contractor (as is the consignee or indorsee under a bill of lading) the case falls outside the rationale of the rule. The original building owner will not be entitled to recover damages for loss suffered by others who can themselves sue for such loss. I would therefore hold that Corporation is entitled to substantial damages for breach by McAlpine's of the building contract . . .

Darlington Borough Council v. Wiltshier Northern Ltd

COURT OF APPEAL

(1994) 41 Con LR 122

The appellant (Darlington) wished to build a recreational centre, known as the Dolphin Centre, on land which Darlington already owned. Owing to statutory restrictions on local authority borrowing, Darlington arranged for the scheme to be financed by a merchant bank. The Dolphin Centre was built in two phases, under two separate contracts under seal in the JCT 63 form executed on 29 October 1979 and 1 December 1981 respectively. The respondent (Wiltshier) entered into each of these contracts as 'the contractor', but 'the Employer' was the merchant bank (Morgan Grenfell). Collaterally to the building contract for phase 1, Darlington and Morgan Grenfell entered into a covenant agreement dated 29 October 1979, which was replaced by a subsequent agreement covering both phases of the project made on 1 August 1980 (the 1980 agreement).

Clause 3(4) of the agreement provided comprehensively for Morgan Grenfell to assign to Darlington all rights it had against, among others, Wiltshier. It read as follows:

> At the end of the Construction Period or whenever called upon so to do the Company will at the request and cost of the Council assign to the Council the benefit of any rights against the Contractors (or any of them) the Architect the Consultant Structural Mechanical and Electrical Engineers or any other Consultant to which the Company may then be or become entitled. If any cause of action accrues to the Company against the Contractors or any of them the Architect the Consultant Structural Mechanical and Electrical Engineers or any other Consultant and the Council gives written notice to the Company of its wish to pursue the same the Company shall at the cost of the Council be obliged to assign and the Council shall be obliged to take an assignment of the benefit of such rights or cause of action and any agreement entered into between the Company and a Contractor the Architect the Consultant Structural Mechanical and Electrical Engineers or any other Consultant as the case may be.

Clause 4 (5) provided:

> The Council agrees that the Company shall not be liable to the Council for any liability cost claim demand loss damage or expense of any kind or nature caused directly or indirectly by out of or by the use of any part or the whole of any of the Building or the Landscaping or for any incompleteness thereof or any inadequacy thereof for

> any purpose or any deficiency or defect therein or the use or maintenance thereof or any repairs servicing or adjustments thereto or any delay in providing or failure to provide any such or any interruption or loss of service or use thereof or any loss of business or any damage whatsoever and howsoever caused. The Council agrees to indemnify and hold the Company harmless from and against all and any such liabilities costs claims demands losses damages and expenses.

When executing the building contracts, Morgan Grenfell, Wiltshier and Darlington entered into contemporaneously a tripartite deed which gave Darlington direct contractual rights against Wiltshier for any liquidated damages for late completion.

Darlington claimed that there were serious defects in the Dolphin Centre due to bad workmanship or other breaches of contract by Wiltshier and on 22 August 1991, Morgan Grenfell assigned to Darlington by deed all rights and causes of action which Morgan Grenfell might then have or become entitled to against Wiltshier.

Preliminary issues were directed to be tried before Judge John Newey QC who ruled against Darlington who appealed. The live issues on appeal were:

(1) Did Darlington as assignee have a valid claim against Wiltshier for damages other than nominal damages for breach of contract?
(2) If so, upon what principles are such damages to be assessed? Judge Newey answered (1) 'No' and (2) 'not applicable'.

It was common ground between the parties that Darlington as assignees of Morgan Grenfell could not recover any damages from Wiltshier beyond those which Morgan Grenfell could have recovered from Wiltshier if there had been no assignment.

Held (allowing the appeal):

Darlington as assignee had a valid claim against Wiltshier for substantial damages which should be assessed on the normal basis as if Darlington had been the employer under the building contract. Although the general rule is that damages are compensatory and a third party cannot sue for damages on a contract to which he is not a party, there are exceptions to the general rule. This case fell within those exceptions. It was foreseeable that damage would cause loss to Darlington and not merely to Morgan Grenfell, and the parties had entered into the contracts on the footing that Morgan Grenfell would be entitled to enforce contractual rights for the benefit of those who suffered from defective performance, i.e. Darlington, but who, under the terms of the contract could not acquire any right to hold Wiltshier liable for breach.

> DILLON LJ: The key passage giving the ratio in *St Martin's Property Corp Ltd* v. *Sir Robert McAlpine Ltd* (the McAlpine case) is in the speech of Lord Browne-Wilkinson, with which all other members of the House agreed . . .
>
> The present case is, in my judgment, a fortiori since so far from there being a prohibition on the assignment of Morgan Grenfell's rights against Wiltshier under the building contracts, the covenant agreement, of which Wiltshier was aware, gave Darlington the right to call for an assignment of such rights. The argument to the contrary, that Lord Browne-Wilkinson's

decision depended on the prohibition on assignments in the building contract in the McAlpine case seems to me to lead to absurdity namely:

(1) Because there was a prohibition on assignments in the McAlpine case it was right to allow Corporation, the original contracting party, to enforce its rights for the benefit of those successors in title who suffered from defective performance.
(2) In the present case there is no justification for allowing Morgan Grenfell, the original contracting party, to enforce its rights for the benefit of Darlington which will obviously suffer from defective performance, because assignment is permitted.
(3) Therefore Morgan Grenfell had no right before the assignment to recover substantial damages for the benefit of Darlington and the permitted assignment therefore fails of its purpose and Darlington cannot get the damages.

One would thus be back at the starting point that in Lord Browne-Wilkinson's words repeating those of Lord Diplock in *The Albazero*:

> It is truly a case in which the rule provides 'a remedy where no other would be available to a person sustaining loss which under a rational legal system ought to be compensated by the person who has caused it'.

Counsel for Wiltshier also sought to distinguish the decision in the *McAlpine* case on the ground that in the present case Morgan Grenfell never acquired or transmitted to Darlington any proprietary interest in the Dolphin Centre. I do not see that that matters as Darlington had the ownership of the site of the Dolphin Centre all along. It was plainly obvious to Wiltshier throughout that the Dolphin Centre was being constructed for the benefit of Darlington on Darlington's land.

Accordingly I would allow this appeal by direct application of the rule in *Dunlop* v. *Lambert* as recognised in a building contract context in Lord Browne-Wilkinson's speech in *McAlpine*'s case . . .

STEYN LJ: Lord Browne-Wilkinson's conclusion was supported by all members of the House of Lords although, it is right to say, Lord Griffiths wished to go further. Relying on the exception recognised in the Linden Gardens case, as well as on the need to avoid a demonstrable unfairness which no rational legal system should tolerate, I would rule that the present case is within the rationale of Lord Browne-Wilkinson's speech. I do not say that the relevant passages in his speech precisely fit the material facts of the present case. But it involves only a very conservative and limited extension to apply it by analogy to the present case. For these reasons I would hold that the present case is covered by an exception to the general rule that a plaintiff can only recover damages for his own loss . . .

Alfred McAlpine Construction Ltd v. Panatown Ltd

HOUSE OF LORDS

(2000) 71 Con LR 1

This action arose out of a contract dated 2 November 1989 for the design and construction of an office building, multi-storey car park and associated external works at Cambridge. This contract was between the employers (Panatown) and the main contractors (McAlpine). The contract sum was approximately £10.44 m.

It was alleged by Panatown that the design and build contract formed part of a chain of contracts under which Panatown were liable to Unex Investment Properties Ltd (UIP) and UIP were liable to Unex Corporation Ltd (UCL). UIP, UCL and Panatown were all associated companies. The site is owned by UCL. The admitted purpose of this alleged chain of contracts was legitimately to avoid incurring liability to VAT on the building.

Panatown alleged that the building was fundamentally flawed. McAlpine argued that even if the building were flawed their only liability to Panatown was for nominal damages, since Panatown were not the owners of the land on which the building was erected. They further argued that Panatown had not established the alleged chain of contracts with UIP and UCL.

The dispute was referred to arbitration. The arbitrator held that the chain of contracts was established by the minutes of a board meeting held on 23 March 1989, attended by Mr Gredley and Mr Helme who were both directors of UIP, UCL and Panatown, the meeting being a board meeting of all three companies held simultaneously. The minutes were only drawn up and signed nearly two years later. The arbitrator further held that Panatown could recover the full loss caused by any defective building work by McAlpine, although they were not the owners of the land.

Held by Judge Anthony Thornton QC:

(1) The arbitrator was wrong in law to treat the minutes of the meeting of 23 March 1989 as conclusive evidence against McAlpine that there were contracts between UIP, UCL and Panatown and as to the terms of those contracts; but

(2) There was other evidence which clearly established that there was some contract between the parties which concerned the building development. UCL had paid £7.5m to Panatown and Panatown had entered into the building contract with McAlpine; however

(3) The contracts arranged between Panatown, UIP and UCL were not identical with those between Panatown and McAlpine. The evidence did not establish that Panatown were required by the contracts to do anything more than to procure a design and build contract with an outside building contractor. Any obligations of Panatown to comply with the terms of the outside building contract only arose after the contract with McAlpine had been entered into.

(4) The general rule was that a contracting party could not recover more than its own loss. There was an exception to this principle arising out of the decision of the House of Lords in *St Martin's Property Corp Ltd* v. *Sir Robert McAlpine & Sons Ltd* (1993) 36 Con LR 1 and the Court of Appeal in *Darlington BC* v. *Wiltshier Northern Ltd* (1994) 41 Con LR 122 but this exception did not apply in the present case as McAlpine had also entered into a direct contract with UIP, though on different terms. This contract excluded the possibility of Panatown recovering substantial damages in their action against McAlpine.

Panatown appealed.

Held by the Court of Appeal:

The appeal should be allowed. The underlying rationale of the decisions in *St Martin's Property Corp Ltd* v. *Sir Robert McAlpine & Sons Ltd* and *Darlington BC* v. *Wiltshier*

Northern Ltd was that the proper construction of the contract was that the parties had agreed that the employer could recover the whole of the loss suffered by others having an interest in the transaction, although the employer was not the owner of the land. The existence of the direct contract between McAlpine and UIP did not prevent the court holding that the parties had the same intention in the present case. The protection given to the owners of the land by the direct warranty was significantly narrower than that given by the building contract and there was no reason to assume that the parties intended those who suffered a loss to be restricted to what they could recover under the direct warranty. This reasoning applied in the same way to claims for liquidated damages as to claims for unliquidated damages.

McAlpine appealed.

Held by the House of Lords (Lords Goff and Millett dissenting):

Since the owner of the land had a direct claim against the builder under the duty of care deed, the employer under the building contract could not recover the loss suffered by the owner of the land.

LORD BROWNE-WILKINSON: In my judgment the direct cause of action which UIPL has under the duty of care deed is fatal to any claim to substantial damages made by Panatown against McAlpine based on the narrower ground. First, the principle in *The Albazero* case as applied to building contracts by the *St Martin's* case is based on the fact that it provides a remedy to the third party 'where no other would be available to a person sustaining loss which under a rational legal system ought to be compensated by the person who has caused it' (see *The Albazero* [1976] 3 All ER 129 at 137, [1977] AC 774 at 847 and the *St Martin's* case 36 ConLR 1 at 28, [1994] 1 AC 85 at 114). If the contractual arrangements between the parties in fact provide the third party with a direct remedy against the wrongdoer the whole rationale of the rule disappears. Moreover, as I have said, both the decision in *The Albazero* case itself and *dicta* in the *St Martin's* case 36 ConLR 1 at 29, [1994] 1 AC 85 at 115 state that where the third party (C) has a direct claim against the builder (B) the promisee under the building contract (A) cannot claim for the third party's damage.

I turn now to the broader ground on which Lord Griffiths decided the *St Martin's* case. He held that the building contractor (B) was liable to the promisee (A) for more than nominal damages even though A did not own the land at the date of breach. He held in effect that by reason of the breach A had himself suffered damage, being the loss of the value to him of the performance of the contract. On this view even though A might not be legally liable to C to provide him with the benefit which the performance of the contract by B would have provided, A has lost his 'performance interest' and will therefore be entitled to substantial damages being, in Lord Griffiths' view, the cost to A of providing C with the benefit. In the *St Martin's* case Lord Keith of Kinkel, Lord Bridge of Harwich and I all expressed sympathy with Lord Griffiths' broader view. However, I declined to adopt the broader ground until the possible consequences of so doing had been examined by academic writers. That has now happened and no serious difficulties have been disclosed. However, there is a division of opinion as to whether the contracting party, A, is accountable to the third party, C, for the damages recovered or is bound to expend the damages on providing for C the benefit which B was supposed to provide. Lord Griffiths in the *St Martin's* case 36 ConLR 1 at 9, [1994] 1 AC 85 at 97 took that view. But as I understand them Lord Goff of Chieveley and Lord Millett in the present case (in agreement with Lord Steyn in *Darlington BC* v. *Wiltshier Northern Ltd* (1994) 41 ConLR 122 at 139, [1995] 1 WLR 68 at 80) would hold that, in the absence of the specific circumstances of the present case, A is not accountable to C for any damages recovered by A from B.

I will assume that the broader ground is sound in law and that in the ordinary case where the third party (C) has no direct cause of action against the building contractor (B) A can recover damages from B on the broader ground. Even on that assumption, in my judgment Panatown has no right to substantial damages in this case because UIPL (the owner of the land) has a direct cause of action under the duty of care deed.

The essential feature of the broader ground is that the contracting party A although not himself suffering the physical or pecuniary damage sustained by the third party C, has suffered his own damage being the loss of his performance interest, i.e. the failure to provide C with the benefit that B had contracted for C to receive. In my judgment it follows that the critical factor is to determine what interest A had in the provision of the service for the third party C. If, as in the present case, the whole contractual scheme was designed, *inter alia*, to give UIPL and its successors a legal remedy against McAlpine for failure to perform the building contract with due care, I cannot see that Panatown has suffered any damage to its performance interests: subject to any defence based on limitation of actions, the physical and pecuniary damage suffered by UIPL can be redressed by UIPL exercising its own cause of action against McAlpine. It is not clear to me why this has not occurred in the present case: but, subject to questions of limitation which were not explored, there is no reason even now why UIPL should not be bringing the proceedings against McAlpine. The fact that the duty of care deed may have been primarily directed to ensuring that UIPL's successors in title should enjoy a remedy in tort against McAlpine is nothing to the point: the contractual provisions were directed to ensuring that UIPL and its successors in title did have the legal right to sue McAlpine direct. So long as UIPL enjoys this right Panatown has suffered no failure to satisfy its performance interest.

The theoretical objection to giving the contracting party A substantial damages for breach of the contract by B for failing to provide C with a benefit which C itself can enforce against B is further demonstrated by great practical difficulties which such a view would entail. Let me illustrate this by postulating a case where, before the breach occurred, UIPL had with consent assigned the benefit of the duty of care deed to a purchaser of the site, X. What if Panatown itself was entitled to, and did, sue for and recover damages from McAlpine? Presumably McAlpine could not in addition be liable to X for breach of the duty of care deed: yet Panatown would not be liable to account to X for the damages it had recovered from McAlpine. The result would therefore be another piece of legal nonsense: the party who had suffered real, tangible damage, X, could recover nothing but Panatown which had suffered no real loss could recover damages. Again, suppose that X agrees with McAlpine certain variations of McAlpine's liability under the building contract. What rights would Panatown then have against McAlpine? The Law Commission in its Report on *Privity of Contract: Contracts for the Benefit of Third Parties* (Law Com no 242) (1996) considered at length questions like these (see in particular paras 11.14, 11.21 and 11.22) and many other problems such as set-off and counterclaims. The Law Commission recommended that in certain defined circumstances third parties should be entitled to enforce the contract. But in the draft Bill annexed to the report and in the Act of Parliament which enacted the recommendations, the Contract (Rights of Third Parties) Act 1999, specific statutory provisions were included to deal with the difficulties arising. Although both the Law Commission's report (paras 5.10 and 5 11) and sections 4 and 6(1) of the Act make it clear that the Act is not intended to discourage the courts from developing the rights of third parties when it is appropriate to do so, in my judgment there is little inducement in a case such as the present where a third party has himself the right to enforce the contract against the contract breaker, to extend the law so as to give both the promisee and the third party concurrent rights of enforcement.

Contribution

In construction contracts a claimant will often have claims against more than one defendant. In such situations, the general rule is that if C has a claim against D1 and D2, he is entirely free to sue either D1 or D2 or both of them. One of the major strategic decisions that C and his lawyers have to make is which defendant to pursue. Such decisions are inevitably largely determined by which defendant has most money or is best insured. The chosen defendant will often wish to explore the possibility of getting other defendants involved. This is governed by the Civil Liability (Contribution) Act 1978.

The most important provisions of this Act are:

1. Entitlement to contribution

(1) Subject to the following provisions of this section, any person liable in respect of any damage suffered by another person may recover contribution from any other person liable in respect of the same damage (whether jointly with him or otherwise).

(2) A person shall be entitled to recover contribution by virtue of subsection (1) above notwithstanding that he has ceased to be liable in respect of the damage in question since the time when the damage occurred, provided that he was so liable immediately before he made or was ordered or agreed to make the payment in respect of which the contribution is sought.

(3) A person shall be liable to make contribution by virtue of subsection (1) above notwithstanding that he has ceased to be liable in respect of the damage in question since the time when the damage occurred, unless he ceased to be liable by virtue of the expiry of a period of limitation or prescription which extinguished the right on which the claim against him in respect of the damage was based.

(4) A person who has made or agreed to make any payment in bona fide settlement or compromise of any claim made against him in respect of any damage (including a payment into court which has been accepted) shall be entitled to recover contribution in accordance with this section without regard to whether or not he himself is or ever was liable in respect of the damage, provided, however, that he would have been liable assuming that the factual basis of the claim against him could be established.

(5) A judgment given in any action brought in any part of the United Kingdom by or on behalf of the person who suffered the damage in question against any person from whom contribution is sought under this section shall be conclusive in the proceedings for contribution as to any issue determined by that judgment in favour of the person from whom the contribution is sought.

(6) References in this section to a person's liability in respect of any damage are references to any such liability which has been or could be established in an action brought against him in England and Wales by or on behalf of the person who suffered the damage; but it is immaterial whether any issue arising in any such action was or would be determined (in accordance with the rules of private international law) by reference to the law of a country outside England and Wales.

2. Assessment of contribution

(1) Subject to subsection (3) below, in any proceedings for contribution under section 1 above the amount of the contribution recoverable from any person shall be such as may be found by the court to be just and equitable having regard to the extent of that person's responsibility for the damage in question.

(2) Subject to subsection (3) below, the court shall have power in any such proceedings to exempt any person from liability to make contribution or to direct that the contribution to be recovered from any person shall amount to a complete indemnity.

(3) Where the amount of the damages which have or might have been awarded in respect of the damage in question in any action brought in England and Wales by or on behalf of the person who suffered it against the person from whom the contribution is sought was or would have been subject to –

(a) any limit imposed by or under any enactment or by any agreement made before the damage occurred;

(b) any reduction by virtue of section 1 of the Law Reform (Contributory Negligence) Act 1945 or section 5 of the Fatal Accidents Act 1976; or

(c) any corresponding limit or reduction under the law of a country outside England and Wales; the person from whom the contribution is sought shall not by virtue of any contribution awarded under section 1 above be required to pay in respect of the damage a greater amount than the amount of those damages as so limited or reduced.

A key concept here is that of 'the same damage' in section 1(1). It is essential to show that the damage in respect of which the contributee is liable is 'the same damage' as that of which the contributor is liable.

Royal Brompton Hospital NHS Trust v. Hammond (No. 3)

HOUSE OF LORDS

(2002) 81 Con LR 1

The highly complex litigation in this case arose out of major building works at the hospital between 1987 and 1990. Disputes broke out between the hospital and Taylor Woodrow, the main contractors which became the subject of an arbitration in which Taylor Woodrow made a claim against the hospital and the hospital made a counter-claim. The arbitration was settled in 1995 on terms that the hospital would pay Taylor Woodrow some £6.2 million.

In 1993 the hospital had started the present action (which proceeded after the settlement of the arbitration) against its professional advisers (architects, project managers, electrical and mechanical engineers and structural engineers). In the present proceedings the architects sought contribution from Taylor Woodrow. (It was part of the settlement between the hospital and Taylor Woodrow that Taylor Woodrow should be indemnified against such proceedings but the agreement was not binding on the architects who were not parties to it.)

Judge John Hicks QC held that even if the architects could show that Taylor Woodrow had been wrongly successful in the arbitration, the primary damage to the hospital would have been the wrongful delay in practical completion whereas the claim against the architects was that their behaviour had weakened the prospects of success in the arbitration. This was not the same damage. The Court of Appeal and the House of Lords agreed.

LORD BINGHAM:

5. It is plain beyond argument that one important object of the 1978 Act was to widen the classes of person between whom claims for contribution would lie and to enlarge the hitherto restricted category of causes of action capable of giving rise to such a claim. It is, however, as I understand, a constant theme of the law of contribution from the beginning that B's claim to share with others his liability to A rests upon the fact that they (whether equally with B or not) are subject to a common liability to A. I find nothing in section 6(1) (c) of the 1935 Act or in section 1(1) of the 1978 Act, or in the reports which preceded those Acts, which in any way weakens that requirement. Indeed both sections, by using the words 'in respect of the same damage', emphasise the need for one loss to be apportioned among those liable.

6. When any claim for contribution falls to be decided the following questions in my opinion arise:

(1) What damage has A suffered?
(2) Is B liable to A in respect of that damage?
(3) Is C also liable to A in respect of that damage or some of it?

At the striking-out stage the questions must be recast to reflect the rule that it is arguability and not liability which then falls for decision, but their essential thrust is the same. I do not think it matters greatly whether, in phrasing these questions, one speaks (as the 1978 Act does) of 'damage' or of 'loss' or 'harm', provided it is borne in mind that 'damage' does not mean 'damages' (as pointed out by Roch LJ in *Birse Construction Ltd* v. *Haiste Ltd (Watson and Anor, third parties)* (1995) 47 Con LR 162 at 170, [1996] 1 WLR 675 at 682) and that B's right to contribution by C depends on the damage, loss or harm for which B is liable to A corresponding (even if in part only) with the damage, loss or harm for which C is liable to A. This seems to me to accord with the underlying equity of the situation: it is obviously fair that C contributes to B a fair share of what both B and C owe in law to A, but obviously unfair that C should contribute to B any share of what B may owe in law to A but C does not.

7. Approached in this way, the claim made by the architect against the contractor must in my opinion fail in principle. It so happens that the employer and the contractor have resolved their mutual claims and counterclaims in arbitration whereas the employer seeks redress against the architect in the High Court. But for purposes of contribution the parties' rights must be the same as if the employer had sued both the contractor and the architect in the High Court and they had exchanged contribution notices. The question would then be whether the employer was advancing a claim for damage, loss or harm for which both the contractor and the architect were liable, in which case (if the claim were established) the court would have to apportion the common liability between the two parties responsible, or whether the employer was advancing separate claims for damage, loss or harm for which the contractor and the architect were independently liable, in which case (if the claims were established) the court would have to assess the sum for which each party was liable but could not apportion a single liability between the two. It would seem to me clear that any liability the employer might prove against the contractor and the architect would be independent and not common. The employer's claim against the contractor would be based on the contractor's delay in performing the contract and the disruption caused by the delay, and the employer's damage would be the increased cost it incurred, the sums it overpaid and the liquidated damages to which it

was entitled. Its claim against the architect, based on negligent advice and certification, would not lead to the same damage because it could not be suggested that the architect's negligence had led to any delay in performing the contract.

LORD STEYN:

VI. A description of the claims

22. The characterisation of the employer's claim against the contractor is straightforward. It is for the late delivery of the building. This is not a claim which the employer has made against the architect. Moreover, notionally it is not damage for which the architect could be liable merely by reason of a negligent grant of an extension of time. It is conceivable that an architect could negligently cause or contribute to the delay in completion of works, e.g. by condoning inadequate progress of the work or by failing to chivvy the contractor. In such a case the contractor and the architect could be liable for the same damage. There are, however, no such allegations in the present case.

23. The essence of the case against the architect is the allegation that his breach of duty changed the employer's contractual position detrimentally as against the contractor. The employer's case is that the architect wrongly evaluated the contractor's claim for an extension of time. It is alleged that by negligently giving an extension of time in respect of an unmeritorious claim by the contractor, the architect presented the contractor with a defence to a previously straightforward claim by the employer for breach of contract in respect of delay. The employer lost the right under the contract to claim or deduct liquidated damages for the delayed delivery of the building. The contractor committed no wrong by retaining the money until the extension of time had been set aside in an arbitration. The detrimental effect on the employer's contractual position took place when the extension of time was negligently given. In such a case the employer must go to arbitration in order to restore his position. He has the burden of proof in the arbitration and has to face the uncertain prospect of succeeding in what may perhaps be a complex arbitration. The employer's bargaining position against the contractor is weakened. A reasonable settlement with the contractor may reflect this changed position: a case with a 100% prospect of success may become, for example, a case with only a 70% prospect of success.

Where there are a number of possible defendants the right to contribution may be substantially affected by the insurance position, because this in turn may affect liability.

Co-operative Retail Services Ltd v. Taylor Young Partnership

HOUSE OF LORDS

(2002) 82 Con LR 1

In April 1993 the claimant (CRS) engaged a main contractor (Wimpey) to build a new office headquarters building in Rochdale. Hall were the electrical sub-contractors and CRS, Wimpey and Hall entered into a joint names insurance policy with Commercial Union.

In March 1995, before practical completion, the building was extensively damaged by fire. CRS brought claims against Taylor Young (TYP), the architects, and Hoare Lea and Partners (HLP), the mechanical and electrical engineers.

The defendants sought contribution from Wimpey and Hall. The House of Lords held that even if Wimpey and/or Hall were factually involved, the insurance arrangements meant that they could not be legally liable to CRS and could not therefore be required to contribute.

LORD HOPE OF CRAIGHEAD:

46. There is no doubt that both the main contract and the sub-contract contain provisions which have the effect in the clearest terms of excluding liability for damage to the works, work executed and site materials due to the negligence, breach of statutory duty, omission or default of the contractor and the sub-contractor respectively: see clause 20.3 of the main contract and clause 6.4 of the sub-contract. This has not been disputed by [Counsel for Taylor Young Partnership]. It is also plain that the purpose of the all-risks insurance which the contractor is required to take out and maintain in the joint names of the employer, the contractor and the sub-contractors is to provide funds for the reinstatement of the works in the event of their being damaged up to and including the date when the certificate of practical completion is issued, whatever the cause of the fire. But the contractual scheme does not end there. For an understanding of its true effect it is necessary to pay close attention to the provisions of clause 22A.4, which deal with what is to happen in the event of loss or damage affecting work executed or any site materials occasioned by any one or more of the risks covered by the joint names policy.

47. The effect of clause 22A.4 may be summarised in this way. On the one hand there is the position of the employer. He is not entitled to deduct anything from the sums payable to the contractor under or by virtue of the contract as compensation for any loss and damage which he has sustained due to the fire. This is so even if the fire was caused by the contractor's act or omission or default or by anyone else for whose acts, omissions or defaults he would otherwise be responsible. Clause 22A.4.2 provides that the occurrence of such loss or damage shall be disregarded in computing any amounts payable to the contractor under or by virtue of the contract. On the other hand there is the position of the contractor. Clause 22A.4.3 requires him with due diligence to restore the work that has been damaged by the fire, to replace or repair any site materials that have been lost or damaged by it and to proceed with the carrying out and completion of the works. Clause 22A.4.4 requires him to authorise the insurers to pay all moneys that are payable from the insurance in respect of the fire to the employer, who is required in his turn to use this money for the purpose of paying the contractor and the associated professional fees for the restoration work. Clause 22A.4.5 provides that the contractor is not to be entitled to any payment for the reinstatement work other than the moneys received under the insurance policy. As the contractor is entitled to an extension of time under clause 25, he is not liable to the employer for losses due to any delay caused by the fire in the completion of the works under the contract.

48. The position therefore is that there is no liability to pay compensation on either side. The employer has no claim for compensation against the contractor. All he can do is insist that the contractor must proceed with due diligence to carry out the reinstatement work and must authorise the release to him of the insurance moneys. The contractor has no claim for compensation against the employer. All he can do is insist that the employer must use the insurance moneys for payment of the cost of carrying out the reinstatement work. It makes no difference whether the fire was caused by the negligence of the contractor or one of his sub-contractors or of the employer or of some third party for whose acts or omissions neither of the parties to the contract is responsible. The ordinary rules for the payment of compensation for negligence and for breach of contract have been eliminated. Whatever the cause of

the fire, the obligation of the contractor is to carry out such work as is needed to put the matter right. His obligation is to restore the fire damage at his own cost, except in so far as the cost of doing so is met by sums recovered under the joint names insurance policy.

49. This is not to say that the contractor may not be found liable to the employer for any loss or damage which the employer may sustain due to his failure to take out and maintain the joint names policy, or his failure to fulfil his obligation with due diligence to carry out the reinstatement works under clause 22A.4.3. But this feature of the contractual scheme is of no assistance to Mr Blackburn. Any liability which the contractor may be under to pay compensation to the employer for those breaches of contract is entirely separate and distinct from the liability of those who caused or contributed to the fire. It could not be said in that event that Wimpey were liable to CRS 'in respect of the same damage' within the meaning of section 1(1) of the 1978 Act read together with section 6(1) of that 1978 Act. CRS would not be entitled to compensation from Wimpey for the same harm or the same wrong as that for which TYP and HLP are said to be liable, as the harm for which Wimpey would be liable would be that resulting from its failure to insure or its failure to carry out the reinstatement works.

50. For these reasons I consider that the Court of Appeal were right to dismiss the appeal by TYP and HLP against the answers which the judge gave to the preliminary issue. In my opinion the meaning and effect of the main contract was to exclude Wimpey's liability to CRS for loss and damage caused by the fire in so far as this was due to its breach of contract. [Counsel for Taylor Young Partnership] accepted that the same reasoning must be applied in Hall's case also. So, as Wimpey and Hall are not persons from whom CRS is entitled to recover compensation in respect of the fire damage, it is not open to TYP and HLP to recover contribution from either Wimpey or Hall in respect of the fire damage for which they are said to be liable.

Chapter 13

Prolongation and Disruption Claims

Standard form building contracts contain express provisions enabling the contractor to claim for loss and expense which he incurs as a result of certain specified events. In every case the exact wording of the relevant clause is important. The forms currently in use generally allow additional or alternative claims for breach of contract at common law.

The machinery of JCT contracts for the ascertainment and reimbursement of direct loss and/or expense is not exhaustive of the contractor's remedies. The contractor is not bound to make an application for ascertainment but may prefer to wait until completion of the work and join any claim for damages for breach of obligation with other claims for damages.

London Borough of Merton v. Stanley Hugh Leach Ltd

CHANCERY DIVISION

(1985) 32 BLR 51

The facts of this case are set out on p. 142.

VINELOTT J: It is common ground that insofar as the contract imposes an obligation to be performed by the building owner (for instance, to give possession of the site) and contains no express provisions specifying the consequences of a breach of such an obligation (by, for instance, permitting an extension of time for completion of the works as in clause 21 if and so far as it applies to a delay in giving possession) the breach of such an obligation founds an obligation to pay damages to the contractor for the loss occasioned by the breach. But, it is said save for (the failure of Merton to appoint an architect in succession to the Borough Architect (who retired in April 1980)) the obligations, breaches of which are relied on to found a claim for damages, are obligations which by the terms of the contract, express or implied, are imposed upon the architect. The case for Merton is that the contract provides an exhaustive machinery whereby the contractor and the building owner can protect themselves from loss flowing from any failure of the architect to perform any duty falling to be performed by him or, so far as he is required to make a decision, any adverse decision. Insofar as the architect is required to carry out any duty which does not involve the exercise of discretion (for instance the issue of instructions, drawings and so forth specifically requested by the contractor) the

Powell-Smith and Furmston's Building Contract Casebook, Fifth Edition. Michael Furmston.

contractor's remedy for the failure by the architect to carry out that duty is provided within the four corners of the contract – in the instance given his remedy is to apply for an extension of time and for compensation for any direct loss or expense consequent on the delay. Insofar as the architect exercises a discretion or acts (as it has been said) as a certifier the contractor is entitled to refer the matter to an arbitrator who (under clause 35(3)) will stand in the shoes of the architect and can review the matter in the same manner as if the certificate or decision had not been given. The arbitrator is himself part of the machinery for correcting any failure or mistake on the part of the architect. So, it is said, there is no room for the imposition of any secondary obligation on the part of Merton to pay compensation for any such failure or mistake. . . .

In my judgment under the contract Merton undertook to ensure that there would at all times be a person who would carry out the duties to be performed by the architect and that he would perform those duties with reasonable diligence, skill and care and that where the contract required the architect to exercise his discretion he would act fairly. It is true that the contract contains an elaborate machinery designed to enable the contractor to spell out in detail steps which he requires the architect to take in relation to some specific matter. But the machinery is clearly not exhaustive. An example will make this clear. One of the allegations in the points of claim is that the architect did not perform the duty imposed on him by clause 30(5)(a) to supply Leach with priced bills of variation. It is now too late for bills of variation to be prepared. After the arbitrator had given his award experts were appointed by both parties to agree the *quantum* of the claim under clause 11(4). I understand that a figure has been agreed. What is said by Leach is that although a figure has been agreed the calculation of that sum cannot be treated as equivalent to priced bills of variation and that in the absence of priced bills of variation Leach and its advisers are deprived of information essential to the calculation of other claims: in particular under clauses 11(6) and 24(1).

Moreover there is a clear indication in the contract that the draftsman contemplated that the contractor might have parallel rights to claim compensation under the express terms of the contract and to pursue claims for damages. That arises under clause 24(2) which I have already read and which, of course, expressly provides that the provisions of the conditions are to be without prejudice to other rights and remedies of the contractor. The effect of clause 24(2) (as I understand it) is this. Clause 24(1) specifies grounds upon which the contractor is entitled to make a claim for reimbursement of direct loss or expense for which he would not otherwise be reimbursed by a payment made under the other provisions of the contract. The grounds specified may or may not result from a breach by the architect of his duties under the contract; a claim by the contractor under subparagraph (a) will normally, though not perhaps invariably, arise from a failure by the architect to answer with due diligence a proper application by the contractor for instructions, drawings and the like, while a claim by the contractor under the subclause (b) following a proper instruction requiring the opening up of works under clause 6(3) normally (though not perhaps invariably) will not involve any breach by the architect of any obligation under the contract. In either case the contractor can call on the architect to ascertain the direct loss or expense suffered and to add the loss or expense when ascertained to the contract sum. The contractor will then receive reimbursement promptly and without the expense and delay of a claim for damages. But the contractor is not bound to make an application under clause 24(1). He may prefer to wait until completion of the work and join the claim for damages for breach of the obligation to provide instructions, drawings and the like in good time with other claims for damages for breach of obligations under the contract. Alternatively he can, as I see it, make a claim under clause 24(1) in order to obtain prompt reimbursement and later claim damages for breach of contract, bringing the amount awarded under clause 24(1) into account.

A claim under clause 11(6) by contrast will not normally arise in consequence of any breach of duty on the part of the architect. The architect has power under clause 11(1) to issue instructions requiring variation and under clause 11(2) 'variation' is described in very wide terms apt to cover any desired departure from the work as originally conceived. The purpose of clause 11(4) is to provide remuneration for such additional or varied work calculated in accordance with the principles which govern pricing of the work originally contracted to be carried out and the purpose of clause 11(6) is to supplement that remuneration by reimbursing the contractor for consequential direct loss or expense. However, a claim for damages may arise if, for instance, following application by the contractor the architect fails to ascertain or to instruct the quantity surveyor to ascertain the amount of the direct loss or expense suffered. There is nothing in the contract which excludes such a claim for damages.

Under JCT terms there is no necessary connection between the grant of an extension of time and a claim for direct loss and/or expense. A claim for direct loss and/or expense need not either be preceded by or accompanied by the grant of an extension of time.

H. Fairweather & Co Ltd v. London Borough of Wandsworth

QUEEN'S BENCH DIVISION
(1987) 39 BLR 106

Disputes arose between the parties to a JCT 63 contract, largely concerned with the test adopted by the arbitrator for allocating extensions of time.

Held: Under JCT terms the grant of an extension of time is not a condition precedent to the contractor's right to reimbursement of direct loss and/or expense.

JUDGE JAMES FOX-ANDREWS QC: The grounds relied upon by Fairweather are:

> The arbitrator erred in law in his award in relation to 'the strikes' in holding . . . that, as between the various heads under clause 23 pursuant to which an extension of time may be granted, the extension should be granted in respect of the dominant reason, whereas he should have held that where the reasons for delay correspond with more than one head, the extension may be granted under either or both heads. The applicants will rely upon the decision of His Honour Judge Edgar Fay QC (sitting as a deputy High Court judge) in *Henry Boot Construction Limited* v. *Central Lancashire New Town Development Corporation* (1980).

In his third consideration of Fairweather's claim for extension of time the architect granted an extension of 81 weeks under condition 23(d) by reason of strikes and combination of workmen. The *quantum* of extension was not challenged but Fairweather contended before the arbitrator that eighteen of those 81 should be reallocated under condition 23(e) or (f). The reasoning behind the contention was that only if there was such a reallocation could Fairweather ever recover direct loss and expense under condition 11(6) in respect of those weeks reallocated to condition 23(e) or condition 24(1)(a) in respect of those weeks reallocated to condition 23(f).

The nub of the challenged decision of the arbitrator is to be found in sections 6.11 and 6.12 of his interim award:

> 6.11 It is possible to envisage circumstances where an event occurs on site which causes delay to the completion of the works and which could be ascribed to more than one of the eleven specified reasons but there is no mechanism in the conditions for allocating an extension between different heads so the extension must be granted in respect of the dominant reason.
>
> 6.12 I accept the respondent's contention that faced with the events of this contract, nobody would say that the delays which occurred in 1978 and 1979 were caused by reason of the architect's instructions given in 1975 to 1977. I hold that the dominant cause of the delay was the strikes and combination of workmen and accordingly the architect was correct in granting his extension under condition 23(d).
>
> In 6.14 he said: For the sake of clarity I declare that this extension does not carry with it any right to claim direct loss and/or expense.

. . . It is I think of assistance to consider two passages in the judgment in the *Boot* case referred to in the ground of appeal. Having considered the provision of conditions 23 and 24, the learned judge said:

> The broad scheme of these provisions is plain. There are cases where the loss should be shared, and there are cases where it should be wholly borne by the employer. There are also cases which do not fall within either of these conditions and which are the fault of the contractor where the loss of both parties is wholly borne by the contractor. But in cases where the fault is not that of the contractor the scheme clearly is that in certain cases the loss is to be shared; the loss lies where it falls. But in other cases the employer has to compensate the contractor in respect of the delay, and that category, where the employer has to compensate the contractor, should, one would think, clearly be comprised of the cases where there is fault upon the employer or fault for which the employer can be said to bear some responsibility.

Loss and expense resulting from delay caused by strikes falls on both employer and contractor. The employer *pro tanto* will lose his right to liquidated damages in respect of any extension of time given by the architect under condition 23(d). But since loss and expense suffered by the contractor resulting from strikes is not a matter within condition 24 or the fault of the employer the contractor had to bear his own loss and expense.

Later in his judgment having considered some cases within condition 23 which did and some cases which did not give a right to a contractor to recover loss and expense under condition 24 the learned judge said that logically the same set of facts might fall within two separate paragraphs of condition 23 one giving the contractor the right to loss and expense and the other not. In the event the learned judge found *Boot* was not such a case.

Those features which are the same and those which are different in condition 24(1) and condition 11(6) were considered by Vinelott J in *Leach* v. *Merton* (1985). Fairweather accept that if they have a claim at all for extension of time under 23(e) it can only be in respect of architect's instructions involving a variation under 11(1). But I do not consider that the obtaining of an extension of time under 23(e) is a condition precedent to recovering loss and expense under 11(6).

Assume the following facts: A contract is entered into in this form of contract in May for one year for completion on 31 July of the next. The work is of a tunnelling nature. No tunnelling can be carried out from 1 November to 31 March for seasonal reasons but during that period the contractor will have expensive equipment lying idle. In early April when the works were on course for completion on 31 July the architect issues an instruction under 11(1) requiring a variation the execution of which will add three months to the contract period. At the same time on the contractor's application he grants an extension of time for completion to 31 October. A fortnight before 31 October when the works as varied are on course for completion in due time a strike occurs which continues until 31 March. The contractor recommences work on 1 April but because he had no opportunity to protect his machinery during the six months period it then takes the contractor two months not two weeks to complete. There has been no fault on either party.

If the architect grants an extension of time of eight months only under 23(d) I can see no reason why the contractor under the contract cannot still recover all his direct loss and expense under 11(6). An extension of time under 23(e) is neither expressly nor I find impliedly made a condition precedent to a right to payment under 11(6). There appears to be no requirement that the contractor should first successfully challenge in arbitration the sub-clause under which the architect has extended time.

The question arises whether the position is different so far as a claim under condition 24(1)(a) is concerned. The wording of that sub-clause matches the wording of condition 23(f). In *Leach* the learned judge considered the nature and extent of a contractor's rights under this contract. He said:

> Moreover there is a clear indication in the contract that the draughtsmen contemplated that contractor might have parallel rights to claim compensation under the express terms of the contract and to pursue claims for damages. That arises under clause 24(2) which I have already read and which, of course, expressly provides that the provisions of the conditions are to be without prejudice to other rights and remedies of the contractor. The effect of clause 24(2) (as I understand it) is this. Clause 24(1) specifies grounds upon which the contractor is entitled to make a claim for reimbursement of direct loss or expense for which he would not otherwise be reimbursed by a payment made under the other provisions of the contract. The grounds specified may or may not result from a breach by the architect of his duties under the contract; a claim by the contractor under sub-paragraph (a) will normally, though not perhaps invariably, arise from a failure by the architect to answer with due diligence a proper application by the contractor for instructions, drawings and the like, while a claim by the contractor under sub-clause (b) following a proper instruction requiring the opening up of works under clause 6(3) normally (though not perhaps invariably) will not involve any breach of the architect's real obligation under the contract. In either case the contractor can call on the architect to ascertain the direct loss or expense suffered and to add the loss or expense when ascertained to the contract sum. The contractor will then receive reimbursement promptly and without the expense and delay of a claim for damages. But the contractor is not bound to make an application under 24(1). He may prefer to wait until completion of the work and join the claim for damages for breach of the obligation to provide instructions, drawings and the like in good time in with other claims for damages for breach of obligations under the contract. Alternatively he can, as I see it, make a claim under 24(1) in order to obtain prompt reimbursement and later claim damages for breach of contract, bringing the amount awarded under clause 24(1) into account.

Neither this part of the judgment nor the terms of the contract itself points to an extension of time under condition 23(f) being a condition precedent to recovery of direct loss and expense under condition 24(1)(a). However the practical effect ordinarily will be that if the architect has refused an extension under the former the contractor is unlikely to be successful with the architect on an application under condition 24(1)(a).

If I am correct in my judgment the matter on which the arbitrator was asked to decide was irrelevant. If however I am wrong it would be necessary to consider whether the arbitrator erred in law in his finding.

If the arbitrator's finding had been limited to the first sentence of section 6.12 that would have been a finding of fact which would have disposed of any right of reallocation assuming such a right existed. I think, however, reading the whole of these paragraphs that there was not such a finding of fact.

'Dominant' has a number of meanings: 'Ruling, prevailing, most influential'. On the assumption that condition 23 is not solely concerned with liquidated or ascertained damages but also triggers and conditions a right for a contractor to recover direct loss and expense where applicable under condition 24 then an architect and in his turn an arbitrator has the task of allocating, when the facts require it, the extension of time to the various heads. I do not consider that the dominant test is correct. But I have held earlier in this judgment that that assumption is false. I think the proper course here is to order that this part of the interim award should be remitted to (the arbitrator) for his reconsideration . . .

Under JCT terms the phrase 'direct loss and/or expense' or 'direct loss and/or damage' means that the sums recoverable are equivalent to damages at common law.

Wraight Ltd v. P.H. & T. (Holdings) Ltd

QUEEN'S BENCH DIVISION

(1968) 13 BLR 26

The contractors agreed, on JCT 63 terms, to build properties at Sheerness in Kent. Shortly after work commenced, unsuitable soil conditions were encountered, and the architect properly directed a suspension of the works. The suspension having continued for more than the period stated in the appendix to the contract, the contractors determined their employment under clause 26(1)(c)(iv) of the contract (now JCT 80, clause 28.1.3.). There was no dispute that the employer was bound to pay the contractors an amount as direct expenditure on starting the work, but the employer disputed the contractors' claim for the profit which they would have earned had the contract been completed.

Held: The words 'direct loss and/or damage' in clause 26 must be given the same meaning as they would have in the case of a breach of contract. The loss of profit was the direct and natural consequence of the determination of the contract, and the contractors were entitled to succeed.

MEGAW, J: For the [contractors], it is put forward that the phrase 'direct loss and/or damage' means what is its usual, ordinary and proper meaning in the law: one has to ask whether any particular matter or items of loss or damage claimed has been caused by the particular matter.

> Here, the particular matter is not breach of contract but determination of the contract. If it has been caused by it, then one has to go on to see whether there has been some intervention of some other cause which prevents the loss or damage from being properly described as being the direct consequence of the determination of the contract.
>
> Causation, as is so well known, is a difficult concept with all kinds of metaphors having been used to explain it. It is, I think, for general purposes best regarded in this way: that there may be a very large number of causes, in one sense of the word, to produce any particular result, but the law has to look at the question of causation from the point of view of common sense and it has to eliminate causes which are of little real significance, though they might in one sense be called 'causes', and to find what is the real cause. The same thing applies with regard to intervening causes.
>
> It is said here on behalf of the [contractors] that quite plainly the loss of profit to the [contractors] was, in any ordinary sense of the word, a direct consequence of the determination of the contract. If the contract had not been determined, they would have gone on and done the work, continued their legal obligations thereunder and would have had their legal right to obtain their full contractual remuneration at the end of it, and therefore they have, by this determination, lost the right to receive that sum of money which they would have received if the contract had not been determined: what can be more direct than that?
>
> But, of course, they will also be excused by the determination from their other legal obligations which would otherwise have fallen upon them, and they cannot claim as being a loss or a partial loss or damage, or a part of damage incurred by them directly resulting from the determination, any expenses which they have been saved from incurring by reason of the limitation of their legal obligations as a result of the determination.
>
> In my judgment, there are no grounds for giving to the words 'direct loss and/or damage caused to the contractor by the determination' any other meaning than that which they have, for example, in a case of breach of contract or other question of the relationship of a fault to damage in a legal context. Therefore it follows . . . that the [contractors] are, as a matter of law entitled to recover that which they would have obtained if this contract had been fulfilled in terms of the picture visualised in advance but which they have not obtained . . .

'Direct damage' is damage which flows naturally from the breach or event relied on without other intervening cause.

Saint Line Ltd v. Richardsons, Westgarth & Co. Ltd

KING'S BENCH DIVISION

[1940] 2 KB 99

In a contract for the supply of engines for a ship one clause provided 'nor shall [the manufacturers'] liability ever and in any case . . . extend to any indirect or consequential damages or claims whatsoever'. The manufacturers were in breach of contract and the owners brought an action claiming damages for: (a) loss of profit for the time during which they were deprived of the use of the ship; (b) expenses of wages, stores, etc.; (c) fees paid to experts for superintendence.

Held: All these heads of claim were recoverable as direct damage and were not excluded by the clause.

ATKINSON J: What does one mean by 'direct damage'? Direct damage is that which flows naturally from the breach without other intervening cause and independently of special circumstances, while indirect damage does not so flow. The breach certainly has brought it about, but only because of some supervening event or some special circumstances. The word 'consequential' is not very illuminating, as all damage is in a sense consequential, but there is a definition in the *Oxford English Dictionary* to which both sides have appealed: 'Of the nature, or a consequence, merely; not direct or immediate; eventual'. It cites the definition of 'consequential damages' from Wharton as: 'losses or injuries which follow an act, but are not direct or immediate upon it'. But, apart from that, I have the guidance of the Court of Appeal as to what is meant by 'consequential'. In *Millar's Machinery Co Ltd* v. *David Way & Son* (1934), where the Court was construing a clause of the same class as this, I find this, that Maugham LJ said: 'On the question of damages the word "consequential" has come to mean "not direct"'. Roche LJ agreed 'that the damages recovered by the defendants on the counterclaim were not merely "consequential" but resulted directly and naturally from the plaintiffs' breach of contract'. It is quite clear that the Court there took it for granted that the word 'consequential' meant 'merely consequential' and referred to something which was not the direct and natural result of the breach.

In my judgment, the words 'indirect or consequential' do not exclude liability for damages which are the direct and natural result of the breaches complained of . . . What the clause does do is to protect the respondents from claims for special damages which would be recoverable only on proof of special circumstances and for damages contributed to by some supervening cause. I am satisfied that it does not protect them from the claims which are made in this case.

Croudace Construction Ltd v. Cawoods Concrete Products Ltd

COURT OF APPEAL

(1978) 8 BLR 20

Croudace were main contractors for building a school, and contracted with Cawoods for the supply and delivery of masonry blocks. Cawoods sought to limit their liability by providing in the supply contract that:

> We are not under any circumstances to be liable for any consequential loss or damage caused or arising by reason of late supply or any fault, failure or defect in any materials or goods supplied by us or by reason of the same not being of the quality or specification ordered or by reason of any other matter whatsoever.

Croudace claimed damages for breach of contract, alleging late delivery and defects. The claim included items for loss of productivity, additional costs of delay in executing the main contract works, and an indemnity against a claim by another subcontractor for delay to his programme.

Held: All the losses directly and naturally resulted in the ordinary course of events from Cawoods' alleged breach and were not excluded as 'consequential loss or damage'.

MEGAW LJ: Counsel for [the defendants] submitted that the word 'consequential' could have three possible meanings. The first meaning that it could have is, as he put it, 'all resulting damage'. He would have been prepared to argue in this case that that was the meaning here of 'consequential', so that the exclusion of all consequential damage for delay would have the effect of excluding all damage resulting from delay, were it not that he was precluded from putting forward that argument because of [another clause in the contract . . .].

The second possible meaning which [he] said could be given to 'consequential' in the context of this contract was one which he defined in this way:

> 'Consequential' is all damage other than the normal or ordinary damage which is recoverable in the case of delay of delivery in a contract for the sale of goods.

That was the meaning which [he] suggested was the appropriate meaning here. His submission would involve, if that meaning were accepted, that the damages to which I have referred as being heads of damage in the particulars of the statement of claim would here not be recoverable, because, he contended, such things as extra costs on site as a result of men and materials being idle as a result of delay, inflated price costs and the claim by the sub-contractor would be within this definition of 'consequential': they would be damages other than the normal or ordinary damages which resulted from delay in a sale of goods case. His submission as to what were the normal or ordinary damages which result from delay in a sale of goods case was, in effect, the difference in market value of the relevant goods – the late-delivered goods – as between the date when the goods should have been delivered under the contract and the date when they were in fact delivered. Those, [he] submitted, were the normal, or ordinary, damages which result from delay in delivery in a sale of goods case: any damages which fall outside that are not 'normal' damages; and, not being normal damages, they come within the meaning of 'consequential' damages in this clause, and hence the defendants are excused from liability in respect of them.

The third possible meaning of 'consequential' damage as put forward by [counsel] was, as he put it, some meaning referring to 'second-stage' consequences – consequences which did not follow as a first stage, or immediately, from the breach, but which were, nevertheless, damages which, in the absence of this clause, would have been recoverable on ordinary principles of the law as to damages.

[He] accepted that one way of expressing that third possible meaning was the meaning given to 'consequential' in *Millar's Machinery Co Ltd* v. *David Way and Son* (1934). In the report it is recorded that:

> Roche LJ agreed that the damages recovered by the defendants on the counterclaim were not merely 'consequential', but resulted directly and naturally from the plaintiffs' breach of contract.

So in essence the third meaning which [he] submitted might possibly be given to the word 'consequential' in the context of the clause with which we are concerned is that damages were not consequential if they resulted directly and naturally from the breach of contract. But [he] submitted that, in the context of this clause, the second suggested meaning ought to be preferred.

[Counsel] has put forward (if he will allow me to say so) a clear, closely-reasoned and interesting argument in support of his submission as to the meaning of the clause in this contract. He will, I am sure, understand that it is in no way disrespect to the clarity and the interesting nature of his argument if I deal with the matter quite shortly. I do that for this reason, that to my mind the decision of this court in *Millar's Machinery Co Ltd* v. *David Way & Son* is a decision, the *ratio decidendi* of which is directly applicable to the present case and which is binding on this court.

In these circumstances the references which [he] gave us to passages in textbooks, and in other cases not directly concerned with the issue with which we are concerned, are not really of assistance. It is clear that the word 'consequential' can be used in various senses. It may be difficult to be sure in some contexts precisely what it does mean. But I think that the meaning given to the word in *Millar*'s case is applicable to the present case. It is binding on us in this case. Even if strictly it were not binding, we ought to follow it. That case was decided in the year 1934. It has stood, therefore, now for more than 43 years. So far as I know it has never been adversely commented upon.

'Direct loss and/or expense' includes a claim for managerial and supervisory expenses directly attributable to matters which have materially affected progress of the works.

Tate & Lyle Food & Distribution Ltd v. Greater London Council

QUEEN'S BENCH DIVISION

[1982] 1 WLR 149

Although this case eventually went to the House of Lords and is reported in [1983] 2 WLR 649, the issues discussed in the judgment set out below did not arise in the House of Lords. The judgment casts doubt on the legitimacy of charging a 'head office percentage' to cover managerial time spent consequent on disruption and delay and in particular on the use of formulae to calculate such a percentage in these circumstances.

GLC constructed two new piers in the river Thames, causing heavy deposits of silt which interfered with the use of the plaintiffs' jetties. As a result, the plaintiffs incurred heavy dredging costs. It was found that the engineers who designed the piers should have foreseen this and adopted a different design.

Held: The GLC were liable. In principle the plaintiffs could properly recover damages for managerial and supervisory expenses directly attributable to GLC's wrongful act, but the claim failed because the plaintiffs could not prove their loss by proper records or otherwise. It was not permissible to allow a percentage of the other items of the claim.

FORBES J: [The] plaintiffs claim that 'they have expended managerial and supervisory resources in attending to the problems created by the infringement of their rights'. Originally, they asserted that such expenditure could not be quantified, but . . . they now claim such expenditure at 2.5 per cent of the total loss and damage. [Counsel] for the plaintiffs, accepts that there is no direct evidence, other than passing references, of the fact that such managerial and supervisory resources were expended or of the extent to which managerial time was deployed on attending to the problems involved. He takes his stand on established [Admiralty] practice . . . Apparently, it is usual in Admiralty collision cases for the judge to decide the issue of liability and for all questions of damage to be subsequently referred to the arbitrament of the registrar sitting with 'merchants'. One of the items usually allowed, without proof, is known as 'agency'. I quote from Marsden, *British Shipping Law*, 11th edition (1961), volume 4, paragraph 520:

> In cases of damage to vessels, the most usual causes of loss to the claimant (apart from detention which will be considered later) are out of pocket expenses. These may be recovered as damages from the wrongdoer. They may be classified as follows: . . . (6) Agency, that is to say certain miscellaneous expenses and work based on the disturbance and extra work which a collision causes in the ship-owner's office and those of his agents in ports, e.g. making fresh arrangements for the use of the vessel, paying accounts, etc. Agency is allowed as a single sum without strict proof, and is assessed as far as possible on the probable expense incurred and work done which would not have been necessary if there had been no collision. Underwriters' agency is not, however, recoverable in so far as differing from owner's agency. Superintendence may also be recovered.

[Counsel] tells me that the usual figure is taken as 1 per cent of the collision damage because any exact computation of the expense due to the 'disturbance and extra work which a collision causes in the shipowner's office' is almost impossible. He also referred me to a passage in *The Liesbosch* (1933), where Lord Wright said:

> . . . compensation for disturbance and loss . . . including in that loss such items as overhead charges, expenses of staff and equipment, and so forth thrown away . . .

I do not think, however, that *The Liesbosch* helps [him]. I think Lord Wright in the passage quoted was dealing with the expenses of retaining staff and hired equipment for the period for which they could not usefully be employed because there was no dredger available to carry on with the dredging contract, and not with the cost of staff time involved in remedying the injury.

[The defendant's counsel] suggest that an item of damages under this head and calculated in this way is unknown in a Queen's Bench action; in addition, [they maintain] that managerial expenses could only be recovered as loss of profit, and that they might be recovered if quantified by acceptable evidence, but that there is no such evidence here.

The problem, it seems to me, resolves itself into two constituents: (a) Is there any warrant for suggesting that managerial time, which otherwise might have been engaged on the trading activities of the company, had to be deployed on the initiation and supervision of remedial work (excluding anything which might properly be regarded as preparation for litigation)? And (b) if so, could this reasonably have been the subject of evidence, or is it so difficult to quantify that the application of some suitable rule of thumb is justified?

I think there is evidence that managerial time was in fact expended on dealing with remedial measures. There was a whole series of meetings, in which the plaintiffs' top managerial people took a leading role, the object of which was to find out what could be done to remedy the situation and to persuade the defendants to do something about it. In addition, while there is no evidence about the extent of the disruption caused, it is clear that there must have been a great deal of managerial time involved in dealing with the dredging required and in rearranging berthing schedules and so on . . . to enable the delivery of material to the refinery to proceed without interruption. I have no doubt that the expenditure of managerial time in remedying an actionable wrong done to a trading concern can properly form the subject matter of a head of special damage. In a case such as this it would be wholly unrealistic to assume that no such additional managerial time was in fact expended. I would also accept that it must be extremely difficult to quantify. But modern office arrangements permit of the recording of the time spent by managerial staff on particular projects. I do not believe that it would have been impossible for the plaintiffs in this case to have kept some record to show the extent to which their trading routine was disturbed by the necessity for continual dredging sessions. In the absence of any evidence about the extent to which this occurred the only suggestion [counsel] can make is that I should follow Admiralty practice and award a percentage on the total damages. But what percentage? [Counsel] claims 2.5 per cent but tells me that the usual Admiralty figure is 1 per cent, and this appears to cover the work of port agents as well as that in the shipowner's office.

While I am satisfied that this head of damage can properly be claimed, I am not prepared to advance into an area of pure speculation when it comes to *quantum*. I feel bound to hold that the plaintiffs have failed to prove that any sum is due under this head.

The essence of a claim for overheads is that because of some circumstance for which the contractor is entitled to make a claim, the contract has not made the contribution to the contractor's overheads which he is entitled to expect. The most obvious case is where the contractor is working at 100% of capacity and has to turn other work away because of extra time devoted to the contract. Conversely, if the contractor is working well below capacity it may be difficult to show that extra work at head office actually causes loss. These factors go to whether there is a recoverable loss. Once this is established it is necessary to quantify the loss. In principle the previous case shows the loss should be established by evidence, but the courts have allowed the use of formulae in some cases.

J.F. Finnegan Ltd v. Sheffield City Council

QUEEN'S BENCH DIVISION

(1988) 43 BLR 124

SIR WILLIAM STABB QC: It is generally accepted that, on principle, a contractor who is delayed in completing a contract due to the default of his employer, may properly have a claim for head office or off-site overheads during the period of delay, on the basis that the work-force, but for the delay, might have had the opportunity of being employed on another contract which would have had the effect of funding the overheads during the overrun period. This principle was approved in the Canadian case of *Shore & Horwitz Construction Co Ltd* v. *Franki of Canada* [1967] SRC 589, and was also applied by Mr Recorder Percival QC, in the unreported case of *Whittal Builders Company Limited* v. *Chester le Street District Council*. Furthermore, in *Hudson's Building Contracts*, at page 599 of the tenth edition, a simple formula is set out to determine the amount of the loss of funding of overheads and profit during the period of overrun.

In the present case, however, the plaintiffs assess this sum on a more complicated basis by taking a percentage of the total sum of a notional contract on which the plaintiffs' disrupted work-force could have been employed had it not been for the overrun. That notional contract sum is arrived at, as I understand it, by applying a constant of 3.51 to the percentage of the company's sales value which represents site-based resources. In this way they avoid any calculation based upon the company's direct labour cost during the period of overrun, which was in fact approximately one third of the total cost, since the subcontracted work, which constituted approximately the remaining two thirds, was not disrupted in the sense that the sub-contractors made no claims on the contractor for loss due to delay.

It seems clear to me, however, that this claim in respect of overheads is not a claim for actual loss but a claim for funding overheads during the period of overrun, for which the contract sum had not made provision. The contract sum provided 5% to cover overheads and profit for a twenty-five weeks contract. The contract, in the event, ran for fifty-nine and a half weeks and any overheads expenditure incurred during the overrun period of thirty-four and a half weeks was not funded by the percentage allowed for in the contract sum.

As I understand it, the defendants contend that any sum recoverable for overheads and profit during the period of overrun should be based upon the proportion which the plaintiffs' direct labour force bore to the total labour force engaged, since it was only the contractor's labour force that was disrupted and not that of the subcontractors, that is to say, one third of the total. The defendants maintain that the plaintiffs' claim is purely speculative and bears no relation to the actual loss. That is true, in one sense, but this claim for overheads during the period of overrun is not related to actual loss but is assessed by allowing the contractor's average yearly percentage for overheads on turnover for the period of overrun, provided of course that expenditure on overheads was incurred in that period. It is a notional figure in the sense that it applies the average values of the company's working figures to the period of overrun, as if the company had been able to deploy its work-force on another contract during that period.

On this contract, valued at £630,000, one-third direct labour and two-thirds subcontract labour were engaged and 5% was allowed to fund overheads and provide profit. If the figure for overheads is based on a percentage of annual turnover, which I assume includes the value of sub-contracted work, I can see no reason why the calculation for unfunded overheads in the overrun period of this contract should be restricted to the one-third disrupted direct labour only. It is the period of delay which is the relevant matter, not who or how many were disrupted. Overheads are incurred in respect of direct work as well as sub-contracted work and it seems to me that the overheads during this overrun period were related to both direct and sub-contracted work. If overheads are incurred in respect of both, then the value of funding overheads for the period of overrun should be based upon both.

The work carried out during the period of overrun must have consisted of both direct and sub-contracted labour and both must have incurred expenditure on overheads. It is this unfunded expenditure which is the subject of this part of the plaintiffs' claim. The plaintiffs do not seek to recover their actual overheads expenditure for the overrun period but to substitute the funding of overheads which a notional other contract would have provided.

However, I confess that I consider the plaintiffs' method of calculation of the overheads on the basis of a notional contract valued by uplifting the value of the direct cost by the constant of 3.51 as being too speculative and I infinitely prefer the *Hudson* formula which, in my judgment, is the right one to apply in this case, that is to say, overhead and profit percentage based upon a fair annual average, multiplied by the contract sum and the period of delay in weeks, divided by the contract period.

The company's accounts over the period 1984 to 1987 show a gradual improvement from overall net loss to net profit. The plaintiffs have used the figure of 8.6%, which was the gross profit for the year ending 30 September 1986. I should have thought that that was on the optimistic side. The figure for the previous year was 6.88%. The figure allowed for in the tender was 5%. I should have thought, as Mr Uff submitted, that 6% for overheads and profit was generous and, on this basis and on the basis of my finding on issue no. 1, the formula would be 6/100 multiplied by £632,080.86 multiplied by $34^1/_2/25$, which produces a figure, according to my arithmetic, of in the region of £53,000. This figure, as indicated, includes profit.

The plaintiffs' percentage figure of 8.60% was a gross profit figure and clearly should not have related to overheads only . . .

Property and Land Contractors Ltd v. Alfred McAlpine Homes North Ltd

QUEEN'S BENCH DIVISION

(1995) 47 Con LR 74

JUDGE HUMPHREY LLOYD QC: 'The *Emden* formula (see Emden's *Construction Law* vol. 1, para. 921) is one of a number of methods conventionally applied in an attempt to arrive at an approximation of the damages supposedly incurred by a contractor when there has been delay to the progress of the works whereby completion is similarly delayed. The theory is that because the period of delay is uncertain and thus the contractor can take no steps to reduce its head office expenditure and other overhead costs and cannot obtain additional work there are no means whereby the contractor can avoid incurring the continuing head office expenditure, notwithstanding the reduction in turnover as a result of the suspension or delay to the progress of the work. The reduced activity no longer therefore pays its share towards the overhead costs. This type of loss (sometimes called a claim for 'unabsorbed overheads') is however to be contrasted with the loss that may occur if there is a prolongation of the contract period which results in the contractor allocating more overhead expenditure to the project than was to have been contemplated at the date of contract. The latter might perhaps be best described as 'additional overheads' and will, of course, be subject to proof that the additional expenditure was in fact incurred. Furthermore, the *Emden* formula, in common with the *Hudson* formula (see *Hudson on Building Contracts*, (11th edn, 1995) paras 8–182 et seq) and with its American counterpart the *Eichleay* formula, is dependent on various assumptions which are not always present and which, if not present, will not justify the use of a formula. For example the *Hudson* formula makes it clear that an element of constraint is required (see *Hudson* para. 8.185) i.e. in relation to profit, that there was profit capable of being earned elsewhere and there was no change in the market thereafter affecting profitability of the work. It must also be established that the contractor was unable to deploy resources elsewhere and had no possibility of recovering cost of the overheads from other sources, e.g. from an increased volume of the work. Thus such formulae are likely only to be of value if the event causing delay is (or has the characteristics of) a breach of contract. The findings and reasoning of the arbitrator as set out in para. 22.2 to 22.5 of his award show that the pre-conditions for the use of the Emden formula could not be satisfied for the purposes of clause 26 of the JCT conditions . . .

Loss of profit is a permissible head of claim under JCT 98 clause 26 [JCT 63, clauses 11(6) and 24(1)], but it seems that the contractor must establish proof of profitability elsewhere, i.e. that he could have employed his resources elsewhere had he been free to do so. See now JCT 2005, clause 4.23 to 4.26.

Peak Construction (Liverpool) Ltd v. McKinney Foundations Ltd

COURT OF APPEAL

(1970) 1 BLR 111

The facts of this case are set out on p. 325.

SALMON LJ: Loss of profit can only be awarded for such a period of delay as the official referee concludes was caused by the defendants.

The way in which the claim for loss of profit was dealt with below had caused me some anxiety. The basis upon which the claim was put in the pleadings was that for 58 weeks no work had been done on this site. Accordingly, a large part of the plaintiffs' head office staff, and what was described as their site organisation, was either idle or employed on non-productive work during this period, and the plaintiffs accordingly suffered considerable loss of gross profit. It was said that what I can shortly describe as the land bank – which was a quantity of land owned by the plaintiffs which they could develop if there were no other private contracts on hand – could not avail them, because they did not know during any part of the 58 weeks when they would be able to re-commence on the East Lancashire Road site, and accordingly they had to remain ready to do so and could not undertake any further work.

[At first instance] the matter was put rather differently. The case was put on the basis that in the time during 1966 and 1967 when they were engaged in completing the construction of the East Lancashire Road project they were unable to take on any other work, which they would have been free to do had the East Lancashire Road been completed on time, and they lost the profit which they would have made on this other work. When the case was argued in this court it seemed to me that the plaintiffs were a little uncertain about which basis they were opting for. In the end, however, I think they came down in favour of the second basis: that is, the one that was argued before the deputy official referee. I think when there is a new trial every issue under this head should be open for the official referee to consider. It might be of some help to him not to be left only with the evidence of the plaintiffs' auditor on this point: possibly some evidence as to what the site organisation consisted of, what part of the head office is being referred to, and what they were doing at the material times, could be of help. Moreover, it is possible, I suppose, that an official referee might think it useful to have an analysis of the yearly turnover from, say, 1962 right up to say, 1969, so that if the case is put before him on the basis that work was lost during 1966 and 1967 by reason of the plaintiffs being engaged upon completion of this block, and therefore, not being free to take on any other work, he would be helped in forming an assessment of any loss of profit sustained by the plaintiffs. However, that is a matter for the parties.

'Direct loss and/or expense' under JCT terms is the loss and/or expense which directly and naturally results in the ordinary course of events from the matters specified. It covers financing charges where appropriate, such being a constituent part of the loss and/or expense.

F.G. Minter Ltd v. Welsh Health Technical Services Organisation

COURT OF APPEAL

(1981) 13 BLR 1

Minter was main contractor on a hospital project. The contract was in JCT 63 form. Many variations were made as the contract progressed, and these affected regular progress of the works as a whole and a nominated sub-contractor's work was materially affected by lack of necessary instructions. Claims were made and paid under clauses 11(6) and 24(1)(a) (now JCT 80, clause 26.1.) The wording of clauses 11(6) and 24(1) had been amended by the parties so as to substitute 'not more than 21 days' for the standard 'within a reasonable time'. Minter claimed that finance charges which they had incurred as a result of being kept out of their money were 'direct loss and/or expense'. WHTSO argued that these charges were not 'direct' and were 'a naked claim for interest'.

Held: The loss was 'direct'. Minter was entitled under the contract terms to recover financing charges.

JCT 80, clause 26, has been re-worded so that only one written application is required to be made by the contractor in respect of loss and/or expense caused by the happening of one of the prescribed events.

> STEPHENSON LJ: I approach the question of construction which we have to answer on a certain amount of common ground. It is agreed that clauses 11(6) and 24(1) must be read together and the words 'direct loss and/or expense' in them – and indeed in clause 34(2) – ought to mean the same thing. It is also agreed that if the subclauses can be construed without reference to other provisions of the contract there are no grounds for giving to those words any other meaning than that which they have in a case of breach of contract in a legal context. That was the opinion of Megaw LJ of the words 'direct loss and/or damage' in clause 26(2)(b)(vi) (which I have read) in *Wraight Ltd* v. *P.H. & T. (Holdings) Ltd* (1968). He there considered that loss or damage caused by the determination of the contract was direct if it would be recoverable as damages flowing naturally in the usual course of things from a determination in breach of contract, and he held that loss of 10 per cent for establishment charges and profit which the contractor would have earned if the contract had not been determined and of $12^1/_2$ per cent for a proportion of the overhead costs of his business attributable to the contract were recoverable as direct loss or damage.
>
> It is agreed that accordingly the court should apply to the interpretation of what loss or expense is direct the distinction between direct and indirect or consequential which was discussed by Atkinson J in *Saint Line Ltd* v. *Richardsons, Westgarth & Co* (1940), in construing an exclusion of liability clause and should at least recognise that loss of profit and expenses thrown away on wages and stores may be recoverable as direct loss or expense, as he there held. So this court held that the cost of men and materials being kept on a site without work was recoverable as damages and not excluded as 'consequential' in *Croudace Construction Ltd* v. *Cawoods Concrete Products Ltd* (1978). It had held the same in *B. Sunley & Co* v. *Cunard White Star Ltd* (1940), making there also a small award of interest on money invested and wasted.
>
> It is further agreed that in the building and construction industry the 'cash flow' is vital to the contractor and delay in paying him for the work he does naturally results in the ordinary course of things in his being short of working capital, having to borrow capital to pay wages and hire charges and locking up in plant, labour and materials capital which he would have invested elsewhere. The loss of the interest which he has to pay on the capital he is forced to borrow and on the capital which he is not free to invest would be recoverable for the employer's breach of contract within the first rule in *Hadley* v. *Baxendale* (1854) without resorting to the second, and would accordingly be a direct loss, if an authorised variation of the works, or the regular progress of the works having been materially affected by an event specified in clause 24(1), has involved the contractor in that loss.
>
> On reaching this point the claimants might be thought to be nearly home. Why, [ask the claimants] should they not be entitled to be repaid what they have lost as a natural as well as

contemplated result of what has happened to their contract through happenings which are in no way their responsibility, if not the respondents' faults or the fault of their architect acting for them? If the respondents choose to vary the contract, or if their architect fails to give them the necessary instructions and they have to keep men and machinery idle and have to pay for them and their subcontractors, why should they be allowed the cost of the extra work but be deprived of what is part of the same loss? Why should the contract sum be adjusted to include the one but not the other? [They submit] that the whole risk of the claimants' loss, or the risk of the whole of their loss, naturally resulting from the happenings necessary to activate clauses 11(6) and 24(1), was contractually assumed by the respondents. Yet if the judge is right, the claimants are worse off as a result of actions for which the respondents are responsible and that unjust result is reached by limiting the natural meaning of direct loss and/or expense to exclude losses which were just as clearly within that meaning as the sums which the architect has allowed. There is, [they submit], nothing in the machinery provided by the two clauses or its operation in this case to break the chain of causation or to halt the operation of the root cause of these losses, even if there is any concurrent or supervening cause, or to turn an obviously direct business loss into a loss that is indirect or consequential.

In support of [this submission] we were reminded of what eminent judges had said about causation in *HMS London* (1914), *The Edison* (1932), and *Smith Hogg & Co* v. *Black Sea & Baltic General Insurance Co* (1970). [The claimants] conceded that the language of clauses 11(6) and 24(1), particularly clause 11(6), created a difficulty in the way of interpreting direct loss and/or expense to include the losses [they claim], in particular the use of the past tense in 'has involved' and 'having been incurred' in clause 11(6) and 'has been involved' in clause 24(1). But [they] asked us to treat the applications required by both clauses as what [they] called 'foot in the door' applications to give warning to the architect of losses both past and future.

It is the language of these clauses in which the machinery for paying the contractor's loss and expense is described, on which the respondents principally rely to defeat the claim.

His lordship then considered and rejected the respondents' argument that interest was not recoverable because of the decision in *London Chatham & Dover Railway Co* v. *S.E. Railway Co* (1893), holding that JCT terms contained an implied promise to pay interest as a constituent part of 'direct loss and/or expense'. He continued:

In the context of this building contract and the accepted 'cash flow' procedure and practice I have no doubt that the two kinds of interest claimed here are direct loss and/or expense unless there is something in the contractual machinery for paying direct loss and or expense which excludes this loss and expense of interest by the claimants. It was not obviously indirect, like the loss of a generous productivity bonus paid by the contractor to his men. It was . . . interest reasonably paid on capital required to finance variations or work disrupted by lack of necessary instructions and as such direct loss or expense in which the claimants had been involved.

To see how the contractual machinery appears to work it may help to set out the temporal sequence in chronological order . . . In a clause 11(6) situation first comes (1) a variation, then (2) a direct loss or expense, then (3) a written application . . . [within a reasonable time after it has been incurred], then (4) the architect's opinion that it would or would not be reimbursed by payments in respect of a valuation made in accordance with clause 11(4): then (5) if his opinion is that it would not, his ascertainment of the amount of the loss or expense, then (6) its inclusion in the next monthly interim certificate; finally (7) its payment within 14 days.

The only difference in a clause 24(1) situation is that it begins with (1) the architect's failure to instruct, followed by (2) it becoming apparent that the regular progress of the works has been materially affected by it. Then follow the same five steps or stages.

If the clauses mean what they appear to say, the loss or expense must come before the application, not just some loss or expense but the loss or expense to be included in the certificate. For clause 11(6) requires the application to be made 'within [a reasonable time] of the loss or expense having been incurred', which must mean the loss in which the variation 'has involved the contractor' and it is 'the amount of such loss' which the architect has to ascertain and certify. The matter is not quite so clear in clause 24(1) because there is in that clause no intermediate reference to the loss having been incurred between the loss in which 'the contractor has been involved' by the progress of the works being affected and 'the amount of such loss' which the architect has to ascertain and certify. But both [parties] agree that the clauses must be read together and given the same interpretation, so the choice would seem to be between (1) reading 'the loss or expense' in clause 11(6) as 'some loss or expense' or adding 'or being incurred' after 'having been incurred' and (2) reading into clause 24(1) after 'the work has been affected as aforesaid' some such words as 'so as to involve the contractor in such loss or expense', or 'so that the loss or expense has been incurred'.

I can find no justification for (1) which involves rectifying clause 11(6), not construing it. I prefer (2) which does no violence to the language of clause 24(1) and merely reads into it what is already included in the natural meaning of its express language. I therefore proceed to apply that construction of the two clauses to the three periods (a), (b) and (c) in the question stated in the case. Period (a) is the interval between (2) the incurring of the loss or expense and (3) the making of an application. Period (b) is the interval between (3) the application and (5) the ascertainment of the loss or expense. Period (c) is the interval between (5) ascertainment and (6) certification.

(a) I cannot see why interest lost or expended before the application, a loss or expense necessarily not more than 3 weeks old, is not direct loss or expense. The judge held that it was not direct within clause 11(6) because it was 'directly due to the fact that they (the Claimants) have chosen to wait before making application', and they were not direct within clause 24(1) because they are 'losses resulting from the fact that the effect of the interruption was not immediately ascertainable'. I find the reasoning difficult to follow and impossible to accept. Of the clause 11(6) loss I think the judge is saying that it would be direct were it not for the intervening (and avoidable?) delay of up to 21 days which makes it indirect. I regret that I cannot agree. It would not only be reasonable for the claimants to wait 21 days; the contract contemplates that they may do so. If the loss is direct when the 21 days start to run, it is, in my opinion, still direct throughout those 21 days. Of clause 24(1) he seems to say that no loss can result from the interruption of the progress of the works by a specified event unless the loss has been incurred at the time at which it becomes apparent that the works have been affected, so that interest charges between that time and the application result directly from the 21 days (or less) delay and not from the interruption. Again with respect I cannot agree. If the charges are incurred because of the interruption during the short period allowed by the contract for making the application, why are they not a direct result of the interruption?

[As to] (b) and (c) I agree with the judge's construction of clauses 11(6) and 24(1) and with his conclusion that the architect can only ascertain and certify the amount of interest charges lost or expended at the date of the application. It is these charges which are the subject of the application; it is these charges which he has power to investigate, ascertain and certify. I respectfully agree with the judge that the architect would be exceeding his powers were he to take into account further financial charges or other losses accruing during these two periods, however long, and such further charges and losses would be recoverable only, if at all, under a subsequent application or subsequent applications – although he might obtain the respondents' approval to waiving the required applications or extending the time for making them. But again I feel bound to differ from the judge's conclusion that they are therefore not direct losses or expenses and are therefore not recoverable at all. I appreciate that a series of successive applications may increase the burden of the architect's labours, and that burden may be further increased by delay on the part of contractors like the claimants in providing information which he reasonably requires either for forming his opinion and for ascertaining amounts under those clauses. But contractors have a duty under clause 30(1)(b) of the main contract (not altered by an additional duty imposed by an amendment to clause 30) to send him 'all documents necessary for the purposes of the computations required' by the conditions of the contract; and I should not have thought, in spite of the judge's contrary opinion, that the investigation and computation of interest charges at three-weekly intervals would make his burden so intolerable as to be outside the contemplation of the parties to the contract, particularly if the charges are likely to be recurring charges assessed and paid in the same amounts at the same intervals of time. No doubt the substitution for 'a reasonable time' of a period which is less than one month will not reduce the architect's labours in doing his duty to his employers and the contractors under these clauses during these two different periods of 21 days and one month. It is, I suspect, to meet the difficulties created by that amendment and by the need for successive applications that the 1980 edition of the Standard Contract has been redrafted (clause 26.1) to require applications to be made 'as soon as it has become, or should reasonably have become apparent to him (the contractor) that the regular progress of the works or any part thereof has been or is likely to be affected' by specified events (including instructions requiring variation) and to state 'that he has incurred or is likely to incur direct loss and or expense'; and also to require the architect to ascertain 'the amount of such loss and or expense which has been or is being incurred by the contractor'.

The architect under all forms of this contract has to investigate and compute values and expenditure from time to time and to adjust the contract price by adding certified amounts as a consequence of action taken or not taken by himself and/or his employer. It is only if the duties which these two clauses on their true construction put upon him are so unreasonable, if they cover investigating and ascertaining and certifying interest charges of this kind, as to have gone beyond the contemplation of the parties to this contract that a court would be driven to hold that they are no part of the claimants' direct loss or expense.

It seems to have been the opinion of the judge that they are so unreasonable. His conclusion was: '. . . the facts that the contract provides for direct losses and expenses and that the losses and expenses here sought to be recovered are losses and expenses which arise only because the machinery for ascertainment and certification provided for by the contract involve delay, in my judgment indicates that they are indirect. The position in the end is that the parties have provided a machinery which will enable payment of the primary loss to be recovered with reasonable speed. They have, notwithstanding that both must have been well aware that the machinery would involve some delay, not provided for interest or other payment in respect of the delay. They could easily have so provided and they have not done so notwithstanding that they have amended the standard form in a number of respects.' I say 'seems to have been' because what he said . . . was that charges or losses incurred during period (b) and made the subject of a further application 'would in my judgment be demonstrably not direct loss and expense resulting from the variation. They would be indirect losses so resulting assuming the period was reasonable and if the period was unreasonable the chain of causation would be completely broken. This might give rise to a claim against the architect . . .' – indeed the respondents – 'but the charges are not in my judgment recoverable under the contract. If they were, it would follow that no claim under clause 11(6) or 24(1) could ever be completely dealt with. For as each claim was investigated, ascertained, certified and paid, a further claim would necessarily arise.'

Again I regret to say that I cannot follow the learned judge or his reasoning, if he is, as I understand that passage, saying that a direct loss can become indirect by lapse of time in applying for it to be paid or in its becoming payable, or perhaps – compared with what he said of period (a) – that a loss cannot be direct if it is not applied for when a direct loss of the same kind was applied for. What this seems to me to overlook or underestimate is that the losses and expenses claimed here are one continuous loss and expense, incurred as it were by instalments but continuing to flow from the variations or specified events in the ordinary course of things without interruption from any delay by contractor or architect. And direct loss and expense can continue and recur for an indefinite period, as many a party to an action for damages for personal injuries knows to his cost. For a continuing and recurring loss of instalments of interest charges like those claimed the machinery provided by clauses 11(6) and 24(1) is very ill-fitting. It is the continuation of the loss and expense which prevents finality – not for ever, but until completion of the contract works as varied. But I do not consider the machinery fits these charges so ill that I am driven to conclude that they must be treated for the purpose of these clauses as indirect loss and expense when they appear to be direct.

An application under JCT 98, clause 26 (JCT 1963 Clauses 11 and 24) must make it clear, if that is the case, that an element of the claim is for financing charges. The claim must be presented within a reasonable time but an employer may be precluded by his conduct from objecting to late claims. Financing charges run up until the last application before the issue of a relevant certificate. Financing charges cannot be recovered for periods when non payment arose from an independent cause which was not direct loss or expense. In calculating interest which compensates a contractor for having a bigger overdraft than he would otherwise have had, it is proper to take account of the fact the banks charge compound interest with periodic rests.

Rees & Kirby Ltd v. Swansea Corporation

COURT OF APPEAL

(1985) 5 Con LR 34

The plaintiffs were engaged to construct a housing estate for the defendants for a fixed price under a contract on JCT 1963 terms dated 1 February 1972. The date fixed for practical completion was 6 July 1973. There were a large number of variations ordered under Clause 11 and there were a number of delays in issuing instructions within clause 24(1). Practical completion in fact took place on 4 July 1974. In June 1979 an extension of time under the contract of 52 weeks was agreed.

During 1972 there was a very sharp escalation in wage rates in the construction industry and it soon appeared that this would convert the present contract into a substantial loss maker. Business relationships between the plaintiffs and the relevant offices of the Council were at this stage excellent and it appeared that the Council would approach the plaintiff's difficulties sympathetically. Further, in October 1973 the Minister for Housing issued a statement that councils would be entitled to consider making *ex gratia* payments to contractors who suffered losses on fixed price contracts. In the light of this the plaintiff's contractual claims were put on one side while the parties negotiated for an *ex gratia* settlement. However, by the end of 1976 the plaintiffs had concluded that no such settlement was possible and reverted to their claim under the contract. They also indicated to the council that their losses were significantly increased by interest charges arising from their having been out of their money for so long. The plaintiff's full claim when submitted in 1978 was for £178,780.67 exclusive of interest. Interim certificates were issued by the architect for £60,000 and £25,000 respectively on 14 February 1979 and 2 April 1979 and a 'final interim certificate' for £156,762.19 on 30 August 1979. None of these sums included any element of interest.

The plaintiffs claimed interest on the sum of £156,762.19 from the date of practical completion until the date of payment under the final interim certificate (12 September 1979) and compound interest on the accrued interest so claimed until the date of judgment.

Held: (1) The contractors' application for payment under clauses 11(6) and 24(1)(a) of JCT 1963 must contain a reference sufficient to make it clear that it includes some loss or expense incurred by reason of his being out of pocket.

(2) Such notice must ordinarily under the terms of these clauses be given within a reasonable time of the loss or expense having been incurred. However, the defendants could not rely on the delay between practical completion in 1974 and the formal application of June 1978 since they were estopped from enforcing their strict legal rights.

(3) Under the structure of the clauses there is no cut-off point at the date of practical completion and no reason why financing charges should not continue to constitute direct loss or expense until the last written application before the relevant certificate was issued.

(4) However, the delay in payment between the date of practical completion and 11 February 1977, when it became clear that the respondent's claim must be advanced under the terms of the contract, was not direct loss or expense since it arose from an independent clause.

(5) The plaintiffs were entitled to financing charges to cover the period from 11 February 1977 to 10 August 1979 when they appended their signature to the draft final account.

(6) Since the plaintiffs were operating on the basis of a substantial overdraft at their bank and since banks calculate interest with periodic rests, it was proper to take this into account.

ROBERT GOFF LJ: The sums claimed by the [contractors] as due and owing under clause 11(6) or clause 24(1)(a) were, as I have said, claimed by them under the principles stated by this Court in *Minter*'s case. That case was also concerned with a claim under clauses in the same form, except that the period for the making of written applications under the clauses was not within a reasonable time of the loss or expense having been incurred but within 21 days of the loss or expense having been incurred. In *Minter*'s case, the contractors were paid amounts under the two clauses in respect of direct loss and/or expense in which they had been involved;

but they claimed that the amounts so paid to them were insufficient. Their claim was that, since the amounts in question had not been certified and paid until long after the time when they were involved in the relevant loss or expense, the sum certified ought to have included amounts in respect of the loss and/or expense in which they had been involved by way of finance charges and/or being stood out of their money for those long periods. This claim raised the question whether loss or expense of this kind was recoverable under these clauses. This court, reversing the decision of Parker J. (as he then was), held that they were so recoverable. Judgments were delivered by both Stephenson LJ and Ackner LJ; Sir David Cairns agreed. In relation to the meaning of the word 'direct' in the expression 'direct loss and/or expense', the court concluded (following the decision of Megaw J (as he then was) in *Wraight Ltd* v. *P.H. & T. Holdings Ltd* (1968), that loss or expense should be regarded as direct, for the purposes of both clauses, if it would be recoverable as damages flowing naturally in the usual course of things, within the meaning of those words as used in the first rule in *Hadley* v. *Baxendale* (1854). This was common ground between the parties. It was also common ground that since, in the construction industry, cash flow is vital to contractors and delay in payment results in the ordinary course of things in the contractor being short of working capital, the loss of interest which he has to pay on capital which he has borrowed and the loss of interest on capital which he is not free to invest does fall within the first rule in *Hadley* v. *Baxendale*, and so is direct loss and/or expense. Furthermore, it made no difference that the law has declined to award, by way of damages, interest on debts which have been paid late. That has always been regarded as an anomalous exception to the general principles on which damages are assessed (see *London Chatham & Dover Railway Co* v. *South Eastern Railway Co* (1893); and the existence of this anomalous exception to the recovery of damages should not inhibit the court from holding that such loss or expense is direct loss or expense when considering the meaning of those words used in the two clauses. As Ackner LJ put it, the contractors were seeking to claim not interest on a debt, but a debt which had, as one of its constituent parts, interest charges which had been incurred. The court, in holding that the contractors' claim was recoverable, also rejected certain arguments (which had impressed Parker J) that the two clauses contained provisions which were, upon examination, inconsistent with such a claim being recoverable under either of the clauses.

Now the claim of the [contractors] in the present case is of a similar kind to, though substantially greater in amount than, the claim successfully advanced by the contractors in *Minter*'s case. In order to explain how the claim arose in the present case, it is now necessary for me to set out the history of the matter in some detail.

[His lordship then set out the history of the contract and the attempts made to reach an *ex gratia* settlement and continued:]

It was submitted to the judge, on behalf of the appellants, that no written application within clauses 11(6) and 24(1)(a) was made within a reasonable time, as required by both clauses; and that on that ground alone the respondents' claim must fail. The judge rejected that contention. He accepted the [contractors'] submission that the claim need not be in any set form, and concluded that in all the circumstances the claim had been made within a reasonable time. He was satisfied that the claim fell within the principle in *Minter*'s case, and allowed it in full. On the alternative claim for damages for breach of contract, he held that the architect plainly failed to comply with the time fixed in the contract for measuring, valuing and issuing certificates. He rejected a submission that there had been acquiescence by the [contractors] in the delay which had occurred. He furthermore held that, as the council and the Town Clerk were to all intents and purposes dictating and controlling the exercise by the architect of his duty, the [Council] were responsible to the [contractors] for the architect's breach of the contract. However, he dismissed the claim to interest as damages on the ground that he was precluded from awarding interest as damages by the decision of the House of Lords in *London Chatham & Dover Railway Co* v. *South-Eastern Railway Co.*

Before this court, [Counsel for Swansea] accepted that we were bound by the decision in *Minter*'s case; accordingly, his argument was directed towards challenging the conclusion of the judge that his clients were liable, under the principles stated in that case, in respect of financing charges from the date of practical completion. He submitted (1) that the respondents had failed to make the requisite written application within a reasonable time, as required both by clause 11(6) and clause 24(1)(a), and that for that reason alone their claim should fail; (2) that in any event there was a 'cut-off date' as at the date of practical completion, so that financing charges incurred after that date were not recoverable; (3) that, in the circumstances, the financing charges were incurred not by reason of the variations and delays in giving instructions, but by reason of other events; and, (4) that the claim should be calculated not on the basis of compound interest, but of simple interest.

In considering these submissions, we start from the fact that it was decided by the Court of Appeal in *Minter*'s case that financing charges are recoverable under the clauses in respect of the period between 'the loss and/or expense being incurred and the making of the written application for reimbursement of the same as required by the Contract': per Ackner LJ. However, the application of the clauses in the circumstances of the present case has been very much complicated by a number of matters. First, almost immediately after intimating that they were making claims under clauses 11(6) and 24(1)(a), the [contractors] entered upon prolonged negotiations with the [Council] for an *ex gratia* payment, so that the whole matter of proceeding with contractual claims was, so to speak, put on ice. Second, even when those negotiations had collapsed, the parties took a little time to pick up the threads of proceeding with claims under the contract; and, since by then the date of practical completion had long since passed, in practical terms the matter which then fell to be considered was the amount of money which was to be included in the final certificate in respect of those claims. Finally, *Minter*'s case was not decided by the Court of Appeal until March 1980, which was six months after payment had been made by the [Council] under the final interim certificate; so, understandably, although the [contractors] were expressing periodic concern about the delay which was continuing and about the fact that they were suffering by being out of their money, their claim, in so far as it was formulated, was expressed primarily as a claim to interest on monies outstanding rather than as a claim to financing charges as direct loss and/or expense; though, with some prescience, it was so formulated on 17 July 1979.

Faced with these somewhat unusual circumstances, we have, I think, to regard with some caution submissions which are founded upon suggestions that the [contractors] failed strictly to comply with the contractual provisions.

With these preliminary observations in mind, I turn to [the] first submission, which was that [Rees & Kirby] failed to make the requisite written application within a reasonable time.

There was some argument before us about the form which a written application under the two clauses should take. [Counsel for the contractors] adopted as part of his argument a passage from *Keating's Building Contracts* (4th edn, at p. 56) in which the learned author has this to say about the contents of an application under clause 11(6):

> Contents of the application. A consideration of the effect of the *Minter* case raises some questions about what an application should contain in order to be valid under clause 11(6). No form is required, but it must, it is submitted, be expressed in such a way, or made in such circumstances, as to show that the architect is being asked to form the opinion referred to in the sub-clause and identify the variation or provisional sum work relied on. It is thought that great particularity is not contemplated by the sub-clause. The architect must

> know the variation of work relied on; he must also either be told in the application, or must be taken to know from his knowledge of the circumstances, sufficient to enable him to form the opinion that the contractor has been involved in direct loss and/or expense for which he would not be reimbursed under sub-clause (4). He does not have to have at the stage of forming his opinion sufficient details to ascertain the amount; it is sufficient if he has enough to form the view that there must be some loss. Thus, it is thought, providing the principles just suggested are met, an application is valid to found a claim at the ascertainment stage that interest should be included even though the application itself did not expressly refer to interest.

I have much sympathy with the general approach adopted in this passage. It seems to me that, in the ongoing relationship that exists between a contractor and an architect carrying out their functions under a contract in this form, a sensible and not too technical attitude must be adopted with regard to the form of such an application. But, having regard to the form which clauses 11(6) and 24(1)(a) take in the contract presently before us, I do not feel able to say that an application need not expressly refer to what Mr Keating calls 'interest'. This is because the Court of Appeal's decision in *Minter*'s case proceeded on the basis that a claim to financing charges cannot be regarded simply as parasitic to the primary expense incurred by a contractor by reason of, for example, a variation; and that financing charges are recoverable under either of the clauses between the date when the loss or expense is incurred and the date when the written application for reimbursement is made, so that, in respect of a continuing loss or expense such as financing charges, successive applications will have to be made from time to time. On that basis, it is difficult to avoid the conclusion that the application in question, if it is to be read as relating to financing charges, must make some reference to the fact that the contractor has suffered loss or expense by reason of being out of his money. I observe however in parenthesis that the clauses in question have since been revised to allow for applications in respect of loss or expense which the contractor has incurred or is likely to incur; and it may be (though I express no opinion upon it) that, under the clauses as so revised, there need be no express reference to interest as financing charges. But, for the purposes of the present case we must, I consider, proceed on the basis that some reference is necessary. Even so, I do not consider that more than the most general reference is required, sufficient to give notice that the contractor's application does include loss or expense incurred by him by reason of his being out of pocket in respect of the relevant variation or delayed instruction, or whatever may be the relevant event giving rise to a claim under the clause.

Even so, I feel unable to accept [the Council's] submission that the [contractors'] claim must fail in the present case because they failed to give the requisite notice within a reasonable time of the loss or expense having been incurred, as required by the clause. In so far as the loss or expense in question consists of the primary loss or expense incurred by the relevant variations or delayed instructions, it appears to me that the [Council] are estopped from enforcing their strict legal right to have such a notice given within a reasonable time of such loss or expense having been incurred, on the principle stated by Lord Cairns LC in *Hughes* v. *Metropolitan Railway Co* (1877). The passage in Lord Cairns' speech is so well known, that I trust that I will be excused from setting it out in this judgment. The present case seems to me to be a classic example of parties entering 'upon a course of negotiation which has the effect of leading one of the parties to suppose that the strict rights arising under the contract will not be enforced'. Here, as soon as the [contractors] raised the question of a claim under either clause 11(6) or clause 24(1)(a), the parties entered into the negotiations for an *ex gratia* settlement, negotiations which lasted over a period of years. Even after those negotiations had come to an end, the [Council] expressed no objection to the [contractors] instructing their own quantity surveyor to prepare their claim, nor did they give any notice to the effect that they intended to resume their right under either clause to require a written application within any particular period of time. Furthermore, once the [contractors] had submitted their claim . . . on 30 June 1978, they had given what I regard to be a sufficient intimation of the fact that they were out of pocket by reason of delay in payment of their claims under the two clauses, and they continued periodically thereafter to give further intimations of that fact almost up to the date of the issue of the final interim certificate. In these circumstances, I am satisfied that the first submission of the Appellants is not well-founded.

Nor, in my judgment, is there any substance in the second of the [local authority's] submissions, which is that there is a 'cut-off point' at the time of practical completion. As I read the clauses, given that (on the clauses in the form which they take in the contract now before us) successive applications are made at reasonable intervals, I can see no reason why the financing charges should not continue to constitute direct loss or expense in which the contractor is involved by reason of, for example, a variation, until the date of the last application made before the issue of the certificate issued in respect of the primary loss or expense incurred by reason of the relevant variation. At the date of the issue of the certificate, the right to receive payment in respect of the primary loss or expense merges in the right to receive payment under the certificate within the time specified in the contract, so that from the date of the certificate the contractor is out of his money by reason either (1) that the contract permits time to elapse between the issue of the certificate and its payment, or (2) that the certificate has not been honoured on the due date. But I can for my part see no good reason for holding that the contractor should cease to be involved in loss or expense in the form of financing charges simply because the date of practical completion has passed.

There is however in my judgment much more force in the third submission of the [Council], that part of the delay in payment in respect of the primary loss or expense incurred by the [contractors] in the present case is attributable to an independent cause, so that financing charges incurred by them during the period when such independent cause operated should not be recoverable by them under either clause 11(6) or clause 24(1)(a). As I see it, the whole of the period of delay which occurred by reason of (1) the negotiations for an *ex gratia* settlement irrespective of the parties' strict contractual rights, and (2) the proposal by the [contractors] that the contract might be converted from a fixed price contract to a fluctuating contract under the NEDO formula, must be regarded as delay attributable to those matters, so that financing charges incurred by [them] during such period cannot be direct loss or expense in which [they] were involved by reason of the variations or delayed instructions. The [contractors] have made no claim in respect of financing charges incurred by them before practical completion: and the period of delay resulting from the above two matters must have run on past practical completion until the architect's letter dated 11 February 1977, after which it became clear that [their] claim must be advanced under the terms of the contract. However, from the date of that letter, I can discern no cause of the [contractors] incurring the financing charges which form the subject matter of their claim, other than the fact that they incurred the primary loss or expense incurred by them by reason of the variations or delayed instructions under one or other of the two clauses. It is true that [their] detailed claim was not submitted until 30 June 1978; but I do not consider that the time which passed in the meanwhile should be regarded as the product of any other cause so independent of the primary loss or expense as to relieve the [Council] from liability in respect of financing charges incurred by the [contractors] during that period. Once [the contractors' quantity surveyor's] report had been delivered to the [Council], the financing charges must have continued to constitute direct loss or expense up to the date of the issue of the final interim certificate on 30 August 1979, subject to the requirement, established in *Minter*'s case, that such loss or expense is only recoverable in respect of a period ending with a written application made under the relevant clause. In my judgment, written applications containing a sufficient reference to such loss or expense were made not only by the delivery of [the quantity surveyor's] report, with its reference to the fact that the [contractors] had been out of funds, and that they had 'a just contractual entitlement to settlement of the value we have assessed, plus interest'; but also by the letters claiming interest, sent at periodical intervals, on 5 December 1978, 13 March 1979 and 17 July 1979, and by their signature on the draft final account on 10 August 1979 reserving their right to interest. It follows, in my judgment, that [the contractors] have established a right to recover financing charges under the clauses in respect of the period from 11 February 1977 to 10 August 1979, on the principle in *Minter*'s case.

There remains to be considered the question whether [they] are entitled to recover their financing charges only on the basis of simple interest, or whether they are entitled to assess their claim on the basis of compound interest, calculated at quarterly rests, as they have done. Now here, it seems to me, we must adopt a realistic approach. We must bear in mind, moreover, that what we are here considering is a debt due under a contract; this is not a claim to interest as such, as for example a claim to interest under the Law Reform Act, but a claim in respect of loss or expense in which a contractor has been involved by reason of certain specific events. [Rees & Kirby], like (I imagine) most building contractors, operated over the relevant period on the basis of a substantial overdraft at their bank, and their claim in respect of financing charges consists of a claim in respect of interest paid by them to the bank on the relevant amount during that period. It is notorious that banks do themselves, when calculating interest on overdrafts, operate on the basis of periodic rests: on the basis of the principle stated by the Court of Appeal in *Minter*'s case, which we here have to apply, I for my part can see no reason why that fact should not be taken into account when calculating the [contractors'] claim for loss or expense in the present case. It follows that, in order to calculate [their] contractual claim it will be necessary to calculate it with reference to the total sum of £156,762.19 in relation to the period I have indicated, viz. 11 February 1977 to 10 August 1979, taking into account (1) the two payments made on account during that period; (2) the rates of interest charged by the bank to [them] at various times over that period; and (3) the periodic rests on the basis of which the bank from time to time added outstanding interest to the capital sum outstanding for the purposes of calculating interest thereafter . . .

Finally, I must briefly mention an alternative basis on which the [contractors] sought to recover their financing charges as damages for breach of contract. The breach of contract alleged by them was that there was a failure by the architect to measure and to make valuations and to issue certificates in respect of the relevant variations and delayed instructions at the times specified in the contract (other than in the final interim certificate itself). In my judgment, however, this claim must fail. The time which passed between the [contractors'] letter of 18 December 1973, in which they first advanced the claim under clauses 11(6) and 24(1)(a), and the issue of the final certificate on 30 August 1979, can be sub-divided into the following periods:

(1) 18 December 1973–24 December 1976: negotiations over a possible *ex gratia* payment.
(2) 24 December 1976–11 February 1977: considering the [contractors'] suggestion that the contract should be converted into a fluctuating contract on the NEDO formula.
(3) 11 February 1977–8 August 1977: a period of quiescence.
(4) 8 August 1977–1 November 1977: time occupied by the [contractors] in instructing a quantity surveyor.
(5) 1 November 1977–30 June 1978: preparation of the [contractors' claim].

(6) 30 June 1978–30 August 1979: consideration by the [Council of the claim], and the issue by them of the final certificate.

None of the periods of delay can possibly be attributed to the [Council], except that it can be argued that they took too long to consider the [contractors'] claim during the period 30 June 1978 to 30 August 1979. But, bearing in mind the time allowed in the contract, under clause 30(5) and (6) and the Appendix to the contract, for the preparation and issue of a final certificate after practical completion, I do not consider that it would be right, in all the circumstances of this case, to hold that there was any breach of contract by the [Council] in this respect. It follows, in my judgment, that the [contractors'] alternative claim to damages for breach of contract must fail . . .

Any interest allowed as part of 'direct loss and/or expense' should be assessed at a rate equivalent to the cost of borrowing, disregarding any special position of the contractor.

Tate & Lyle Food & Distribution Ltd v. Greater London Council

QUEEN'S BENCH DIVISION

[1982] 1 WLR 149

For the facts and holdings, see p. 281.

FORBES J: Despite the way in which Lord Herschell LC in *London, Chatham and Dover Railway Co* v. *South Eastern Railway Co* (1893) stated the principle governing the award of interest on damages, I do not think the modern law is that interest is awarded against the defendant as a punitive measure for having kept the plaintiff out of his money: I think the principle now recognised is that it is all part of the attempt to achieve *restitutio in integrum*. One looks, therefore, not at the profit which the defendant wrongfully made out of the money he withheld – this would indeed involve a scrutiny of the defendant's financial position – but at the cost to the plaintiff of being deprived of the money which he should have had. I feel satisfied that in commercial cases the interest is intended to reflect the rate at which the plaintiff would have had to borrow money to supply the place of that which was withheld. I am also satisfied that one should not look at any special position in which the plaintiff may have been; one should disregard, for instance, the fact that a particular plaintiff, because of his personal situation, could only borrow money at a very high rate or, on the other hand, was able to borrow at specially favourable rates. The correct thing to do is to take the rate at which plaintiffs in general could borrow money. This does not, however, to my mind, mean that you exclude entirely all attributes of the plaintiff other than that he is a plaintiff. There is evidence here that large public companies of the size and prestige of these plaintiffs could expect to borrow at 1 per cent over the minimum lending rate, while for smaller and less prestigious concerns the rate might be as high as 3 per cent over the minimum lending rate. I think it would always be right to look at the rate at which plaintiffs with the general attributes of the actual plaintiff in the case (though not, of course, with any special or peculiar attribute) could borrow money as a guide to the appropriate interest rate. If commercial rates are appropriate I would take 1 per cent over the minimum lending rate as the proper figure for interest in this case.

In wholly exceptional circumstances it may be possible to adopt a 'broad-brush' approach to the assessment of individual heads of claim where proof of each and every individual item is very difficult, but this is probably only the case where the contract machinery has broken down.

Penvidic Contracting Co Ltd v. International Nickel Co of Canada Ltd

SUPREME COURT OF CANADA

(1975) 53 DLR (3d) 748

Contractors agreed to lay ballast and track for a railroad. The employer was in breach of its obligation to facilitate the works. The contractors had agreed to do the work for a certain sum for each ton of ballast. They claimed as damages the difference between that sum and the larger sum which they would have demanded had they foreseen the problems caused by the employer's breaches of the contract. There was evidence that the figure claimed was a reasonable estimate.

Held: Damages should be assessed on the basis claimed. Where proof of the actual additional costs caused by the breach was difficult, it was permissible to assess damages broadly.

SPENCE J: Viscount Haldane LC in *British Westinghouse Electric & Manufacturing Co Ltd* v. *Underground Electric Railways Co of London Ltd* (1912) said:

> The quantum of damage is a question of fact, and the only guidance the law can give is to lay down general principles which afford at times but scanty assistance in dealing with particular cases. The judges who give guidance to juries in these cases have necessarily to look at their special character, and to mould, for the purposes of different kinds of claim, the expression of the general principles which apply to them, and this is apt to give rise to an appearance of ambiguity.
>
> Subject to these observations I think that there are certain broad principles which are quite well settled. The first is that, as far as possible, he who has proved a breach of a bargain to supply what he contracted to get is to be placed, as far as money can do it, in as good a situation as if the contract had been performed.

The difficulty in fixing an amount of damages was dealt with in the well-known English case of *Chaplin* v. *Hicks* (1911), which had been adopted in the Appellate Division of the Supreme Court of Ontario in *Wood* v. *Grand Valley Railway Co* (1913), where Meredith CJO, said:

> There are, no doubt, cases in which it is impossible to say that there is any loss assessable as damages resulting from the breach of a contract, but the Courts have gone a long way in holding that difficulty in ascertaining the amount of the loss is no reason for not giving substantial damages, and perhaps the furthest they have gone in that direction is in *Chaplin* v. *Hicks* (1911). In that case the plaintiff, owing, as was found by the jury, to a breach by the defendant of his contract, had lost the chance of being selected by him out of fifty young ladies as one of twelve to whom, if selected, he had promised to give engagements as actresses for a stated period and at stated wages, and the action was brought to recover

damages for the breach of the contract, and the damages were assessed by the jury at £100. The defendant contended that the damages were too remote and that they were unassessable. The first contention was rejected by the Court as not arguable, and with regard to the second it was held that 'where it is clear that there has been actual loss resulting from the breach of contract, which it is difficult to estimate in money, it is for the jury to do their best to estimate; it is not necessary that there should be an absolute measure of damages in each case:' per Fletcher Moulton LJ.

When *Wood* v. *Grand Valley Railway Co* reached the Supreme Court of Canada, judgment was given by Davies J and . . . the learned Justice said:

It was clearly impossible under the fact of that case to estimate with anything approaching to mathematical accuracy the damages sustained by the plaintiffs, but it seems to me to be clearly laid down there by the learned Judges that such an impossibility cannot 'relieve the wrongdoer of the necessity of paying damages for his breach of contract' and that on the other hand the tribunal to estimate them whether jury or Judge must under such circumstances do 'the best it can' and its conclusion will not be set aside even if 'the amount of the verdict is a matter of guess work'. [emphasis added]

I can see no objection whatsoever to . . . using the method suggested by the plaintiff of assessing the damages in the form of additional compensation per ton rather than attempting to reach it by ascertaining items of expense from records which, by the very nature of the contract, had to be fragmentary and probably mere estimations.

London Borough of Merton v. Stanley Hugh Leach Ltd

CHANCERY DIVISION

(1985) 32 BLR 51

The facts of this case are set out on p. 142. One of the many points at issue was, if the contractor proved all the alleged breaches of contract, he was entitled to remuneration on a *quantum meruit* basis in respect of some or all of the items. The arbitrator's interim award gave the contractor a 'partial *quantum meruit*'. Vinelott J treated the issue as hypothetical but, as the following passage indicates, it is unlikely that such a claim would succeed in England if the contract was fully performed.

VINELOTT J: This issue relates primarily to the claim that the architect failed to supply Leach with priced bills of variation within the period specified in clause 30(5), that the bills of variation delivered with the points of defence were not prepared in accordance with that subclause and that in view of the delay that has occurred and the failure by Merton to appoint anybody answering the description of the architect it is not now practicable for proper bills of variation to be prepared.

The case for Leach, in essence, is that the machinery for quantifying additions to the contract sum has broken down and that the arbitrator can and should award in lieu of a sum calculated in accordance with the contract reasonable remuneration for the work actually carried out by it . . .

[After the award] agreement was given in respect of the sum payable under clause 11(4), though I would stress that Leach do not agree that that agreement represents a bill of variations and, as I have said, stress the absence of a bill of variations as a matter which impedes the proper ascertainment of the sum due in accordance with the machinery of the contract.

After referring to (amongst others) the decisions of the House of Lords in *Sudbrook Trading Estate Ltd* v. *Eggleton* (1983) and *Lodder* v. *Slowey* (1904), the arbitrator concluded:

> In the case [Leach] has achieved practical completion of the works and so there is no question of valuing partially completed work as was the case in *Lodder* v. *Slowey*. Although I will be unable to determine the point until I hear evidence in Part II of this arbitration, I share [Merton's expert's] anticipation that the 'bill of variations' may prove a useful starting point for my ultimate assessment of the adjusted contract sum. I also anticipate that there may be several items in the 'bill of variations' that are acceptable to [Leach] and which are capable of being readily valued in accordance with the contract machinery. I do not consider it right to interfere with such assessments and I therefore hold that the [Leach's] entitlement to recover on the basis of *quantum meruit* is limited to items of work other than those included in the bills of variation and which either (1) have been agreed between the parties or (2) are readily capable of being valued in accordance with the contract machinery.

After the hearing had concluded Merton made further submissions founded on the then recent decision of the Court of Appeal in British Columbia in *Morrison-Knudsen & Co Inc* v. *British Columbia Hydro and Power Authority (No. 2)* (1978) and concluded that his decision 'takes into account the principles of this Canadian case by restricting assessment on *quantum meruit* to those items which cannot now be valued under the contract machinery'.

It became apparent in the course of the argument that the claim advanced by Leach before the arbitrator is an alternative to the claim to be entitled to damages in respect of the breaches alleged . . . The claim in essence is that if Leach cannot claim to be entitled to be paid a sum calculated otherwise than in accordance with the machinery provided by the contract and if that machinery has broken down by reason of default on the part of Merton they claim they must be entitled to a *quantum meruit* either for the whole of the work or to reflect additional expense which they would, apart from that default, have been entitled to recover by virtue of the machinery in the contract. This issue therefore seems to me in the light of my conclusions on other matters to be a hypothetical one and although I have been anxious to assist the parties by dealing as comprehensively as possible with the issues formulated by them I do not think it is necessary or desirable that I should venture further into this field.

Where a contractor's claim depends on 'an extremely complex interaction in the consequences of various denials, suspensions and variations, it may be difficult or even impossible to make an accurate apportionment of the total extra cost between the several causative events' it is permissible to assess the claim on a global basis. This applies to claims for both time and money.

London Borough of Merton v. Stanley Hugh Leach Ltd

CHANCERY DIVISION

(1985) 32 BLR 51

The facts are set out on p. 142.

VINELOTT J: The case for Merton is shortly as follows. It is said that under the terms of clause 11(6) and 24(1) upon written application by the contractor the architect must form an opinion whether for the reason there set out the contractor has been involved in direct loss or expense in consequence of one of the causes there mentioned; if the architect is of the opinion that he has, the architect must ascertain or instruct the quantity surveyor to ascertain the amount of direct loss stemming from that clause. It is said that under those provisions the architect cannot make an award unless he is in a position to ascertain the direct loss stemming from a specific cause identified in the application and cannot therefore make an award if the loss stemming from the two different causes cannot be separated and each separate part identified as the direct loss stemming from each cause. In this respect the arbitrator cannot be in any better position than the architect: although his function is not purely arbitral (he has power to open up and review any decision of the architect) nonetheless to the extent that he does so he stands in the shoes of the architect.

This broad submission is clearly and admittedly inconsistent with the decision of Donaldson J in *Crosby* v. *Portland UDC* (1967). In *Crosby* the arbitrator rolled up several heads of claim arising under different heads and indeed claims for which the contract provided different bases of assessment. The question accordingly is whether I should follow that decision. I need hardly say that I would be reluctant to differ from a judge of Donaldson J's experience in matters of this kind unless I was convinced that the question had not been fully argued before him or that he had overlooked some material provisions of the contract or some relevant authority. Far from being so convinced, I find this reasoning compelling. The position in the instant case is, I think, as follows. If application is made (under clause 11(6) or 24(1) or under both sub-clauses) for reimbursement of direct loss or expense attributable to more than one head of claim and at the time when the loss or expense comes to be ascertained it is impracticable to disentangle or disintegrate the part directly attributable to each head of claim, then, provided of course that the contractor has not unreasonably delayed in making the claim and so has himself created the difficulty the architect must ascertain the global loss directly attributable to the two causes, disregarding, as in *Crosby*, any loss or expense which would have been recoverable if the claim had been made under one head in isolation and which would not have been recoverable under the other head taken in isolation. To this extent the law supplements the contractual machinery which no longer works in the way in which it was intended to work so as to ensure that the contractor is not unfairly deprived of the benefit which the parties clearly intend he should have.

In the two previous cases the question of 'global claims' arose in connection with possible review of arbitrators' awards. The question may also arise at the earlier stage of pleading. Defendants in construction disputes often argue that claimants should particularise in their pleadings the precise effect of each alleged cause of delay or disruption. Often this is difficult if not impossible to do, at least until the evidence has been heard.

Some commentators thought that the advice of the Privy Council in* Wharf Properties Ltd *v.* Eric Cumine Associates *(1991) 29 Con LR 84 supported such claims, although that advice was expressly said to be subject to the very special facts of the case.

The correct position is stated in the two following decisions of the Court of Appeal.

GMTC Tools and Equipment Ltd v. Yuasa Warwick Machinery Ltd

COURT OF APPEAL

(1994) 44 Con LR 68

LEGGATT LJ: These submissions, however, appear to me to miss the point so far as the plaintiffs' presentation of their damages claim is concerned. They criticise the formulation imposed upon them by the judge on the ground that it presupposes that the plaintiffs' production process is so flexible and instantaneously reactive to a period of downtime, that the plaintiffs ought to have been able to link each incident of downtime with a particular purchase of precise number of blanks or finished cutters which replaced the lost production. It also presupposes that the plaintiffs ought to have been able to purchase precise numbers of blanks or finished cutters to match the loss of production.

The plaintiffs, however, put their case quite simply in relation to each of the heads of claim. First, they contend that after May 1988 the lathe has been capable of producing the whole of their requirement of blanks, from which they argue that had it not been defective it could and would have produced the whole of their requirement of blanks during the period of defect between August 1985 and May 1988. In relation to Rolls Royce, the claim is for loss of profit from a business opportunity which would have existed in the period between 1986 and 1988 had the lathe been functioning properly and reliably.

I have referred to the fact that according to Mr Storey's submission an official referee or a commercial judge, when dealing with the interlocutory stages of an action, is entitled to prescribe the way in which *quantum* of damage is to be pleaded and proved. I disagree. No judge is entitled to require a party to establish causation and loss by a particular method, especially when the method proceeds, as happened here, on what can only be regarded as an imperfect understanding of the commercial realities of the plaintiffs' manufacturing processes. With the benefit of hindsight, it is apparent that the plaintiffs might have fared better if at an earlier stage they had objected to the way in which they were being obliged to plead their claim. Understandably, they tried so far as possible to do as a matter of expedience what was being demanded of them by judge and defendants alike. When they proved unable to accomplish this satisfactorily, it was said against them that they were unable to particularise their claims sufficiently, and so those claims were struck out. In my judgment, they ought never to have been put in that predicament in the first place. The plaintiffs were entitled to put their claims in the rational way which has been explained to us by Mr Reese. That is not to say that either of the claims will necessarily succeed at trial, whether wholly, mainly or in part. But the judge was wrong to determine that issue at this stage, especially when the approach involved stipulating the manner in which proof of *quantum* was to be attempted, and then striking out the pleadings when the plaintiffs failed to comply with it.

I have come to the clear conclusion that the plaintiffs should be permitted to formulate their claims for damages as they wish, and not forced into a straightjacket of the judge's or their opponents' choosing. This is not the stage at which to try the efficacy of these claims. It will be for the trial judge to determine whether the plaintiffs can establish that the loss claimed under each of their two main headings was caused by the breaches of contract that the judge has found.

Since, in exercising his discretion as he purported to do, this judge was plainly wrong, it follows that in my judgment the appeal should be allowed, the re-amended statement of claim should stand, and the case should be remitted to another judge to determine the outstanding issues of quantum . . .

British Airways Pension Trustees Ltd v. Sir Robert McAlpine & Sons Ltd

COURT OF APPEAL

(1994) 45 Con LR 1

SAVILLE LJ: The judge described this part of the pleadings as embarrassing, though not sufficiently so on its own to justify striking out the claims. This conclusion is not challenged on this appeal, in the sense that it is accepted that the pleading as it stands is open to a request for further and better particulars which the plaintiffs offered to supply during the course of the hearing. The various defects alleged by the plaintiffs might not all be attributable to all the defendants, the cost of remedying the individual defects was not given and no attempt was made to ascribe to each defect the amount by which it contributed to the alleged diminution in value. At the same time I have some difficulty in seeing how the defendants could fairly be said to be seriously prejudiced by these omissions. The pleading alleges that the defects respectively attributable to McAlpine's and [the architects] each caused the alleged diminution in value. The alleged defects themselves were set out in some detail; McAlpine's and [the architects] had been on site for a considerable time after practical completion and so had their own means of knowledge of the alleged defects. Thus it seems to me that it can hardly be said that these defendants were in any real fashion placed in a position where they were unable to know what case they had to meet or were facing an unfair hearing. They could, in my view, each prepare to deal with the allegation that they were responsible for the defects and could each assess the cost of remedying any particular defect. They could also investigate with their own experts to what extent (if at all) any particular alleged defect or class of defects would diminish the sale value of the building. It is true that the pleading does not seek to apportion liability between the active defendants, but this is because it was the plaintiffs' case (good or bad does not matter in this context) that the defects attributable to each defendant caused the whole of the diminution in value. In any event it seems to me that the defendants themselves, if they were minded to make any offer to settle the proceedings or to pay money into court, could calculate without difficulty their respective responsibility for defects and a proportionate amount of the diminution in value (if any) attributable to those defects. The decision to make a payment in or an offer of settlement does not depend so much on what the plaintiff is claiming as on what the defendants calculate the claim is worth.

The basic purpose of pleadings is to enable the opposing party to know what case is being made in sufficient detail to enable that party properly to prepare to answer it. To my mind it seems that in recent years there has been a tendency to forget this basic purpose and to seek particularisation even when it is not really required. This is not only costly in itself, but is calculated to lead to delay and to interlocutory battles in which the parties and the court pore over endless pages of pleadings to see whether or not some particular point has or has not been raised or answered, when in truth each party knows perfectly well what case is made by the other and is able properly to prepare to deal with it. Pleadings are not a game to be played at the expense of the litigants, nor an end in themselves, but a means to the end, and that end is to give each party a fair hearing. Each case must of course be looked at in the light of its own subject matter and circumstances. Thus general statements to the effect that global or composite claims are embarrassing and justify striking out, to be found for example in *Hudson on Building and Engineering Contracts* (11th edn, 1994) para. 8–204 are not automatically applicable to every case. With regard to the particular pleadings in question, I remain unpersuaded that either McAlpine's or [the architects] were put to any sort of material unfair disadvantage by the way the matter had been set out by the plaintiffs . . .

The position was very fully and helpfully reviewed in the following case.

John Doyle Construction Ltd v. Laing Management (Scotland) Ltd

COURT OF SESSION (OUTER HOUSE)

(2002) 85 Con LR 98

LORD MACFADYEN:

(c) Discussion

33. This case is not concerned with whether a global claim for loss and expense may relevantly be advanced by a contractor under a construction contract. The debate proceeded on the basis that it was common ground that such a claim could in principle relevantly be made (*Merton London Borough* v. *Stanley Hugh Leach Ltd* (1985) 32 BLR 51; *Wharf Properties Ltd* v. *Eric Cumine Associates (No 2)* (1991) 52 BLR 1; *John Holland Construction & Engineering Pty Ltd* v. *Kvaerner RJ Brown Pty Ltd* (1996) 82 BLR 81). Nor is it in issue in this case at this stage whether the circumstances are such as to permit a claim to be made in that form. The pursuers aver (at page 32 of the closed record) 'Despite the Pursuers' best efforts, it is not possible to identify causal links between each such cause of delay and disruption, and the cost consequences thereof'. That averment having been made, the defenders accept that the pursuers are in principle entitled to advance a global claim. I prefer to reserve my opinion on whether such an averment is essential to the relevancy of a global claim, on what the pursuers need do to establish that averment, and on what the consequences would be if they failed to do so.

34. The pursuers attribute their global loss to a number of causal factors. One of these is the delaying and disruptive effect on WP2011 of delay, between 22 January and 12 February 1996, in completion of WP2010. As I have already held, that is not a relevant basis for an extension of time in relation to WP2011, and is in any event a factor on which the pursuers are contractually barred from founding for that purpose. It is equally not a relevant basis for a claim for loss and expense. The contractual bar also operates, in my opinion in the context of a claim for loss and expense. The issue which arises is whether the fact that the global loss and expense is said to have been caused, *inter alia*, by that factor for which the defenders have no contractual responsibility is fatal to the relevancy of the global claim. (The defenders also point to a short period when delay is said to have been caused by snow, which is also a cause which does not relevantly found a loss and expense claim.)

35. Ordinarily, in order to make a relevant claim for contractual loss and expense under a construction contract (or a common law claim for damages) the pursuer must aver (1) the occurrence of an event for which the defender bears legal responsibility, (2) that he has suffered loss or incurred expense, and (3) that the loss or expense was caused by the event. In some circumstances, relatively commonly in the context of construction contracts, a whole series of events occur which individually would form the basis of a claim for loss and expense. These events may inter-react with each other in very complex ways, so that it becomes very difficult, if not impossible, to identify what loss and expense each event has caused. The emergence of such a difficulty does not, however, absolve the pursuer from the need to aver and prove the causal connections between the events and the loss and expense. However, if all the events are events for which the defender is legally responsible, it is unnecessary to insist on proof of which loss has been caused by each event. In such circumstances, it will suffice for the pursuer to aver and prove that he has suffered a global loss to the causation of which each of the events for which the defender is responsible has contributed. Thus far, provided the pursuer is able to give adequate specification of the events, of the basis of the defender's responsibility for

each of them, of the fact of the defender's involvement in causing his global loss, and of the method of computation of that loss, there is no difficulty in principle in permitting a claim to be advanced in that way.

36. The logic of a global claim demands, however, that all the events which contribute to causing the global loss be events for which the defender is liable. If the causal events include events for which the defender bears no liability, the effect of upholding the global claim is to impose on the defender a liability which, in part, is not legally his. That is unjustified. A global claim, as such, must therefore fail if any material contribution to the causation of the global loss is made by a factor or factors for which the defender bears no legal liability. That point has been noted in *Keating on Building Contracts* (7th edn, 2001) para 17–18, in *Hudson's Building and Engineering Contracts* (11th edn, 1995) para 8–210, more clearly in *Emden's Construction Law* (8th edn, 1999), III, para 231, in the American cases, and most clearly by Byrne J in *John Holland Construction & Engineering Pty Ltd* v. *Kvaerner RJ Brown Pty Ltd* (1996) 82 BLR 81 at 85 and 86. The point has on occasions been expressed in terms of a requirement that the pursuer should not himself have been responsible for any factor contributing materially to the global loss, but it is in my view clearly more accurate to say that there must be no material causative factor for which the defender is not liable.

37. Advancing a claim for loss and expense in global form is therefore a risky enterprise. Failure to prove that a particular event for which the defender was liable played a part in causing the global loss will not have any adverse effect on the claim, provided the remaining events for which the defender was liable are proved to have caused the global loss. On the other hand, proof that an event played a material part in causing the global loss, combined with failure to prove that that event was one for which the defender was responsible, will undermine the logic of the global claim. Moreover, the defender may set out to prove that, in addition to the factors for which he is liable founded on by the pursuer, a material contribution to the causation of the global loss has been made by another factor or other factors for which he has no liability. If he succeeds in proving that, again the global claim will be undermined.

38. The rigour of that analysis is in my view mitigated by two considerations. The first of these is that while, in the circumstances outlined, the global claim as such will fail, it does not follow that no claim will succeed. The fact that a pursuer has been driven (or chosen) to advance a global claim because of the difficulty of relating each causative event to an individual sum of loss or expense does not mean that after evidence has been led it will remain impossible to attribute individual sums of loss or expense to individual causative events. The point is illustrated in certain of the American cases. The global claim may fail, but there may be in the evidence a sufficient basis to find causal connections between individual losses and individual events, or to make a rational apportionment of part of the global loss to the causative events for which the defender has been held responsible.

39. The second factor mitigating the rigour of the logic of global claims is that causation must be treated as a common sense matter (*John Holland Construction & Engineering Pty Ltd* v. *Kvaerner RJ Brown Pty Ltd* (1996) 82 BLR 81 at 84, per Byrne J). That is particularly important, in my view, where averments are made attributing, for example, the same period of delay to more than one cause.

40. The particular issue that arises in this case is whether the pursuers' reliance on a causal factor (or factors) which can be held, as a matter of relevancy, not to involve liability on the part of the defenders should result in dismissal of the loss and expense claim, or whether in the circumstances a proof before answer should be allowed. In my judgment the defenders' submission on the point comes very close to success. I am, however, persuaded that a proof before answer should be allowed by a combination of two considerations. The first is that in

my view on a fair reading of the pursuers' pleadings they do aver concurrent causes of delay in respect of the period between 22 January and 12 February 1996. Although there is reference in Schedule C to the summons to WP2010 and snow being 'critical' causes, I do not consider that the averment on page 10 – that before and after 22 January 1996 the pursuers were in any event delayed by late issue of drawings and information – can properly be ignored. How each of the concurrent causes ought to be viewed in determining whether the causes for which the defenders had no liability played a material part in causing the global loss is a matter that is, in my view, best left for consideration at the conclusion of a proof before answer. The second consideration which leads me to allow a proof before answer on the global claim is that it would be wrong to exclude, at this stage, the possibility that the evidence properly led at proof before answer will afford a satisfactory basis for an award of some lesser sum than the full global claim.

41. Before leaving this aspect of the debate, I would make two further observations. The first is that the risk that the pursuers' global claim will fail because a material part of the causation of the loss and expense was an event for which the defenders are not liable, if the evidence discloses no rational basis for the award of any lesser sum, remains a live one. Secondly, the allowance of a proof before answer does not afford the pursuers *carte blanche* to attempt to prove their loss and expense in any way they choose. Their pleadings remain the measure of what they are entitled to prove by way of computation of loss and expense. If a lesser claim is to be made out, that must be done on the basis of evidence which is properly led within the scope of the existing pleadings.

Petromec Inc v Petroleo Brasileiro SA (Petrobras) and another (Petromec Inc and others, Part 20 defendants)

COURT OF APPEAL

(2007) 115 ConLR 11, [2007] EWHC 1589 (Comm), [2007] EWCA Civ 1371

In this case the dispute arises at the case management stage. The parties were not agreed as to how *Petromec* should plead its case and the Court of Appeal was giving guidance.

MAY LJ:

The judge referred to a number of authorities and textbooks (at [42]) which tend to disapprove global claims. Rather, if there is a claim under a construction contract for multiple breaches or multiple variations, there is a need to show the consequences of complying with instructions in terms of work done and cost reasonably incurred. The causal nexus must be spelt out in intelligible form. The terms of cll 12.1 and 12.2 did not obviate this need. The judge accepted that 'an adequate indemnity was given' but that indemnity was circumscribed by the terms of cll 12.1–12.3. The judge accepted that the court should not dictate to a party to litigation the case that the party must advance so as to put the party into a straightjacket of the court's devising. If, therefore, Petromec's method had been sufficient to prove its case, the judge would not have debarred it from pleading and proving it by that method.

In summary, therefore, the judge held (at [48]), that, on the proper construction of the supervision agreement, the sum due to Petromec under cll 12.1 and 12.2 cannot be ascertained by calculating the difference in the manner which Petromec proposed. Petromec must specify the instructions, the work required to comply with those instructions (or with the amended specification under cl 11) and the cost attributable to that work. The changes and causal nexus must be pleaded. Petromec can contend that the work done and the cost is reasonable. By one means or another, it must plead with sufficient particularity the work done and its cost by reference to the amended specification or the instructions given. The judge, however, said (at [49]) that it was not necessary to ascertain separately sums due under cll 12.1 and 12.2. All that was necessary was to establish the total of the additional costs referred to in cll 12.1 and 12.2. What mattered was the total reasonable extra cost payable in consequence of complying with Petrobras's instructions under cl 10, as reflected cll 11.1 and 12.1 or 12.2.

Merely because an item is omitted from the bills or specification does not necessarily mean that the contractor has a claim for extras. The employer is not bound to pay for 'things that everybody must have understood are to be done but which happen to be omitted from the quantities': Channell J in* Patman & Fotheringham Ltd *v.* Pilditch *(1904).

Williams v. Fitzmaurice

COURT OF QUEEN'S BENCH
(1858) 3 H & N 844

The plaintiff contracted to build a house 'to be completed and dry and fit for Major Fitzmaurice's occupation by 1 August, 1858'. In the specification, prepared on the defendant's behalf, no mention was made of floorboards. It stated that 'the whole of the materials mentioned or otherwise in the foregoing particulars, necessary for the completion of the work, must be provided by the contractor'. The plaintiff signed the specification and agreed 'to do all the works of every kind mentioned and contained in the foregoing particulars, according, in every respect, to the drawings furnished or to be furnished, for the sum of £1,100'. The plaintiff brought floorboards on site, but refused to lay them without extra payment.

Held: The plaintiff was not entitled to recover for the flooring as an extra, because it was included in the contract, though not mentioned in the specification.

Default by a nominated sub-contractor or nominated supplier does not give rise to any claim by the contractor against the employer.

Kirk & Kirk v. Croydon Corporation

QUEEN'S BENCH DIVISION
[1956] JPL 585

The plaintiffs contracted to build a school for the defendants, the contract being in RIBA form (1939 edition, revised 1948). The contract included a clause providing for nominated suppliers. After making enquiries of a brick manufacturer as to delivery dates, the architect instructed the plaintiffs to place an order with the manufacturer.

The manufacturer delayed in supplying the bricks, thereby causing delays and involving the plaintiffs in extra costs.

Held: The defendants were not liable to the plaintiffs.

Where the employer has himself undertaken to provide goods or carry out work or procure someone else to do so, a term may be implied into the contract that such goods will be supplied at the proper time and the contractor will have an action for damages for breach of that term:* Thomas Bates & Son Ltd *v.* Thurrock Borough Council *(1975), p. 178.

The employer is not liable to the contractor for acts of his architect which are not within the scope of the architect's authority.

Stockport Metropolitan Borough Council v. O'Reilly

QUEEN'S BENCH DIVISION

[1978] 1 Lloyds Rep 595

O'Reilly contracted to build 105 houses, garages and ancillary works for the council. The contract was in JCT 63 form. Disputes between the parties were referred to arbitration. The main points at issue were (a) whether the contract had been properly determined or repudiated and (b) the extent to which the works had been varied. The arbitrator failed in his interim award to find the extent to which the works had been varied. He confused variations under the contract with orders given by the architect which were not authorised by the contract.

Held: The interim award of the arbitrator should be set aside.

> JUDGE EDGAR FAY QC: An architect's *ultra vires* acts do not saddle the employer with liability. The architect is not the employer's agent in that respect. He has no authority to vary the contract. Confronted with such acts, the parties may either acquiesce in which case the contract may be *pro tanto* varied and the acts cannot be complained of or a party may protest and ignore them. But he cannot saddle the employer with responsibility for them.

The employer may be liable to the contractor for the architect's failure properly to operate the terms of the contract.

Rees & Kirby Ltd v. Swansea City Council

QUEEN'S BENCH DIVISION

(1985) 25 BLR 129

Although this case was reversed on appeal, the following extract from the judgment at first instance appears to be good law.

KILNER BROWN J: The more interesting point is whether the architect was agent for the defendants making them vicariously responsible. There are two authorities which are in point, namely *Burden* v. *Swansea Corporation* (1957) and *Sutcliffe* v. *Thackrah* (1974). As I understand the law, an architect is usually and for the most part a specialist exercising his special skills independently of his employer. If he is in breach of his professional duties he may be sued personally. There may, however, be instances where the exercise of his professional duties is sufficiently linked to the conduct and attitude of the employer that he becomes the agent of the employers so as to make them liable for his default. In the instant case the employers through the behaviour of the Council and the advice and intervention of the Town Clerk were to all intents and purposes dictating and controlling the architect's exercise of what should have been his purely professional duty. In my judgment this was the clearest possible instance of responsibility for the breach attaching to the employers.

In practice a very common ground of complaint by contractors is that the employer and his professional advisers have been late in providing information. In a great many building projects work starts before the planning process has been completed. This creates a risk, which often comes to pass, that the necessary information will be provided too late and that delay will thereby be caused.

Royal Brompton Hospital NHS Trust v. Hammond (No. 8)

QUEEN'S BENCH DIVISION (TECHNOLOGY AND CONSTRUCTION COURT)
(2002) 88 Con LR 1

The procedural history of this case has been outlined above (p. 267). The present hearing involved the substantive consideration of the issues. One of these concerned the late provision by the mechanical and electrical services engineer (AA) of the drawings for the mechanical and electrical services. AA had said, untruthfully, that the bulk of the design information was ready to issue. The judge held that this was a breach of AA's contract with the hospital. He also held that the project manager (PMI) was in breach of contract for failing adequately to supervise AA. He held that the architect (WGI) was not in breach of contract.

JUDGE HUMPHREY LLOYD QC:
64. In this contract the provisions of the bills and of the M & E specification to which I have referred make it clear that the co-ordination drawings were to be provided to the contractor in accordance with whatever might be agreed at the pre-contract meeting and thereafter in accordance with the master programme. Those arrangements would ordinarily meet the requirements of clause 5.4 in establishing what was 'necessary' and 'reasonably necessary' and, if the information did not appear until late, TW [Taylor Woodrow] would be able to rely on them as specific applications for the purposes of clauses 25 and 26. Even discounting the possible effect of clause 2.2.1 these provisions did not purport to deal comprehensively with clause 5.4 as it applied to co-ordination drawings, nor was it likely that operational arrangements of this kind were intended to have any such contractual effect. So [counsel for some of the defendants] arguments based on Roberts would fail.

65. It is necessary for the contractor to have the drawings a certain number of weeks before starting the relevant part of the installation. The relevant start date is that on the contractor's current programme, adjusted as appropriate for any delays that have occurred in the meantime that will result in the date being put back. The contractor's own expectation of the amount of time required is a relevant factor to be taken into account when determining the necessary period, but it is not decisive. For example, as the decision of the Court of Appeal establishes, the mere fact that a contractor requests information for a particular date does not make that date the date by which the information should have been provided. It is, on the other hand, evidence that the contractor did not require it earlier than that date. Equally, the contractor may have been led to think that the drawings will be of such a quality that he will be almost able to build straight off them and so leave himself relatively little time for preparing his own drawings. That will also be a material factor. Any question of timing may not therefore be divorced from that of the quality of the drawings and vice versa. On the other hand, the contractor may have little confidence in the architect or the engineer, and has to plan for the worst. The architect's or engineer's obligations to provide drawings in time can therefore be determined by the subjective understanding of the contractor – hence the desirability of making specific applications or notices as required, for example by clause 26.2.1 so as to establish what might otherwise have been appreciated by the consultant. Furthermore the period required by a contractor, after receipt of co-ordination drawings, for preparation of installation drawings and for site mobilisation depends on many factors within his own control, including in particular: (a) the quantity and quality of drawing resources, and (b) the planned relationship between the drawing period and the mobilisation period. In this instance it is material that clause B1:03:08 of the M & E specification provided that TW had to 'Give due consideration to detailed co-ordination drawings where provided with the specification' and 'No deviation from the positions indicated on the Engineer'. Thus RBH submitted, correctly, that TW could reasonably wait until it had AA's drawings before starting on its own drawing programme – a point accepted by AA. Accordingly the times required by the contractor may be earlier than the consultant may have expected. The employer is nevertheless bound to comply with them, unless they are quite unreasonable, i.e. not based on the contractor's real requirements. Hence the prudent employer requires advance notice by way of pre-contract notification and thorough programming, as found in TW's contract with RBH. As the contract proceeds what is 'necessary' from time to time may depend upon the contractor's actual state of progress. But the test is not whether the contractor has reached the position where it is ready to make use of the information, or whether it is at that time necessary to what the contractor is doing, for either would be to defer it to the latest possible date and the contractor could not be sure that it would arrive. Under clause 5.4 the contractor has to have the information earlier than he may in fact be able to use it. Obviously, if the information arrives later, but still earlier than the time when the contractor actually needs it, there will be no consequences either in terms of breach of contract or delay or disruption (still less potential delay) qualifying for consideration under clauses 25 or 26.

66. I also do not accept AA's case that the employer can avoid its responsibilities under clause 5.4 by relying on the absence of an application or notice from the contractor although that may be in fact a material factor in determining the temporal necessity for the information (or whether it was reasonably necessary for the contractor to have that information, i.e. that it was needed to explain or amplify the contract drawings or to enable the contractor to carry out and complete the works). First, there is no such limitation in clause 5.4. The JCT conditions are not notable for brevity: indeed the purposes of the repetition of refrains and of seeming prolixity are to ensure that those who actually use the form do not have to be lawyers and do not have to embark on processes of interpretation and cross-referencing where none

is indicated. If therefore it had been intended that the explicit and clear requirements in clauses 25 and 26 were to condition clause 5.4 it would have been made plain. Secondly, clause 26.6 is very clear:

> 26.6 Reservation of rights and remedies of Contractor
>
> The provisions of clause 26 are without prejudice to any other rights and remedies which the Contractor may possess.

In my judgment that means that, whatever the rights the contractor may (or may not) have under clause 26, if any of the circumstances described in clause 26 as entitling the contractor to apply for the reimbursement of loss and expense occasioned by it is also a breach of contract by the employer then it remains actionable by the contractor. That was the view of Vinelott J in *Merton London Borough* v. *Stanley Hugh Leach Ltd* (1985) 32 BLR 51 at 108 in relation to the comparable provisions of the 1963 edition of the JCT form and, so far as I am aware, it has stood unchallenged. It has been accepted as correct by the leading commentators (see *Hudson's Building and Engineering Contracts* (11th edn, 1995) para 8-070 and *Keating on Building Contracts* (7th edn, 2001) paras 18-82, 18-318, 18-341) and it has been followed (see *Fairclough Building Ltd* v. *Vale of Belvoir Superstore Ltd* (1990) 28 Con LR 1). [Counsel for some of the defendants] did not argue to the contrary. I therefore accept the submissions of [counsel for the claimants] and reject those for AA.

67. Thirdly, non-compliance with clause 5.4, which delayed completion and which rendered the contractor liable to liquidated damages would not make time at large. That proposition assumes, incorrectly, that the liquidated damages payable under clause 24 is a penalty from which the contractor may seek relief. That is a dated view of liquidated damages provisions in construction contracts. The modern and better view treats the provisions as the agreed measure of the damages payable (see *Keating on Building Contracts* (7th edn, 2001) paras 9–10 to 9–12, 9–26). Although the figure may have been inserted by the employer it is priced for and accepted by the contractor as the limit of its liability. Hence, even if the provision becomes inoperative, the employer cannot recover more than the amount that would have been recoverable had the clause operated. Clause 25 and the list of relevant events provide an apparently comprehensive code whereby time may or may not be extended for all the common types of occurrence (and many rare events). Late information, in other words, non-compliance with clause 5.4 is, after variations, probably the most common cause of delay. Lack of proper pre-planning has been (and remains) the bane of construction contracts, mainly where there has been competitive tendering and where the fees payable to consultants are not conducive to getting information out on time, whether for pre- or post-contract purposes. In order to remind the contractor of the need to establish the circumstances which justify the time or money sought, both clauses 25 and 26 require proof of the necessity for the late information, whether in terms of time or content, by way of a specific application or notice. That does not affect the contractor under clause 26 since its absence does not prevent a claim for damages for breach of contract. The contractor is similarly not prejudiced in relation to an extension of time since the architect is obliged to consider whether or not to grant an extension of time once he becomes aware of any relevant event. This was also established by the judgment of Vinelott J in *Merton London Borough* v. *Stanley Hugh Leach Ltd* (1985) 32 BLR 51 at 97 to 98 and 104. Before issuing a certificate that the contractor has failed to complete the works by the completion date, as provided by clause 24.1, the architect has to consider whether it should be issued. In doing so the architect has thus to consider any grounds which might excuse the contractor for failing to complete the works by the completion date. Thus the architect will

carry out the same exercise as that required under clause 25.3.3. Clause 25.3.3.1 requires the architect to fix a completion date which is –

> . . . fair and reasonable having regard to any of the Relevant Events, whether upon reviewing a previous decision or otherwise and whether or not the Relevant Event has been specifically notified by the contractor under clause 25.2.1.1 [my emphasis].

The JCT conditions do not make the issue of a certificate under clause 24.1 dependent upon the grant of any extension of time (or refusal to do so) under clause 25 so in law there is no connection even though in practice the same ground will be covered. Once the contractor has given notice under clause 25.2.1.1 the architect is obliged to consider the circumstances (and may also have been obliged to do so if he became aware of them earlier) even if no specific application had been made as provided for by clause 25.4.6.

68. For present purposes, the more important question, in my view, is the date by which the architect or consultant is obliged under the contract with the client to provide the information. In my judgment, as I have indicated, that date cannot be the date by which it turns out that the contractor actually needed the information. A target is not to be determined *ex post facto*. That is to confuse whether the employer can prove any loss resulting from the breach of duty with the duty itself. Obviously if clause 5.4 is fulfilled then a duty based upon its fulfilment will necessarily be discharged to the extent the contractor has made no claim against the employer so the employer should not be liable to the contractor and thus should have no complaint against the consultant responsible. The duty is primarily contractual and, although its breach may also give rise to an action in negligence, the employer's architect or consultant has to provide the information by the date when it is expected. That date will depend on a number of factors. The terms of the construction contract will be one of them but in the pre-contract period the duty will have to be framed by reference to what is likely to happen. Once the contract is made, the discharge of the duty may then be refined by the production of programmes by the contractor or other yardsticks by which the performance of the architect or consultant may be measured, all the more so if the dates are agreed (as envisaged by the bills and specification in this contract). If the programmes are not met then the consultant may no longer be obliged to provide the information by the programmed date, although, as I have said, it would be misleading to equate the consultant's obligations with the employer's obligations under a provision such as clause 5.4. Thus, for example, it was pointed out by Vinelott J in *Merton London Borough* v. *Stanley Hugh Leach Ltd* (1985) 32 BLR 51 at 88 that the date when otherwise information might be required may be postponed if the contractor falls behind his programme. It does not necessarily follow that the consultant is thereby relieved of the duty, but, as already stated, it may mean there will be no consequences to the client if the information does not arrive at a time when the contractor is not in a position to make use of it. Thus RBH's case stems from liability to TW for non-compliance with clause 5.4 and TW's claim. The client has a distinct interest in avoiding not only claims but the possibility of claims and in my judgment there is ample justification for requiring the consultant to comply with its original obligations if to do so would prevent the contractor claiming. Thus under clause 23.1 of the JCT conditions the contractor has a distinct obligation to proceed 'regularly and diligently' and the employer is under a corresponding obligation not to hinder or prevent that obligation – the references to *Roberts* v. *Bury Improvement Comrs* (1870) LR 5 CP 310 made by [counsel for the claimants] and [counsel for some of the defendants] are apposite here, as is [counsel for some of the defendants'] citation from *McCarrick* v. *Liverpool Corp* [1946] 2 All ER 646, [1947] AC 219. In *Roberts* Kelly CB and Blackburn and Mellor JJ said ((1870) LR 5 CP 310 at 326) –

> . . . a principle well established at common law, that no person can take advantage of the non-fulfilment of a condition the performance of which has been hindered by himself . . . and also that he cannot sue for a breach of contract occasioned by his own breach of contract, so that any damages he would otherwise have been entitled to for the breach of contract to him would immediately be recoverable back as damages arising from his own breach of contract.

69. The timing of further information required under clause 5.4, like the release of information generally, must not disrupt the contractor's progress even if it does not lead to delay in completion for otherwise there may be liability under clause 26. In practice therefore that may mean that the consultant will be obliged under the contract with the client to produce information before the time when the contractor can make use of it. The contractor will however then know that it exists and will be able to plan to catch up with confidence. The word 'necessary' in the first part of clause 5.4 is not limited to providing the information 'just in time' nor is it to be measured by reference to whether the contractor suffers critical delay to planned completion. Delay to completion, and, perhaps even more important in practice, disruption to progress may therefore be minimised or averted, especially if the delay in progress is the contractor's responsibility. That is in the interests of the client and the contractor and their avoidance is a proper objective for the consultant. Unless the client agrees, the terms of the construction contract and the performance of the contractor cannot in law modify the terms of the consultant's contract. Provisions such as clause 5.4 are not to be used to diminish the terms or effect of the consultant's contract. . . .

99. RBH authorised a start on 2 March 1987. In a 'Preliminary Contract Programme' dated 16 February 1987 PMI used November 1987 as the date for the start of M & E first fix. Accordingly I do not accept AA's case that PMI must have known that AA's response to item 65 of TW's requirements 'as previously discussed' meant that a number of single service drawings were still in a fluid state or otherwise undrawn, as AA either had been or still were waiting for client confirmation or information from other parties and would or might not be ready on commencement. In addition to the summary in para 8.3.5 of his report [expert witness for AA] also drew attention to changes being made by RBH which affected the production of AA's drawings but in my view his opinion relates to the general state of AA's preparation.

100. I thus conclude that, as the co-ordination drawings were not ready for issue by mid-February 1987, the lack of many single service drawings meant that AA did not have 'the bulk of the design information ready for issue', as [a partner in AA] agreed and that to that extent the statement was untrue and made negligently by AA and contrary to the terms of AA's retainer. The bulk for this purpose obviously did not include co-ordination drawings. It remains unclear why the statement stood. Clearly AA had an interest in not revealing how far behind it was. Of the possibilities available AA in my view knew that there was a real risk of not being able to provide the drawings to TW on time. I do not consider that AA honestly believed that in some way which was not explained it could and would meet TW's requirements. If so, it could not have held that view if it had taken care to justify it. The absence of staffing records inhibited [expert witness for RBH] from expressing a view. I have to reach a conclusion on the available evidence which includes an assessment of the witnesses from AA and the other parties. Probably AA thought that somehow it would be able to avoid delaying the contractor because RBH would not stop changing its mind. AA certainly decided not to advertise (as RBH put it) how far behind they were. Even if AA had not told PMI that the bulk of the drawings were ready for issue, it ought to have corrected the statements in PMI's letter of 18 February and advised PMI (or even WGI) that the co-ordination drawings would or might not be ready on time.

101. Although AA is liable to RBH equally PMI has at least a case to answer. [A partner in PMI] was there to find out what was going on and to get to the bottom of the alternatives which I have set out, although [a partner in AA] said that AA did not see much of him and that its main contact was through [W]. I conclude, having seen and heard [a partner in PMI] (and allowing for the fact that he was probably not asked about the events until some ten years afterwards) that he almost certainly did not make any proper and thorough investigation and was too easily persuaded by AA not to do so. [A partner in PMI] did not appear to me to be someone who would make things up. He said that he would not have been able to give [an employee of PMI] of RBH the information that he gave him if it was true that the drawings had not been changed or developed or worked on in six months. He therefore accepted that, if that was the state of affairs, he should have discovered it and therefore would not have reported as he did. That was the purpose of his visit. I do not therefore accept evidence [of a partner in AA] that the situation (other than in relation to co-ordination drawings) was well known by other members of the project team from discussions between them. If it had been known I have no doubt that someone would have intervened to warn RBH that the start contemplated was dependent on an assumption which might turn out to be incorrect.

102. As regards the co-ordination drawings the impression was that AA would produce them to meet TW's requirements and it attracts the same or similar criticisms and conclusion. AA's case was that in mid-February 1987, notwithstanding the uncertainties and the considerable work that remained to be done, it thought that it would be able to produce the co-ordination drawings by the time that the contractor needed them. It did so since, as [an expert witness for AA] said: 'gut feeling told me we would have done it . . . This is not the first time that we have been in this sort of situation'. The experts . . . also agreed that –

> . . . a reasonably competent consulting engineer could anticipate commencing co-ordination drawings for level 1 first and would not need a contractor's programme to know this on a project such as this.

In answer to [counsel for some of the defendants], [expert witness for RBH] considered that there was reason for someone in the position of [a partner in PMI] to have concluded that AA should be able to get their co-ordination information (as he put it) out within three or four months, and the level 1 drawings within about five to seven weeks from a planned start. But were AA's forecasts and alleged feelings based on reality? In my judgment [a partner in PMI] either did not investigate whether the drawings would be ready or was dissuaded by AA from doing so. As a result AA managed to conceal from everybody else that it was not really ready.

103. In my view [a partner in PMI] ought to have had a programme by which he and PMI could have seen the progress of the production by AA of its drawings. Some such programme had been used in the previous year. It would have enabled him to test AA's statements as it would have had to demonstrate how it would produce the drawings. [A partner in PMI] agreed that such programmes were desirable and that 'in order to monitor information flow you have to have some sort of programme against which to monitor it'. He also accepted that if there had been a clearly established programme of the release dates for the co-ordination drawings, it was likely that things would have gone better than they in fact did. [An expert witness] agreed with [a partner in PMI]. He also agreed that [the partner in PMI] should have got, if not a programme, then a fairly firm plan from AA as to when it would issue co-ordination drawings by certain dates which it considered would be more than adequate for TW. [The expert witness] baulked at the conclusion that if that was not done then PMI did not act with

reasonable competence, but in my view only because there was no record of what [the partner in PMI] had done, the absence of which he did regard as unsatisfactory. In my judgment there is no record because [the partner in PMI] did not take those prudent steps to obtain for RBH a reliable basis for advising it to proceed and that accordingly PMI also failed in its duty to RBH as result of which RBH was misinformed on 16 February (and, of course, the misleading statement was not corrected by PMI, as AA did not do so).

104. It would be kind to find that PMI thought that everybody else was aware of AA's actual position (as opposed to AA's unreliability), but I cannot do so as there is no real evidence to support it (although there can be no doubt that no one could have thought that the co-ordination drawings were then ready). Such a finding would mean that PMI knew that its advice to RBH was incorrect and unjustified as it did not tell the whole story. PMI had many failings but not of that gravity. On the other hand AA obviously knew that the co-ordination drawings would probably not be ready within four months. They certainly ought to have been ready as AA had enough time to prepare them and, as [an expert witness for RBH] said they could easily have been prepared much before they actually were (autumn 1987). However [the expert witness for RBH], in a typically careful answer, did not say that AA ought to have warned PMI and RBH of the likelihood that they would not be ready on time but only because he had insufficient information about AA's staffing to justify such an opinion. (That does not affect the time by which they ought to have been issued.) As already set out I think that AA believed that it would in some way be delivered from its predicament. [A partner in AA] said that it was AA's 'considered position' that there was no reason to suppose that M & E information would not be available to the contractor as and when he needed it and: 'It was a difficult task but we thought it could be achieved'. [An expert witness for AA] said that AA was confident that it could feed the information to the contractor on time. Perhaps AA thought that there would be more changes which would excuse everybody. In his second witness statement [a partner in AA] said:

> Further, I took the view that if no more changes were made and a tight control over the contractor was taken by PMI and the Supervising Officer then there was a chance that the scheme could be completed in the manner required.

He certainly thought that the risk being run was not 'minimal'.

105. If PMI had known of AA's situation, I am sure that, unpalatable though it would have been to RBH and the DHSS, either TW would not have been allowed to start on site when it did, or measures would have been taken to ensure that the risk of delay was minimised. The money allocated had been used up; the contingency figure was very low; I do not believe that any project manager, even PMI, would have permitted its client to have courted such a danger at that stage (see [an expert witness'] view). I am certain that had [an employee of RBH] and [RBH's former general manager] been given the full picture they would not have been permitted by the DHSS to go ahead. It is true that [the employee of RBH's] evidence was that only 'something very serious' or 'a disaster' would have caused RBH to pause, but in my view the situation at that time, if known, would have been so regarded. In such circumstances the likelihood of TW being able to make a claim for the late delivery of co-ordination drawings should have been eliminated or reduced to an acceptable risk (i.e. in the context of this project, a minimal one). In terms of 'loss of a chance' I would assess the chance of RBH going ahead, had it been correctly advised by AA or PMI, as no higher than 5%.

106. RBH also pleaded that had AA advised it of the true state of affairs then RBH would have contracted with TW to produce the co-ordination drawings. The case was not formally

withdrawn but I am sure that TW would not have agreed to produce the co-ordination drawings, not least since, as will become apparent, there was a shortage of competent engineers and draughtsmen. In addition it would have taken some time to negotiate the change so any saving of time would have evaporated and TW would naturally have taken advantage of its position as the preferred contractor so the price increase would not be controllable by RBH. TW had been paid for the delayed start and hoped for savings had not materialised. [An employee of RBH] said that in January RBH wanted to start promptly although the reason never became clear from his evidence. This part of RBH's case was not made out.

Under a contract in JCT 63 terms, the architect's failure to ascertain, or instruct the quantity surveyor to ascertain, the amount of the direct loss and/or expense suffered or incurred by the contractor, as required by clauses 11(6) and 24(1) of the contract is a breach of contract for which the employer is liable in damages.

A contractor's claim under the contract for direct loss and/or expense cannot be maintained by legal action in the absence of an architect's certificate. However, sums may be so claimed by way of an action for damages for breach of contract, provided it is established that the contractor has suffered some damage. Unless the employer establishes that there are no matters in respect of which the contractor is entitled to claim under clauses 11(6) and 24(1), it necessarily follows that the contractor must have suffered some damage as a result of the failure to ascertain amounts due. See JCT 80 and JCT 98, clause 2.6.

Croudace Ltd v. London Borough of Lambeth

COURT OF APPEAL
(1985) 6 Con LR 70

On 12 October 1979 Croudace Ltd (Croudace) contracted with the Council (Lambeth) in JCT 63 terms, to erect houses and ancillary buildings at Stockwell. The contract provided that 'the architect' and 'the quantity surveyor' meant the chief architect and chief quantity surveyor of Lambeth respectively, though these officers delegated their day-to-day duties under the contract to outside firms. From the date of the contract until his retirement on 31 May 1983, Lambeth's chief architect was Mr W. Jacoby.

Croudace commenced work in July 1979. A mains electrical cable was discovered on 17 December 1979, and the architect instructed Croudace to revise their order to the London Electricity Board to provide for the diversion of the cable. The diversion was not completed until 11 June 1980. Croudace gave notice of loss and/or expense and also made application for an extension of time, alleging that the delay was Lambeth's responsibility, though this was later denied by Lambeth. Croudace also alleged that they were delayed by Lambeth's failure to conclude their negotiations with a scaffolding contractor who was in possession of part of the site.

Although the contract completion date was 19 October 1981, practical completion was not achieved until 22 July 1982. On 2 February 1983, the chief architect awarded extensions of time under clause 23(e), totalling 31 weeks and 3 days, and also issued his certificate of delay under clause 22, ending his letter by stating that 'it will be necessary for you to submit a formal claim for any loss or expense that you may consider is due to you, to be ascertained by the quantity surveyor'.

On 12 February 1983 Croudace submitted a draft claim to the outside quantity surveyors, who replied that 'all payments relating to loss and expense' claims must be approved by Lambeth, to whom they had forwarded a copy of the submission. Despite letters chasing the money, no payment was in fact made, and on 25 May 1983 Croudace submitted a final claim for loss and expense in the amount of £624,544.54. On the following day there was a meeting between the outside quantity surveyors and a representative of Croudace, and the quantity surveyors agreed that a payment of £100,000 would be fair and reasonable as an interim payment on account.

On 31 May 1983 Mr Jacoby retired and Lambeth did not appoint anyone to succeed him. Although Croudace continued to press Lambeth both orally and in writing, they received no further payment. They issued a writ on 22 February 1984 and sought summary judgment against Lambeth for £100,000 plus interest, or alternatively for damages and interest to be assessed. Lambeth sought to stay the action under section 4 of the Arbitration Act 1950. The trial judge found in favour of Croudace and ordered that they be at liberty to enter judgment for damages to be assessed and that Lambeth should make an interim payment of £100,000 to Croudace. He dismissed Lambeth's application for a stay of proceedings. Lambeth appealed against the trial judge's orders.

Held (dismissing the appeal): Lambeth's acts and omissions after 2 February 1983, including but not limited to their failure to appoint a successor architect, amounted to failure by them to take such steps as were necessary to enable Croudace's claim to be ascertained, and as such amounted to a breach of contract.

> BALCOMBE LJ: Although he did not formally concede the point, [counsel] for Lambeth did not seriously contest the judge's finding that there was implied in the contract an obligation on the part of Lambeth to nominate a successor to Mr Jacoby when he retired as Chief Architect, and that their failure to do so constituted a breach of contract on their part. In my judgment Lambeth's acts and omissions after 2 February 1983, including but not limited to their failure to appoint a successor to Mr Jacoby, amounted to a failure by them to take such steps as were necessary to enable Croudace's claim for loss and expense to be ascertained, and as such amounted to a breach of contract on their part: cf: *Smith* v. *Howden Union & Anor* (1890).
>
> The real issues in this case are whether Croudace can establish that it has suffered any damage resulting from that breach of contract and, if so, whether the amount of that damage should be determined by the court or by arbitration.
>
> Unless it can be successfully maintained by Lambeth that there are no matters in respect of which Croudace are entitled to claim for loss and expense under clauses 11(6) and 24(1)(e), it necessarily follows that Croudace must have suffered some damage as a result of there being no one to ascertain the amount of their claim. Without going into the details of the many matters raised by Croudace, I am satisfied that they can establish that they have suffered some damage in this connection . . . Further, when damages come to be assessed it will be open to Lambeth to take the point that Croudace's applications under clauses 11(6) and 24(1) were not made within a reasonable time, since the judge's statement of his view that 'the application made under both conditions was made within a reasonable time' was not a finding necessary for the orders which he made. In my judgment, therefore, the judge was entitled to make the order . . . for damages to be assessed, unless it was a case for a stay of proceedings pending arbitration.

Chapter 14

Liquidated Da

Liquidated damages

Liquidated damages are a
the damages payable in th
uidated damages are gene
A provision for liquidated
of the loss likely to be cau
agreed sum is extravagant
Liquidated damages are re

Dunlop Pneumatic Tyre Co Ltd v. New Garage Motor Co Ltd

HOUSE OF LORDS
[1915] AC 79

Dunlop contracted to sell tyres and other accessories to New Garage on terms designed to ensure that the tyres, etc. were not sold below the manufacturer's list price. New Garage agreed to pay Dunlop 'the sum of £5 for each and every tyre . . . sold or offered in breach of this agreement, as and by way of liquidated damages and not as a penalty'. New Garage sold tyres in breach of the agreement.

Held: In the circumstances of the case, the sum of £5 was liquidated damages and not a penalty.

LORD DUNEDIN: I shall content myself with stating succinctly the various propositions which I think are deducible from the decisions which rank as authoritative: (i) Though the parties to a contract who use the words penalty or liquidated damages may *prima facie* be supposed to mean what they say, yet the expression used is not conclusive. The court must find out whether the payment stipulated is in truth a penalty or liquidated damages. This doctrine may be said to be found *passim* in nearly every case. (ii) The essence of a penalty is a payment of money stipulated as *in terrorem* of the offending party; the essence of liquidated damages is

Powell-Smith and Furmston's Building Contract Casebook, Fifth Edition. Michael Furmston.

ge: *Clydebank Engineering Company* v. *Yzquierdo* whether a sum stipulated is penalty or liquidated n to be decided upon the terms and inherent circum-, judged of as at the time of the making of the contract, not lic *Works Comr* v. *Hills* (1906) and *Webster* v. *Bosanquet* (1912). nstruction various tests have been suggested, which, if applicable eration, may prove helpful or even conclusive. Such are: (a) It will be the sum stipulated for is extravagant and unconscionable in amount in the greatest loss which could conceivably be proved to have followed from ustration given by Lord Halsbury LC, in the *Clydebank* Case (1905). (b) It will e a penalty if the breach is only in not paying a sum of money, and the sum stipu- sum greater than the sum which ought to have been paid: *Kemble* v. *Farren* (1829). though one of the most ancient instances, is truly a corollary to the last test. Whether it d its historical origin in the doctrine of the common law that, when A. promised to pay B. a sum of money on a certain day and did not do so, B. could only recover the sum with, in certain cases, interest, but could never recover further damages for non-timeous payment, or whether it was a survival of the time when equity reformed unconscionable bargains merely because they were unconscionable – a subject which much exercised Jessel MR, in *Wallis* v. *Smith* (1882) – is probably more interesting than material. (c) There is a presumption (but no more) that it is a penalty when

> a single lump sum is made payable by way of compensation, on the occurrence of one or more or all of several events, some of which may occasion serious and others but trifling damages.

per Lord Watson in *Lord Elphinstone* v. *Monkland Iron and Coal Co* (1886). On the other hand: (d) It is no obstacle to the sum stipulated being a genuine pre-estimate of damage that the consequences of the breach are such as to make precise pre-estimation almost an impossibility. On the contrary, that is just the situation when it is probable that pre-estimated damage was the true bargain between the parties: *Clydebank* Case (1905), per Lord Halsbury; *Webster* v. *Bosanquet* (1912), per Lord Mersey.

If there is a liquidated damages clause, it does not matter whether the actual loss is greater or less. The liquidated damages are recoverable whether or not the employer can prove that he has in fact suffered loss, or even if, in the event, there is no loss at all.

Clydebank Engineering & Shipbuilding Co v. Castaneda and Others

HOUSE OF LORDS

[1905] AC 6

The Spanish government ordered four torpedo boats from Clydebank. The contract provided a 'penalty' for late delivery at the rate of £500 per week. The boats were delivered late, and the Spanish government claimed £67,500 at £500 for each week of late delivery. One of the arguments put forward by Clydebank was that after the delivery date the greater part of the Spanish fleet had been sunk by the American fleet. In effect, they argued that they had done the Spanish government a favour by being late in delivery.

Held: The sum of £500 a week was recoverable as liquidated damages. Clydebank's contentions were rejected.

LORD HALSBURY LC: Then comes the question whether, under the agreement, the damages are recoverable as an agreed sum, or whether it is simply a penalty to be held over *in terrorem*, or whether it is a penalty so extravagant that no court ought to enforce it. It is impossible to lay down any abstract rule as to what might or might not be extravagant without reference to the principal facts and circumstances of the particular case. A great deal must depend on the nature of the transaction. On the other hand, it is an established principle in both countries to agree that the damages should be so much in the event of breach of agreement. The very reason why the parties agreed to such a stipulation was that, sometimes, the nature of the damage was such that proof would be extremely difficult, complex and expensive. If I wanted an example of what might be done in this way, I need only refer to the argument of counsel as to the measure of damage sustained by Spain through the withholding of these vessels. Suppose there had been no agreement in the contract as to damages, and the Spanish government had to prove damages in the ordinary way, imagine the kind of cross-examination of every person connected with the Spanish administration. It is very obvious that what was intended by inserting these damages in the contract was to avoid a minute, difficult and complex system of examination which would be necessary if they had attempted to prove damage in the ordinary way.

It was also suggested that there could be no measure of damage in the case of a warship which had no commercial value at all. It is a strange and somewhat bold assertion to say that, in the case of a commercial ship, the damages could be easily ascertained, but that the same principle could not be applied to a warship as it earned nothing. The deprivation of a nation of its warships might mean very serious damage, although it might not be very easy to ascertain the amount. But is that a reason for saying they were to have no damages at all? It seems to me hopeless to advance such a contention. It is only necessary to state the assertion to show how absurd it is.

But there was a more startling suggestion still. It was argued for the appellants that, if they had acted up to their contract, and these vessels had been sent out at the specified date, they would have shared the same fate as the other Spanish warships at the hands of the American Navy. Therefore, the respondents, instead of suffering damage, had, by the action of the appellants, been saved the loss of these ships, which otherwise would have been at the bottom of the Atlantic. After considerable experience, I do not think that I have ever heard such an argument before, nor do I think that I am likely to hear it again. Nothing could be more absurd, and to give effect to it would be a striking example of defective jurisprudence. If your Lordships look at the nature of the transaction, it is hopeless to contend that the penalty was intended merely to be *in terrorem*. Both parties recognised that the question was one in which time was the main element of the contract. I have come to the conclusion that the judgment of the court below was perfectly right. There is no ground for the contention that the sum in the contract was not the damages agreed on between the parties for very good and excellent reasons at the time at which the contract was entered into.

A clause is not a penalty simply because excessively ingenious speculation can show that in some improbable event the liquidated damages will exceed the loss.

Philips Hong Kong Ltd v. The Attorney General of Hong Kong

JUDICIAL COMMITTEE OF THE PRIVY COUNCIL

(1993) 61 BLR 41

LORD WOOLF: Except possibly in the case of situations where one of the parties to the contract is able to dominate the other as to the choice of the terms of a contract, it will normally be insufficient to establish that a provision is objectionably penal to identify situations where the application of the provision could result in a larger sum being recovered by the injured party than his actual loss. Even in such situations so long as the sum payable in the event of non-compliance with the contract is not extravagant, having regard to the range of losses that it could reasonably be anticipated it would have to cover at the time the contract was made, it can still be a genuine pre-estimate of the loss that would be suffered and so a perfectly valid liquidated damage provision. The use in argument of unlikely illustrations should therefore not assist a party to defeat a provision as to liquidated damages. As the Law Commission stated in Working Paper No 61 (page 30):

> The fact that in certain circumstances a party to a contract might derive a benefit in excess of his loss does not . . . outweigh the very definite practical advantages of the present rule upholding a genuine estimate, formed at the time the contract was made of the probable loss.

A difficulty can arise where the range of possible loss is broad. Where it should be obvious that, in relation to part of the range, the liquidated damages are totally out of proportion to certain of the losses which may be incurred, the failure to make special provision for those losses may result in 'liquidated damages' not being recoverable. (See the decision of the Court of Appeal on very special facts in *Ariston SRL* v. *Charly Records Ltd* (1990), *The Independent*, 13 April 1990.) However, the court has to be careful not to set too stringent a standard and bear in mind that what the parties have agreed should normally be upheld. Any other approach will lead to undesirable uncertainty especially in commercial contracts . . .

To conclude otherwise involves making the error of assuming that, because in some hypothetical situation the loss suffered will be less than the sum quantified in accordance with the liquidated damage provision, that provision must be a penalty, at least in the situation in which the minimum payment restriction operates. It illustrates the danger which is inherent in arguments based on hypothetical situations where it is said that the loss might be less than the sum specified as payable as liquidated damages. Arguments of this nature should not be allowed to divert attention from the correct test as to what is a penalty provision – namely is it a genuine pre-estimate of what the loss is likely to be? – to the different question, namely are there possible circumstances where a lesser loss would be suffered? Here the minimum payment provision amounted to about 28% of the daily rate of liquidated damages payable for non-completion of the whole works by Philips. The Government point out that if there is delay in completion it will continue inevitably to incur expenses of a standing nature irrespective of the scale of the work outstanding and that those expenses will continue until the work is completed. This being a reasonable assumption and there being no ground for suggesting that the minimum payment limitation was set at the wrong percentage, its presence does not create a penalty . . .

Some commentators discerned in the *Phillips* case a significant move away from the reasoning in the *Dunlop* case. This was denied by the Court of Appeal in the following case.

Jeancharm Ltd v. Barnet Football Club Ltd

COURT OF APPEAL

(2003) 92 Con LR 26

The parties made an agreement for the supply of football kit by Jeancharm to Barnet Football Club. The contract contained a clause under which, if the buyer failed to pay within 45 days of the invoice date, interest should be payable at the rate of 5% per week.

The Court of Appeal held that the sum was far in excess of any conceivable loss flowing from late payment and was therefore a penalty.

JACOB J:

11. [Counsel for the respondents] suggested that following the *Philips* case and the Australian decision (referred to) the law had moved on from what was stated by Lord Dunedin to the extent that it had virtually abandoned it. In particular, he suggested that one should look at the contract as a whole, look at the risks being undertaken by both sides and ask whether the clause was an appropriate clause, having regard to the risk undertaken by the opposite party. Here, for instance, he said that his clients were at very considerable risk if they were in late delivery, having regard to the 20 pence per garment per day clause, and that should be balanced against the interest for late payments.

12. I can find nothing in the *Philips* case that suggests a departure of that gigantic nature from the law as laid down by Lord Dunedin. Lord Dunedin indicated that the question of whether or not a clause was a penalty clause depended upon whether it could be regarded as a genuine pre-estimate of the damage caused if there was a breach. Mr Kay's formulation abandons that entirely.

Alfred McAlpine Capital Projects Ltd v. Tilebox Ltd

JACKSON J.

(2005) 104 ConLR 39; [2005] EWHC 281 (TCC)

In 1998 the defendants (Tilebox) purchased a leasehold interest in Onslow House in Guildford. They planned to refurbish it as a top-of-the-range office building. In 2000 they had discussions with a possible funder, Standard Life, and on 12 February 2001 Tilebox and Standard Life entered into a development funding agreement (the DFA). The scheme of the DFA was that Tilebox would surrender its lease to Standard Life, would enter into a building contract for the development and engage relevant professionals. Standard Life would fund the development, subject to a maximum, and Tilebox would arrange for the building to be let to one or more tenants for a term of at least 15 years. Tilebox would receive a management fee of £225,000 and a development completion payment (DCP) based on a formula set out in the contract.

Tilebox then entered into negotiations for a building contract with the claimants (McAlpine) and a contract on JCT terms with contractor's design (1998 edn) was entered into on 27 April 2001. This provided for liquidated damages at the rate of £45,000 per week or part thereof. This figure had been the subject of negotiations.

The contract provided for completion on 9 August 2002. There were very substantial delays and McAlpine accepted that there was a substantial period of delay for which it cannot be granted any extension of time. McAlpine initiated the present proceedings seeking a declaration that the figure of £45,000 a week was a penalty.

JACKSON J:

Let me now stand back from the authorities and make four general observations, which are pertinent to the issues in the present case.

(1) There seem to be two strands in the authorities. In some cases judges consider whether there is an unconscionable or extravagant disproportion between the damages stipulated in the contract and the true amount of damages likely to be suffered. In other cases the courts consider whether the level of damages stipulated was reasonable. Mr Darling submits, and I accept, that these two strands can be reconciled. In my view, a pre-estimate of damages does not have to be right in order to be reasonable. There must be a substantial discrepancy between the level of damages stipulated in the contract and the level of damages which is likely to be suffered before it can be said that the agreed pre-estimate is unreasonable.

(2) Although many authorities use or echo the phrase 'genuine pre-estimate', the test does not turn upon the genuineness or honesty of the party or parties who made the pre-estimate. The test is primarily an objective one, even though the court has some regard to the thought processes of the parties at the time of contracting.

(3) Because the rule about penalties is an anomaly within the law of contract, the courts are predisposed, where possible, to uphold contractual terms which fix the level of damages for breach. This predisposition is even stronger in the case of commercial contracts freely entered into between parties of comparable bargaining power.

(4) Looking at the bundle of authorities provided in this case, I note only four cases where the relevant clause has been struck down as a penalty. These are *Public Works Comr* v. *Hills* [1906] AC 368, [1904–7] All ER Rep 91.9, *Campbell Discount Co Ltd* v. *Bridge* [1962] 1 All ER 385, [1962] AC 600, *Workers Trust and Merchant Bank Ltd* v. *Dojap Investments Ltd* [1993] 2 All ER 370, [1993] AC 573, and *Ariston SRL* v. *Charly Records Ltd* (1990), *Independent*, 13 April. In each of these four cases there was, in fact, a very wide gulf between (a), the level of damages likely to be suffered, and (b), the level of damages stipulated in the contract.

On the basis of the conclusions set out in parts 6 and 7 of this judgment, cl 24.2 of the building contract was an entirely reasonable pre-estimate of damages. Indeed, when one adds together Tilebox's own losses and Tilebox's liability in damages to Standard Life, it can be seen that £45,000 per week is too low.

There is, however, a separate question which I must address. Suppose that I am wrong in my interpretation of cl 6.8.1 of the DFA. Suppose that this clause does not impose upon McAlpine a back-to-back liability to Standard Life in the event of delay by McAlpine. In those circumstances, would cl 24.2 of the building contract become a penalty clause?

On this hypothesis, Tilebox's foreseeable losses flowing from delay would fall under two heads, namely: (i) diminution of the DCP, and (ii) Tilebox's own direct losses. For the reasons set out in part 7 of this judgment, the amount of Tilebox's future losses under each of those two heads would, in April 2001, have been difficult to quantify. A weekly loss somewhere in the region of £30,000 could have been foreseen under head (i), although the actual figure might turn out to be higher or lower than that. A weekly loss somewhere between £5,000 and £10,000 could have been foreseen under head (ii). From the viewpoint of April 2001, it was most unlikely, although just conceivable, that the total weekly loss would be as high as £45,000. Against that background, should cl 24.2 be struck down as a penalty? In my judgment, it should not, for five reasons.

(1) The figure of £45,000 was at or slightly above the top of the range of possible weekly losses flowing from delay. Whether one takes the top of the range or the middle of the range of possible future losses as the yardstick, it seems to me that the gap between that yardstick and £45,000 was not nearly wide enough to warrant characterising this clause as a penalty.

(2) Mr Hutley did make a genuine attempt to estimate the losses which would flow from future delay. Mr Hutley, unlike me, did not have the good fortune to listen to two days of legal argument about what that loss was likely to be. Instead, perfectly sensibly, he took as his yardstick a conservative estimate of rental value. I accept, of course, that if Mr Hutley's estimate was substantially wrong, then the genuineness of his efforts cannot save the clause. Nevertheless, this is a relevant factor.

(3) The difficulty which was inherent in the exercise of estimating future losses makes it particularly sensible in this case for the parties to have agreed upon a weekly figure (see *Clydebank Engineering and Shipbuilding Co Ltd* v. *Yzquierdo y Castaneda* [1905] AC 6 at 1.1, [1904–7] All ER Rep 251 at 253. See also Lord Dunedin's speech in *Dunlop Pneumatic Tyre Co Ltd* v. *New Garage and Motor Co Ltd* [1915] AC 79 at 87–88, [1914–15] All ER Rep 739 at 742 (sub-para (d)).

(4) This court, following the lead set by higher courts, is predisposed where possible to uphold contractual terms which fix the level of damages. This predisposition is somewhat stronger in the present case for the following reason: the building contract dated 27 April 2001 is a commercial contract made between two parties of comparable bargaining power.

(5) During the course of the pre-contract negotiations, the level of liquidated damages was the subject of specific debate. A figure of £45,000 was considered not only by the parties, but also, as can be seen from the documents, by their legal advisors. The fact that cl 24.2 and its appendix survived such scrutiny is further evidence that, as at April 2001, the liquidated damages provision was reasonable.

Where the employer is wholly or partly responsible for the contractor's failure to complete on time, he cannot recover liquidated damages, unless the contract provides otherwise.

Peak Construction (Liverpool) Ltd v. McKinney Foundations Ltd

COURT OF APPEAL

(1970) 1 BLR 111

Peak contracted with Liverpool Corporation to construct multi-storey flats. The contract was in Liverpool Corporation's own form, described by Salmon LJ as containing 'the most one-sided, obscurely and ineptly-drafted clauses in the United Kingdom', with a completion period of 24 months. The corporation's Director of Housing was appointed as architect-in-charge. Clause 22 provided that 'time shall be considered as of the essence of the contract' and provided for liquidated damages at the rate of '25 s per dwelling for each and every week or part of a week' of delay. Clause 23 empowered the architect to grant extensions of time for, *inter alia*, extras and additions, *force majeure* and 'other unavoidable circumstances'. The defendants were nominated subcontractors for the piling. They completed their work and left site. Subsequently, the piles were found to be seriously defective.

At a site meeting, it was agreed that work should cease pending investigation and at a later meeting it was further agreed that the problem should be submitted to one of three named consulting engineers. Everyone, including the architect, agreed to accept his opinion. A month later, the corporation's housing committee resiled on this decision and merely agreed to consider the consultant's report 'without prejudice'. The corporation did not appoint the engineer for several months and a further period elapsed whilst he completed his report. There was a further delay before the corporation authorised the report and the defendants were not able to recommence until 58 weeks after work had ceased.

Held: (1) Since part of the total delay was attributable to the corporation's fault, there should be a new trial to assess how much of that period was attributable to the defendants' breach.

(2) So far as the employers were concerned, they were not entitled to recover any liquidated damages against the plaintiffs under the terms of the main contract. The corporation could not recover liquidated damages for delay for which it was partly to blame.

SALMON LJ: A clause giving the employer liquidated damages at so much a week or month which elapses between the date fixed for completion and the actual date of completion is usually coupled, as in the present case, with an extension of time clause. The liquidated damages clause contemplates a failure to complete on time due to the fault of the contractor. It is inserted by the employer for his own protection; for it enables him to recover a fixed sum as compensation for delay instead of facing the difficulty and expense of proving the actual damage which the delay may have caused him. If the failure to complete on time is due to the fault of both the employer and the contractor, in my view, the clause does not bite. I cannot see how, in the ordinary course, the employer can insist on compliance with a condition if it is partly his own fault that it cannot be fulfilled: *Wells* v. *Army & Navy Co-operative Society Ltd* (1902); *Amalgamated Building Contractors* v. *Waltham Urban District Council* (1952) and *Holme* v. *Guppy* (1838). I consider that unless the contract expresses a contrary intention, the employer, in the circumstances postulated, is left to his ordinary remedy; that is to say, to recover such damages as he can prove flow from the contractors' breach. No doubt if the extension of time clause provided for a postponement of the completion date on account of delay caused by some breach or fault on the part of the employer, the position would be different. This would mean that the parties had intended that the employer could recover liquidated damages notwithstanding that he was partly to blame for the failure to achieve the completion date. In such a case the architect would extend the date for completion, and the contractor would then be liable to pay liquidated damages for delay as from the extended completion date.

The liquidated damages and extension of time clauses in printed forms of contract must be construed strictly *contra proferentem*. If the employer wishes to recover liquidated damages for failure by the contractors to complete on time in spite of the fact that some of the delay is due to the employer's own fault or breach of contract, then the extension of time clause should provide, expressly or by necessary inference, for an extension on account of such a fault or breach on the part of the employer. I am unable to spell any such provision out of . . . the contract [clause] in the present case. In any event, it is clear that, even if [the] clause had provided for an extension of time on account of the delay caused by the contractor, the failure in this case of the architect to extend the time would be fatal to the claim for liquidated damages. There had clearly been some delay on the part of the corporation. Accordingly, as the architect has not made 'by writing under his hand such an extension of time', there is no date under the contract from which the defendants' liability to pay liquidated damages for delay could be measured. And therefore none can be recovered: see *Miller* v. *London County Council* (1934).

M&J Polymers Ltd v Imerys Minerals Ltd

BURTON J

(2008) 117 ConLR 88, [2008] EWHC 344 (Comm)

The claimant (Polymers) was a company supplying dispersants; that is, chemicals used in the breakdown of clay and other materials. It had been supplying dispersants to the defendants (Imerys Minerals) (previously known as English China Clays) since 1991. The present contract signed in January 2005 was for the supply of four dispersants: Jaypol 1183 (1183), Jaypol BTC 2 (BTC 2), Jaypol 1160 (1150) and Jaypol 1140. The contract was for a three-year term and contained provisions for minimum purchases of products and an undertaking that the defendants would 'take or pay' for the minimum quantities. There were provisions as to the quality of the goods and art 1.6 gave the defendants an option to terminate if there were a failure to meet significant specification requirements.

The defendants purported to terminate in May 2006 and on 19 May 2006 the claimants accepted this as an unlawful repudiation. There were three issues before the court:

(1) whether the 1160 dispersant delivered in 2005 was fit for purpose and whether the defendant was entitled ro refuse to accept deliveries after 31 August 2005;
(2) whether the 'take and pay' provision was a penalty;
(3) whether there was a shortfall of orders for 1183 in January 2005.

BURTON J:

On the facts of this case, I am entirely satisfied that the take or pay clause was *commercially justifiable*, did not amount to *oppression*, was negotiated and freely entered into between *parties of comparable bargaining power*, and did not have the *predominant purpose of deterring* a breach of contract nor amount to a provision '*in terrorem*'. The evidence was wholly clear.

> The negotiations took place between extremely well-qualified, able and savvy commercial men against a very significant commercial background, including a background of previous dealings. At the time when they were negotiating there was, on the one hand, an extreme scarcity of acrylic acid, a willingness on the part of the claimant to commit supply to the defendant, notwithstanding the requirements of their other customers, but in return expecting an absolute commitment to take a minimum quantity of the product to be manufactured with that acid, and, on the other hand, a clear recognition of the difficult commercial position in which the defendant found itself, a desperate need to secure supplies of the product and a desire to keep the contract as short as possible; with the result that there was give and take on both sides, a shorter term than the claimant desired and a longer one than the defendant desired, but so far as the defendant was concerned, the significant advantage, as it then saw it, of a get-out clause via the meet or release provision in art 5.6.

An employer is not entitled to liquidated damages if by his acts or omissions he has prevented the contractor from completing the works by the completion date. This general rule may be amended by the express terms of the contract and is so amended by JCT terms.

Percy Bilton Ltd v. Greater London Council

HOUSE OF LORDS

(1982) 20 BLR 1

Bilton contracted with the GLC to erect a housing estate. The contract was dated 25 October 1976, and was substantially in JCT 63 form. The original completion date was 24 January 1979. The nominated sub-contractor for mechanical services went into liquidation on 28 July 1978, being already 40 weeks behind programme, and Bilton determined their employment on 31 July 1978. Bilton had asked for a new sub-contractor to be nominated on 7 July 1978, but the sub-contractor nominated as a replacement withdrew before starting work. Ultimately a third firm was nominated on 31 October 1978, the sub-contract being eventually agreed on 22 December 1978. The ultimate sub-contractor's programme was such that the work would not be completed until 25 January 1980.

Various extensions of time were granted, and eventually the extended completion date became 1 February 1980, part of the extension of time being awarded under clause 23(f). Bilton had not completed by that date and thereafter GLC deducted liquidated damages from money certified due to Bilton.

Bilton contended that the architect's clause 22 certificate of delay was invalid and argued that GLC were not entitled to deduct liquidated damages, time having become 'at large' because of GLC's failure to re-nominate so as not to delay them.

Held: Withdrawal of a nominated sub-contractor is not a fault or breach of contract on the part of the employer, nor is it covered by clause 23 of JCT 63. GLC's delay in making a timely renomination fell within JCT 63, clause 23(f) (now JCT 80, clause 25.4.6.), and Bilton was entitled to an extension of time to cover that delay. The architect's clause 22 certificate was valid and GLC was entitled to deduct liquidated damages.

LORD FRASER OF TULLYBELTON: The true position is . . . correctly stated in the following propositions:

(1) The general rule is that the main contractor is bound to complete the work by the date for completion stated in the contract. If he fails to do so, he will be liable for liquidated damages to the employer.
(2) That is subject to the exception that the employer is not entitled to liquidated damages if by his acts or omissions he has prevented the main contractor from completing his work by completion date – see for example, *Holme* v. *Guppy* (1838), and *Wells* v. *Army & Navy Co-operative Society* (1902).
(3) These general rules may be amended by the express terms of the contract.
(4) In this case, the express terms of clause 23 of the contract do affect the general rule. For example, where completion is delayed '(a) by *force majeure*, or (b) by reason of any exceptionally inclement weather,' the architect is bound to make a fair and reasonable extension of time for completion of the work. Without that express provision, the main contractor would be left to take the risk of delay caused by *force majeure* or exceptionally inclement weather under the general rule.
(5) Withdrawal of a nominated sub-contractor is not caused by the fault of the employer, nor is it covered by any of the express provisions of clause 23. Paragraph (g) of clause 23 expressly applies to 'delay' on the part of a nominated sub-contractor but such 'delay' does not include complete withdrawal . . .
(6) Accordingly, withdrawal falls under the general rule, and the main contractor takes the risk of any delay directly caused thereby.
(7) Delay by the employer in making the timeous nomination of a new subcontractor is within the express terms of clause 23(f) and the main contractor, the appellant, was entitled to an extension of time to cover that delay. Such an extension has been given.

Where the reason for the contractor's failure to complete on time is wholly or partly the fault of the employer, time may become 'at large' and the contractor's obligation is then to complete 'within a reasonable time'.

Wells v. Army & Navy Co-operative Society Ltd

KING'S BENCH DIVISION
(1902) 86 LT 764

A building contract provided that certain matters causing delay and 'other causes beyond the contractor's control' were to be submitted to the decision of the directors of the employer. The directors were 'to adjudicate thereon and make due allowance therefor if necessary, and their decision shall be final'. There was a liquidated damages clause. There was a year's delay in completion, and the directors allowed a three-month extension for delays caused by sub-contractors. Other breaches were established, including failure by the employer to give possession of the site and failing to provide plans and drawings in due time.

Held: The words 'other causes beyond the contractor's control' did not extend to delay caused by the interference of the employers or their architect. Liquidated damages could not therefore be deducted and the contractor's obligation was to complete within a reasonable time.

Where the contract stipulates that work is to be done within a stated period, if by his conduct the employer renders it impossible or impracticable for the contractor to do his work within the stipulated time, then the employer can no longer insist upon strict adherence to the time for completion and he cannot enforce any liquidated damages.

Dodd v. Churton

COURT OF APPEAL

[1897] 1 QB 562

A contract provided for the whole of the works to be completed by 1 June 1892, with liquidated damages of £2 per week for every week that any part of the work remained unfinished after that date. There was a provision that any authority given by the architect for any alteration or addition in or to the works was not to vitiate the contract. There was no provision for extending the time for completion if additional work was ordered. Additional works were ordered involving delay in the completion of the works beyond the specified date. The works were not completed until 5 December 1892. A fortnight was a reasonable time for the doing of the additional work, and the employer, allowing a fortnight's additional time for the completion of the works, claimed £2 per week in respect of the delay of 25 weeks.

Held: By giving the order for the additional works the employer had waived the stipulation for liquidated damages in respect of non-completion of the work by 1 June.

> LORD ESHER MR: Where one party to a contract is prevented from performing it by the act of the other, he is not liable in law for that default; and, accordingly, a well recognised rule has been established in cases of this kind, beginning with *Holme* v. *Guppy* (1838), to the effect that, if the building owner has ordered extra work beyond that specified by the original contract which has necessarily increased the time requisite for finishing the work, he is thereby disentitled to claim the penalties for non-completion provided for by the contract.

Under a contract in JCT 98 terms where clause 24 is incorporated and the parties complete the relevant appendix entry, either by stating a rate at which liquidated damages are to be calculated or by stating that the sum is nil, that constitutes an exhaustive agreement as to the damages which are or are not to be payable by the contractor in the event of his failure to complete the works on time.

Temloc Ltd v. Errill Properties Ltd

COURT OF APPEAL

(1987) 12 Con LR 109

On 12 March 1984 the respondent (Temloc) entered into in JCT 80 (Private, Without Quantities version) with the defendant (Errill) for the construction of a large shopping development at Plympton, Devon. The contract sum was £840,000, and the contract completion date was 28 September 1984. The contract overran, and the architect certified practical completion as having taken place on 20 December 1984, though when he reviewed extensions of time under clause 25.3.3 on 2 May 1985 he wrote to the contractor giving a revised completion date of 14 November 1984.

Clause 24 (the liquidated damages clause) was left in the printed conditions of contract as executed by the parties and the relevant Appendix entry was filled in as '£nil'. Temloc issued writs against Errill claiming sums due on certificates issued by the architect. There were also claims and counterclaims between Errill and third parties who had agreed to occupy parts of the development, which resulted in Temloc being brought in as a third party, arising out of the delay in completion. The following preliminary issues were before the court, on Errill's appeal against the determination of them in Temloc's favour.

> Issue
> Whether upon a true construction, the Contract made between Temloc and Errill and dated 12 March 1984 entitles Errill to claim any relief as a result of any failure by Temloc to complete the contract works either by the date specified in the contract or any other later date certified or within a reasonable time and if so precisely what relief and in what circumstances?
>
> Sub-issues
> (I) Does the inclusion of '£nil' nevertheless permit a claim for (a) damages at large or (b) for an indemnity?
> (II) If the effect of the £nil entry is to preclude a claim (whether for damages or an indemnity) as a result of a failure to complete by the date fixed by the contract or a date extended in accordance with the contract, can Errill nevertheless have a claim in respect of a failure to complete within a reasonable time?

Held (dismissing the appeal):

(1) The employer was not entitled to claim any relief as a result of the contractor's failure to complete the contract works either by either the date specified in the contract or any later date fixed under the contract or within a reasonable time because clause 24 was exhaustive of the employer's remedies for the breach of late completion.

(2) The effect of the '£nil' entry in the Appendix was to provide a negative figure for liquidated damages. The fact that the parties agreed that there should be no damages also excluded any other claims which clause 24.2 would have excluded, including a claim for damages for failure to complete within a reasonable time.

NOURSE LJ: I think it clear, both as a matter of construction and as one of common sense, that if (1) clause 24 is incorporated in the contract and (2) the parties complete the relevant part of the appendix, either by stating a rate at which the sum is to be calculated or, as here, by stating that the sum is to be nil, then that constitutes an exhaustive agreement as to the damages which are or are not to be payable by the contractor in the event of his failure to complete the works on time. In my view the submissions of [counsel for the defendant] for the employer have proceeded on an overvalue of the significance of the words 'as liquidated and ascertained damages' in clause 24.2.1. The damages payable in respect of late completion of the works are one head of the general damages which may be recoverable by an employer for the contractor's breach of a building contract. Their character is not in any way altered according to whether the rate at which they are payable is agreed by the parties in advance, so that they become liquidated, or determined by the court after breach, so that they remain unliquidated until so determined.

Turning to the wording of clause 24, I observe, first, that it is headed 'Damages for non-completion'. It does not refer to liquidated damages for non-completion. That may not be a very large point. But then clause 24.1 provides for the architect to issue a certificate if the contractor fails to complete the works by the completion date. That failure necessarily involves a breach of contract giving rise to a claim for damages. If clause 24.1 stood alone, the damages would be unliquidated. What clause 24.2.1 then does is to provide that the contractor shall, as the employer may require, pay or allow to the employer the whole or part of 'a sum calculated at the rate stated in the Appendix', being a sum payable or allowable by way of damages for the breach of contract. The agreement as to rate necessarily makes them liquidated, as is recognised by the following words 'as liquidated and ascertained damages'. They nevertheless remain damages for the breach established by the architect's certificate under clause 24.1.

Viewing the clause in this way. I find it impossible to attribute to parties who complete the appendix in one way or the other an intention that the employer shall have the option of claiming damages of precisely the same character but in an unliquidated amount.

J-Corp Pty Ltd v. Mladenis and another

SUPREME COURT OF WESTERN AUSTRALIA COURT OF APPEAL

(2008) 131 Con LR 188, [2009] WASCA 157

The appellant entered into a contract with the respondent to build a house on the respondent's land for $311,484.12. The contract was on a standard form used by the appellant. Clause 1'7 of the contract dealt with commencement, completion and delays. Clause LL.9 provided:

> If the Builder breaches sub-clause 11.1, it shall be liable to pay the Proprietor liquidated damages at the rate of NIL DOLLARS ($00.00) per day for each day beyond the due date for practical completion until practical completion is deemed to have taken place.

A preliminary question of law was ordered:

> Does clause 77.9 of the Contract on its proper construction have the effect of excluding the [respondents'] right to claim damages at common law for losses suffered by the [respondents] by reason of the [appellant's] breach of the Contract (assumed for the purpose of this question), namely the failure to reach practical completion of the Works by the Due Date for Practical Completion?

The primary judge answered the question in the negative and held that cl 71.9 only excluded a claim to liquidated damages. The appellant appealed.

BUSS JA:

The critical issue in the appeal is whether the parties, by specifying the rate of liquidated damages in cl 11.9 of the contract as 'NIL DOLLARS ($00.00) per day' intended that the respondents/proprietors should not be entitled to recover any damages (either liquidated damages at an agreed rate or unliquidated common law damages) for breach of the appellant's/builder's obligation under cl 11.1 to complete the works by a specified date.

It is well-established that an intention to exclude the common law right to damages for breach of contract must be expressed in clear and unambiguous terms. See, for example, *Concut Pty Ltd* v. *Worrell* [2000] HCA 64, (2000) 176 ALR 693 at [23], (2000) 74 ALJR 312 at [23] (Gleeson CJ, Gaudron and Gummow JJ) and *Décor Ceilings Pty Ltd* v. *Cox Constructions Pty Ltd (No 2)* [2005] SASC 483, (2007) 23 BCL 347 at [66] (Besanko J).

It is true that a claim by a proprietor against a builder for late completion under a building contract, whether for liquidated damages at an agreed rate or for unliquidated common law damages, involves a single cause of action for the same loss: see *Temloc Ltd* v. *Errill Properties Ltd* (1987) 12 Con LR 109 at 115–116 (per Nourse LJ). In other words, a claim for liquidated damages at an agreed rate and a claim for unliquidated common law damages in respect of the same loss are not separate heads of damages.

However, this point of principle does not preclude the parties to a building contract from agreeing that the proprietor will be confined to the common law right to damages for any breach-by the builder of its obligation to complete the works by a specified date The intention of the parties in this respect turns on the proper construction of the particular contract.

The construction of a written contract is concerned with ascertaining what a reasonable person would have understood the parties to mean. Consideration should ordinarily be given not only to the language of the document, but also to the surrounding circumstances known to the parties, and the apparent purpose and object of the transaction: see *Toll (FGCT) Pty Ltd* v. *Alphapharm Pty Ltd* [2004] HCA 52, (2004) 219 CLR 165 at [40] (Gleeson CJ, Gummow, Hayne, Callinan and Heydon JJ). Where the parties have entered into a building contract, the apparent purpose and object of the transaction may include the evident allocation between the parties of the risk of loss or damage arising from a particular occurrence or contingency. The statutory context (if any) in which the parties contracted may also be relevant to issues of construction.

As Giles J noted in *Baese Pty Ltd* v. *RA Bracken Building Pty Ltd* (1990) 5 BCL 137 at 140, a clause in a building contract providing for liquidated damages at an agreed rate in the event of the builder's failure to complete the works on time is normally included for the benefit of the proprietor. It relieves the proprietor from the difficulty and expense of proving that he or she has suffered actual damage as a result of the delay.

I agree with Newnes JA that on a proper construction of the contract between the appellant/builder and the respondents/proprietors, the parties did not express in clear and unambiguous term an intention to exclude the common law right to damages for breach of the appellant's/builder's obligation to complete the works on time. Clause 11.9, in the context of the contract as a whole, manifests an intention that the standard form liquidated damages clause should, in substance, be excluded, but does not, in my view, manifest an intention that any and all remedies (whether under the contract or at common law) should be excluded.

The architect's certificate of delay under clause 24.1 of JCT 98 is a condition precedent to the contractor's liability to pay liquidated damages. Once such certificate has been issued, the employer may deduct liquidated damages from amounts due to the contractor under interim certificates.

J.F. Finnegan Ltd v. Community Housing Association Ltd

COURT OF APPEAL (CIVIL DIVISION)

(1995) 47 Con LR 25

In December 1986 the respondent J.F. Finnegan Ltd (Finnegan) entered into a written agreement with the appellant, Community Housing Association Ltd (CHA), for the construction of 18 flats at 46–47 Coram Street, London WC1, for the sum of just over £750,000. The date for possession of the site was 1 December 1986 and the date for completion was 1 March 1988. Liquidated and ascertained damages for non-completion under clause 24 were stated to be at the rate of £2,500 a week or part of a week.

Clause 24 of the contract provided:

> Certificate of Architect.
> 24.1 If the Contractor fails to complete the Works by the Completion Date then the Architect shall issue a certificate to that effect.
> Payment or allowance of liquidated damages.
> 24.2.1 Subject to the issue of a certificate under clause 24.1 the Contractor shall, as the Employer may require in writing not later than the Final Certificate, pay or allow to the Employer the whole or such part as may be specified in writing by the Employer of a sum calculated at the rate stated in the Appendix as liquidated and ascertained damages for the period between the Completion Date and the Date of Practical Completion and the Employer may deduct the same from any monies due or to become due to the Contractor under this Contract (including any balance stated as due to the Contractor in the Final Certificate) or the Employer may recover the same from the Contractor as a debt.
> 24.2.2 If, under clause 25.3.3, the Architect fixes a later Completion Date the Employer shall pay or repay to the Contractor any amounts recovered, allowed or paid under clause 24.2.1 for the period up to such later Completion Date.

Clause 25.3.3 provided:

> 25.3.3 Not later than the expiry of 12 weeks from the date of Practical Completion the Architect shall in writing to the Contractor either
> 3.1 fix a Completion Date later than that previously fixed if in his opinion the fixing of such later Completion Date is fair and reasonable having regard to any of the Relevant Events . . .

On 9 March 1988 the architect certified under clause 24.1 that Finnegan had failed to complete the works by the completion date. There was a delay to the project of 19 weeks and CHA deducted £47,500 (i.e. 19 weeks × £2,500) from the certified amount, accompanying their cheque by a note which contained the following details:

Cert. 17	£61,518.00
LAD Damages	£47,500.00
	£14,018.00

Finnegan commenced proceedings against CHA challenging the deduction, and arguing that the £2,500 figure was a penalty, and also that CHA was not entitled to deduct liquidated damages. The trial judge (Judge Maurice Carr QC) held in favour of Finnegan on the ground that CHA had failed to comply with clause 24.2.1 in that it did not serve on Finnegan a statement in writing as required by that clause. On the facts the judge held that the note accompanying the cheque was not a compliance with clause 24.2.1, and that the deduction of £47,500 was in breach of contract. He therefore declared that CHA was not entitled to deduct that sum.

CHA appealed and argued (i) that the requirement in writing under clause 24.2.1 was not a condition precedent to deduction; (ii) that if the requirement was a condition precedent, it need only require the contractor to pay or allow the liquidated and ascertained damages and, if part only was claimed, to specify what part the contractor was to pay or allow, and no more; (iii) that it had in any event satisfied the condition precedent.

Held (allowing the appeal):

(1) The requirement in writing by the employer is a condition precedent to the operation of clause 24.2.1 in the sense that without such requirement there can be no lawful deduction. The language of the clause makes it clear that the employer's entitlement to deduct or recover liquidated and ascertained damages is dependent on the employer indicating that he does require a payment or allowance of the whole or part of the liquidated damages. The clause was inoperable without such requirement.

(2) From the language of clause 24.2.1 there are only two matters which must be contained in the written requirement. One is whether the employer is claiming a payment or deduction in respect of liquidated and ascertained damages. The other is whether the requirement relates to the whole or part (and, if so, what part) of the sum for liquidated and ascertained damages.

(3) The note accompanying the cheque sent on 28 September 1988 was a requirement in writing sufficient for the purpose of clause 24.2.1 and thus satisfied the condition precedent.

PETER GIBSON LJ:

Issue (1)

There can, in my opinion, be no doubt but that a requirement in writing by the employer is a condition precedent to the operation of clause 24.2.1 in the sense that without such requirement there can be no lawful deduction. The language of the clause makes clear that the entitlement of the employer to deduct or recover LADs [liquidated and ascertained damages] is dependent on the employer indicating by such requirement that he does require a payment or allowance of the whole or part of the LADs. In short the clause is inoperable without such requirement. That a requirement in writing is a condition precedent was the view of Judge John Newey QC in *A. Bell & Son* v. *CBF* (1989) 16 Con LR 62 at 65–66, 46 BLR 102 at 107. To the extent that Judge James Fox-Andrews QC in *Jarvis Brent Ltd* v. *Rowlinson Construction Ltd* (1990) 6 Const LJ 292 was of a different view, I prefer the contrary view of Judge Newey and of the judge in the present case . . .

Similarly, a little later he said that he had already 'concluded that clause 25.2.1 is a condition precedent which must be complied with before a deduction can be made'. The judge explained his thinking in the sentence that followed that which I cited first, that 'to allow an employer to deduct without making a requirement in writing means that the words "may require in writing" are superfluous'. The judge in my view is not to be understood as making a temporal point in using the word 'before'.

Issue (2)
From the language of clause 24.2.1 I consider that there are only two matters which must be contained in the written requirement. One is whether the employer is claiming a payment or a deduction in respect of LADs. The other is whether the requirement relates to the whole or a part (and, if so, what part) of the sum for the LADs. I stress the words 'the sum' because it is quite clear, in my view, from the language of clause 24.2.1 that what is required is a statement of the sum in respect of LADs . . .

For my part, I do not see that clause 30.1.1.3 really assists the contractor in the present case. The reason for the deduction can be made plain by the employer by referring to the fact that what he is claiming is LADs. That in itself imports that there has been a failure to complete on the completion date. Nor do I understand the difficulty said to face the contractor. The contractor will know at the time of the requirement in writing the length of the overrun and the contractor will know the existing completion date and the date of practical completion if that has already been certified. Counsel for the plaintiff said that the employer may not know the completion date because a completion date fixed by the architect subsequent to the original contractual completion date is not required to be notified to the employer under the provisions of the contract. For my part, I regard the notion that the employer will not know the revised completion date as wholly unreal. The architect is the employer's agent. In my judgment, it really cannot be assumed that the draftsman of clause 24.2.1 would have contemplated the possibility that the employer did not know of the actual completion date in operation at the time of a requirement in writing.

Until another completion date is fixed, the employer is entitled to claim LADs for the whole of the period of the overrun, repaying under clause 24.2.2 the excess when a later completion date is fixed by the architect under clause 25.3.3. But the employer, if he knows that a later completion date is likely to be fixed by the architect, may sensibly prefer not to claim the whole of the LADs which are available to him at the time of the requirement, but he may not know what later completion date is to be fixed. I do not see why the clause should be construed as requiring the employer to specify the precise dates of the overrun, the LADs for which he is making the subject of a requirement, so long as the LADs which he is requiring to be paid or allowed do not exceed the whole of the LADs to which he is entitled at the time of the requirement. No doubt it will be good practice for the employer to make its position clear beyond doubt. No doubt in certain circumstances, where the case is particularly complicated, it may be highly desirable that explanations should be given. But I would be reluctant to import into this commercial agreement technical requirements which may be desirable but are not required by the language of the clause and are not absolutely necessary and any breach of which would defeat the intended entitlement of the employer. Indeed, I would say that in the present case there is no ambiguity in the contractual language to leave me in any doubt as to what information clause 24.2.1 requires to be given to the contractor . . .

Extensions of time

Unless there is an extension of time clause, the contractor is under a strict obligation to complete on time, unless he is prevented from doing so by acts or breaches of the employer or by operation of the law.

If the contractor is so prevented from completing on time, any liquidated damages clause ceases to apply, even to subsequent delays which are the contractor's fault.

The contract time may be extended by the employer because of his own delay only if there is an express provision for extension of time and the terms of the clause must be strictly complied with.

Under JCT terms, in considering whether or not to grant an extension of time, it is the effect on completion which is critical. If the contractor is ahead of programme, the architect is entitled to take that fact into account.

London Borough of Hounslow v. Twickenham Garden Developments Ltd

CHANCERY DIVISION

(1970) 7 BLR 81

The facts of this case are set out on p. 383.

> MEGARRY J: Provided the contractor has given written notice of the cause of the delay, the obligation to make an extension appears to rest on the architect without the necessity of any formal request for it by the contractor. Yet he is required to do this only if in his opinion the completion of the works 'is likely to be or has been delayed beyond the date for completion', or any extended time for completion previously fixed. If a contractor is well ahead with his works and is then delayed by a strike, the architect may nevertheless reach the conclusion that completion of the works is not likely to be delayed beyond the date for completion . . . The contractor is under a double obligation: on being given possession of the site, he must 'thereupon begin the works and regularly and diligently proceed with the same', and he must also complete the works 'on or before the date for completion', subject to any extension of time. If a strike occurs when two-thirds of the work has been completed in half the contract time, I do not think that on resuming work a few weeks later the contractor is then entitled to slow down the work so as to last out the time until the date for completion (or beyond, if an extension of time is granted) if thereby he is failing to proceed with the work 'regularly and diligently'.

Under JCT terms the architect may grant an extension of time even after completion of the works. JCT 98, clause 25.3.3, makes express provision for the architect to review the position as regards extensions within 12 weeks of practical completion. See now JCT 2005, clauses 2.26 to 2.29.

Amalgamated Building Contractors Ltd v. Waltham Holy Cross Urban District Council

COURT OF APPEAL

[1952] 2 All ER 452

Clause 18 of the then current RIBA Contract provided:

> If in the opinion of the architect the works be delayed (i) by *force majeure*, or (ii) by reason of exceptionally inclement weather . . . or (ix) by reason of labour and material not being available as required . . . then in any such case the architect shall make a fair and reasonable extension of time for completion of the works . . .

The date for completion was 7 February 1949, and the contractors were liable to liquidated damages in the event of delay at the rate of £50 a week. On 19 January 1949, the contractors sought a 12 months' extension of time because of labour and materials difficulties, repeating their request on 27 January. The architect merely gave a formal acknowledgment. The work was completed on 28 August 1950 and it was not until 20 December 1950 that the architect wrote extending the time for completion to 23 May 1949.

Held: The retrospective extension of time was valid.

DENNING LJ: The contractors say that the words in clause 18

> the architect shall make a fair and reasonable extension of time for completion of the works

mean that the architect must give the contractors a date at which they can aim in the future, and that he cannot give a date which has passed. I do not agree with this contention. It is only necessary to take a few practical illustrations to see that the architect, as a matter of business, must be able to give an extension even though it is retrospective. Take a simple case where the contractors, near the end of the work, have overrun the contract time for six months without legitimate excuse. They cannot get an extension for that period. Now suppose that the works are still uncompleted and a strike occurs and lasts a month. The contractors can get an extension of time for that month. The architect can clearly issue a certificate which will operate retrospectively. He extends the time by one month from the original completion date, and the extended time will obviously be a date which is already past. Or take a cause of delay, such as we have in this case, due to labour and materials not being available. That may cause a continuous delay operating partially, but not wholly, every day, until the works are completed. The works do not stop. They go on, but they go on more slowly right to the end of the works. In such a case, seeing that the cause of delay operates until the last moment, when the works are completed, it must follow that the architect can give a certificate after they are completed. These practical illustrations show that the parties must have intended that the architect should be able to give a certificate which is retrospective, even after the works are completed . . . *Miller* v. *London County Council* (1934) is distinguishable. I regard that case as turning on the very special wording of the clause which enabled the engineer 'to assign such other time or times for completion as to him may seem reasonable'. Those words, as du Parcq J said, were not apt to refer to the fixing of a new date for completion *ex post facto*. I would also observe that on principle there is a distinction between cases where the cause of delay is due to some act or default of the building owner, such as not giving possession of the site in due time, or ordering extras, or something of that kind. When such things happen the contract time may well cease to bind the contractors, because the building owner cannot insist on a condition if it is his own fault that the condition has not been fulfilled.

The process by which the length of extension of time to be granted is calculated is usefully considered in the next case.

Balfour Beatty Building Ltd v. Chestermount Properties Ltd

QUEEN'S BENCH DIVISION

(1993) 32 Con LR 139

By a contract dated 16 June 1988, which was substantially in JCT 80 Form, private edition with approximate quantities, the appellant (the contractors) agreed to carry out the construction of a seven-storey office block at 126–137 Houndsditch, London EC3, for the respondent (the employers).

The works comprised the construction of the shell and core of the building, together with certain elements of developer fit-out. Under clause 2.2 of the contract, the employers could elect by a particular date to confine the contract to the shell and core works only. Work commenced on 18 September 1987, prior to the execution date, and the date for completion was 17 April 1989. In March 1988, the employers gave notice confining the scope of the work to shell and core only. On 11 October 1988 the architect granted an extension of time for completion of 15.5 days, taking the completion date to 9 May 1989.

The work had not been completed by 9 May 1989 and on 8 June 1989 the architect issued a certificate of non-completion under clause 24.1 of the contract. The shell and core works remained uncompleted in January 1990 and the anticipated completion date was the beginning of July 1990. Between 31 January 1990 and 9 February 1990, there were discussions between the parties and correspondence, the agreed effect of which was to entitle the employers to reinstate the fit-out works as part of the works, but not varying any of the other provisions of the contract.

Pursuant to that agreement, between 12 February and 12 July 1990 the architect issued variation instructions for the carrying out of the fit-out works, at which time the contractors were in culpable delay. The shell and core works were practically completed on 12 October 1990 and the fit-out works reached practical completion on 25 February 1991. Two further extensions of time were granted, one on 18 December 1990, which fixed the revised completion date as 12 September 1989, and a second and final one on 14 May 1991 which fixed the revised completion date at 24 November 1989, which was some 2½ months before the issue of the first of the architect's instructions for the carrying out of the fit-out works as variations on the contract.

On 15 May 1991 the architect issued a further and final certificate of non-completion under clause 24.1, which gave rise to a claim by the employers to be entitled to deduct from payments to the contractors under the contract the sum of £3.84m as liquidated and ascertained damages.

The contractors challenged the employers' entitlement to deduct that amount. They contended that the consequence of the issue of variation instructions by the architect during the period of culpable delay was to set time 'at large', so that no liquidated damages were recoverable. Alternatively, they contended that, even if the structure of the completion date and liquidated damages was intact, the architect should have given an extension of time starting from the date when the variation instruction was given, calculating the additional time which ought to be allowed by reason of the instruction, and then projecting forward from the date of the instruction by adding on that additional period to fix a new completion date after the date of the instruction, called the 'gross basis'.

The employers contended that the correct approach was that the revised completion date should be calculated by taking the date currently fixed and adding on to it the period of time which the architect regarded as fair and reasonable in respect of the consequences of the variations instructed, called 'the net basis', which is what the architect had purported to do.

These and other disputes were referred to arbitration. By an interim award dated 11 May 1992 the arbitrator, Mr Christopher J. Willis, FRICS, FCIArb, dealt with two questions put to him as preliminary issues.

To the question:

> Does clause 25 confer upon the architects jurisdiction to grant an extension of time for completion of the Works in respect of a relevant event occurring during a period of culpable delay?

the arbitrator answered that clause 25 did confer such jurisdiction on the architect.

In relation to the question:

> In granting an extension of time in respect of a relevant event occurring during a period of culpable delay, ought the architect to order a 'gross' extension (that is one that re-fixes the completion date at the calendar date upon which the work would reasonably be expected to be completed, having regard to the calendar date upon which it is instructed), or ought it to be a 'net' extension (that is one which calculates the revised completion date by taking the date currently fixed and adding the number of days which the architect regards as fair and reasonable)?

the arbitrator answered:

> In granting an extension of time in respect of a relevant event occurring during a period of culpable delay, the architect ought to award a 'net' extension of time, that is, one which calculates the completion date by taking the date currently fixed and adding the number of days which the architect regards as fair and reasonable.

The contractors appealed in relation to these two questions of law and, in the event that the court found that the arbitrator was wrong in relation to the first question, the court was asked to indicate the answer to a third question, which was not answered by the arbitrator because of his answer to the first question. The third question was:

> If the architect had no jurisdiction to extend time and in consequence there could be no completion date fixed by reference to the variation instructions and no liquidated and ascertained damages calculated by reference to such new completion date,

how does one calculate the time for completion of the works? Is there any residual entitlement for the employers to levy liquidated and ascertained damages on account of late completion? And does the architect have any power to issue certificates of non-completion under the contract?

Held (dismissing the appeal):

(1) The arbitrator was entirely correct in holding that clause 25 conferred on the architect jurisdiction to grant an extension of time for completion of the works in respect of relevant events occurring during a period of culpable delay.

(2) The arbitrator was also correct in holding that where a relevant event occurred during a period of culpable delay, the revised completion date should be calculated on a net basis, that is by taking the date currently fixed and adding to it the number of days which the architect regarded as fair and reasonable in respect of the consequences of the relevant event. It would be wrong in principle to apply the 'gross' method, and the 'net' method represented the correct approach.

COLMAN J: The concept that clause 25.3.3.1 gives the architect a power to consider that a variation instruction fairly and reasonably justifies an extension of time which involves putting back the completion date from a point of time before the instruction was given to a later point of time also before the instruction was given strikes one at first sight as distinctly peculiar. It gives rise to a new completion date by which the additional works were supposed to be finished which precedes those additional works becoming part of the contract works at all! Does not that 'flout business commonsense'?

In order to answer this question, it is right to examine the underlying contractual purpose of the completion date/extension of time/liquidated damages regime. At the foundation of this code is the obligation of the contractor to complete the works within the contractual period terminating at the completion date and on failure to do so to pay liquidated charges for the period of time by which practical completion exceeds the completion date. But superimposed on this regime is a system of allocation of risk. If events occur which are non-contractor's risk events and those events cause the progress of the works to be delayed, in as much as such delay would otherwise cause the contractor to become liable for liquidated damages or for more liquidated damages, the contract provides for the completion date to be prospectively or, under clause 25.3.3, retrospectively, adjusted in order to reflect the period of delay so caused and thereby reduce *pro tanto* the amount of liquidated damages payable by the contractor.

Likewise, if the works are reduced by an omission instruction by the architect, it may be fair and reasonable to reduce the contract period for completion prospectively or retrospectively and therefore to advance the completion date. In view of the inherent difficulties in predicting with precision the impact on the progress of the works of non-contractor's risk events, the architect is given a power of retrospective adjustment of the completion date. The underlying objective is to arrive at the aggregate period of time within which the contract works as ultimately defined ought to have been completed, having regard to the incidence of non-contractor's risk events and to calculate the excess time, if any, over that period, which the contractor took to complete the works. In essence the architect is concerned to arrive at an aggregate period for completion of the contractual works, having regard to the occurrence of non-contractor's risk events and to calculate the extent to which the completion of the works has exceeded that period.

This objective is clearly reflected in the provisions of clauses 25.3.2 and 25.3.3.2 which deal with architect's instructions to omit part of the original Works. The architect is by clause 25.3.2 empowered to bring forward the completion date, if, in his opinion, that is fair and reasonable having regard to instructions issued by him to omit part of the works as previously defined. This power cannot be exercised until after an initial exercise of the power of extension under clause 25.3.1 and must be exercised before the completion date previously fixed. Obviously there is nothing to stop the architect in an appropriate case from refixing the completion date at a point of time which is not only before the date on which he exercises the power but is also before the date on which he issued the omission instruction. This would have the effect of retrospectively imposing on the contractor an obligation to complete the works at a date before the event on the basis of which the completion date was advanced. That would arise in a case where the contractor had fallen well behind the clock in the progress of the rest of the works.

Then one comes to clause 25.3.3.2, which provides the architect with what on the wording is an identical power, save that it can be exercised after the completion date, if this occurs before practical completion. It expressly applies to omission instructions issued by the architect 'after the last occasion on which the architect fixed a new completion date'. There is nothing in the words to limit this power to omission instructions issued before as distinct from after the last-fixed completion date. Indeed, there is every reason for not limiting the power in this way. If it may be fair and reasonable after that completion date to reduce the contract period by reason of an omission instruction issued before that completion date, why might not an omission instruction issued after that completion date equally justify a reduction in the contract period and the consequent advancement of the completion date, if necessary to a point of time before the omission instruction was issued? If an architect were not so empowered, there would follow the anomalous result that, if the contractor had fallen behind the clock on the rest of the work and had already overshot the completion date, he could be compelled to pay more liquidated damages for his delay if the omission instructions happened to have been issued before the completion date but the same amount of liquidated damages if the omission instructions were issued after the completion date, notwithstanding that in both cases the works, and therefore the amount of time needed to complete them, would be reduced by the same amount.

This cannot have been the intention of the parties and the words of clause 25.3.3.2 do not suggest any such limitation. If there is no such limitation in respect of omission instructions under that clause, there could hardly be such a limitation in respect of relevant events, which include architect's variation instructions (clause 23.4.5.1) under clause 25.3.3.1, unless the words of the latter clause expressly suggested it. They contain no such suggestion.

Moreover, if one tests the contractor's construction of clause 25.3.3.1 by reference to its legal and commercial consequences, a very striking picture emerges. If variation instructions under clause 14.2 can only be treated as relevant events if such instructions are given before the completion date, it must follow that any variation instruction which follows the completion date would be an act of prevention by the employer. It was common ground that if the contract failed to provide for power to grant an extension of time on account of delays caused by an act of prevention, the effect of the act of prevention was to prevent the employer relying on the completion date/liquidated damages provisions in the contract. The obligation to complete the works was to be performed within a reasonable time, there could be no extensions on account of relevant events and the employer's only hope of compensation would be to recover unliquidated damages for delay (see *Peak Construction (Liverpool) Ltd* v. *McKinney Foundations Ltd* (1970) 1 BLR 111). The remarkable consequences of the application of this principle could therefore be that if, as in the present case, the contractor fell well behind the clock and overshot the completion date and was unlikely to achieve practical completion until far into the future, if the architect then gave an instruction for the most trivial variation, representing perhaps only a day's extra work, the employer would thereby lose all right to liquidated damages for the entire period of culpable delay up to practical completion or, at best, on the respondents' submission, the employer's right to liquidated damages would be confined to the period up to the act of prevention. For the rest of the delay he would have to establish unliquidated damages. What might be a trivial variation instruction would, on this argument, destroy the whole liquidated damages regime for all subsequent purposes.

So extreme a consequence for the future operation of the contract could hardly reflect the common intention, particularly having regard to the very specific distribution of risk provisions which are agreed to be applicable in respect of relevant events occurring before the completion date. It is certainly a construction which is most improbable in the absence of some other express provision supporting it . . .

In conclusion, therefore, on the first question, in my judgment the construction for which the contractor contends involves legal and commercial results which are so inconsistent with other express provisions and with the contractual risk distribution regime applicable to pre-completion date relevant events that, in the absence of express words compelling that construction, it cannot be right. In this respect, the contract is not ambiguous or so unclear as to call for application of the *contra proferentem* rule or the resolution of nicely-balanced issues of construction in favour of the employer for whose benefit the liquidated damages regime is introduced. The apparently anomalous consequence of the application of the arbitrator's construction that the architect could refix a completion date before the issue of the variation instruction is in my view entirely consistent with the basic purpose of the liquidated damages regime for reasons which I have already explained. Moreover, the retrospective postponement of the completion date to a date before the event causing delay was an eventuality contemplated with equanimity by Lord Denning MR in *Amalgamated Building Contractors Ltd* v. *Waltham Holy Cross UDC* [1952] 2 All ER 452 at 454. On this first question, the arbitrator was entirely correct . . .

Failure by the employer properly to grant an extension of time may have the effect of setting time at large. Whether the completion date is set at large by delay in granting an extension of time depends on all the circumstances.

Fernbrook Trading Co Ltd v. Taggart

SUPREME COURT OF NEW ZEALAND

[1979] 1 NZLR 556

Taggart contracted to carry out road construction works for Fernbrook. The contract period was 15 weeks, to commence 14 days after the date of the engineer's order to commence, and liquidated damages of $500 a week were payable by Taggart for delay. The extension of time clause provided for time to be extended for, *inter alia*, 'delays caused by other contractors on the site, delays in installation of services or exceptional circumstances'.

The completion date was 30 December 1975. Taggart applied for an extension of time on 18 November because, as well as being required to carry out additional work, there were delays by other contractors on the site. He received no acknowledgment until 2 March 1976 when the engineer granted an extension to 24 March. On 29 April the engineer granted a further extension to 1 June. The delays by other contractors caused a summer job to run into winter and exceptionally bad weather caused further delays; as did a drivers' strike. In addition Fernbrook was in breach of contract in failing to make progress payments for a total of 23 weeks, causing Taggart to reduce labour on the site because of financial difficulties. There was also a failure by the engineer to issue progress payment certificates for five separate months. The work was completed 46 weeks after the final extended completion date.

Held: The exercise of the engineer's discretion to extend time after the completion date was reasonable in the circumstances. Accordingly, the extensions were validly given, and the completion date was not set at large. However, the engineer had no jurisdiction to extend time on account of Fernbrook's breach of contract in failing to make progress payments. This breach set the completion date at large, with the effect that Taggart was not liable to pay liquidated damages.

Note: This judgment is quoted at length as it represents an admirable summary of the law.

ROPER J: [Taggart's counsel] submitted that to be effective an extension of time must be granted within the original completion period, or within a validly granted extension of that period; or in other words, that a retrospective extension was invalid and set the completion time at large. As the author of *Hudson's Building and Engineering Contracts* (10th edn, 1970) says at page 643, 'Regrettably, the law on this subject [i.e. the time for exercise of extensions] appears to have become unnecessarily complicated'. The differing circumstances and contractual provisions under review make for difficulty in reconciling the authorities.

The decision of our Court of Appeal in *Anderson* v. *Tuapeka County Council* (1900) is the logical starting point. In that case the contract provided for extensions of time if extra works were ordered. Extras were ordered on two occasions, one being after the contract completion date, and on neither occasion was reference made to an extension of time beyond the contract date. In June, some months after the date originally fixed for completion, the engineer purported to deduct penalties indicating that the time for completion had been extended from 15 November to a date in April. Stout CJ said:

> This is not a case in which the engineer is appointed an arbitrator, as in many building cases, to decide on amounts due at the end of contract. Were it such a case, he might, at the completion of the contract, have had power to determine the amount due by either party to the other. Here he has a function to perform – namely, to grant such an extension of time as he shall think adequate if 'alterations, deviations, additions, or extra work' are ordered. It must, I think, be admitted that until 20 June, when the time was extended – if the words I have quoted can be deemed an extension – the penalties were at large. No time for finishing the contract was then in existence. The 15 November was no longer the date, and no date had been substituted for the 15 November. I do not think that once the penalties were at large they could be reimposed by the engineer. He was, in fact, creating on the 20 June a new contract not in existence, and antedating it back to the first week in May. I can find no authority for his so doing. The words 'as he shall think adequate' seem to me

to imply that the moment to fix the extended time is when the extra works have not been done but have to be done, and that therefore it was before the contractors began to execute the extra works that the time had to be fixed.

Williams J said:

> If no date is specified within which the works are to be completed, how is it possible for the contractor to complete the works by a specified date? Or how can he have broken a contract to complete on a specified date if he did not know beforehand what the date was on which he was under an obligation to complete? A proviso which was intended to preserve to the contractee the right to recover penalties in an event which, had it not been for the proviso, would have deprived him of that right, should be expressed in clear and unambiguous terms. If it had been intended to allow the engineer to decide *ex post facto* whether there was a breach of contract to complete, the contractor never having been made aware of when he was required to complete, it should have been very plainly stated. The language of the proviso seems to be altogether unadapted to carry out such an intention.

In Anderson the contract provided for completion on 15 November:

> Provided nevertheless that in the event of any alterations deviations additions or extra work being required the Engineer shall allow such an extension of time as he shall think adequate in consequence thereof and any sum to become payable by way of damages for non-completion as aforesaid shall be computed from the expiration of such extended time instead of from the day hereinbefore mentioned for the completion of the works.

Anderson's case was referred to by du Parcq J in *Miller* v. *London County Council* (1934). There the engineer granted an extension of time in November 1932, the works having been completed in July of that year, the completion date thereby being extended from 15 November 1931 to 7 February 1932. The proviso relating to extensions of time in that particular contract read:

> Provided always that if by reason of additional work or enlargements of the work or for any other just cause arising with the council or the engineer or in consequence of any strikes or combination of workmen or for want or deficiency of any orders, drawings or directions, or by reason of any difficulties, impediments, obstructions, oppositions, doubts, disputes or differences whatsoever, and howsoever occasioned the contractor shall in the opinion of the engineer have been unduly delayed or impeded in the completion of the work or any part of it, it shall be lawful for the engineer, if he shall think fit, to grant from time to time, and at any time or times by writing under his hand such extension of time for completion of the work, and that either prospectively or retrospectively, and to assign such other time or times for completion as to him may seem reasonable without thereby prejudicing or in any manner affecting the validity of the contract.

Du Parcq J held that the words 'either prospectively or retrospectively' did not confer on the engineer a right to fix the extension of time *ex post facto* when the work was completed, but merely empowered him to wait till the cause of delay had ceased to operate, and then 'retrospectively' with regard to the cause of delay to assign to the contractor a new date to work to. In the result he concluded that the extension not having been granted in time no liquidated damages were payable. It appears that the cause of the delay in completion in that case was extra work ordered by the engineer.

Miller v. *London County Council* was applied and followed by Burt J in *MacMahon Construction Pty Ltd* v. *Crestwood Estates* (1971). The particular contract under review contained two clauses relating to extensions of time, the one applicable to delays from such causes as bad weather, strikes or shortage of materials, and the other where variations in the work had been ordered by the engineer. The latter read:

> Clause 40.2: Modification of Obligations and Guarantees. If, in the opinion of the contractor, any variation under clause 40.1 prevents him from fulfilling any of his obligations or guarantees under the contract, he shall notify the engineer in writing as soon as possible in what manner his obligations or guarantees may be affected. The engineer shall notify the contractor forthwith whether he confirms or withdraws the order for variation.
>
> If the engineer confirms the order for variation, the contractor's obligations and guarantees shall be varied in so far as the variation under clause 40 shall in the opinion of the engineer require.

The contract completion date was 3 December 1969. The engineer ordered a number of variations and the contractor wrote some seven letters requesting extensions for each variation, the last being dated 3 December 1969. It was not until 17 April 1970 that the engineer replied and on 23 April he purported to grant extensions. The work was completed in fact on 29 May 1970. Burt J said:

> An 'extended time', should it exist, must be the product of the proper exercise of a power appropriate to the circumstances to be found in the contract and by a 'proper exercise' I mean that if, upon the proper construction of the power to extend, it should appear that the power must be exercised within a period of time either fixed or reasonable, then a purported exercise outside that time is ineffective and there then being no date from which liquidated damages can run, the building owner loses the benefit of that provision: see [*Hudson's Building Contracts*], 6th edition page 359, and *Miller* v. *London County Council* (1934) per du Parcq J.

And further:

> The plaintiff, on the other hand, submits that the power is to be found within clause 40 and more specifically within clause 40.2. The submission is that if the impact of a variation order upon the contractor's time schedule for the carrying out of the works is such that the contractor is unable to complete 'the works' as that term is defined so as to include the variations within the stipulated time, then in the terms of clause 40.2 the 'variation . . . prevents him from fulfilling . . . his obligations . . . under the contract'. This submission I accept. The condition controlling the operation of the clause can, in this way, be satisfied and it was so satisfied and the engineer was notified. Confirmation of the order is, I think, a necessary implication from the facts. But no opinion was expressed by the engineer that the contractor's obligations should be varied by the time for completion being extended, and that power not having been exercised at that time cannot, in my opinion, be exercised later.
>
> In the result there is no 'extended time' for the purposes of 40.1 and, therefore, no liquidated damages can be recovered.

His lordship then referred to *Amalgamated Building Contractors Ltd* v. *Waltham Holy Cross UDC* (1952), and cited the passage from the judgment of Denning LJ quoted on p. 338, and continued:

> In the instant case it was competent for the engineer to grant extensions for 'extra or additional work of any kind or other special circumstances' (clause 11.4 general conditions); and 'for delays caused by other contractors on the site, delays in installation of services or exceptional circumstances' (addendum to special conditions). The delay with which we are concerned (for which it is said the granted extensions were retrospective and therefore invalid), arose from delays caused by other contractors, namely the sewer contractor. We are therefore not dealing with a cause of delay attributable to Fernbrook. It appears that one of Lord Denning's reasons for distinguishing the *Miller* case was that in *Miller* the delay stemmed from an act of the owner, whereas in the *Amalgamated Building* case it did not.
>
> In *Hudson's Bulding and Engineering Contracts* (10th edn, 1970) pages 644–645 the three possible interpretations of extension of time clauses are dealt with as follows:
>
>> In the first place, the contract may contemplate that the power should be exercised at once upon the occurrence of the event causing delay. This construction may be appropriate to non-continuing causes of delays, such as the ordering of extras. Secondly, the contract may contemplate that the power should be exercised when the full effect upon the contract programme is known. This is appropriate to continuing causes of delay, such as strikes, withholding of the site, and so on, or to cases, like some extras, where precise estimation is difficult or impossible. Or, thirdly, the contract may contemplate exercise of the power at any time before issue of the final certificate. Since the case of *ABC Ltd* v. *Waltham Holy Cross UDC* (1952) a decision on the then current RIBA form of contract, which distinguished *Miller* v. *LCC* (1934) on somewhat slender grounds, it is suggested that this latter interpretation will normally prevail in the absence of clear language to the contrary, particularly as the ambit of most modern extension of time clauses usually comprehends delays due to causes of many different kinds.

The above passage from *Hudson* was considered by Casey J in *New Zealand Structures & Investments Ltd* v. *McKenzie* (1979) where he said:

> In reply to the defendants' submissions that extensions could not be given after completion date, Mr Hicks referred me to page 644 of *Hudson*, where the cases cited are reviewed. I agree that the defendants have totally misunderstood the comments at the top of page 645, in claiming they supported the view expressed in *Miller* v. *London County Council* (1934) – namely, that the exercise of the power after completion was too late under that contract. On the contrary, *Hudson*'s comments are clearly intended to support the last of three possible interpretations of extension clauses – that the contract may contemplate the exercise of the power at any time before issue of the final certificate (in this case the maintenance certificate of 10 June 1977). That was the effect of the judgment of the Court of Appeal in England in *Amalgamated Building Contractors Ltd* v. *Waltham Holy Cross UDC* (1952) which specifically departed from the reasoning in those earlier cases to the contrary, where they suggest that the time had to be fixed before completion date so that the contractor could know his new date in time to achieve it. (This was also the approach taken by Stout CJ in the Court of Appeal in *Anderson* v. *Tuapeka County Council* (1900).) In modern building practice such clauses are not generally used for this purpose; their

> principal function is to enable a date to be fixed for the calculation of damages for delay – liquidated or otherwise. As this case so clearly demonstrates, in a major contract it is virtually impossible to gauge the effect of any one cause of delay while it is still proceeding, let alone assess the consequences of concurrent or overlapping causes. Finally, any need to have a prompt decision loses some force as a factor in interpreting such a clause, when one considers the normal review and arbitration procedures, which can take a decision well beyond the final certificate date. In my view of this contract, adopting the reasoning in the *Amalgamated Building* case, and bearing in mind modern building practice, the engineer may grant the extension at any time up to the stage when he becomes *functus officio*, and he is not limited by the completion date.

The provisions of the contract in the instant case are not appropriate to the fixing of a new date of completion *ex post facto* regardless of the cause of the delay. The extension clauses themselves could hardly be said to contain, in *Hudson*'s words, 'clear language to the contrary' so that the decision in *Amalgamated Building* did not apply, but clause 11.5.1 of the general conditions obviously contemplates extensions being given, at least for some causes of delay, before the issue of the final certificate. Clause 11.5.1 provides:

> Should the rate of progress of the works or any part thereof be at any time in the opinion of the engineer too slow to ensure completion of the works by the prescribed time or extended time for completion the engineer shall so notify the contractor in writing and the contractor shall thereupon take such steps as he may think necessary and as the engineer may approve in order to expedite progress so that the works may be completed by the prescribed time or any extended time.

It is significant I think that in the *Anderson*, *Miller* and *MacMahon* cases the sole cause of the delay requiring an extension appeared to be the ordering of extra work, where there could be no justification for delaying a decision as to the appropriate extension, whereas in the *Amalgamated Building* case the causes of the delays did not rest with the employer and were of a nature which made the duration of the delay uncertain.

In my opinion no one rule of construction to cover all circumstances can be postulated and the best that can be said on the present state of the authorities is that whether the completion date is set at large by a delay in granting an extension must depend upon the particular circumstances pertaining.

I think it must be implicit in the normal extension clause that the contractor is to be informed of his new completion date as soon as is reasonably practicable. If the sole cause is the ordering of extra work then in the normal course the extension should be given at the time of ordering so that the contractor has a target for which to aim. Where the cause of delay lies beyond the employer, and particularly where its duration is uncertain then the extension order may be delayed, although even there it would be a reasonable inference to draw from the ordinary extension clause that the extension should be given a reasonable time after the factors which will govern the exercise of the engineer's discretion have been established. Where there are multiple causes of delay there may be no alternative but to leave the final decision until just before the issue of the final certificate.

Applying that approach to the present case I think the exercise of the engineer's discretion to extend time after the completion date was reasonable in the circumstances and I therefore agree with the arbitrator's conclusion that the extensions were validly given, and that the completion date was not set at large by that circumstance.

I therefore reject [counsel for T's] first submission.

His second submission was that the contract completion time had been set at large, and the liquidated damages clause invalidated, by Fernbrook's breach of the contract terms. The breach relied upon was Fernbrook's delay of 23 weeks (as found by the arbitrator) in meeting progress payments, and allied to that the engineer's admitted failure to issue progress payment certificates in five separate months although that failure was at a much later date in the contract. Clause 18(c) of the arbitrator's memorandum . . . refers to Taggart's difficulties because of delayed payments. He was in effect deprived of what has been referred to as the life blood of commerce, an expression which has particular significance in the contracting world. Another consequence was that the contract extended into the winter months and clause 18(b) of the arbitrator's memorandum deals with that aspect.

[Fernbrook] argued that any delays that might have been occasioned by delayed payments, or failure to issue progress certificates, could have been met by extensions of time granted by the engineer, but as Taggart had not applied for them it was not open to the arbitrator to review the matter.

On this score I am satisfied that the arbitrator fell into error on the law, and that the authorities are against Taggart's submission.

The general conditions provide for extensions of time upon the happening of certain specified events 'or other special circumstances of any kind whatsoever', and the addendum to the special conditions refers to delays caused by other contractors, or in the installation of services, 'or exceptional circumstances'.

In *Perini Pacific Ltd* v. *Greater Vancouver Sewerage and Drainage District* (1966), the British Columbia Court of Appeal held that where a building contract contains a clause providing that the owner may extend the time for completion upon application by the builder, by reason of extras or delays occasioned by strikes, lockouts, *force majeure* 'or other cause beyond the control of the contractor', the concluding words must be construed narrowly and with reference to the preceding specific causes of delay and so as not to include defaults of the owner which would unreasonably result in making him the judge of the extent of his own default.

In that case the employer, in default under the implied terms of the contract, had delivered defective machinery for incorporation in the sewage disposal plant being erected by the contractor. The necessary repairs to the machinery delayed the contractor's completion date. The court held that the contractor was thereby left at large as to the date of completion and was released of his liability to pay liquidated damages even though the contractor's own tardiness may have caused some of the delay. Davey JA said:

> The contract was an entire one providing for the construction of a number of units making up the plant, and the plaintiff had to finish all parts of the work by the due date in order to perform the contract in due time. Even if the plaintiff had not been otherwise in default, the repairs to the engines would have prevented the plaintiff from completing the power house by 10 January 1963, and so fulfilling its contract. In my opinion the defendant's default, although preventing the completion by the due date of only one unit of several covered by the entire contract, disentitles the defendant to liquidated damages for the plaintiff's failure to complete the whole plant by the due date: *Holme* v. *Guppy* (1838); *Roberts* v. *Bury Improvement Com'rs* (1870); *Dodd* v. *Churton* (1897); *Russell* v. *Viscount Sa da Bandeira* (1862). Some of those cases speak of the owner's having made it impossible for the contractor to complete, or having prevented completion by the due date. Other language in those cases, sometimes by the same judge in the same case, speaks of the owner's having delayed completion as the thing that exonerates the contractor. In *Holme* v. *Guppy*, Parke B said the contractor was thereby left at large as to the date of completion.

> As I understand those authorities, neither the fact that the work would not have been completed in time if there had been no delay by the defendant, nor the fact that the plaintiff's tardiness or wrongful acts caused some of the delay, prevents the plaintiff being released of its liability for the liquidated damages through the defendant's substantial defaults that delayed it in +2completing the contract on time.

And Bull JA said:

> The case in England of *Wells* v. *Army & Navy Co-operative Society Ltd* (1902) has, I suggest, considerable value with respect to the matter in discussion. In this case, which is singularly similar to the one at bar, an owner claimed liquidated damages for a contractor's failure to construct certain buildings by a date which, under the contract procedures, had been extended by three months by the owner by reason of certain delays. The contractor failed to complete by the extended time claiming that it was prevented from so doing by the defaults of the owner. It was found that material delays were caused by the action or fault of the owner with respect to part of the project, but that there were other delays, including those caused by the contractor, present. Wright J, the trial Judge, said with reference to various causes of delay.
>
> > It is very difficult to determine how far any particular defaults of this kind on the part of the defendants would entitle the plaintiffs to relief from penalties, especially when, as in this case, there were other and more important causes of delay which would not be grounds of relief in this action; but on the whole I think that the delays in giving details not merely contributed to the delay of completion, but were such as even in the absence of the other causes of delay would have prevented completion in due time, and in my view have to a great extent increased the delay of completion.
>
> and again:
>
> > On the whole, I think that the conclusion must be that the defaults of the defendants were such that in their cumulative effect they were inconsistent with their claim to insist on completion within the stipulated time. The defaults were, in my opinion, sufficiently substantial to cast upon the defendants the burden of showing that the defaults did not excuse the delay. It is true that, apart from their defaults, the plaintiffs had, by the default of subcontractors, been delayed to an extent which might of itself have involved them in penalties, but in the absence of the further defaults by the defendants, it is impossible to say to what extent the liability to penalties might not have been reduced.
>
> On appeal, where the decision was affirmed, Vaughan Williams LJ with whom the other Lord Justices expressed agreement, dealt specifically with the matter here concerned after discussing the submission of counsel for the owners that despite the fact that the owners had delayed the work it could not be said that that prevented the completion in time because a contributory cause was the fault of the contractor, and said:
>
> > In law, I wholly deny the proposition Mr Bray put forward which was this really in effect: 'Never mind how much delay there may be caused by the conduct of the building owner, the builder will not be relieved from penalties if he too has been guilty of delay in the execution of the works.' I do not accept that proposition in law.

Accordingly, it is my respectful view that it can be fairly said that the respondent acted in a manner in breach of its contract that actually prevented or made it impossible for the appellant to perform its obligation at the proper time. To hold that despite this absolute prevention, the appellant under the circumstances could not have so performed, would, I suggest with respect, have the effect of permitting the respondent to take full advantage of its own wrong, or, to put it as Mathew J did in *Wells* v. *Army & Navy Co-operative Society*, 'the defaults of the defendants were such that they were inconsistent with their claim to insist upon completion within the stipulated time'.

The present case is not one where it was a question of whether the engineer should have granted an extension of time for his employer's breach of the contract as the arbitrator found (although in fairness to the arbitrator I should say that the *Perini* case was not cited to him) for in my opinion there was no jurisdiction in the engineer so to do, but rather a question of whether Fernbrook's breach relieved Taggart of liability under the liquidated damages clause. On the authority of *Perini* I conclude that there was no jurisdiction in the engineer to extend time on account of Fernbrook's breach, and that that breach set the completion date at large with effect that Taggart was not liable to pay liquidated damages in the sum of $2500, as found by the arbitrator, nor in any other sum.

Interpretation of provisions

Although in general liquidated damages and extension of time clauses in standard form contracts are to be construed strictly contra proferentem, ***this does not apply to negotiated contracts such as the JCT forms.***

Tersons Ltd v. Stevenage Development Corporation

COURT OF APPEAL

(1963) 5 BLR 54

In a dispute arising out of the ICE conditions (2nd edition), it was argued that since the conditions had been included in the invitation to tender sent by SDC to the contractors, they should be interpreted *contra proferentem* the employer and in favour of the contractors. The court rejected this argument.

PEARSON LJ: [Counsel] has contended that the maxim *verba fortius accipiuntur contra proferentem* should be applied in this case in favour of the contractor against the corporation on the ground that the general conditions were included in the invitation to tender sent by the corporation to the contractor. In my view, the maxim has little, if any, application in this case. The general conditions are not a partisan document or an 'imposed standard contract' as that phrase is sometimes used. It was not drawn up by one party in its own interests and imposed on the other party. It is a general form, evidently in common use, and prepared and revised jointly by several representative bodies including the Federation of Civil Engineering Contractors. It would naturally be incorporated in a contract of this kind, and should have the same meaning whether the one party or the other happens to have made the first mention of it in the negotiations.

The relationship between the JCT extension of time and direct loss and/or expense provisions (JCT 98, clauses 25 and 26; JCT 63, clauses 23 and 24) is illustrated and explained by the following case.

Henry Boot Construction Ltd v. Central Lancashire New Town Development Corporation

QUEEN'S BENCH DIVISION

(1980) 15 BLR 1

The facts are not relevant for present purposes.

JUDGE EDGAR FAY QC: Part of the contractual framework is relevant . . . The contract provided for liquidated damages in case of the completion date not being met. In the case of such delay the contractors were to pay by way of liquidated damages the sum, among other things, of £7 per week per dwelling. That is a factor to be borne in mind in considering the background of this case. Now if there is delay in the carrying out of a construction contract there is of course loss. There is in the first instance loss to both parties. To the employer, the owner, there is loss of the return upon his investment. The day when he starts getting a return by way of rent from his property is postponed and he may well have an extended period of expenditure upon supervision and the like. Equally, there is loss upon the contractor, owing to the prolongation of the period for which he has to supply matters falling within overheads as well as other expenditure, and indeed possibly idle time as well. If the delay is the fault of the contractor, then the damages for delay, the liquidated provision which I have referred to, comes into operation. He is not entitled to any relief and he has to pay the damages at £7 per week per dwelling in this instance. But if the delay is not the fault of the contractor, or he alleges that it is not, then under the terms of the contract there are two provisions which enable him to obtain relief from the loss which he otherwise would suffer.

Those provisions are condition 23 and condition 24 . . .

The broad scheme of these provisions is plain. There are cases where the loss should be shared, and there are cases where it should be wholly borne by the employer. There are also those cases which do not fall within either of these conditions and which are the fault of the contractor, where the loss of both parties is wholly borne by the contractor. But in the cases where the fault is not that of the contractor the scheme clearly is that in certain cases the loss is to be shared: the loss lies where it falls. But in other cases the employer has to compensate the contractor in respect of the delay, and that category, where the employer has to compensate the contractor, should, one would think, clearly be composed of cases where there is fault upon the employer or fault for which the employer can be said to bear some responsibility.

The scheme can be illustrated by provisions which are not in issue in this case. Under condition 23 the first two of the paragraphs which invoke the condition are (a) *force majeure*, or (b) by reason of any exceptionally inclement weather. If the delay is caused by those matters then under condition 23 the contractor is relieved of his penalty payment by reason of not meeting the completion date, but he gets no compensation himself. The loss, in other words, lies where it falls, as one would expect from something over which neither party has any control, such as *force majeure* or exceptionally inclement weather.

Those provisions are not repeated in condition 24. There is no question in those cases of any compensation being paid by the employer to the contractor.

PART III

DISCHARGE OF THE CONTRACT

Part III is concerned with cases where the contractual obligations of the parties come to an end. In some cases the contract itself is brought to an end; in other cases one of the parties is relieved from further performance.

Powell-Smith and Furmston's Building Contract Casebook, Fifth Edition. Michael Furmston.

Chapter 15
Release

The parties may agree to release each other from further performance. This presents no problem where the contract is still partly (or wholly) unperformed on both sides. More difficulty arises where one party has wholly performed and it is alleged that there is agreement that the other may not perform in full. The most obvious example in a building contract would be for the builder to have completed the work and then agreed to accept less than full payment. There is no problem if there is a genuine dispute as to whether the work is up to contract standard.

If the contractor agrees to accept part payment and to release the employer from payment of the balance, the agreement will not be binding unless at least the contractor accepts it voluntarily and (perhaps) unless the agreement is under seal or supported by fresh consideration.

D. & C. Builders Ltd v. Rees

COURT OF APPEAL

[1966] 2 QB 617

The plaintiffs were a small firm of jobbing builders who had done work for the defendant to the undisputed value of £482.13s.1d. Six months later, the money not having been paid and the plaintiffs being desperate for cash, they were offered £300 in full settlement, being told in effect that if they did not accept it, they would get nothing. The plaintiffs accepted the cheque and issued a receipt stating that it was 'in completion of the account'. They then sued for the balance.

Held: The action succeeded.

LORD DENNING MR: [The principle of equitable estoppel] has been applied to cases where a creditor agrees to accept a lesser sum in discharge of a greater. So much so that we can now say that, when a creditor and a debtor enter upon a course of negotiation, which leads the debtor to suppose that, on payment of the lesser sum, the creditor will not enforce payment of the balance, and on the faith thereof the debtor pays the lesser sum and the creditor accepts it as satisfaction: then the creditor will not be allowed to enforce payment of the balance when

Powell-Smith and Furmston's Building Contract Casebook, Fifth Edition. Michael Furmston.

it would be inequitable to do so. This was well illustrated during the last war. Tenants went away to escape the bombs and left their houses unoccupied. The landlords accepted a reduced rent for the time they were empty. It was held that the landlords could not afterwards turn round and sue for the balance: see *Central London Property Trust Ltd* v. *High Trees House Ltd* (1947). This caused at the time some eyebrows to be raised in high places. But they have been lowered since. The solution was so obviously just that no one could well gainsay it.

In applying this principle, however, we must note the qualification: The creditor is only barred from his legal rights when it would be inequitable for him to insist upon them. Where there has been a true accord, under which the creditor voluntarily agrees to accept a lesser sum in satisfaction, and the debtor acts upon that accord by paying the lesser sum and the creditor accepts it, then it is inequitable for the creditor afterwards to insist on the balance. But he is not bound unless there has been truly an accord between them.

In the present case, on the facts as found by the judge, it seems to me that there was no true accord. The debtor's wife held the creditor to ransom. The creditor was in need of money to meet his own commitments, and she knew it. When the creditor asked for payment of the £480 due to him, she said to him in effect: 'We cannot pay you the £480. But we will pay you £300 if you will accept it in settlement. If you do not accept it on those terms, you will get nothing. £300 is better than nothing.' She had no right to say any such thing. She could properly have said: 'We cannot pay you more than £300. Please accept it on account.' But she had no right to insist on his taking it in settlement. When she said: 'We will pay you nothing unless you accept £300 in settlement,' she was putting undue pressure on the creditor. She was making a threat to break the contract (by paying nothing) and she was doing it so as to compel the creditor to do what he was unwilling to do (to accept £300 in settlement): and she succeeded. He complied with her demand. That was on recent authority a case of intimidation: see *Rookes* v. *Barnard* (1964) and *J.T. Stratford & Son Ltd* v. *Lindley* (1964). In these circumstances there was no true accord so as to found a defence of accord and satisfaction: see *Day* v. *McLea* (1889). There is also no equity in the defendant to warrant any departure from the due course of law. No person can insist on a settlement procured by intimidation.

In my opinion there is no reason in law or equity why the creditor should not enforce the full amount of the debt due to him. I would, therefore, dismiss this appeal.

WINN LJ: In my judgment it is an essential element of a valid accord and satisfaction that the agreement which constitutes the accord should itself be binding in law, and I do not think that any such agreement can be so binding unless it is either made under seal or supported by consideration. Satisfaction, viz. performance, of an agreement of accord, does not provide retroactive validity to the accord, but depends for its effect upon the legal validity of the accord as a binding contract at the time when it is made: this I think is apparent when it is remembered that, albeit rarely, existing obligations of debt may be replaced effectively by a contractually binding substitution of a new obligation.

The previous law was probably substantially modified by the next case although the Court of Appeal did not explicitly recognise that they were changing the law. In practice they stretched the traditional doctrine of consideration so far that an employer who agrees to pay a contractor (or a contractor a sub-contractor as in this case) more for completing the works will nearly always be liable to pay unless the party receiving extra payments has been guilty of economic duress.

Williams v. Roffey Bros & Nicholls (Contractors) Ltd

COURT OF APPEAL

[1990] 1 All ER 512

The defendants, a firm of building contractors, had entered into a contract to refurbish a block of 27 flats. They sub-contracted the carpentry work to the plaintiff. Halfway through the contract the plaintiff found that he was in financial difficulties. This was partly because he had contracted at too low a price and partly because he had failed to supervise his men properly so that they worked economically. The defendants were anxious that the plaintiff would keep going and finish the work, since if he had stopped and they had to engage another carpentry contractor there would have been a substantial risk that they would be late on the contract and would have to pay liquidated damages. The defendants therefore agreed with the plaintiff that they would pay the plaintiff an extra £10,300 at the rate of £575 per flat on completion. The plaintiff carried on work and completed eight further flats after this modification of the original agreement. When he sued for payment the defendants argued that the second agreement was not supported by any consideration and accordingly they were not liable.

Held: That the defendants were liable to pay the extra money promised.

> GLIDEWELL LJ: Accordingly, following the view of the majority in *Ward* v. *Byham* and of the whole court in *Williams* v. *Williams* and that of the Privy Council in *Pao On* v. *Lau Yiu* the present state of the law on this subject can be expressed in the following proposition: (i) if A has entered into a contract with B to do work for, or to supply goods or services to, B in return for payment by B and (ii) at some stage before A has completely performed his obligations under the contract B has reason to doubt whether A will, or will be able to, complete his side of the bargain and (iii) B thereupon promises A an additional payment in return for A's promise to perform his contractual obligations on time and (iv) as a result of giving his promise B obtains in practice a benefit, or obviates a disbenefit, and (v) B's promise is not given as a result of economic duress or fraud on the part of A, then (vi) the benefit to B is capable of being consideration for B's promise, so that the promise will be legally binding . . .

Chapter 16
Frustration

The obligations of the parties may be brought to an end by the doctrine of frustration. Originally this doctrine applied to unexpected events which made continued performance of the contract impossible. An early example is Appleby v. Myers, p. 137, where a factory was destroyed by fire. Later cases have applied the doctrine to developments which, without making performance strictly impossible, have made it fundamentally different from that contemplated at the time the contract was made.

Metropolitan Water Board v. Dick, Kerr & Co Ltd

HOUSE OF LORDS
[1918] AC 119

In July 1914, the respondents contracted to build a reservoir for the appellants within six years. The contract gave the engineer typical powers to award extra time. In February 1916 the Ministry of Munitions, acting under wartime powers, ordered the respondents to cease work, which they did. Granted the possibility of extensions of time it was not impossible to perform, but the House of Lords held that a contract resumed after the wartime interruption would be fundamentally different from that before the interruption.

LORD DUNEDIN: On the whole matter I think that the action of the government, which is forced on the contractor as a *vis major*, has by its consequences made the contract, if resumed, a work under different conditions from those of the work when interrupted. I have already pointed out the effect as to the plant, and, the contract being a measure and value contract, the whole range of prices might be different. It would in my judgment amount, if resumed, to a new contract; and as the respondents are only bound to carry out the old contract and cannot do so owing to supervenient legislation, they are entitled to succeed in their defence to this action.

Powell-Smith and Furmston's Building Contract Casebook, Fifth Edition. Michael Furmston.

However, it is not sufficient to show that the contract has turned out more difficult and expensive for one party to perform than he expected.

Davis Contractors Ltd v. Fareham Urban District Council

HOUSE OF LORDS

[1956] AC 696

The plaintiffs contracted to build 78 council houses for the authority within a period of eight months and for a fixed price. Through no fault of either party, there was a scarcity of skilled labour and the work took 22 months to complete. The plaintiffs argued that because of the shortage of labour, the contract had been brought to an end by frustration, and that they were entitled to recover a sum well in excess of the fixed price on the basis of a *quantum meruit* (see p. 162).

Held: The claim must fail. The contract had not been frustated and the plaintiffs had not been released from its terms as regards price.

> LORD RADCLIFFE: Lord Loreburn ascribes the dissolution to an implied term of the contract that was actually made. This approach is in line with the tendency of English courts to refer all the consequences of a contract to the will of those who made it. But there is something of a logical difficulty in seeing how the parties could even impliedly have provided for something which *ex hypothesi* they neither expected nor foresaw; and the ascription of frustration to an implied term of the contract has been criticised as obscuring the true action of the court which consists in applying an objective rule of the law of contract to the contractual obligations that the parties have imposed upon themselves. So long as each theory produces the same result as the other, as normally it does, it matters little which theory is avowed (see *British Movietonews Ltd* v. *London and District Cinemas Ltd* (1951), *per* Viscount Simon). But it may still be of some importance to recall that, if the matter is to be approached by way of implied term, the solution of any particular case is not to be found by inquiring what the parties themselves would have agreed on had they been, as they were not, forewarned. It is not merely that no one can answer that hypothetical question: it is also that the decision must be given 'irrespective of the individuals concerned, their temperaments and failings, their interest and circumstances' . . . The legal effect of frustration 'does not depend on their intention or their opinions, or even knowledge, as to the event'. On the contrary, it seems that when the event occurs 'the meaning of the contract must be taken to be, not what the parties did intend (for they had neither thought nor intention regarding it), but that which the parties, as fair and reasonable men, would presumably have agreed upon if, having such possibility in view, they had made express provision as to their several rights and liabilities in the event of its occurrence' (*Dahl* v. *Nelson* (1881), *per* Lord Watson).
>
> By this time it might seem that the parties themselves have become so far disembodied spirits that their actual persons should be allowed to rest in peace. In their place there rises the figure of the fair and reasonable man. And the spokesman of the fair and reasonable man, who represents after all no more than the anthropomorphic conception of justice, is and must be the court itself. So perhaps it would be simpler to say at the outset that frustration occurs whenever the law recognizes that without default of either party a contractual obligation has become incapable of being performed because the circumstances in which performance is called for would render it a thing radically different from that which was undertaken by the contract. *Non haec in foedera veni.* It was not this that I promised to do . . .

I am bound to say that, if this is the law, the appellants' case seems to me a long way from a case of frustration. Here is a building contract entered into by a housing authority and a big firm of contractors in all the uncertainties of the post-war world. Work has begun shortly before the formal contract was executed and continued, with impediments and minor stoppages but without actual interruption, until the 78 houses contracted for had all been built. After the work had been in progress for a time the appellants raised the claim, which they repeated more than once, that they ought to be paid a larger sum for their work than the contract allowed; but the respondents refused to admit the claim and, so far as appears, no conclusive action was taken by either side which would make the conduct of one or the other a determining element in the case.

Whether a particular event is a frustrating event must depend not only on the event but also on the precise terms of the contract.

Wong Lai Ying v. Chinachem Investment Co Ltd

PRIVY COUNCIL

(1979) 13 BLR 81

The respondents, who were building two blocks of flats in Hong Kong, entered into 24 contracts to sell flats as yet unbuilt to the appellants. The contracts were entered into between March and November 1971. Work began in December 1971, the contractual date for completion was 17 May 1973, and time was made of the essence but with provision in certain circumstances for extension for not more than one year.

In June 1972, part of the hillside above the building site slipped, taking with it a 13-storey block of flats, the debris of which landed on the site, obliterating the building works already completed. Work stopped and, since it could not be recommenced within three months, the respondents' building permit came to an end. A new permit was not issued until November 1975.

These facts would appear to present a classic example of frustration but the appellants argued that the contract was not frustrated (and that they were therefore entitled to specific performance) because of clause 22 of the contract which provided:

> It is further agreed that notwithstanding anything herein contained . . . should any unforeseen circumstances beyond the vendor's control arise whereby the vendor becomes unable to sell the said undivided share and apartment to the purchaser as hereinbefore provided, the vendor shall be at liberty to rescind the agreement forthwith and to refund to the purchaser all instalments of purchase price paid by the purchaser hereunder without interest or compensation . . .

Held: Clause 22 should not be read as applying to the kind of unforeseen disaster which had happened. The contract was frustrated.

LORD SCARMAN: It was, however, urged that the language of clause 22 is wide enough to cover the event which happened. So it is. But the question is whether the general words of the clause are sufficient to support the inference that the parties must be presumed to have made provision for the event. In answering the question, the concurrent findings of fact by four judges, all of whom would be well aware of conditions in Hong Kong, must be respected by the board. The event was, admittedly, an 'unforeseen natural disaster', and [both the trial judge and the

Court of Appeal] spelt out its consequences for the contract . . . All of them were prepared to characterise it as a 'frustrating event': they differed only in their construction of clause 22.

The unforeseen character of the event is not in dispute. Clause 22 cannot, in their Lordships' opinion, be construed as making provision for the possibility of this particular unforeseen contingency. The clause, coming at the end of the contract, replete with specific provisions and time limits, was plainly intended to confer upon the vendor a remedy of rescission if a dispute arose or it became clear he could not complete in accordance with the contract, provided he acted 'forthwith' to terminate the contract. It does not follow from the provision of a summary remedy avoiding litigation in such circumstances that the parties must have agreed that their contract would continue after an unforeseen natural disaster having the consequences analysed and assessed by the judges below. The Board agrees with the view expressed by Huggins J in the Court of Appeal that:

> the inclusion of a clause such as clause 31 in the Bank Line charter-party [*Bank Line* v. *Capel* (1919)] or clause 22 of the agreement in this case is . . . not inconsistent with the operation of the doctrine of frustration and does not show an intention that the doctrine shall not apply

Strictly, of course, the issue is not whether the doctrine of frustration is excluded but whether provision was made for an event causing the circumstances of performance to be radically different from that undertaken by the contract. The word 'forthwith' is an indication, if any other than the context and circumstances of the contract is needed, that this 'frustrating event' was not contemplated; for a most significant feature of this event was that as a result of it 'the position of the parties was clouded in uncertainty' (Briggs CJ). Another indication is the bizarre consequences of holding the event covered by the clause. If the vendor should fail to act 'forthwith', and completion should become indefinitely delayed, the purchaser, exercising his right under clause 3(3), could wait as long as he pleased, collecting all the time his interest at the rate of one per cent per month. A further indication is the presence in the contract of clause 20. Requisition, if not provided for, could well be a frustrating event: indeed it is a classic instance in some circumstances of frustration. The parties by clause 20 made specific provision for it.

Codelfa Construction Proprietary Ltd v. State Rail Authority of New South Wales

HIGH COURT OF AUSTRALIA

(1982) 149 CLR 337

The appellants, a construction company, contracted with the respondents, a railway authority, for the construction of an underground railway. The contract required the work to be completed within a fixed period and the appellants started work on a three shift a day, seven days a week basis. However, the work was noisy and third parties affected by the work obtained injunctions restraining evening and Sunday work.

The appellants commenced an arbitration claiming an addition to the contract price, either on the basis that there was an implied term in the contract that if the contractor was restrained by injunction from working the planned shifts, the authority would indemnify it against the additional costs or on the basis that the contract had been frustrated by the grant of the injunction.

The arbitrator found as a fact that the parties had entered into the contract on the basis that work was to be carried out on a seven-day, three-shift basis and in the shared belief that no injunction would be granted to restrain noisy work in the evenings or at weekends.

The High Court of Australia held that no term was to be implied but (by a majority of 4–1) that the contract had been frustrated by the grant of the injunction.

MASON, J: It is not surprising that the cases commonly throw up situations of supervening impossibility caused by a change in the law – they are the more common instances of the unforeseen or unexpected occurrence. But in principle there is no reason why a mutual assumption arising from a mistaken view that an activity is immune from injunctive relief should not attract the principle of frustration. No doubt it is more difficult in such a case to show that the grant of injunctive relief was not foreseen or could not reasonably have been foreseen, but if that can be shown then the doctrine of frustration should apply. The injunction is a supervening event though it does not stem from any alteration in the law.

An unusual element in the present case is that the parties appear to have received, accepted and acted on erroneous legal advice that the contract work could not be impeded by the grant of an injunction to restrain noise or other nuisance, advice which was based on an erroneous interpretation of section 11 of the City and Suburban Electric Railways Act 1915 (NSW). One might have expected the parties and their advisers to have had reservations about the correctness of the advice and to have given consideration to the possibility that, despite the advice, an injunction might be granted. However, the findings do not reflect the existence of any reservations; indeed, they record Codelfa's acceptance of the representations made by the Authority. Codelfa is a wholly owned subsidiary of an Italian company and this may explain Codelfa's willingness to accept and act on the representation made by the Authority.

The doctrine of frustration is closely related to the concept of mutual mistake. However, in general, relief on the ground of mutual mistake is confined to mistakes of fact, not of law. If the common contractual assumption is of present fact it is a case of mutual mistake; if the assumption is of future fact it is a case of frustration (*Bell* (14), *per* Lord Atkin), the distinction being that in one case the contract is void *ab initio* and in the other it is binding until the assumption is falsified. Here the mistake is not one of present fact; it is either a mistake as to future fact or a mistake of law. Even if it be a mistake of law, this is not, I think, fatal to the application of the doctrine of frustration. The unsatisfactory distinction between a mistake of fact and one of law has not so far been carried over into frustration and I see no reason to further complicate the doctrine by invoking this distinction.

The critical issue then is whether the situation resulting from the grant of the injunction is fundamentally different from the situation contemplated by the contract on its true construction in the light of the surrounding circumstances. The contract itself did not require that the work be carried out on a three shift continuous basis six days a week without restriction as to Sundays. But it required completion of the works within 130 weeks. And Codelfa with its tender had submitted a construction programme which involved a three shift continuous basis six days a week. By clause S.6 of the specifications Codelfa was required to submit a revised programme of work to the Engineer for his determination within thirty calendar days of the issue of a notice to proceed under the contract. This Codelfa did. Again it made provision for the method of operation already mentioned. It was accepted by the Engineer.

It is in this contractual setting that the findings of the Arbitrator have special significance. The relevant findings are set out in pars. 14, 15, 16, 18 and 19 of his award.

14. The Parties to Contract ESR 1005 each entered into such Contract on the common and mutual understanding and on the basis that:

(a) the works the subject of the Contract should and would be carried out by the Contractor on a 3-shift continuous basis six days per week and without restriction as to Sundays, and

(b) the work to be performed was inherently of a noisy and disturbing nature and the work or substantial parts thereof was to be carried out in close proximity to areas of residential neighbourhood, and

(c) no Injunction or other Restraining Order could or would be granted against the Contractor in relation to noise or other nuisance arising out of the carrying on of the said works on such basis.

15. The matter mentioned in paragraph 14(c) was represented by the Principal to the Contractor and was accepted as the situation by the Contractor prior to and at the time of entering into the Contract. . . .

16. The said works could not in fact be carried out by the Contractor in accordance with methods and programmes agreed to by the parties and in acccordance with the contractual stipulations as to time of performance unless the works were carried out on the basis mentioned in paragraph 14(a) hereof. . . .

18. Restraining Orders and Injunctions were in fact issued by the Court on grounds of noise and other nuisance arising out of the carrying out of the works by the Contractor on the basis mentioned in paragraph 14(a) hereof the effect of which was (inter alia) to prohibit and preclude the works from being carried out by the Contractor on the said basis and to cause the Contractor to incur additional cost in the carrying out of the works.

19. The said works could not in fact be carried out by the Contractor in accordance with methods and programmes agreed to by the parties without substantial noise and other disturbances arising there from or in connection therewith by reason of the inherently noisy and disturbing nature of the work.

The submission of the proposed programme of work with the tender, its supersession by the revised programme pursuant to clause S.6 of the specifications, together with the very provisions of clause S.6 itself dealing with the construction programme, provide a link between the contract and the antecedent discussions so as to enable us, subject to a consideration of specific provisions in the specifications, to say that the contract contemplated that completion would be achieved within the time stipulated by the method of work already mentioned, it being assumed that it could not be disturbed by the grant of an injunction.

I reject the Authority's argument that clause S.6 is inconsistent with the notion that the contract looked to this method of work as the mode by which the work was to be completed. Certainly clause S.6(5) envisaged that a major change to the work diagram as determined by the Engineer might be required by 'revisions to the programme'. It seems that the responsibility for initiating such a change lay with Codelfa. But I do not think that this is necessarily inconsistent with Codelfa's case on frustration nor do I think that clause G.44(7) of the contract, which provides for the grant of an extension of time in case of delays 'owing to causes beyond the control or without the fault or negligence of the Contractor', covers the position. Delay due to the grant of an injunction on the ground of nuisance committed by Codelfa scarcely answers this description, even though it results from performance of the work in the only manner which will enable completion to take place within the time stipulated.

Clause S.8(2)(c) however, poses a greater obstacle. It provides:

> The operation of all plant and construction equipment shall be such that it does not cause undue noise, pollution or nuisance. This may require the use of sound insulated compressor and air tools, silencers on ventilating fans and restrictions on the working hours of plant or such other measures as approved by the Engineer. The Contractor shall not be entitled to additional payment if the Engineer requires that measures be taken to reduce noise and pollution.

Once the injunctions were granted the engineer gave notices reflecting the provisions of the injunctions, restricting the hours of work so as to prohibit work at night and thereby inhibited Codelfa from continuing with its three shift operation under the contract.

The first paragraph of clause S.8(2)(c) contains a promise by Codelfa that it will not operate plant and equipment so as to cause a nuisance. The second specifically looks to the possibility of a restriction on working hours of plant. And the third denies additional remuneration if the engineer requires measures to be taken to reduce noise and pollution.

Do these provisions support the view that Codelfa was undertaking in any event to perform the contract work, even though the method contemplated by the parties might prove to be unlawful or impossible by reason of its amounting to a nuisance and its being restrained by injunction? I do not think that clause S.8(2)(c) has such a wide-ranging effect. It involves no subtraction from the language of the provisions to say that it is quite consistent with the contemplated method of work being an essential element of the contract. Indeed, there would be no inconsistency between these provisions and an explicit provision for termination of the contract in the event that the method of work was restrained by injunction. There was plenty of scope for an exercise of the Engineer's power under the second paragraph so long as it did not displace the continuation of that method of work.

I come back then to the question whether the performance of the contract in the new situation was fundamentally different from performance in the situation contemplated by the contract. The answer must, I think, be in the affirmative. Paragraphs 14, 15, 16, 18 and 19 of the Arbitrator's award go a long way towards establishing this answer. The finding contained in para. 16 proceeds on the footing that the contract work could not be carried out as contemplated by the contract once injunctions were granted, the effect of which was to prohibit the continuous three shift a day operation six days a week. Performance by means of a two shift operation, necessitated by the grant of the injunctions, was fundamentally different from that contemplated by the contract.

There is, of course, no inconsistency between the conclusion that a term cannot be implied and the conclusion that events have occurred which have brought about a frustration of the contract. I find it impossible to imply a term because I am not satisfied that in the circumstances of this case the term sought to be implied was one which parties in that situation would necessarily have agreed upon as an appropriate provision to cover the eventuality which has arisen. On the other hand I find it much easier to come to the conclusion that the performance of the contract in the events which have occurred is radically different from performance of the contract in the circumstances which it, construed in the light of surrounding circumstances, contemplated.

Chapter 17
Illegality

Illegality does not bring the contract to an end, but it may prevent one or both parties from enforcing it. A building contract might be illegal because the building owner intended to use the building for an improper purpose; for example, as a brothel. The building owner could not then enforce the contract, nor could the builder if he knew of the improper purpose.

By far the most common practical examples arose under the building licence scheme in force during the 1939–45 war and for a period thereafter, under which contracts to carry out unlicensed work were illegal.

Bostel Brothers Ltd v. Hurlock

COURT OF APPEAL

[1949] 1 KB 74

The defendant had obtained a licence to carry out repairs to his house and engaged the plaintiffs to do the work. The plaintiffs innocently exceeded the amount of the licence without having applied for a supplementary licence. The plaintiffs claimed the balance in excess of the amount of the licence.

Held: The claim failed, as the prohibition on doing unlicensed work was absolute and did not depend on the builder's state of mind.

Note: This reasoning produces a result which seems unfair, the defendant having the benefit of the work without having to pay for it. The theoretical justification for the rule is that it is necessary to be unfair to individual builders in order to encourage builders as a class to comply with the rules. However where it is the duty of the employer to obtain the licence, it was held that he had given a collateral warranty that he would obtain the necessary licence or stop the work.

Powell-Smith and Furmston's Building Contract Casebook, Fifth Edition. Michael Furmston.

Strongman (1945) Ltd v. Sincock

COURT OF APPEAL

[1955] 2 QB 525

The defendant, an architect owner, engaged the plaintiffs to undertake building work and agreed to obtain the necessary licences. In fact, he allowed work considerably in excess of the licence figure to be done and then refused to pay for it.

Held: Although the builders could not maintain a direct action for the balance, they could sue the defendant upon his undertaking to obtain the necessary licences unless they were themselves morally to blame or culpably negligent. Since it was agreed that as between architect and builder the primary duty to obtain licences lay on the architect, it was held that they were not.

DENNING LJ: The third question is whether the plaintiffs were guilty of negligence. I can well see that if there was culpable negligence on the part of the persons seeking damages, he might not be entitled to recover. In *Askey* v. *Golden Wine Co Ltd* (1948) some merchants had been guilty of culpable negligence in not taking proper steps to see whether a liquor was safe for consumption and they were not allowed to recover. I said: 'If they were allowed to be negligent and yet recover damages, it would offer an inducement to them to turn a blind eye to contamination.' So, in any of these licensing cases, if the builder were negligent and yet were allowed to recover damages, it would be an inducement to him to turn a blind eye to the statute. That is what I had in mind in *J. Dennis & Co Ltd* v. *Munn* (1949) and the other cases to which [counsel] referred us. When a builder is doing work for a lay owner – if I may so describe him – the primary obligation is on the builder to see that there is a licence. He ought not simply to rely on the word of the lay owner. He ought to inspect the licence himself. If he does not do so, it is his own fault if he finds himself landed in an illegality. But in this case there was not a lay owner. The owner was the architect, and he himself said in evidence: 'I agree that where there is an architect it is the universal practice for the architect and not the builder to get the licence.' No fault, it seems to me, can, in these circumstances, be attributed to the builder.

It was contended before us . . . that on the facts of this case there must have been negligence and that in point of law the official referee ought to have found it. I think not. The official referee found: 'I do not consider that the plaintiffs in the present case have done an immoral act, nor were they negligent in not insisting on the production of supplementary licences. They had done a great deal of other work for the defendant without any question being raised with regard to the sufficiency of the licences.'

As I said at the beginning of this case, it comes very ill from the mouth of the defendant to raise this point as against the plaintiffs. His attitude was well shown by an observation which he made to the solicitor . . . He said: 'If the Nicholls can be bluffed they deserve to lose their money.' In other words, he was saying: 'If they were fools enough to trust in me, they ought to lose their money.' That is a very wrong attitude for a professional man, an architect, to take up. It shows quite clearly that on his own admission he has misled them and now seeks to turn it into his own advantage. In my judgment, his objection fails. On the findings of the official referee, the plaintiffs were entirely innocent people who were led into this unfortunate illegality by the representation of the architect, amounting to a collateral contract, that he would get the licences. That contract not having been fulfilled, I see no objection in point of law to the plaintiffs recovering the damages, and I think that the appeal should be dismissed.

The next case provides a modern example – perhaps all too common in practice but usually concealed from the public gaze.

Taylor v. Bhail

COURT OF APPEAL

(1995) 50 Con LR 70

The defendant was the headmaster of a school at Hounslow. The school suffered damage as a result of gales and there were other building works to be done at the school. The plaintiff, a building contractor, was engaged to do this work. There were three contracts but no live issue survived as to contracts two and three.

Contract one related to works of repair which were covered by insurance. The defendant told the plaintiff that if he inflated the price by £1000, and gave him the £1000, he would see that the plaintiff got the work. The plaintiff agreed. The trial judge held that the quality of the illegality affecting the transaction was not such that the plaintiff should not be permitted to succeed on his claim for payment. The defendant appealed.

Held: The plaintiff's claim for payment should fail, as the plaintiff could not present his case in any way which did not depend on the contract having been procured by the payment of the additional £1000. Appeal allowed.

MILLETT LJ: The question in this appeal is whether the plaintiff, who is a builder, can enforce a contract which was intended to be used to practise a fraud on the defendant's insurers. The plaintiff was aware of the intended fraud and he was a willing participant in it. The question is whether the building contract can be identified and enforced separately from the fraudulent arrangements of which it formed an integral part. In my judgment it cannot.

This was a conspiracy to defraud. It is quite unrealistic to regard it as a building contract for the sum of £12,480 with a separate and independent agreement to defraud the defendant's insurers superimposed upon it. The plaintiff was willing to do the work for £12,480. The defendant agreed to give the work to the plaintiff provided that the plaintiff would provide him with an inflated estimate for £13,480 and assist him in deceiving the insurers into believing that this was the true price for the work . . .

In many contexts it maybe important to analyse a transaction in order to determine whether it consisted of a single contract or two contracts. But illegality is a question of substance, not form. Whether the arrangements between the plaintiff and the defendant comprised a single contract or two separate contracts is, in my judgment, immaterial; they constituted a single, indivisible arrangement tainted by fraud, neither component of which was ancillary or subsidiary to the other, and neither of which is severable so as to leave the other enforceable.

It is important to bear in mind that the law refuses to enforce not only contracts which are in themselves illegal, but also contracts which are ex facie legal but which, to the knowledge of the parties, have an illegal purpose or are intended to be performed in an illegal manner. In *Pearce* v. *Brooks* (1866) LR 1 Exch 213, [1861–73] All ER Rep 102 the plaintiffs agreed to supply the defendant with a brougham on hire-purchase. The defendant was a prostitute. The jury found that the plaintiffs knew that she intended to use the brougham in the course of plying her trade. The plaintiffs were unable to recover the cost of the hire. This is a well-known example of a contract which was legal on the face of it but was unenforceable because it was intended for an illegal or immoral purpose.

In *Miller* v. *Karlinksi* (1945) 62 TLR 85 the plaintiff was employed on terms that he was to make inflated returns for expenses which in truth represented disguised remuneration. The purpose of these arrangements was to defraud the Inland Revenue. The plaintiff later brought an action against his employer for arrears of salary and travelling expenses. He was not allowed to recover. The case has frequently been followed. Employees who connive with their employers to defraud the Inland Revenue have been unable to recover compensation for unfair or wrongful dismissal or to enforce their contracts of employment in any way. In such cases the contract of employment is tainted by fraud by the illegal use for which it was intended to be put. These cases cannot be distinguished in principle from the present.

In all such cases there is an underlying, lawful contract – to supply a brougham on hire, to engage a workman at a wage, or to build a wall – and in all of them that underlying contract has been rendered unenforceable by the illegal manner in which it was intended to be performed or by the illegal use to which it was intended to be put . . .

Chapter 18
Forfeiture Clauses, Repudiation and Determination

Building contracts usually contain clauses (sometimes called forfeiture clauses) entitling one party to bring the contract to an end in certain circumstances. Independently of such clauses, contracting parties have power to determine the contract in certain circumstances under the general law of contract. In practice there is a significant area of overlap between these two possibilities.

Under the general law one party is entitled to terminate in certain circumstances if the other party has broken the contract. It is not every breach which has this result, however. Basically, the party seeking to determine must show that the other party has either committed a breach which goes to the root of the contract or has repudiated his obligations under it. These concepts are often difficult to apply in practice and, as the following cases illustrate, it sometimes happens that A commits a breach which B wrongly thinks entitles him to terminate and by taking this view commits a repudiatory breach entitling A to terminate.

Lubenham Fidelities & Investments Co Ltd v. South Pembrokeshire District Council & Wigley Fox Partnership

COURT OF APPEAL

(1986) 6 Con LR 85

There were two contracts in JCT 63 form. The original contractors went into liquidation, and the plaintiffs, who were bondsmen, elected to perform the contract themselves rather than pay on the performance bond. They engaged sub-contractors to carry out the work. Because of various building defects, Wigley Fox (the architect under the contract) wrongfully advised the council to deduct liquidated damages from interim certificates. The plaintiffs thereupon wrote terminating the contracts for alleged breach, and ordered the sub-contractors to cease work. The council thereupon served notice of determination under clause 25(1), on the ground that the plaintiffs were in default by wholly suspending the carrying out of the works without reasonable cause. Simultaneously, the plaintiffs served notices under clause 26(1) alleging that the council were in default by failing to make payment under the interim certificates; they then issued a writ.

Powell-Smith and Furmston's Building Contract Casebook, Fifth Edition. Michael Furmston.

Held: The plaintiffs had repudiated the contracts by not complying with their terms and were in breach, even though the council were wrong in deducting liquidated damages at that stage. By withdrawing the sub-contractors from site and so suspending work without reasonable cause the plaintiffs had acted contrary to clause 25(1). By serving invalid notices of termination and thereafter issuing a writ, the plaintiffs had indicated an intention not to be bound by the contracts and since the council had no alternative but to accept the repudiation, the contracts were brought to an end.

Alkok v. Grymek

SUPREME COURT OF CANADA

(1968) 67 DLR (2d) 718

The facts sufficiently emerge from the judgment.

SPENCE J: The appellant is a contractor in the City of Toronto. He entered into a contract with the respondents dated 14 May 1956, whereby he agreed to:

(a) provide all the materials and perform all the work shown on the drawings and described in the specifications entitled 'proposed residence for Mr I. Grymek' which have been signed in duplicate by both the parties and which was prepared by Edward I. Richmond, MRAIC, acting as and hereinafter entitled 'the architect' and
(b) do and fulfill everything indicated by this agreement, the specifications, and the drawings.

The contract provided that the owner would pay to the contractor $57,500 and

(b) Make payment on account thereof upon the architect's certificate (when the architect is satisfied that payments due to sub-contractors have been made), as follows:

i) Upon completion of the sub-floor $6,000.00;
ii) Upon completion of the roof $12,000.00;
iii) Upon completion of the brown coat of plaster $8,000.00;
iv) Upon completion of the white coat of plaster, including all plumbing and electrical work $12,000.00;
v) Upon completion of trim $8,000.00;
vi) the sum of eleven thousand, five hundred dollars ($11,500.00) . . .

The appellant commenced work of erecting the house in accordance with the said contract and from time to time the respondents required the addition of certain extras which the appellant added and which the Court of Appeal for Ontario found were of the value of $950. The respondents paid to the appellant the whole of the first two instalments of $6,000 and $12,000, respectively, and one-half of the third instalment, i.e. $4,000, and did so without requiring that the appellant comply with the provisions of para. (b) aforesaid by satisfying the architect that the payments due to the subcontractors had been made.

Later when the appellant pressed for the payment of the balance of the third instalment, the respondents required the appellant to comply with the provisions of the contract and to so satisfy the architect. After several conferences, the appellant failed to so satisfy the architect and in addition there were complaints from the respondents and from their architect . . . that there were defects in the construction and that the construction was delayed. The solicitor for the respondents who had drafted the original contract and who had been present at the various conferences when an attempt was made to satisfy the architect that the sub-contractors had been paid, wrote to the appellant on 10 November 1956, complaining of the progress of the work and of certain defects expressing the fear that mechanics' liens would be registered against the property and concluded with this paragraph:

> Unless all of the building infractions, which are your responsibility, have been remedied, and the work carried on at a proper pace, my clients shall have no alternative than to employ their own specific trades to complete your portion of the uncompleted work, and any moneys or expenses incurred by my clients in employing tradesmen for either work done or materials supplied shall be deducted from the contract price herein.

Six days later, on 16 November 1956, the said solicitor wrote again to the appellants in which he said:

> Further to the above matter, in confirming my conversation with you yesterday, I hereby advise you on behalf of my clients Issie Grymek and Yetta Grymek that they are terminating their contract with you as of today's date.
>
> They intend to complete the lands and premises in the manner in which they believe the work should be done.

In accordance with that letter the appellant ceased work on the contract though the architect for the respondents has testified that the appellant was willing and anxious to continue it. The respondents proceeded to complete the building themselves through the intervention of other contractors and materialmen.

It was the view of the learned Master that the respondents were entitled to terminate the contract at the time and in the fashion aforesaid.

The Court of Appeal for Ontario, however, came to the conclusion that sufficient grounds for the termination of the contract had not been established. McGillivray JA in his reasons for judgment, quoted *Anson on Contract*, 21st edition, page 424, as follows:

> The question to be answered in all these cases of incomplete performance is one of fact; the answer must depend on the terms of the contract and the circumstances of each case. The question assumes one of two forms – Does the failure of performance amount in effect to a renunciation on his part who makes default? Does it go so far to the root of the contract as to entitle the other to say, 'I have lost all that I cared to obtain under this contract; further performance cannot make good the prior default'?

That proposition needs no support by citation from judgments and I accept it as expressing the proper test which the Court must apply here. As pointed out by McGillivray JA, in his reasons, the contract as between the parties was for the contractor to build a house and for the defendants to pay for it. The contractor had proceeded with the building although not in accordance with the pace at which the owners believed he should be proceeding and had been guilty of what the Court of Appeal for Ontario has found were certain minor defects in construction. The owners having paid two instalments and part of the third were refusing to pay the balance of the third instalment or those which would become due thereafter, and in so doing were relying upon the provision of the contract which required the appellant as contractor to satisfy the respondents' architect that the payments to the sub-contractors had been made. They were entitled to require that the appellant continue his work upon the contract and to refuse to pay him until he did satisfy that provision. If the appellant had refused to proceed on that basis then, of course, he would have been in breach of the provision of the contract going to the root thereof and the respondent would have been entitled to terminate the contract. The appellant did not indicate by word or conduct an intention to so act or not to be bound in every way by the contract. The appellant, therefore, did not give to the respondents the opportunity to terminate the contract on the ground that the appellant had been in breach of a term going to the root of it. It was true that he was in breach of the term requiring him to satisfy the respondents' architect that the subcontracts had been paid, but, with respect, I agree with McGillivray JA's view that this was a mere ancillary term which could be enforced perfectly by the respondents simply refusing to make payments until they were satisfied, as indeed the respondents were refusing. I am, therefore, in accord with the view of McGillivray JA, that the respondents have not shown sufficient grounds to support their termination of the contract.

R.B. Burden Ltd v. Swansea Corporation

HOUSE OF LORDS

[1957] 3 All ER 243

The facts sufficiently appear from the speech of Lord Tucker.

LORD TUCKER: My Lords, the short point for decision in the appeal is whether, on the findings of the learned official referee, the respondent corporation 'interfered with or obstructed the issue of a certificate' within the meaning of clause 20 in a building contract made between the appellants as contractors and the respondents as employer, dated 20 December 1949, for the construction of a primary school at Swansea. Under the contract, the borough architect was appointed architect and surveyor for the purposes of the contract. He was subsequently replaced as surveyor by the firm of Messrs. O who were appointed in his place by the respondents as they had power to do under the contract. H was the representative of the firm who acted pursuant to this appointment.

Clause 24(a) of the contract provides that, at the period of interim certificates (i.e. monthly), interim valuations shall be made whenever the architect considers them necessary and that, subject to clause 21 (which deals with sub-contractors), the architect shall issue a certificate stating the amount due to the contractor from the employer, and the contractor shall be entitled to payment therefor within the period named in the appendix. Clause 24(b) provides that the amount so due shall, subject to clause 21(a) and to any agreement between the parties as to stage payments, be the total value of the work properly executed and of the materials and goods delivered on the site for use in the works up to, and including, a date not more than seven days before the date of the said certificate, less the amount to be retained by the employer and less any instalments previously paid. Clause 27 provides for arbitration in case of any dispute or difference between the employer or the architect on his behalf and the contractor, either during the progress or after completion or abandonment of the works, as to the construction of the contract, or arising thereunder, or in connexion therewith (including any matter or thing left to the discretion of the architect), or the withholding by the architect of any certificate to which the contractor may claim to be entitled or the measurement and valuation mentioned in clause 19 (which deals with variations) or the rights and liabilities of the parties under clause 19, clause 20 or clause 25. Clause 20(1), which is the clause directly in question, reads, so far as relevant, as follows:

> If the employer within the period which is named in the appendix to these conditions and thereafter for seven clear days after written notice from the contractor does not pay to the contractor the amount due on any certificate, or if the employer interferes with or obstructs the issue of any such certificate . . . the contractor may, without prejudice to any other rights or remedies, thereupon by notice by registered post to the employer or architect determine the employment of the contractor under this contract.

It has been found that, in the early stages of the contract, it was the practice for persons in the respondents' surveyor's department from time to time to meet a representative of the appellants to agree figures and valuations for the purpose of the issue by the architect of interim certificates. As time went on, owing to pressure of work it was found impossible for anyone to attend the site on behalf of the respondents, and the appellants' figures were checked by the office staff so far as was practicable. When the work was nearing completion, some of the respondents' officials suspected that this practice had resulted in considerable overpayments, and this was one of the reasons for the appointment, shortly before No. 21 interim certificate was due, of H of the independent firm of Messrs. O who had prepared the original bills of quantities. It is, I think, clear that H was appointed not only to act as surveyor for the limited purposes expressly laid down in the contract but also to carry out the valuations and measurements which had previously been done, with the architect's consent, by members of the respondents' permanent staff with a view to assisting the architect to obtain the necessary materials for the issue of his certificates.

The official referee has found as a fact that 'in the case of certificate 21 the architect decided that an interim valuation should be made and instructed Messrs. O to make it'. H accordingly made what he claimed to be a valuation, as a result of which the architect issued a certificate for £1,287 as against £5,785 claimed by the appellants. The official referee has found that this valuation was not properly made, and the question for decision is whether the defects and errors found constitute obstruction or interference by the employer with the issue of a certificate.

The official referee considered H an unsatisfactory and unreliable witness, but your Lordships are only concerned with his actual ultimate findings as to what was done, or left undone, by H. He found that the amount of £1,287 recommended by H was what he then thought was due to the appellants under clause 24(b), that his valuation had been made in too great a hurry without visual valuation on the site, that it omitted to include approximately £1,500 for hard core, which had been bought and used by the appellants, and that he failed to inform the architect of this omission when it was brought to his notice. H had not included certain excess payments made by the appellants during the certificate period consequent on price variations, and had made certain miscalculations with regard to preliminaries. Finally he found that, as a result of a mistake in the respondents' surveyor's department, the certificate as issued provided for payment to sub-contractors of a sum of £650 or thereabouts out of the total certified. This sum was not, in fact, payable and the appellants never paid it but retained the full sum. It is unnecessary to refer further to this last finding, since counsel for the appellants did not contend that this mistake would, by itself, have justified determination under clause 20.

By reason of these matters, the official referee stated he had come to the conclusion, not without some hesitation, that, by producing that so-called valuation, H obstructed the issue of the 'sort of certificate' the appellants were entitled to and, consequently, they could determine their employment under clause 20.

The Court of Appeal allowed the respondents' appeal. Denning LJ, I think, assumed without deciding that H's acts and omissions amounted to obstruction or interference, but held that the respondents were not liable for the acts of the surveyor within the sphere of duties assigned to him under the contract. Morris LJ considered the respondents not liable because H was employed as an independent consultant not subject to their control. He refers to him 'conducting himself erroneously', but expresses no view whether such erroneous conduct constituted obstruction or interference with the issue of a certificate. Parker LJ said he was prepared to accept that H was guilty of interference or obstruction, but absolved the respondents from responsibility on grounds similar to those expressed by Denning LJ.

His lordship continued in the words set out on p. 220, and went on:

My Lords, one of the difficulties of this case as it has come to this House is that your Lordships do not yet know whether or not there had been over-payments on previous certificates, and whether, in fact, any sum was due when certificate No. 21 was issued. An interlocutory order was made, by consent, postponing the ascertainment of these figures until after the decision on obstruction. It would be a curious result if your Lordships were to hold that the issue of 'the sort of certificate' to which the appellants were entitled has been obstructed or interfered with so as to justify their determining the contract, if it should eventually be found that they were not at the date of the certificate entitled to any payment.

These considerations tend to confirm me in the view that errors due to negligent valuation were never intended to give rise to the remedy afforded to the appellants by clause 20, but fall to be dealt with only by arbitration under clause 27. Holding as I do that the acts and omissions found against H do not amount to obstruction or interference within clause 20, it follows that the letter from the town clerk of 29 October 1951, in which he asserted that a proper valuation executed strictly in accordance with the contract had been made, cannot of itself constitute obstruction although, in fact, the valuation was not correct or properly made.

For these reasons I would dismiss the appeal.

Canterbury Pipe Lines Ltd v. Christchurch Drainage Board

COURT OF APPEAL OF NEW ZEALAND

(1979) 16 BLR 76

The facts sufficiently emerge from the judgment of Cooke and Woodhouse JJ delivered by Cooke J. (McMullin J delivered a separate judgment differing on some points).

COOKE J: Nineteen years ago, on 27 April 1960, the Christchurch Drainage Board accepted the tender of Canterbury Pipe Lines Limited for the laying of about 63 chains of sewer pipes in the Sparks Road area and incidental work. Subject to variations or extras, the contract price was £29,228 (omitting shillings and pence). The specified date for the completion of the works was nine months from acceptance of the tender, namely 27 January 1961. The contract provided that in a number of important respects the engineer of the board should be left with the responsibility of making decisions.

Work started and carried on until September 1960. With the acquiescence of the board there was then a delay pending delivery to the contractor of a piece of heavy duty trenching equipment called a Parsons trenchliner. Once available this machine was expected to enable the contract to be completed more expeditiously. In the event its advantage in that respect proved to be accompanied by the disadvantage that more backfill had to be imported. The machine having been obtained by the contractor, work began again after an interval of about five months. Progress was then quite rapid until mid-April. Then the board's engineer took the view that an amount of some £926 allowed by the contractor in its tender for restoration of the road surface was inadequate; and conversely he maintained that other items had been treated by the contractor too generously in arriving at the total tendered price. The engineer thought that £3,200 was likely to be needed for the restoration. A difference of opinion arose between the engineer and the governing director of the contracting company as to the extent of the company's responsibility in the matter. The amount of restoration needed appears to have been much in excess of anything contemplated on either side at the time of the tender. Against that immediate background and certain difficulties that had arisen between the two men regarding other contracts being carried out by the company for the board, the engineer decided that the £3,200 should be built up in effect in the board's hands before any further monthly progress payments were made to the contractor. Normally, the next progress payment would have been made in May and would have been for about £1,600; but for the reason just mentioned the engineer did not certify for it.

Other differences arose between the parties at about the same time. They arose from claims by the contractor to be paid for the extra backfill necessitated by the machine and to be paid for some extra work at more than schedule rates. The history of the dispute and the correspondence are dealt with in more detail in the judgment of McMullin J, which we have had the advantage of reading. For our purposes it is enough to say that the contractor stopped work and that in reliance on a clause in the general conditions of the contract the engineer then took the work out of the contractor's hands. It was completed partly by two other contractors engaged by the board, partly by the board's own employees . . .

It is unnecessary to decide whether the unjustified withholding of progress certificates by the engineer would have entitled the contractor to rescind its contract with the board . . . [The contractor] elected to keep the contract on foot. What the contractor did was to suspend the work. The question is whether there was any right to do that.

Mersey Steel and Iron Co v. *Naylor, Benzon & Co* (1884) concerned a contract for the sale of steel to be delivered by instalments, the instalments to be separately paid for. It was argued that a contract of that kind was made on the assumption that the parcel already delivered would be paid for punctually and in time to put the manufacturer in funds to provide for the manufacture and delivery of the next parcel. The House of Lords held, however, that payment for a previous delivery was not a condition precedent to the right to claim the next delivery; and further that, although a refusal to pay could in some circumstances amount to repudiation entitling the other party to rescind, the buyers in that case had not repudiated by merely postponing payment on erroneous legal advice that the leave of court was required because the sellers were facing a winding-up petition. Ever since then it has been taken to be the law of England that except in the event of the buyer's insolvency payment for previous instalments is not normally a condition precedent to the liability to deliver, although it can be made so by express provision (*Ebbw Vale Steel Co* v. *Blaina Iron Co* (1901)), or no doubt by necessary implication in the particular contract . . .

Obviously the argument which failed in the *Mersey Steel* case could have been advanced with no less force by a building contractor from whom progress certificates or payments had been wrongly withheld. The *Mersey Steel* decision goes far to explain the apparent absence in English building contract law of any recognition of a common law right to suspend work for wrongful withholding of a progress certificate or payment, as distinct from a right to rescind for a breach going to the root of the contract. If the contract provides for payment on interim certificates from an architect or engineer, the issue of a certificate will normally give rise to a debt payable by the employer (subject, however, to the other terms of the contract). And in some cases of wrongful failure to certify the contractor is able to sue the employer without a certificate. Apart from suing for interim payments, or requiring arbitration where that is provided for, the remedy – and apparently the only remedy – which the contractor is recognised as having at common law is rescission if a sufficiently serious breach has occurred. If he chooses not to rescind, his own obligations continue. He is bound to go on with the work. All the available English and Commonwealth textbooks on building contracts state the law consistently with this view . . .

There is an established principle, illustrated in the building contract field by *Roberts* v. *Bury Commissioners* (1870), which is put in various ways: that no person can take advantage of the non-fulfilment of a condition the performance of which has been hindered by himself; that a party is exonerated from the performance of a contract when the performance is rendered impossible by the wrongful act of the other contracting party; or, more emotively, that a party cannot take advantage of his own wrong. Failures by an employer to make the site available or to supply plans are illustrations. All the cases cited in argument in this connection, and all others that we have found, were cases of physical impossibility or hindrance. No case was cited where the failure was solely in the payment of money.

We think this court should hesitate before seeking to introduce into the building contract law of New Zealand a new general rule as to the right to suspend work. There can be no greater need now for such a rule than in the past. Building contracts have been traditionally a fertile source of disputes. Yet, as has been seen, the rule has not evolved in England or New Zealand or (apparently) Australia. *Kamlee Construction Ltd* v. *Town of Oakville* (1960) suggests that it may not have evolved in Canada either. (But support for it is evidently afforded by . . . *Hoskins* v. *Barber* (1879).) On the other hand such a rule does appear to be recognised in the United States as regards building contracts and some other contracts.

One would have to bear in mind also that building contracts are commonly made in standard forms. For many years the practice has been to provide expressly for a right of suspension if that is intended. For instance the standard forms issued by the Joint Contracts Tribunal in England (the JCT forms) in the revisions reproduced in the fourth edition of *Keating's Building Contracts* (1978) give the contractor a qualified right to determine the building contract, on certain notice, for nonpayment of the amount due on any certificate . . . but give a subcontractor against the contractor a right of suspension of sub-contract works after nonpayment and notice . . .

The significance of these precedents in the present context is twofold. In the first place the parties know where they stand if the rule is that there is no right to suspend work unless it is expressly given in the contract. Customarily the period of delay and the requisite notice are then specified. Some uncertainty would be inevitable if on this pattern there were now superimposed a common law right of suspension. But, bearing in mind that a tendering contractor has little bargaining-power as to the clauses in the contract, that objection need not be treated as decisive. In the second place, however, there is a much more formidable point. Where, as is common, a certificate by an architect or engineer is a condition precedent to a progress payment, we have come across no instance of a New Zealand or English form of contract clearly giving the contractor a right of suspension for non-payment when no certificate had been issued. No doubt the reason is that such a right may be seen as subverting the scheme of the contract in placing the architect or engineer in charge.

This case does not call for a decision on whether New Zealand law recognises a general right in the contractor to suspend work (as distinct from rescinding) by reason of substantial default by the employer in paying a certified progress payment or a progress payment that has fallen due under a contract not requiring certificates. For the reasons already given, it seems to us as well to leave that question open. Whatever the answer to it, we are against recognising such a right when the architect or engineer has declined to issue the certificate stipulated for by the contract as a condition precedent to the employer's duty to pay. It would disrupt the scheme of these contracts. It could encourage contractors to take the law into their own hands. They might stop work which in the public interest needs to be done promptly. In such cases, if the contractor cannot or does not wish to rescind and cannot prove impossibility or its equivalent, he will be left with whatever remedies regarding the recovery of progress payments may be available to him under the contract. As already indicated we do not exclude the view that a case of impossibility or its equivalent might be made out by proving that for want of money the contractor could not carry on or could not reasonably be expected to do so. The present case was fought in a way not truly raising this issue . . .

In our opinion it should be held in the light of these authorities that in certifying or acting under clause 13 here the engineer, though not bound to act judicially in the ordinary sense, was bound to act fairly and impartially. Duties expressed in terms of fairness are being recognised in other fields of law also, such as immigration. Fairness is a broad and even elastic concept, but it is not altogether the worse for that. In relation to persons bound to act judicially fairness requires compliance with the rules of natural justice. In other cases this is not necessarily so. But we do not think that it can be confined to procedure. Its use in the authorities in combination with 'impartiality' suggests that it is not meant to be a narrow concept . . .

The result, as we see it, is that the engineer was acting less than fairly and the notice was accordingly invalid. On this view it follows that, while the contractor was in the wrong in suspending the work, the employer was also in the wrong in purporting to act under clause 13 by taking the works out of the contractor's hands and reletting to other contractors. This unusual impasse is something which the parties could not reasonably have foreseen when they made their contract. To hold the parties to further performance after reletting would alter the fundamental nature of the contract. Indeed it would not have been physically possible. They are both to blame. The question is what should be done to bring the dispute to a just end.

The board had the benefit of more work from the contractor than has been paid for. The parties are agreed that at contract rates the amount unpaid is $3,336. In addition the board is holding retention monies amounting to $4,897 and the deposit of $1,172 which was required with the tender. It is reasonable that the contractor should recover these sums. [Counsel] contends, however, that the case should be remitted to the Supreme Court to allow the contractor to pursue its counterclaim for loss of profit or *quantum meruit*. He cites *Lodder* v. *Slowey* (1904). But there the contractor was not himself at fault.

Where one party is entitled to determine the contract at common law, he will also be entitled to damages to compensate him for his loss. Where a party is terminating under an express provision of the contract, he will only be entitled to such further remedy as the contract gives him.

Thomas Feather & Co (Bradford) Ltd v. Keighley Corporation

QUEEN'S BENCH DIVISION

(1953) 52 LGR 30

Builders contracted to erect houses for the corporation. The contract contained the usual prohibition against sub-contracting without the employers' consent but also expressly provided for a right to determine in the event of sub-contracting without consent. The contractors did sub-contract without consent. The corporation determined the contract and employed other builders to finish the work at an extra cost of £21,000. An arbitrator awarded this sum to the corporation and the builders applied to the court for a review of this decision.

Held: As the breach was of a kind which would not have entitled the corporation to determine but for the express provision of the contract, the corporation were not entitled to the extra cost of completing the houses.

LORD GODDARD CJ: Much as I respect the opinion of the arbitrator in this case I have come to the conclusion that the view taken by him is wrong. I think the meaning of this clause is clear enough. Here is a particular term of a contract which, ordinarily, would not be regarded as one going to the root of the contract or being of the essence of the contract, so that a breach of it would not have given the corporation power to determine the contract or to treat that particular breach as a repudiation by the builder. It seems to me that it would be impossible to say that, in a contract of this magnitude and description, the mere fact that the contractor might put out some small portion of the work, or even a large portion of the work, to a sub-contractor would be a repudiation of the contract. He is not by so doing saying that he does not intend to perform the contract. But what he is doing is breaking a collateral term which says that he is not to sub-contract, and the fact of his subcontracting might not cause a pennyworth of damage to the corporation. But so anxious are the corporation that there should be no sub-contracting that they have put in this special provision, that this particular requirement, namely, that there is to be no sub-contracting, is to be regarded as of the essence of the contract, which it would not otherwise have been, in my opinion, but for that provision. Then it is provided that if there is a breach of it the corporation can determine the contract, or they can claim £100. In my opinion, it is perfectly clear what it means. It means that the corporation can determine the contract, and can determine the contract without the contractor being able to claim any damages against them for so doing. But I cannot see that they need that to preserve what may be called the common law liability. If the contractor repudiates a contract, and shows by his conduct that he does not mean to be bound by it, then, of course, the building owner can recover any extra cost to which he may be put by way of damages. That was the position in the case to which my attention has been called, *Marshall* v. *Mackintosh* (1898), a case in which the contractor did nothing.

I think that the clear meaning of this contract is that if the builder breaks one of these conditions, or certainly the condition against sub-contracting – whether it applies to the earlier part of the clause, the 'Fair Wage' clause, or not, I need not decide, but I am inclined to think it does, though I am only concerned with this matter of sub-contracting – the corporation have certain rights, and that is that they can put an end to the contract once and for all. I would have expected to find, if it was intended that, in those circumstances, the contractor would be liable for damages, that there would have been an express provision put in to that effect. I think that this provision simply gives the corporation a right to terminate the contract, which they would not otherwise have had, and that it gives them nothing more.

Deciding whether a party is entitled to terminate will often involve a very careful analysis of the facts.

Reinwood Ltd v. L Brown & Sons Ltd

HOUSE OF LORDS

(2008) 116 Con LR 1, [2008] UKHL 12

The employer (Reinwood) made a contract with the contractor (Brown) for the construction of 59 apartments in Manchester on the JCT standard form (with quantities) (1998 edition), incorporating amendments 1 to 3.

The date for possession of the site was 14 July 2003 and the completion date 18 October 2004. Liquidated damages were agreed at £13,000 per week or part thereof.

There were considerable delays. On 7 December 2005 Brown applied for an extension of time. On 14 December 2005 the architect issued a certificate of non-completion under cl 24.1. On 11 January 2005 the architect issued interim certificate 29 in a sum of £187,988 with a final date for payment of 25 January.

On 17 January Reinwood issued two notices, one under cl 24.2 of intention to deduct liquidated damages and the other stating that it proposed to pay £126,359 (£187,988 *less* £61,629 liquidated damages). This sum was paid on 20 January. On 23 January the architect granted an extension of time until 10 January 2005. Brown wrote to Reinwood on 24 January stating that the amount payable by way of liquidated damages was now £12,325 and the amount payable under interim certificate 29 was £175,662.

Reinwood did not make any further payment before 26 January and on that day Brown served a notice of default under cl 28.2.1.1. Reinwood paid £49,303 on 1 February.

On 28 June Reinwood should have paid £39,981, but did not do so. On 4 July Brown served a notice of determination relying *inter alia* on the notice of 26 January as a previous default. On 6 July Reinwood wrote to Brown accepting this notice as a repudiatory breach.

Reinwood issued proceedings, claiming a declaration that Brown had unlawfully terminated the contract. His Honour Judge Gilliland QC held that Brown had validly terminated the contract. Reinwood appealed.

LORD NEUBERGER:

There is no doubt that, if the January extension had been granted before 11 January 2006, the employer would not have been entitled to deduct the LADs resulting from the December non-completion certificate, as that certificate would have been 'cancelled' under cl 24.1 as a result of the January extension, by the time of the issue of the interim certificate. Equally, there is no doubt that, if the January extension had been granted after 25 January 2006, the employer's deduction of the LADs based on December non-completion certificate would have been unassailable, as that certificate would not have been cancelled under cl 24.1 by the January extension until after the 'final date for payment' under the interim certificate. The difficulty in this case arises from the fact that the January extension was granted after the date of issue of the interim certificate, but before the 'final date for payment' thereunder.

The contractor's case is that, although the employer was apparently entitled to rely on the December non-completion certificate when it served the preliminary notice and the withholding notice on 17 January, and when it actually paid under the interim certificate, the fact that that entitlement did not exist by the final date for payment deprives the employer of the right to rely on the December non-completion certificate in relation to payment under the interim certificate.

At least on the face of it, the effect of cl 24.2 appears clear. Provided the two pre-conditions in what I have characterised as paras [a] and [b] are satisfied, cl 24.2.1.2 entitles the employer to give notice under cl 30.1.1.4. Although not spelt out, it must, in my view, follow (save, perhaps, in special circumstances) that, where the two pre-conditions are satisfied and the employer has served a withholding notice under cl 30.1.1.4, both parties should be entitled to proceed on the basis that payment will, and can properly, be made in accordance with that notice. Any other conclusion would fly in the face of commercial common sense: the detailed procedures set out in cl 24.2 would appear to have little point if they did not, at least normally, govern the parties' subsequent rights and obligations. That point is reinforced by the fact that the notices referred to in cll 24.2.1[b] and 24.2.1.2 are, respectively, the notices required by ss 110 and 111 of the 1996 Act, and part of the purpose of those sections is to enable parties to a construction contract to know where they stand (see eg *Melville Dundas Ltd (in receivership)* v. *George Wimpey UK Ltd* [2007] UKHL 18 at [19], (2007) 112 ConLR 7 at [19], [2007] 1 WLR 1736, per Lord Hoffmann).

If that analysis applies in this case, it would follow that the employer was entitled to make the deduction from the sum due under the interim certificate that it made in respect of the LADs based on the December non-completion certificate. The issue of that certificate satisfied the first pre-condition, in cl 24.2.1[a], the service of the preliminary notice satisfied the second pre-condition, in cl 24.2.1[b], and the service of the withholding notice complied with cl 24.2.7.2. (The fact that the preliminary notice was served later than the date identified in cl 30.1.1.3 – and in s 110(2) – does not invalidate it, not least because there is no reference to such date in cl 2a.2.1[b].)

So the question is whether, in accordance with the submission of the contractor, the issue of the January extension, after the employer had both served the withholding notice and paid on the assumption that he had the right to rely on the December non-completion certificate, deprived the employer of that right. In my judgment, in agreement with the Court of Appeal, the issue of the January extension did not have that effect.

The decision of the House of Lords was not the end of the matter.

Reinwood Ltd v. L Brown & Sons Ltd (No. 2)

COURT OF APPEAL, CIVIL DIVISION

(2008) 121 ConLR 1, [2008] EWCA Civ 1909

The contract sum of £6.2m was exclusive of VAT. The work was partly zero-rated (residential) and part standard-rated (commercial). The contract required the contractor to make and serve the employer separate provisional assessments of VAT.

The employer did not pay the VAT which was the subject of the first provisional assessment. On 12 May 2005 the contractor gave notice of a specified default in respect of this non-payment. Later the parties agreed on a smaller sum which the employer paid.

The contractor now sought to justify the termination on the combination of the notices of 12 May 2005 and 28 June 2006.

Held: The default notice dated 12 May 2005 was valid because the letter dated 11 April 2005 was a valid provisional assessment and the employer's e-mail of 14 April 2005 was not a valid objection.

The interim assessment of VAT must state the basis on which the certified amount was calculated.

An objection for this purpose required a clear indication of reliance on cl 1.2.2. Merely to ask for more detail was not an objection.

The ineffective reliance by Brown on the non-payment in January 2006 did not prevent it relying on the default notice of 12 May 2005. This was not an example of waiver but an application of the principle that a contracting party who relied on a bad reason for termination, was not barred from relying on a good reason.

LLOYD LJ:

The appellant's contention is inconsistent with the general principle of contract law that if a party refuses to perform a contract, giving a reason which is wrong or inadequate, or giving no reason at all, or terminates a contract under a contractual provision to that effect, the refusal or termination may nevertheless be justified if there were at the time facts in existence which would have provided a good reason for the refusal: see *Chitty on Contracts* (29th edn, 2004), vol. 1, para. 24–014. That principle is often used in relation to facts unknown to the party refusing at the time of its refusal, but there is no reason why it should not be used in relation to facts which were known to that party at that time. 'Waiver can apply to qualify that principle, but only in cases of, in effect, estoppel.'

Accordingly it seems to me that neither of the waiver grounds of appeal can succeed, and that the appeal ought therefore to be dismissed. In my judgment the judge was correct to hold that the notice of determination under cl 28.2.4 was valid, because of the valid notice of default dated 12 May 2005, despite the invalidity of the notice of default served in January 2006.

The fact that one party is contractually entitled to terminate the contract on specified grounds does not preclude him from treating the contract as discharged by repudiatory breach unless the contract itself expressly or implied provides that it can only be determined by exercise of the contractual right.

Architectural Installation Services Ltd v. James Gibbons Windows Ltd

QUEEN'S BENCH DIVISION

(1989) 16 Con LR 68

The plaintiffs were specialist labour-only sub-contractors installing window units for the defendants as main contractors. The plaintiffs sued the defendants for moneys alleged to be due and for damages for wrongful determination of the contract. The defendants contended that they were entitled to determine the contract and counterclaimed for, among other things, damages for defective work. The sub-contract between the parties, made on 22 July 1985, was in the defendants' standard form, and contained an express provision (clause 8) providing for its termination by notice on specified grounds, including the default of the sub-contractor wholly suspending the works or failing to proceed with them expeditiously and remaining in default for 7 days after being given written notice of the default by the main contractor. It was ordered that two preliminary issues be tried, each of which assumed that the plaintiffs had wholly suspended the works and/or failed to proceed with the works expeditiously, in breach of clause 20 of the subcontract. These issues were:

(1) On those assumptions, was the defendants' telex of 21 August 1986 a rightful termination of the sub-contract pursuant to clause 8 of the subcontract or an unlawful repudiation of the same?
(2) On those assumptions and that the plaintiffs acted in such a manner as to evince an intention no longer to be bound by the contract between the parties, was the defendants' telex of 21 August 1986 a rightful termination of the sub-contract at common law or an unlawful repudiation of it?

Held: (1) On the assumptions set out in the preliminary issues the defendants' telex of 21 August 1986 was not a rightful termination of the subcontract pursuant to clause 8 because there was no sensible connection between the defendants' notice of default dated 20 September 1985 and the telex notice of determination of 21 August 1986 in terms of either content or time.
(2) However, on the like assumptions, the defendants' telex of 21 August 1986 was a rightful termination of the sub-contract at common law since clause 8 did not exclude common law rights expressly or by implication but existed side by side with the common law right to terminate.

JUDGE PETER BOWSHER QC: It seems clear to me that the reason for the inclusion of condition 8 in this contract is the same as the inclusion of similar terms in many other contracts. When a contract is going adrift an employing party is faced with a number of problems. Above all, he does not want to terminate the contract or subcontract because that will produce a whole new set of problems. But, if he has to threaten termination to secure compliance with the contract, he knows that if his threat is contested he will later have to satisfy a court or an arbitrator of two things, firstly that the employed party was in breach, and secondly that the party in breach evinced an intention no longer to be bound by the contract. To be certain of being able to prove breach of contract is difficult enough, to be confident that it would be possible to convince the court or arbitrator that the party in breach had repudiated the contract is a difficult state of mind to achieve. The party alleged to be in breach has a similar difficulty in assessing the likely future assessment of the quality of its actions by any tribunal. So the two notice procedure is introduced into the contract to make it easier for both parties to assess their legal liabilities towards each other and to govern their conduct accordingly. I stress, easier for both parties. It seems to me to be totally clear that that procedure applies to each of the three sub-divisions of sub-condition 8(c) and I see no ground for suggesting that it applies to only one of them. It obviously does not apply to condition 8(a) or (b).

For those reasons, I find that on the assumptions set out in the first of the preliminary issues before me, the defendants' telex of 21 August 1986 was an unlawful repudiation of the contract. Condition 8 required two sensibly connected notices. There was not two sensibly connected notices. Therefore, the defendants cannot rely on condition 8 in support of their contention that the sub-contract was rightfully terminated.

[Counsel for the plaintiffs] points to the fact that condition 8 is not prefaced by the words 'without prejudice to other rights and remedies' and submits that the absence of those words tends to support the proposition that other rights and remedies are excluded. I am not impressed by that submission and I would be sorry if draftsmen of contracts felt it necessary to include such legal verbiage in order to avoid unintended results of their drafting. Construction contracts are already sufficiently complicated when the draftsmen seek to state what they do mean. They should not be burdened with the additional task of stating what they do not mean. When someone has obviously gone to a great deal of trouble to draft a contract, and two commercial parties have agreed to a contract on those terms, the court should be very reluctant to step in and suggest that those two parties also agreed something which was not written down in the agreement between them.

I see no reason at all for any implication of any term to the effect that condition 8 of the contract is to be the only machinery for terminating the contract, to the exclusion of common law rights of termination. Indeed, if it were the case that condition 8 provided the only means of terminating for breach, that would produce a nonsensical result in a situation which may not be common but is not unknown. If the plaintiffs are right and condition 8 excludes any common law right to terminate, if the plaintiffs were to say to the defendants 'We are leaving the site and we shall not return and we shall not change our minds about this', the defendants would still have to go through the two notice procedure in order to terminate the contract. That seems to me to be contrary to any business sense. For the reasons set out earlier in this judgment, I find that the provisions of condition 8 of the contract exist side by side with the common law rights to terminate the contract.

There is often serious dispute as to whether one party has in fact done or failed to do that which would entitle the other party to determine the contract. This problem has caused particular difficulty where the employer wishes to determine the contract and turn the contractor off the site. In the first of the two cases below, Megarry J held that where the employer sought the aid of the Court by way of interim injunction ordering the contractor to leave the site he must show a clear entitlement to determine under the contract. That view has been widely criticised and was not followed in the second case.

London Borough of Hounslow v. Twickenham Garden Developments Ltd

CHANCERY DIVISION

(1970) 7 BLR 81

The dispute arose under a contract in JCT 63 terms. The architect purported to give notice to the contractors under clause 25(1) stating that in his opinion they had failed to proceed regularly and diligently with the works and that unless there was an appreciable improvement within 14 days the council would be entitled to determine their employment. By a subsequent letter, the council purported to determine the contractors' employment under the contract. The contractors alleged that the notice of determination was wrongful and amounted to a repudiation of the contract. They refused to accept the alleged repudiation and elected to proceed with the works.

The council claimed an injunction to restrain the contractors from remaining on the site.

MEGARRY J: [In] this case the contract is one for the execution of specified works on the site during a specified period which is still running. The contract confers on each party specified rights on specified events to determine the employment of the contractor under the contract. In those circumstances, I think that there must be at least an implied negative obligation of the borough not to revoke any licence (otherwise than in accordance with the contract) while the period is still running . . .

I hold that without establishing the efficacy of the notices given under condition 25, the borough is not entitled to the injunctions sought . . .

In the event, my conclusion is that although the borough has established some sort of a case for having validly determined the contract, that case falls considerably short of any standard upon which, in my judgment, it would be safe to grant this injunction on motion. What is involved is the application of an uncertain concept to disputed facts. As is so often the case on motion, a court is faced with a choice of evils. I fully accept the importance to the borough, on social grounds as well as others, of securing the due completion of the contract, and the unsatisfactory nature of damages as an alternative. But the contract was made, and the contractors are not to be stripped of their rights under it, however desirable that may be for the borough. A contract remains a contract, even if (or perhaps especially if) it turns out badly. The borough may indeed be able at the trial to make out a formidable case in support of its contentions, just as the contractor may be able to make out a formidable case in reply; and with the full procedure of the trial, and in particular with the advantage of seeing and hearing the witnesses, and the testing process of cross-examination (not least on any expressions of expert opinion), the court may well be able to reach a firm conclusion on one side or the other. But I lack these advantages, and with so much turning on disputed questions of fact, and inferences from the facts, I cannot say that the borough has made out its case for the injunction that it seeks.

Mayfield Holdings Ltd v. Moana Reef Ltd

SUPREME COURT OF NEW ZEALAND

[1973] 1 NZLR 309

Disputes arose between the parties to a building contract. The employer refused to make further payments until alleged defects were remedied. The contractor retaliated by slowing down work, and the employer called in another contractor who was denied access by the first contractor. The owner sought an injunction against the first contractor.

MAHON J: The question at issue in the *Twickenham Garden* case was whether an interlocutory injunction in favour of the owner should be granted pending trial of the action. The plaintiff council, as owners of the building land engaged the defendant contractor to construct roads, car parks and other services for a proposed housing estate, the work to occupy a period of four years. The contract was in the standard Royal Institute of British Architects form and it contained a condition to the effect that in the event of specified default of the contractor, and in particular in the event of the contractor failing to proceed diligently with the works, the architects of the plaintiff council could give notice specifying the default and that unless the default was remedied the plaintiff could give notice determining the contract. At one stage work was halted in consequence of a long strike which itself was not a default under the contract but work on the site was continued thereafter at a pace which the plaintiff's architect thought was too slow. He gave notice of default in terms of the contract. A month later the council gave notice that the default had not been remedied and that the employment of the contractor was therefore terminated. The council then issued a writ claiming damages for trespass in failing to vacate the site, and they also asked for a perpetual injunction. They then moved for an interlocutory injunction to restrain the contractor before trial from entering, remaining or otherwise trespassing on the site or interfering with the plaintiff council's possession of the site. There were two main issues:

(1) Disregarding the question of validity of the notice were the council, as owners, entitled to determine the licence of the contractor to enter and remain on the site?
(2) Were the notices valid and effective so as to determine the contractor's employment?

Megarry J found in favour of the contractor. There appear to have been two grounds for his decision:

(1) The contract was for the execution of specified works over a period which was still continuing when the notices were given and the learned Judge held that there must have been an implied negative obligation on the part of the Borough Council not to revoke the contractor's licence (otherwise than in accordance with the contract) while the contract period was still running, and that a Court of Equity would not assist the council to break its contract.
(2) Although the Borough Council had established 'some sort of a case' for having validly determined the contract, that case fell considerably short of any standard upon which it would be safe to grant an interlocutory injunction.

In respect of the first ground of decision, Megarry J reviewed the relevant authorities in relation to licences to enter upon land and held that this was a contractual licence containing the implied negative covenant above referred to with the result that equity would restrain the Borough Council from revoking the licence in breach of the contract of which it formed part. It followed from this that the contractor could successfully resist the issue of a mandatory injunction ordering the contractor from the site unless and until the Borough Council established a material breach of contract carrying with it the right of revocation of the licence. As to the second ground, because there were substantial questions at issue as to whether any breach of contract had been committed, it therefore followed that the Borough Council was unable in the interlocutory proceedings to prove the material breach of contract which alone would have justified forfeiture of the contract and concurrent revocation of the contractor's licence. Many points of argument were presented to Megarry J but so far as the revocability of the licence was concerned there were two principal submissions made on behalf of the contractor, firstly that the licence was coupled with an interest, and secondly, in the alternative, that the contract contained an implied negative covenant on the part of the corporation not to revoke the licence in breach of the contract. Megarry J accepted the second of these submissions and did not come to a conclusion on the first. I must examine both these arguments as they are each available to the contractor in the case before me. Thus the crux of the decision on this part of the case was the finding of an implied negative covenant in the terms expressed. To this finding there are two possible objections – (a) that the existence of such an implied term might not be thought necessary to give the contract business efficacy, (b) that even if it did exist it could not be specifically enforced, with the result that the defence to the injunction claimed, being in effect a claim for specific performance, could not be maintained. I will deal with these objections in the order named.

The proposed implication of a term in a contract, in particular a commercial contract, calls for careful consideration. A court is not entitled to reconstruct an agreement on equitable principles, or on a view of what the parties should have contemplated. The implication of the unexpressed term must be necessary to give to the transaction such business efficacy as the parties must have intended. 'The implication must arise inevitably to give effect to the intention of the parties' – *Luxor (Eastbourne) Ltd* v. *Cooper* (1941) *per* Lord Wright. With these observations in mind, what would be the presumed common intention of the parties to an ordinary building contract in the event of the development between them of a dispute amounting to an impasse relating to construction of the works? The consequences of the impasse have to be considered. The works have been part completed, and the owner claims to be entitled to terminate the contract for reasons of faulty workmanship, deviation from plans, slow progress, or the like. The owner stops progress payments and instructs his architect not to certify for further work. In theory the contractor can complete the next section of work and bring an action, with its consequential delays, for the progress payment due in respect of that work, raising default by the owner as an answer to the defence that there is no architect's certificate. But in practice, the delay involved would be insurmountable. The contractor would have to lay off his employees, or transfer them to another job. In the absence of progress payments he probably could not pay his sub-contractors. The architect's authority to co-operate in the construction would be withdrawn. Under these circumstances, it seems impossible to say that the contractor can obtain any benefit by remaining on the job. He has a claim for unpaid work to date, and a claim for loss of profit on the whole contract, and instead of working on under difficulties and suing for unpaid progress payments he surely would take the alternative course of leaving the job and suing to enforce his claims, thus recovering by action, if not in default under the contract, the moneys to which he would have been entitled if the works had been completed by him pursuant to the contract. The contractor's only interest in the transaction is to complete the specified works and at a profit to himself. The owner's interest is to secure completion in accordance with his plans and specifications. Each object will still be attained even if the owner, in breach of his contract, procures the dismissal of the contractor from the site and completes the specified works by contract with another builder. Judged from the standpoint of business efficacy, I can for myself see no difficulty in implying a term between builder and owner that in the event of a complete breakdown in their contractual relationship, the builder will surrender the site and the works, leave the owner to complete as and when he will, and enforce by action his remedy against the owner for breach of the building contract. I quote in this context the observations of Knight Bruce LJ in *Garrett* v. *Banstead and Epsom Downs Railway Co* (1864), referred to in *Hudson's Building and Engineering Contracts* (10th edition) page 712:

> To suppose in a case like this, where, if the company are wrong, ample compensation in damages may be obtained by the contractor, that the company are to have a person forced

> on them to perform these works whom they reasonably or unreasonably object to (whereas there would be no reciprocity if the wrong were on the other side), for the purpose of compelling the performance of the works, is more than I am able to do.

The opinion expressed by Knight Bruce LJ can be seen to be inspired by the lack of mutuality of remedy but it might also be justified upon a further ground. The practical exigencies of performance of a building contract must be kept in mind. The purported forfeiture by the owner may be based on alleged departure from the contract specifications. Must the owner be compelled to stand by while his action for breach awaits trial, and watch the building being completed in a manner which may ultimately be decided to have been in breach of the contract? Certainly if the departure from the contract terms were clear and unanswerable I should imagine that nothing in the *Twickenham Garden* case would prevent the issue of an interlocutory injunction dismissing the contractor from the site, subject to correct notices of forfeiture being given, but whenever the supposed breach was a matter of genuine controversy, as it so often is, the *Twickenham Garden* decision would maintain the contractor not only in possession of the site but in continuation of the works on a disputed basis. It is difficult to accept that either contractor or owner would ever have agreed, at the formulation of their contract, to any express term carrying with it such drastic consequences. For the reasons which I have expressed I cannot agree, with respect, with the view taken by Megarry J in the *Twickenham Garden* case. I am fortified to some extent in the view which I have taken by the editorial criticism of the *Twickenham Garden* decision in 87 Law Quarterly Review 309 where it is said, at the end of the note on the case:

> The contractor's sole object is to make a profit; if he gets it, and suffers no loss, why should not the owner be entitled to change his mind and exclude him from his property?

I am clear that this is the sense of the unexpressed term which the law must imply where an impasse between owner and building contractor is envisaged. A continuation of the impasse could only be to the detriment of both parties. Surely the contractor would say: 'If a final breakdown in our relationship occurs, or if you decide to cancel the contract without due cause, then I agree that you go ahead and finish the job yourself, and I will sue you for my lost profit.' I am therefore not prepared to hold in the case before me that there was a covenant implied in the contract binding the owner not to revoke the contractor's licence in breach of contract.

Chapter 19
Limitation

'Limitation' is the name given by lawyers to the rules prescribing within what periods actions must be started. Failure to start the action in time does not theoretically extinguish the obligation but the court will not give its assistance to the prosecution of the claim. (So if the defendant pays up after the limitation period, the payment is valid and cannot be recovered.)

The principal statute is currently the Limitation Act 1980, the most important sections of which in relation to building contracts are:

The Limitation Act 1980

Section 2 Time limit for actions founded on tort
An action founded on tort shall not be brought after the expiration of six years from the date on which the cause of action accrued.

Section 5 Time limit for actions founded on simple contract
An action founded on simple contract shall not be brought after the expiration of six years from the date on which the cause of action accrued.

Section 8 Time limit for actions on a specialty
(1) An action upon a specialty shall not be brought after the expiration of twelve years from the date on which the cause of action accrued.
(2) Subsection (1) above shall not affect any action for which a shorter period of limitation is prescribed by any other provision of this Act.

Section 32 Postponement of limitation period in case of fraud, concealment or mistake
(1) Subject to subsection (3) below, where in the case of any action for which a period of limitation is prescribed by this Act, either

- (a) the action is based upon the fraud of the defendant; or
- (b) any fact relevant to the plaintiff's right of action has been deliberately concealed from him by the defendant; or
- (c) the action is for relief from the consequences of a mistake;

Powell-Smith and Furmston's Building Contract Casebook, Fifth Edition. Michael Furmston.

the period of limitation shall not begin to run until the plaintiff has discovered the fraud, concealment or mistake (as the case may be) or could with reasonable diligence have discovered it.

References in this subsection to the defendant include references to the defendant's agent and to any person through whom the defendant claims and his agent.

(2) For the purposes of subsection (1) above, deliberate commission of a breach of duty in circumstances in which it is unlikely to be discovered for some time amounts to deliberate concealment of the facts involved in that breach of duty.

(3) Nothing in this section shall enable any action:

(a) to recover, or recover the value of, any property; or
(b) to enforce any charge against, or set aside any transaction affecting, any property;

to be brought against the purchaser of the property or any person claiming through him in any case where the property has been purchased for valuable consideration by an innocent third party since the fraud or concealment or (as the case may be) the transaction in which the mistake was made took place.

(4) A purchaser is an innocent third party for the purposes of this section:

(a) in the case of fraud or concealment of any fact relevant to the plaintiff's right of action, if he was not a party to the fraud or (as the case may be) to the concealment of that fact and did not at the time of the purchase know or have reason to believe that the fraud or concealment had taken place; and
(b) in the case of mistake, if he did not at the time of the purchase know or have reason to believe that the mistake had been made.

It is not always easy to establish 'the date on which the cause of action accrued'. In contract cases it will normally be the date on which the contract was broken. The next two cases concern claims in tort and vividly illustrate the difficulties which arise where the defect is covered up.

Anns v. London Borough of Merton

HOUSE OF LORDS

(1978) 5 BLR 1

It was alleged that in 1962 a builder had erected a block of maisonettes with defective foundations and that the defendant council had, through its inspector, either failed to inspect the foundations or had inspected them carelessly. The plaintiffs had taken long leases of some of the maisonettes. In February 1970 structural movement began, resulting in cracks. In February 1972 the plaintiffs issued writs against the builder and against the council. This action was brought to decide whether the proceedings against the council were barred under section 2(1)(a) of Limitation Act 1939. [There is no relevant difference between the Limitation Act 1939 and the Limitation Act 1980.] Obviously if the plaintiffs' cause of action against the council accrued in 1962, it was too late to start an action in 1972. In fact, although the appeal was nominally on the limitation issue, the speeches in the House of Lords were mainly concerned with the more fundamental question of whether an action would lie against the council at all on such facts.

Held: Such an action would lie: it would not be statute barred.

LORD WILBERFORCE: When does the cause of action arise? We can leave aside cases of personal injury or damage to other property as presenting no difficulty. It is only the damage for the house which required consideration. In my respectful opinion the Court of Appeal was right when, in *Sparham-Souter* v. *Town and Country Developments (Essex) Ltd* (1976), it abjured the view that the cause of action arose immediately on delivery, i.e. conveyance of the defective house. It can only arise when the state of the building is such that there is present or imminent danger to the health or safety of persons occupying it. We are not concerned at this stage with any issue relating to remedial action nor are we called on to decide on what the measure of the damages should be; such questions, possibly very difficult in some cases, will be for the court to decide. It is sufficient to say that a cause of action arises at the point I have indicated.

The Limitation Act 1939. If the fact that defects to the maisonettes first appeared in 1970, then, since the writs were issued in 1972, the consequence must be that none of the present actions are barred by the Act.

Pirelli General Cable Works Ltd v. Oscar Faber and Partners

HOUSE OF LORDS

(1982) 21 BLR 99

The defendants, a firm of consulting engineers, advised the plaintiffs on the design and erection of a boiler flue chimney. The chimney was negligently designed and in due course cracks occurred in the internal lining of the chimney. The chimney was built in June and July 1969. The plaintiffs issued their writ in October 1978. The trial judge found that the cracks occurred not later than April 1970, that the plaintiffs did not discover the cracks until November 1977 and that it was not proved that the plaintiffs could with reasonable care have discovered the cracks before October 1972. On these facts the action was time-barred if time started to run when the cracks occurred, but not if it only started to run when the cracks could with reasonable care have been discovered.

Held: That the limitation period began when the cracks occurred and that the action was therefore statute-barred.

LORD FRASER OF TULLYBELTON: Part of the plaintiffs' argument in favour of the date of discoverability as the date when the right of action accrued was that that date could be ascertained objectively. In my opinion that is by no means necessarily correct. In the present case, for instance, the judge held that the plaintiffs as owners of a new chimney, built in 1969, had no duty to inspect the top of it for cracks in spring 1970. But if they had happened to sell their works at that time, it is quite possible that the purchaser might have had such a duty to inspect and, if so, that would have been the date of discoverability. That appears to me to show that the date of discoverability may depend on events which have nothing to do with the nature or extent of the damage.

Counsel for the plaintiffs argued that in *Anns* v. *Merton London Borough* (1978) Lord Wilberforce, and the other members of this House who agreed with his speech, had approved of the observations in *Sparham-Souter* (1976) to the effect that the discoverability date was the date when the cause of action accrued. But I do not so read Lord Wilberforce's speech. He simply narrated the conflict between the cases of *Dutton* (1973) and *Sparham-Souter* (1976) without indicating any preference. He posed the question 'When does the cause of action arise?' and he answered it as follows:

> In my respectful opinion the Court of Appeal was right when, in *Sparham-Souter* v. *Town and Country Developments (Essex) Ltd* (1976) it abjured the view that the cause of action rose immediately on delivery, i.e., conveyance of the defective house. It can only arise when the state of the building is such that there is present or imminent danger to the health or safety of persons occupying it.

The only express approval in that passage is to the Court of Appeal's decision that the cause of action did not arise immediately the defective house was conveyed. His Lordship did not say, nor in my opinion did he imply, that the date of discoverability was the date when the cause of action accrued. The date which he regarded as material (when there is 'present or imminent danger . . . health or safety') was, of course, related to the particular duty resting on the defendants as the local authority, which was different from the duty resting on the builders or architects, but I see nothing to indicate that Lord Wilberforce regarded the date of discoverability of the damage as having any relevance. He was not considering the question of discoverability, no doubt because the main issue in the appeal by the time it reached the House was whether any duty at all was incumbent on the local authority. Three other noble and learned Lords expressed agreement with Lord Wilberforce. Only Lord Salmon delivered a separate reasoned speech and he clearly considered that the cause of action could arise before damage was discovered or discoverable, although he recognised that proof might be difficult. He said:

> Whether it is possible to prove that damage to the building had occurred four years before it manifested itself is another matter, but it can only be decided by evidence. (My emphasis.)

Neither Lord Salmon nor the other Lords seem to have considered that they were dissenting from the majority view on that matter. In these circumstances I do not think that the majority in *Anns* are to be taken as having approved the discoverability test applied in *Sparham-Souter*.

In this case their lordships recommended statutory amendment of the law. This suggestion was adopted by the Latent Damage Act 1986, the full text of which is given below. Note that this Act does not reverse the *Pirelli* case, but in certain circumstances introduces a secondary time limit based on discoverability.

Latent Damage Act 1986

Time limits for negligence actions in respect of latent damage not involving personal injuries

1. The following sections shall be inserted in the Limitation Act 1980 (referred to below in this Act as the 1980 Act) immediately after section 14 (date of knowledge for purposes of special time limits for actions in respect of personal injuries or death):

'Actions in respect of latent damage not involving personal injuries'

Special time limit for negligence actions where facts relevant to cause of action are not known at date of accrual.

14A. (1) This section applies to any action for damages for negligence, other than one to which section 11 of this Act applies, where the starting date for reckoning the period of limitation under subsection (4)(b) below falls after the date on which the cause of action accrued.

(2) Section 2 of this Act shall not apply to an action to which this section applies.

(3) An action to which this section applies shall not be brought after the expiration of the period applicable in accordance with subsection (4) below.

(4) That period is either:

(a) six years from the date on which the cause of action accrued; or
(b) three years from the starting date as defined by subsection (5) below, if that period expires later than the period mentioned in paragraph (a) above.

(5) For the purposes of this section, the starting date for reckoning the period of limitation under subsection (4)(b) above is the earliest date on which the plaintiff or any person in whom the cause of action was vested before him first had both the knowledge required for bringing an action for damages in respect of the relevant damage and a right to bring such an action.

(6) In subsection (5) above 'the knowledge required for bringing an action for damages in respect of the relevant damage' means knowledge both:

(a) of the material facts about the damage in respect of which damages are claimed; and
(b) of the other facts relevant to the current action mentioned in subsection (8) below.

(7) For the purposes of subsection (6)(a) above, the material facts about the damage are such facts about the damage as would lead a reasonable person who had suffered such damage to consider it sufficiently serious to justify his instituting proceedings for damages against a defendant who did not dispute liability and was able to satisfy a judgment.

(8) The other facts referred to in subsection (6)(b) above are:

(a) that the damage was attributable in whole or in part to the act or omission which is alleged to constitute negligence; and
(b) the identity of the defendant; and
(c) if it is alleged that the act or omission was that of a person other than the defendant, the identity of that person and the additional facts supporting the bringing of an action against the defendant.

(9) Knowledge that any acts or omissions did or did not, as a matter of law, involve negligence is irrelevant for the purposes of subsection (5) above.

(10) For the purposes of this section a person's knowledge includes knowledge which he might reasonably have been expected to acquire:

(a) from facts observable or ascertainable by him; or
(b) from facts ascertainable by him with the help of appropriate expert advice which it is reasonable for him to seek;

but a person shall not be taken by virtue of this subsection to have knowledge of a fact ascertainable only with the help of expert advice so long as he has taken all reasonable steps to obtain (and, where appropriate, to act on) that advice.

Overriding time limit for negligence actions not involving personal injuries.
14B. (1) An action for damages for negligence, other than one to which section 11 of this Act applies, shall not be brought after the expiration of fifteen years from the date (or, if more than one, from the last of the dates) on which there occurred any act or omission:

(a) which is alleged to constitute negligence; and
(b) to which the damage in respect of which damages are claimed is alleged to be attributable (in whole or in part).

(2) This section bars the right of action in a case to which subsection (1) above applies notwithstanding that:

(a) the cause of action has not yet accrued; or
(b) where section 14A of this Act applies to the action, the date which is for the purposes of that section the starting date for reckoning the period mentioned in subsection (4)(b) of that section has not yet occurred;

before the end of the period of limitation prescribed by this section.

Provisions consequential on section 1

2. (1) The following section shall be inserted in the 1980 Act immediately after section 28 (extension of limitation period in case of disability on date of accrual of cause of action):

'Extension for cases where the limitation period is the period under section 14A(4)(b)

28A. (1) Subject to subsection (2) below, if in the case of any action for which a period of limitation is prescribed by section 14A of this Act:

(a) the period applicable in accordance with subsection (4) of that section is the period mentioned in paragraph (b) of that subsection;
(b) on the date which is for the purposes of that section the starting date for reckoning that period the person by reference to whose knowledge that date fell to be determined under subsection (5) of that section was under a disability; and
(c) section 28 of this Act does not apply to the action;

the action may be brought at any time before the expiration of three years from the date when he ceased to be under a disability or died (whichever first occurred) notwithstanding that the period mentioned above has expired.

(2) An action may not be brought by virtue of subsection (1) above after the end of the period of limitation prescribed by section 14B of this Act.'

(2) In section 32 of the 1980 Act (postponement of limitation period in case of fraud, concealment or mistake), at the end there shall be added the following subsection:

'(5) Sections 14A and 14B of this Act shall not apply to any action to which subsection (1)(b) above applies (and accordingly the period of limitation referred to in that subsection, in any case to which either of those sections would otherwise apply, is the period applicable under section 2 of this Act).'

Accrual of cause of action to successive owners in respect of latent damage to property

3. (1) Subject to the following provisions of this section, where:

(a) a cause of action ('the original cause of action') has accrued to any person in respect of any negligence to which damage to any property in which he has an interest is attributable (in whole or in part); and
(b) another person acquires an interest in that property after the date on which the original cause of action accrued but before the material facts about the damage have become known to any person who, at the time when he first has knowledge of those facts, has any interest in the property;

a fresh cause of action in respect of that negligence shall accrue to that other person on the date on which he acquires his interest in the property.

(2) A cause of action accruing to any person by virtue of subsection (1) above:

(a) shall be treated as if based on breach of a duty of care at common law owed to the person to whom it accrues; and
(b) shall be treated for the purposes of section 14A of the 1980 Act (special time limit for negligence actions where facts relevant to cause of action are not known at date of accrual) as having accrued on the date on which the original cause of action accrued.

(3) Section 28 of the 1980 Act (extension of limitation period in case of disability) shall not apply in relation to any such cause of action.

(4) Subsection (1) above shall not apply in any case where the person acquiring an interest in the damaged property is either:

(a) a person in whom the original cause of action vests by operation of law; or
(b) a person in whom the interest in that property vests by virtue of any order made by a court under section 538 of the Companies Act 1985 (vesting of company property in liquidator).

(5) For the purposes of subsection (1)(b) above, the material facts about the damage are such facts about the damage as would lead a reasonable person who has an interest in the damaged property at the time when those facts become known to him to consider it sufficiently serious to justify his instituting proceedings for damages

against a defendant who did not dispute liability and was able to satisfy a judgment.

(6) For the purposes of this section a person's knowledge includes knowledge which he might reasonably have been expected to acquire:

(a) from facts observable or ascertainable by him; or
(b) from facts ascertainable by him with the help of appropriate expert advice which it is reasonable for him to seek;

but a person shall not be taken by virtue of this subsection to have knowledge of a fact ascertainable by him only with the help of expert advice so long as he has taken all reasonable steps to obtain (and, where appropriate, to act on) that advice.

(7) This section shall bind the Crown, but as regards the Crown's liability in tort shall not bind the Crown further than the Crown is made liable in tort by the Crown Proceedings Act 1947.

Supplementary: Transitional provisions

4. (1) Nothing in section 1 or 2 of this Act shall:

(a) enable any action to be brought which was barred by the 1980 Act or (as the case may be) by the Limitation Act 1939 before this Act comes into force; or
(b) affect any action commenced before this Act comes into force.

(2) Subject to subsection (1) above, sections 1 and 2 of this Act shall have effect in relation to causes of action accruing before, as well as in relation to causes of action accruing after, this Act comes into force.

(3) Section 3 of this Act shall only apply in cases where an interest in damaged property is acquired after this Act comes into force but shall so apply, subject to subsection (4) below, irrespective of whether the original cause of action accrued before or after this Act comes into force.

(4) Where:

(a) a person acquires an interest in damaged property in circumstances to which section 3 would apart from this subsection apply; but
(b) the original cause of action accrued more than six years before this Act comes into force;

a cause of action shall not accrue to that person by virtue of subsection (1) of that section unless section 32(1)(b) of the 1980 Act (postponement of limitation period in case of deliberate concealment of relevant facts) would apply to any action founded on the original cause of action.

Citation, interpretation, commencement and extent

5. (1) This Act may be cited as the Latent Damage Act 1986.

(2) In this Act:

'the 1980 Act' has the meaning given by section 1; and
'action' includes any proceeding in a court of law, an arbitration and any new claim within the meaning of section 35 of the 1980 Act (new claims in pending actions).

(3) This Act shall come into force at the end of the period of two months beginning with the date on which it is passed.

(4) This Act extends to England and Wales only.

Abbott and another v. Will Gannon & Smith Ltd

COURT OF APPEAL, CIVIL DIVISION

(2005) 103 ConLR 92, [2005] EWCA Civ 198

The claimants were the owners of a hotel in Torquay. In May 1995 they engaged the defendant consulting structural and civil engineers to design work necessary to remedy structural defects in a large bay window at the hotel. Work to the defendant's design was completed by a local builder in March 1997. Cracks in the works appeared in late 1999.

The claimant started proceedings in both contract and tort on 5 September 2003. The defendant raised a limitation defence which was tried as a preliminary issue. It was accepted that the contract claim was time-barred.

The defendants argued that the tort claim started to run on completion of the works in March 1997 and was therefore time-barred. The claimant argued that time began to run on, or shortly before, they first noticed cracking in 1999.

TUCKEY LJ:

So what is the present state of the law of England? With three House of Lords' cases to guide us it ought to be possible to give a clear answer to this question, but I regret that I feel unable to do so with any confidence. *Murphy*'s case establishes that, absent a special relationship a claimant may only sue in tort for personal injury or damage to property caused by a latent defect in a building. But it is not clear whether this extends to damage to the building itself before the defect is discovered. And what is the position where there is a special relationship? It is clear that the duty in such a case extends to taking care not to cause economic loss. But when does such loss occur in a case such as the present and does the duty not to cause physical damage to property constitute a separate cause of action for limitation purposes?

Mr Holwill for the engineers submits that in a case like the present a claimant will only suffer economic loss. In this case that loss occurred at the time when the defectively designed work to the bay window was completed, at which point the claimants suffered economic loss because their building was defective. Mr Holwill further submits that the claimants' cause of action for negligent design was 'single and indivisible' and accrued when damage (economic loss) was first suffered. The fact that the claimants had chosen to claim the cost of carrying out work to remedy the physical damage to their building was not relevant and could not be used to get round the fact that their true claim was time-barred.

I think the simple answer to these submissions is that we are bound by the decisions in *Pirelli General Cable Works Ltd* v. *Oscar Faber & Partners (a firm)* [1983]1 All ER 65, [1983] 2 AC 1 and *Ketteman* v. *Hansel Properties Ltd* to uphold the decision of the district judge. The facts in the *Pirelli* case are indistinguishable from those in the present case. The *Pirelli* case was approved in *Ketteman*'s case and was cited without disapproval in *Murphy* v. *Brentwood DC* (1990) 21 ConLR 1, [1991] 1 AC 398 by the House which included two members (Lord Bridge and Lord Brandon) who were party to the decision in the *Pirelli* case. It has not been expressly overruled and I am not persuaded that this has been done impliedly. Lord Lloyd left open the question as to whether *Pirelli* was still the law in England. It seems to me that only the House of Lords can decide whether it is or not.

PART IV

SUB-CONTRACTS AND SUB-CONTRACTING

Powell-Smith and Furmston's Building Contract Casebook, Fifth Edition. Michael Furmston.

Chapter 20
Assignment and Sub-letting

Building contracts commonly contain clauses restraining assignment and sub-letting. Even without such clauses, a party would not be entitled to assign his obligations without the other party's consent; but he would be entitled to assign his rights. At one time contractors commonly obtained finance by assigning the right to receive payment. If the contract forbids assignment, such an assignment is ineffective to transfer rights to the assignee against the employer.

Linden Gardens Trust Ltd v. Lenesta Sludge Disposals Ltd

St Martins Property Corporation Ltd v. Sir Robert McAlpine & Sons Ltd

HOUSE OF LORDS

(1993) 36 Con LR 1

These two cases were heard together before the House of Lords because they raised overlapping issues. The facts of the second case are given on p. 256.

In *Linden Gardens Trust Ltd* v. *Lenesta Sludge Disposals Ltd* the plaintiffs (Linden Gardens) were leaseholders of part of premises in Jermyn Street, London, having acquired their leasehold interests by assignment from Stock Conversion. Prior to the assignments, the assignors had contracted with the second defendant for the carrying out of alteration works at the premises. The contract was in JCT 63 form, and clause 17(1) provided that:

> The employer shall not without the written consent of the contractor assign this contract.

The third defendants were engaged for remedial work in February 1985. After completion of the works the assignors issued a writ against the first defendants claiming damages for breach of contract and for negligence. Prior to the issue of that writ, the assignors assigned their leasehold interest in three floors of the premises to the plaintiffs, and subsequent thereto assigned to them their interest in another floor.

Powell-Smith and Furmston's Building Contract Casebook, Fifth Edition. Michael Furmston.

On 14 January 1987 the assignors assigned to the plaintiffs their 'right of action as pleaded . . . against Lenesta Sludge Disposals Ltd' and 'all other rights of action currently vested in the assignors which are or were incidental to their leasehold interest'.

On 31 July 1987, by consent, Linden Gardens Trust Ltd were substituted as plaintiffs in the action.

Further asbestos was found in the premises in 1987 and 1988, and on 20 January 1989 leave to add the second and third defendants was granted and the writ was amended accordingly. After service of defences, the following were ordered to be tried as preliminary issues:

> (1) Are the plaintiffs entitled by virtue of the deed of assignment pleaded in paragraph 1F of the amended statement of claim to recover damages against the defendants in respect of the various causes of action and heads of loss pleaded (a) where the loss was incurred by [the assignors] prior to the said deed of assignment, (b) where the loss was incurred by the plaintiffs subsequent thereto?
> (2) Were [the assignors] precluded from lawfully assigning rights of action to the plaintiffs against the second defendants by clause 17(1) of the contract dated 19 July 1979 made between [the assignors] and the second defendants?

The Court of Appeal held (Staughton LJ dissenting) that clause 17(1) of JCT 63 prohibited assignment of the contract itself but it did not prohibit the assignment of benefits arising under the contract. Such benefits included the right to sue for breach of the contract where the rules governing assignment of causes of action were complied with. Since, in the present case, the assignee had both a proprietary and a commercial interest, the assignment of Stock Conversion's causes of action was valid.

Held: allowing the appeal, that clause 17(1) of JCT 63 prohibited the assignment both of the contract itself and of benefits arising under the contract.

> LORD BROWNE-WILKINSON: The argument runs as follows. On any basis, clause 17 is unhappily drafted in that it refers to an assignment of 'the contract'. It is trite law that it is, in any event, impossible to assign 'the contract' as a whole, i.e. including both burden and benefit. The burden of a contract can never be assigned without the consent of the other party to the contract in which event such consent will give rise to a novation. Therefore one has to discover what the parties meant by this inelegant phrase. It is said that the intention of the parties in using the words 'assign this contract' is demonstrated by clause 17(2) which prohibits both the assignment of the contract by the contractor without the employer's consent and the subletting of any portion of the works without the consent of the architect. In clause 17(2), the contractor is only expressly prevented from sub-letting 'any portion of the works'. Yet it must have been the parties' intention to limit the contractor's rights to sublet the whole of the works. Accordingly, the words in clause 17(2) 'assign this contract' have to be read as meaning 'sublet the whole of the works'. If that is the meaning of the words 'assign this contract' in clause 17(2) they must bear the same meaning in clause 17(1), which accordingly only prohibits the employer from giving substitute performance and does not prohibit the assignment of the benefit of the contract.

Like the majority of the Court of Appeal, I am unable to accept this argument. Although it is true that the phrase 'assign this contract' is not strictly accurate, lawyers frequently use those words inaccurately to describe an assignment of the benefit of a contract since every lawyer knows that the burden of a contract cannot be assigned: see e.g. *Nokes* v. *Doncaster Amalgamated Collieries Ltd* [1940] 3 All ER 549 at 551–552, [1940] AC 1014 at 1019–1020. The prohibition in clause 17(2) against subletting 'any portion of the works' necessarily produces a prohibition against the subletting of the whole of the works: any subletting of the whole will necessarily include a subletting of a portion and is therefore prohibited. Therefore there is no ground for reading the words 'assign this contract' in clause 17(1) as referring only to subletting the whole. Decisively, both clauses 17(1) and (2) clearly distinguish between 'assignment' and 'subletting': it is therefore impossible to read the word 'assign' as meaning 'sublet'. Finally, I find it difficult to comprehend the concept of an employer 'subletting' the performance of his contractual duties which consist primarily of providing access to the site and paying for the works.

Accordingly, in my view clause 17(1) of the contract prohibited the assignment by the employer of the benefit of the contract. This, by itself, is fatal to the claim by Investments (as assignee) in the St Martins case.

The majority in the Court of Appeal drew a distinction between an assignment of the right to require future performance of a contract by the other party on the one hand and an assignment of the benefits arising under the contract (e.g. to receive payment due under it or to enforce accrued rights of action) on the other hand. They held that clause 17 only prohibited the assignment of the benefits arising under the contract, in particular accrued causes of action. Therefore, in the Linden Gardens case, where all the relevant breaches of contract by the contractors predated the assignment, an assignment to Linden Gardens of the accrued rights of action for breach was not prohibited. In contrast, in the St Martins case, where all the breaches of contract occurred after the date of the assignment, the majority of the Court of Appeal held that it was a breach of clause 17 to seek to transfer the right to future performance.

This distinction between assigning the right to a future performance of a contract and assigning the benefits arising under a contract was largely founded on a note by Professor Goode 'Inalienable Rights' (1979) 42 MLR 553 on *Helstan Securities Ltd* v. *Hertfordshire CC* [1978] 3 All ER 262. In that case a contract contained a clause prohibiting the contractor from assigning the contract 'or any benefit therein or thereunder'. The contractors assigned to the plaintiffs the right to a liquidated sum of money then alleged to be due to the contractors under the contract. Croom-Johnson J held that the plaintiffs, as assignees, could not sue the employers to recover this sum of money.

In his note Professor Goode rightly pointed out that where a contract between A and B prohibits assignment of contractual rights by A, the effect of such a prohibition is a question of the construction of the contract. There are at least four possible interpretations: (1) that the term does not invalidate a purported assignment by A to C but gives rise only to a claim by B against A for damages for breach of the prohibition; (2) that the term precludes or invalidates any assignment by A to C (so as to entitle B to pay the debt to A) but not so as to preclude A from agreeing, as between himself and C, that he will account to C for what A receives from B: *Re Turcan* (1888) 40 Ch D 5; (3) that A is precluded not only from effectively assigning the contractual rights to C, but also from agreeing to account to C for the fruits of the contract when received by A from B; (4) that a purported assignment by A to C constitutes a repudiatory breach of condition entitling B not merely to refuse to pay C but also to refuse to pay A.

Professor Goode then expressed the view that construction (2) (being the *Helstan* case itself) was permissible and effective but that construction (3) to the extent that it purported to render void not only the assignment as between B and C but also as between A and C was contrary to law.

I am content to accept Professor Goode's classification and conclusions, though I am bound to say that I think cases within categories (1) and (4) are very unlikely to occur. But Professor Goode's classification provides no warrant for the view taken by the majority of the Court of Appeal in the present case: he does not discuss or envisage a case where a contractual prohibition against assignment is to be construed as prohibiting an assignment by A to C of rights of future performance but does not prohibit the assignment by A to C of 'the fruits of performance' e.g. accrued rights of action or debts. Professor Goode only draws a distinction between the assignment of rights to performance and the assignment of rights under the contract in two connections: first in dealing with the effect of a prohibited assignment as between the assignor and the assignee (in categories (2) and (3)); secondly, in dealing with contracts for personal services. In the latter, he rightly points out that, although an author who has contracted to write a book for a fee cannot perform the contract by supplying a book written by a third party, if he writes the book himself he can assign the right to the fee – the fruits of performance. He expressly mentions that such right to assign the fruits of performance can be prohibited by the express terms of the contract.

However, although I do not think that Professor Goode's article throws any light on the true construction of clause 17, I accept that it is at least hypothetically possible that there might be a case in which the contractual prohibitory term is so expressed as to render invalid the assignment of right to future performance but not so as to render invalid assignment of the fruits of performance. The question in each case must turn on the terms of the contract in question.

The question is to what extent does clause 17 on its true construction restrict rights of assignment which would otherwise exist? In the context of a complicated building contract, I find it impossible to construe clause 17 as prohibiting only the assignment of rights to future performance, leaving each party free to assign the fruits of the contract. The reason for including the contractual prohibition viewed from the contractor's point of view must be that the contractor wishes to ensure that he deals, and deals only, with the particular employer with whom he has chosen to enter into a contract. Building contracts are pregnant with disputes: some employers are much more reasonable than others in dealing with such disputes. The disputes frequently arise in the context of the contractor suing for the price and being met by a claim for abatement of the price or cross-claims founded on an allegation that the performance of the contract has been defective. Say that, before the final instalment of the price has been paid, the employer has assigned the benefits under the contract to a third party, there being at the time existing rights of action for defective work. On the Court of Appeal's view, those rights of action would have vested in the assignee. Would the original employer be entitled to an abatement of the price, even though the cross-claims would be vested in the assignee? If so, would the assignee be a necessary party to any settlement or litigation of the claims for defective work, thereby requiring the contractor to deal with two parties (one not of his choice) in order to recover the price for the works from the employer? I cannot believe that the parties ever intended to permit such a confused position to arise.

Again, say that before completion of the works the employers assigned the land, together with the existing causes of action against the contractor, to a third party and shortly thereafter the contractor committed a repudiatory breach? On the construction preferred by the Court of Appeal, the right to insist on further performance, being unassignable, would have remained with the original employers whereas the other causes of action and the land would belong to the assignee. Who could decide whether to accept the repudiation, the assignor or the assignee?

These possibilities of confusion (and many others which could be postulated) persuade me that parties who have specifically contracted to prohibit the assignment of the contract cannot have intended to draw a distinction between the right to performance of the contract and the right to the fruits of the contract. In my view they cannot have contemplated a position in which the right to future performance and the right to benefits accrued under the contract should become vested in two separate people. I say again that that result could have been achieved by careful and intricate drafting, spelling out the parties' intentions if they had them. But in the absence of such a clearly expressed intention, it would be wrong to attribute such a perverse intention to the parties. In my judgment, clause 17 clearly prohibits the assignment of any benefit of or under the contract.

It follows that the purported assignment to Linden Gardens without the consent of the contractors constituted a breach of clause 17. The claim of Linden Gardens as assignee must therefore fail unless it can show that the prohibition of clause 17 was either void as being contrary to public policy or, notwithstanding the breach of clause 17, the assignment was effective to assign the chose in action to Linden Gardens . . .

In the face of this authority, the House is being invited to change the law by holding that such a prohibition is void as contrary to public policy. For myself I can see no good reason for so doing. Nothing was urged in argument as showing that such a prohibition was contrary to the public interest beyond the fact that such prohibition renders the chose in action inalienable. Certainly in the context of rights over land the law does not favour restrictions on alienability. But even in relation to land law a prohibition against the assignment of a lease is valid. We were not referred to any English case in which the courts have had to consider restrictions on the alienation of tangible personal property, probably because there are few cases in which there would be any desire to restrict such alienation. In the case of real property there is a defined and limited supply of the commodity and it has been held contrary to public policy to restrict the free market. But no such reason can apply to contractual rights: there is no public need for a market in choses in action. A party to a building contract, as I have sought to explain, can have a genuine commercial interest in seeking to ensure that he is in contractual relations only with a person whom he has selected as the other party to the contract. In the circumstances, I can see no policy reason why a contractual prohibition on assignment of contractual rights should be held contrary to public policy . . .

Therefore the existing authorities established that an attempted assignment of contractual rights in breach of a contractual prohibition is ineffective to transfer such contractual rights. I regard the law as being satisfactorily settled in that sense. If the law were otherwise, it would defeat the legitimate commercial reason for inserting the contractual prohibition, viz, to ensure that the original parties to the contract are not brought into direct contractual relations with third parties . . .

It is normally permissible to delegate performance of a contract so long as the delegate's performance is exact. In practice, large amounts of most construction contracts are so delegated to sub-contractors. Most construction contracts require consent to sub-contracting. It is not permissible to delegate performance where the contract expressly or impliedly requires personal performance. Normally personal performance is not required in construction contracts, but the next case provides an exception.

Southway Group Ltd v. Wolff

COURT OF APPEAL (CIVIL DIVISION)
(1991) 28 Con LR 109

The plaintiffs (Southway) carry on business in the manufacture of fitted kitchens. In 1989 they owned and occupied a property in Hendon which consisted of a warehouse and a small amount of adjoining land. In January 1989 they contracted to sell the property to Brandgrange Ltd (Brandgrange). Brandgrange was a shell company wholly owned by Initiative Co-Partnership Ltd, which was in turn owned as to 49 per cent by a Mr Ormonde and as to 51 per cent by Initiative Developments Ltd, a company owned by a Mr Obermeister and his wife. Mr Obermeister was an architect; Mr Ormonde was a property developer with particular expertise in devising and financing development schemes, obtaining property ripe for development and obtaining planning permission.

Under the contract of sale completion was to be on 30 April 1990, or earlier on 28 days' written notice, not to be given earlier than 5 December 1989. A notice, which was accepted as good, was given on 17 November 1989 to complete on 5 March 1990.

By a contract dated 21 December 1989, Brandgrange agreed to sell the property to the defendants (the trustees), who were mother and son and the trustees of the Wolff Charity Trust, for £2.9 m. This resale contract contained undertakings by Brandgrange to carry out redevelopment to the property. The redevelopment was described in a specification attached to the contract which was skeletal in the extreme.

Brandgrange failed to complete on 5 March and Southway served notice to complete under clause 22 of the National Conditions of Sale. On 21 March Southway and Brandgrange entered into a deed of assignment by which a new completion date of 17 April was substituted, time being of the essence, and Brandgrange assigned to Southway the benefits of their resale contract with the trustees. Southway gave notice of this assignment to the trustees.

Brandgrange failed to complete on 17 April. On 19 April Southway treated this failure as a repudiation and terminated the contract.

Southway then determined to carry out the redevelopment works under the resale contract themselves. The trustees indicated that this was not acceptable to them and Southway sought a declaration that if they carried out work which complied with the specification within the time provided by the resale contract and tendered a valid transfer of the building, they would be entitled to the purchase price under the contract between Brandgrange and the trustees.

Held: The contract between Brandgrange and the trustees was one which called for personal performance by Brandgrange, and vicarious performance by Southway or contractors employed by Southway was not permitted.

BINGHAM LJ: It is in general permissible for A, who has entered into a contract with B, to assign the benefit of that contract to C. This does not require the consent of B, since in the ordinary way it does not matter to B whether the benefit of the contract is enjoyed by A or by a third party of A's choice such as C. But it is elementary law that A cannot without the consent of B assign the burden of the contract to C, because B contracted for performance by A and he cannot be required against his will to accept performance by C or anyone other than A. If A wishes to assign the burden of the contract to C he must obtain the consent of B, upon which the contract is novated by the substitution of C for A as a contracting party.

It does not, however, follow that in the absence of a novation A must personally perform all the obligations he has assumed under his contract with B. In some classes of contract, as where B commissions A to write a book or paint a picture or teach him to play the violin, it would usually be clear that personal performance by A was required. In other cases, as where A undertakes to repair B's shoes or mend B's watch or drive B to the airport, it may be open to A to perform the contract vicariously by employing the services of C. In this situation the contractual nexus remains unaltered, since A remains liable to B for performance of the contract and no contractual relationship arises between B and C.

Whether a given contract requires personal performance by A, or whether (and if so to what extent) A may perform his contractual obligations vicariously, is in my opinion a question of contractual construction. That does not mean that the court is confined to semantic analysis of the written record of the parties' contract, if there is one. Such is not the modern approach to construction of a commercial contract. It means that the court must do its best, by reference to all admissible materials, to make an objective judgment of what A and B intended in this regard. Where A and B, perhaps with legal advice, have entered into a long and ambitious written contract, the terms of that contract may well be conclusive or almost so. Where a written contract is short and summary, or the contract is made orally, surrounding circumstances are likely to be of much greater significance: a reliable objective assessment of what the parties intended may well require account to be taken of such matters as the type of contract in question, the state of the market, the commercial position of the respective contracting parties, personal relationships between the main protagonists on each side and matters of that kind. But it is of course the joint intention of the parties which matters, not the secret intention of either, so no account may be taken of matters which may influence the mind of one party but are unknown to the other.

The issue in the present appeal does not concern assignment. It concerns vicarious performance. The question is whether, under its contract with the Wolffs, Brandgrange was entitled to perform all its obligations vicariously through the medium of Southway . . .

Assignment involves a transfer of the benefit of a contract from assignor to assignee. It may be desired instead to replace one party by another. This is possible but only by a completely new contract to which all relevant persons are parties. Such a new contract is often called a novation.

Novations are nowadays quite common in certain construction situations. Suppose an employer intends to let a contract on a design-and-build basis under which the contractor will be liable for design (see p. 66). The employer may want to get the designer working before the contractor is appointed, with a view to a novation which will transfer the designer to the contractor's team at a later stage. The following case revealed that this process has hidden difficulties.

Blyth & Blyth Ltd v. Carillion Construction Ltd

COURT OF SESSION (OUTER HOUSE)

(2001) 79 Con LR 142

Blyth & Blyth started working on a project for the employers, THI Leisure, in about January 1997. On 16 March 1998, Blyth & Blyth, THI Leisure and Carillion Construction entered into a three-party novation agreement. In broad terms this was designed to transfer Blyth & Blyth to the Carillion team. In due course Blyth & Blyth sued for unpaid fees and Carillion counterclaimed for defective work. This raised the question of the extent to which Blyth & Blyth were liable to Carillion for work done before the novation. The judge held that Carillion could not recover for their own loss suffered before the novation, because the natural measure of loss in such a case was the loss suffered by THI. (Of course this must turn at least in part on the proper construction of the novation agreement.)

LORD EASSIE: 39. One of the bases upon which counsel for the defenders sought to justify the contention that the deemed alteration of the identity of the parties to the deed of appointment gave rise to a valid claim for the defenders' losses (and not an employers' loss by way of assignation or quasi assignation) was the contention that in view of the terms of the building contract, whereby the defenders undertook responsibility to the employer for design, the employer could never suffer any loss, he having a contractual claim against the defenders. It was said that, since no loss could thus accrue to the employer, there was nothing to be assigned to the contractor. Accordingly, so ran the argument, the relevant provisions of the novation agreement would lack content and lead to a result which the parties could never have intended. The inability of the employer, in the context of the contractual arrangements, to suffer any loss was, said counsel for the defenders, a fundamental misconception in the argument for the pursuers.

40. In addressing this submission, I would observe, as a perhaps minor preliminary, that this argument assumes in the first place that the design undertakings in the building contract are entirely co-extensive with the obligations owed by the pursuers to the employer. I am not persuaded that such is the case. There appear to me to be provisions in the deed of appointment which are different from and go well beyond the scope of the design undertakings in the building contract. I mention, at random, paras 2.2, 2.3, 2.4 and 2.13 of Schedule 1 to the deed of appointment.

41. Secondly, and more importantly, even in relation to the design obligations ultimately undertaken by the defenders so far as co-extensive with the design obligations of the pursuers to the employer, the argument for the defenders – as I understood it – assumed that the existence of a contractual claim by the employer against the defenders for defective design precluded the existence of any claim by the employer against the pursuers. That assumption was disputed by counsel for the pursuers. No authorities on this matter were cited to me, but in principle it appears to me that a valid claim by the employer against the engineer on the basis of defective design is not defeated by the employer's having the possibility of making a valid claim on a similar basis against the contractor by virtue of the design obligations undertaken by the contractor in the building contract. Thus, if defects were to emerge in the leisure centre requiring the execution of repairs and its closing down, and if the defects arose from defective design the genesis of which lay in what was done for the employer before the date of the tender, one would readily think the employer to have a perfectly prestable claim against the designer.

The fact that the terms of the building contract might also give the employer a claim against the contractor by reason of his having effectively underwritten the pre-construction design, would not discharge or destroy the liability of the designer to the employer. Further, applying the litmus test of insolvency of the contractor, the employers' claim against the consultant must persist. There being such liability in solidum it may well be of course that questions of relief or apportionment between the two obligants would arise.

42. I am accordingly not persuaded that the pursuers' argument on the construction of the novation agreement to the effect that the agreement operates as akin to assignation of the claims prestable at the instance of the employer suffers the radical misconception asserted by counsel for the defenders, namely lack of content, on the view that by reason of the contractor's liability under the building contract there can never be a transmissible loss upon which such assignation could operate.

Offer-Hoar and others v. Larkstore Ltd (Technotrade Ltd, Part 20 defendant)

COURT OF APPEAL, CIVIL DIVISION

(2006) 109 ConLR 92, [2006] WECA Civ 1079

In December 1998 Technotrade Ltd (Technotrade) produced a geotechnical report on a site in Hythe in Kent as the result of a contract with Starglade Ltd (Starglade), the owners of the site. In July 1999 Starglade sold the site to Larkstore Ltd (Larkstore). Larkstore came into possession of the Technotrade report and used it without seeking Technotrade's consent. There was no contract between Larkstore and Technotrade.

Larkstore engaged Bess Ltd (Bess), a building and civil engineering contractor, to carry out work on the site. The contract documentation included the Technotrade report.

On 13 October 2001, while works were being carried out on the site by Bess, a landslip occurred which caused damage to properties nearby, uphill of the site. The owners of these properties brought claims. Larkstore had to undertake extensive stabilisation works.

On 23 February 2004 a deed of assignment was entered into between Starglade and Larkstore providing for the assignment of the report and the right to sue in respect of any breach by Technotrade and entitling Larkstore to bring any available claims against Technotrade.

The owners of the affected land commenced proceedings against Larkstore and Bess. Bess was insolvent. On 6 October 2004 Larkstore commenced Part 20 proceedings against Technotrade. On 14 June 2005 an order was made for the trial of three preliminary issues arising out of the Pt 20 claim. The present appeal is concerned with the first of these, which concerns the effect of the assignment.

MUMMERY LJ:

The remedy in damages for breach of contract is not limited to the loss that could have been proved at the date when the breach occurred and the cause of action first arose. Subject to factual and legal issues of causation, remoteness, quantum and limitation of actions, there is a remedy in damages against the contract-breaker for loss which occurs after the cause of action has accrued. A cause of action may arise years before any substantial damage occurs, as, for example, in the case of negligent advice on title. There is no legal principle which protects the contract-breaker by excluding his liability for substantial damage that occurs after the initial breach of contract.

What difference, if any, can an assignment of the cause of action make to the remedies available to the assignee against the contract-breaker? A statutory assignment in writing under s. 136 of the Law of Property Act 1925, of which express notice has been given, as was done in this case, is effectual in law to pass and transfer from the date of notice the legal right to the thing in action and 'all legal and other remedies for the same'. The statutory assignment is expressly made 'subject to equities having priority over the right of the assignee'.

Applying this concise account of the legal principles to the particular circumstances of this case, it is, in my judgment, fallacious to contend that Larkstore cannot recover substantial damages from Technotrade, even if it can prove that Technotrade was in breach of contract and otherwise liable for them.

The contention is based on the propositions that Starglade (the assignor) had only suffered nominal damages at the date of the assignment, because it no longer owned the site, and that Larkstore (the assignee) could not acquire by assignment from Starglade any greater right than Starglade had against Technotrade.

As I see it, that is not the true legal position. What was assigned by Starglade to Larkstore was a cause of action for breach of contract against Technotrade and the legal remedies for it. It was not an assignment of 'a loss', as Mr Friedman described it in his attempt to persuade the court that the amount of the loss recoverable by Larkstore was limited by what loss had been suffered by Starglade, in this case nil. The assignment included the remedy in damages for the cause of action. The remedy in damages for breach of contract is not, in principle, limited to the loss suffered as at the date of the accrual of the cause of action or as at any particular point of time thereafter.

Chapter 21
Domestic Sub-contractors

A major practical problem in relation to sub-contracts is the way in which the main contract and sub-contract fit together. Provisions of the main contract cannot be read into the sub-contract unless they have been expressly incorporated.

Smith and Montgomery v. Johnson Bros Co Ltd

ONTARIO HIGH COURT

(1954) 1 DLR 392

The defendants were main contractors to construct a tunnel sewer for the City of Hamilton through the Hamilton mountain. The plaintiffs, who were miners, undertook a sub-contract for tunnelling 'according to the dimensions and specifications as set forth in the contract between the City of Hamilton and the [defendants]'.

Held: That these words were not apt to incorporate the terms and conditions of the main contract into the sub-contract.

> SCHROEDER J: It is, of course well established that plans and specifications, if not contained in a contract itself but referred to therein or annexed thereto, must be construed therewith when identified. Wherever the plans and specifications are referred to for a specific purpose only, however, they become part of the contract for that purpose only and should be treated as irrelevant for all other purposes. The covenant of the plaintiffs to do the tunnelling 'according to the dimensions and specifications as set forth in the contract between the City of Hamilton and the [Johnson company]' would undoubtedly require the plaintiffs to dig out a tunnel 10ft 6in by 12ft 6in in dimensions, and any other provision affecting the physical characteristics of such tunnel no doubt would be applicable to and binding upon the plaintiffs, but I am not prepared to hold that the reference to the specifications as expressed in this context imports into the contract between the plaintiffs and the defendant, so as to make binding upon the plaintiff-subcontractors, who operate in a very small way, such sweeping and extraordinary powers as are exercisable by the city engineer. Only that part of the general specifications is to be read into the contract in question that relates to the specific purpose for which the same were referred to. Counsel have been unable to refer me to any authority in which all the general provisions of specifications forming part of a main contract became part of a subcontract,

Powell-Smith and Furmston's Building Contract Casebook, Fifth Edition. Michael Furmston.

> unless the latter by its terms made very clear the intention of the parties that all terms and provisions of the principal contract were to form part of the agreement between the principal contractor and the subcontractor. The general clauses under which the city engineer ordered a stoppage of the work and a change in the method of procedure are more in the nature of such provisions as one is likely to find in specifications furnished for the guidance of bidders for the original contract. It does not seem to me to be clear from the situation and the circumstances of the parties that it was their intention that, except in the restricted sense already mentioned, the specifications forming part of the original contract should be a part of the subcontract.

In the absence of express undertakings the main contractor does not necessarily undertake to have the site ready for a sub-contractor by a particular date (but only within a reasonable time).

Piggott Construction Co Ltd v. W.J. Crowe Ltd

ONTARIO COURT OF APPEAL

(1961) 27 DLR (2d) 258

The appellant was appointed main contractor on an extensive government contract. Because of shortages of steel there was no date fixed for the start of plastering work.

The appellant started work on the main contract in September 1955 and the sub-contract was undertaken in the same month. The appellant then expected that it would take about two years to complete the main contract and the respondent expected that the plastering would take about 44 weeks. In September 1956 the respondent was told that the site would be ready for plastering within a few days but shortly afterwards that it would not be ready until the spring of 1957. In April 1957 the appellant informed the respondent that the works were now ready. The respondent then took the position that it would not now do the work unless it was paid extra and claimed that it was so entitled on the ground that the appellant was in breach by not having been ready for the sub-contract work earlier. The trial judge held that there was an implied term in the sub-contract that the building would be ready for plastering not later than 1 January 1957.

Held: There was no such implied term.

> LAIDLAW JA: Looking at the contract between the appellant and the Crown and also the sub-contract between the appellant and the respondent in a reasonable and business way, there is no necessary implication that the appellant agreed that the male infirmary or any other building to be erected by it would be ready for the commencement of the work of lathing and plastering not later than 1 January 1957. While the care and control of the work generally was the responsibility of the appellant and it was provided in the general instructions and requirements made part of the general contract and subcontract that the work was to be carried forward 'in a proper manner and as expeditiously as possible' nevertheless, it is my view that in the circumstances existing at the time the contract was entered into the appellant would not have assumed an obligation to have any of the buildings to be erected by it, ready for commencement of the work of lathing and plastering within any specified period of time. It

would have been utter folly and gross recklessness for it to have done so. It knew that the progress of the work of construction to be done by it was dependent upon the supply of necessary steel and upon other uncertain factors and conditions. It could not foresee whether the delay occasioned by that shortage might continue for a period of weeks or possibly months. Again, the work of 'plumbing and drainage, heating and ventilating, electrical work and elevator and dumb waiter installations' was excluded from the work to be done by the appellant and was provided for under separate contracts. The appellant was not responsible for the commencement or progress of work to be done under those contracts and it realized that commencement of the work of lathing and plastering depended thereon. It is inconceivable that in the particular circumstances any building contractor and especially one possessing the great experience of the appellant, would undertake to have any one part of the work of construction ready for commencement thereof at any particular time or within any prescribed period. Therefore, having these contingent and uncertain matters and conditions clearly and fully in mind, it was the positive and calculated intention of the parties to omit from the terms and conditions of the subcontract an express stipulation and any provision from which it might be implied that the appellant was bound to have any buildings ready for commencement of the work of lathing and plastering at any stated time or within any prescribed period.

The learned trial judge found contrary to the evidence of the respondent that there was no assurance by the appellant that the respondent could start work in September 1956. Likewise, in my opinion, he ought to have concluded from the evidence and the particular circumstances in this case that it was the intention of both the appellant and the respondent that the appellant was not bound to have any part of the premises ready for commencement of lathing and plastering within any particular period of time after the commencement of the work of construction. It is my respectful view that the term implied by Gale J, is directly contrary to the intention of the parties at the time they entered into the subcontract and that an obligation was improperly created and imposed on the appellant by the Court. The authorities to which I have referred make it plain that the paramount consideration governing the decision as to whether or not a term should be implied in a contract is in the last analysis the intention of the parties at the time the contract was entered into and the Court should not add to the contract an obligation which clearly was not and could not have been intended by the parties.

Chapter 22
Nominated Sub-contractors

The use of nominated sub-contractors has been extensive in modern building contracts, but although convenient in some respects, it has generated more than its fair share of legal problems. While lawyers tend to start from the position that a nominated sub-contractor is just another sub-contractor for whom the main contractor is responsible, there is no doubt that, commercially, nominated sub-contractors often look for business to employers and not to contractors and that contractors feel that failures by nominated sub-contractors should be laid at the employer's door. Neither the standard forms nor the cases have fully resolved these tensions. The elaborated provisions for nominated sub-contractors in JCT 98 have disappeared from JCT 2005.

The contractual position between contractor and nominated sub-contractor must be found in the sub-contract.

A. Davies & Co (Shopfitters) Ltd v. William Old Ltd

QUEEN'S BENCH DIVISION

(1969) 67 LGR 395

The defendant was the main contractor for the erection of a new store on JCT 63 terms, which provided for certain work to be sub-contracted to a subcontractor nominated by the architect. The architect obtained a tender for this work from the plaintiff and instructed the defendant to accept it. The defendants sent an order to the plaintiff on their standard printed form, which contained on its reverse printed conditions which included a 'pay when paid' clause. The plaintiffs wrote thanking the defendant for the order and carried out the work. The employer became insolvent before having paid for all the work.

Held: The contract between plaintiff and defendant was on the basis of the defendant's printed conditions, which the plaintiff had accepted. The defendant was only liable to pay for the work in so far as he had himself been paid by the employer.

Powell-Smith and Furmston's Building Contract Casebook, Fifth Edition. Michael Furmston.

The existence of a contract between the employer and a nominated sub-contractor, as is the case where the JCT Standard Form of Employer/Nominated Sub-Contractor Agreement is executed, does not affect the application of the main contract.

George E. Taylor & Co Ltd v. G. Percy Trentham Ltd

QUEEN'S BENCH DIVISION

(1980) 16 BLR 15

Taylor were nominated sub-contractors to Trentham, and also had a contract with the employer whereby they warranted due performance of the subcontract works so that the main contractors should not become entitled to an extension of time. The employer paid Trentham only £7,526,87 against an interim certificate for £22,101, the amount withheld being the balance payable to the sub-contractors after deduction of the main contractor's claim against them for delay.

Held: The employer was not entitled to withhold the money, the contract between employer and nominated sub-contractor being *res inter alios acta.*

JUDGE WILLIAM STABB QC: The architect has issued an interim certificate whereby he has certified that £22,101 is due and payable by the employers to the main contractors, and, by clause 30, the main contractors are entitled to be paid that amount, subject at all time to the employers' common law or equitable rights of set-off (see *Gilbert Ash (Northern) Ltd* v. *Modern Engineering (Bristol) Ltd* (1973)). It is not suggested here by the employers that they have any right of set-off as against the main contractors so as to justify the retention of any part of that sum so certified to be due and payable to the main contractors. What the employers contend is that part of that certified sum is due to be paid by the main contractors to the sub-contractors under the provisions of clause 27(b) of the main contract, and that, since they (the employers) have a claim or counter-claim against the sub-contractors under the separate, so-called 'direct' contract, they should be permitted to withhold payment of that sum, certified to be payable under the RIBA contract between them and the main contractors, in order to satisfy wholly or in part damage which they are counter-claiming from the sub-contractors under the 'direct' contract between themselves and the sub-contractors.

So far as sub-contractors and main contractors are concerned, it has been said that the contract between main contractors and employers is *res inter alios acta*. In my view, but even more plainly, it would seem that, so far as employers and main contractors are concerned, the direct contract between employers and sub-contractors is also *res inter alios acta.*

Most building contracts contain some limited provision for direct payment of nominated sub-contractors by the employer. It has been further held that under JCT 63 and the 1963 NFBTE/FASS form of sub-contract the employer holds retention money on trust for the nominated sub-contractor so that in the event of the contractor's insolvency, the nominated sub-contractor may recover these monies in full. The position is the same under JCT 98.

Re Arthur Sanders Ltd

CHANCERY DIVISION

(1981) 17 BLR 125

The relevant facts are set out in the judgment of Nourse J.

NOURSE J: This is an application by the Greater London Council in the liquidation of a company called Arthur Sanders Limited, which formerly carried on business as a building contractor. Shortly stated, the question is whether the council is entitled to claim a particular set-off under section 31 of the Bankruptcy Act 1914, which is introduced into companies' insolvent winding-up by section 317 of the Companies Act 1948. There is no difficulty over the terms or effect of section 31. The question is what, under two contracts made between the council and the company and those made between the company and six nominated sub-contractors in relation to one of the main contracts, is the extent of the set-off to be made under the section. In other words, the question is more one on the law of building contracts.

As I have said, the council had two contracts with the company, the first dated 9 July 1969 for the execution of work at Wellingborough and the second dated 2 October 1970 for the execution of work at Bletchley. The two contracts were, for present purposes, in identical terms. They incorporated the council's 'Standard Conditions' for use in contracts for works with quantities, which were based on the 1963 edition (as then revised) of the RIBA Standard Conditions applicable to a case where the employer is a local authority. I was told that there has been no material alteration to the RIBA Conditions since 1970 and, further, that there is no material distinction between a case where the employer is a local authority and one where he is a private employer. The six sub-contracts were all in the standard form of sub-contract (1963 edition as then revised) issued under the sanction of the National Federation of Building Trades Employers and the Federation of Associations of Specialists and Sub-Contractors. That is usually known as the FASS sub-contract. Again I was told that there has been no material alteration to that since 1970. It therefore appears that the decision in this case will apply to any building contract for works with quantities which incorporates the RIBA Conditions 1963 edition (as since revised) where there is a FASS sub-contract 1963 edition (as since revised). The particular provisions with which this case is concerned are clauses 25, 27 and 30 of the RIBA Conditions and clause 11 of the FASS sub-contract.

On 13 April 1971 the Company resolved to go into creditors' voluntary winding-up and [a liquidator was appointed]. The effect of that was automatically to determine the employment of the company under both its contracts with the council pursuant to clause 25(2) of the RIBA Conditions. At that date the works under the Wellingborough contract were nearly completed, but those under the Bletchley contract had a long way to go. In neither case was the employment of the company reinstated and the works were completed by other contractors. The result was that the council was left holding net retentions under the Wellingborough contract of £11,086.77, included in which was a sum of £1,374.56 in respect of work carried out by the six nominated sub-contractors, whereas under the Bletchley contract the extra cost to the council for getting the works completed was £98,155. Under the Wellingborough contract the Council owes £11,086.77, but under the Bletchley contract it is owed £98,155. Those figures are not in dispute.

The council contends that it is entitled under section 31 to set off the £11,086.77 owed by it under the Wellingborough contract against the £98,155 owed by the company under the Bletchley contract, and to prove in the liquidation for the balance of £87,068.23. The liquidator accepts that the council is entitled to set off £9,712.21 of the £11,086.77 against the £98,155, but he disputes its right to set off the balance of £1,374.56. He contends that that sum should be paid by the council to him in full, not for the benefit of the general body of creditors, but as trustee for the six nominated sub-contractors. Accordingly, says the liquidator, the council may prove in the liquidation for £88,442.79, but must pay to him the sum of £1,374.56. I should add that the liquidator has so far been able to pay dividends to unsecured creditors amounting to 15p in the £. . . .

In order to analyse the effect of these provisions, it seems to me that I must start with clause 30(4)(a) of the RIBA Conditions. The effect of that is to provide that the employer shall hold the retentions under interim certificates on trust for the contractor, whose beneficial interest therein is expressed to be subject only to a right in the employer to which I shall return later. The first point to note about that provision is that there is no express requirement for the retentions to be funded. However, in *Rayack Construction Ltd* v. *Lampeter Meat Co Ltd* (1979), Vinelott J held that clause 30(4)(a) does impose an obligation on an employer to appropriate and set aside the retentions as a separate trust fund and he made a mandatory order to that effect on motion. That question does not arise in this case, but the employer there was not a local authority but a private company which it was recognised could have got into financial difficulties. The learned judge proceeded on the footing that if no fund were set aside and the employer went into liquidation the contractor would rank as an unsecured creditor, save only for its lien on unfixed goods and materials under the proviso to clause 26(2) of the RIBA Conditions. That may well be so, but I agree with [counsel for the liquidator] that in the case of a solvent employer equity, looking on that as done which ought to be done, will not allow him any advantage from his failure to set aside a fund. Accordingly, it seems to me that I must proceed on the footing that the council did from time to time set aside as a separate fund sums equivalent to the retention money under the interim certificates. I did not understand [counsel for the GLC] to argue to the contrary. Next, it seems to me that the sums thus notionally set aside became held by the company as a trustee. That would seem to me to be a very obvious consequence of the first step, but it was disputed by [the council] on the ground that the provisions of the main contract as a whole render the subject-matter of the trust uncertain. I reject that argument and I do not think that I need say more about it than that . . .

I must now deal with what I have already said appears to me to be the most important question. In my view it is clear that clause 30(4)(a) creates in the first instance a trust of the retention moneys in favour of the contractor. It is then necessary to go to clause 27, which deals with nominated sub-contractors. Although the employer's agent nominates the sub-contractors, it is the contractor who contracts with them. But the proviso to sub-clause (a) erects the two safeguards in favour of the contractor to which I have already referred, the second being that no sub-contractor can be nominated who will not (save where the parties otherwise agree) enter into a sub-contract which provides for the matters set out in the nine paragraphs of sub-clause (a). The only one of those paragraphs with which this case is directly concerned is (viii), which provides that the contractor shall hold a due proportion of the retentions on trust for the sub-contractor. That gives to the sub-contractor a right corresponding to that given to the contractor by clause 30(4)(a) of the RIBA Conditions.

It follows from all this that unless his agent agrees to the contrary the employer will know that any sub-contract will necessarily include a provision of the kind required by (viii) of clause 27(a). Putting it at its lowest it seems to me to be clear that the employer requires and authorises the contractor to enter into a subcontract in that form and that the employer has notice of its contents. The subcontract therefore takes effect as an assignment to the sub-contractor, made with the authority and with the knowledge of the employer, of a due proportion of the contractor's beneficial interest in the retention moneys under clause 30(4)(a) of the main contract. In my judgment there can be no doubt that that is the effect of clause 11(h) of the FASS sub-contract. If authority be needed for that it can be found in the decision of Wynn-Parry J in *Re Tout & Finch Ltd* (1954) . . .

In other words, the position as between employer, contractor and sub-contractor under the present RIBA Conditions and the FASS sub-contract is that the employer holds a due proportion of the retentions on trust for the contractor as trustee for the sub-contractor.

If the foregoing analysis is correct it would seem to follow that as and when the retentions fall to be released the sub-contractor can require the contractor to recover the former's due proportion from the employer. Moreover, whether he is required to do so or not, the contractor can of his own volition take steps to recover all the retentions (including those payable to sub-contractors) from the employer. Applying that to the present case it would mean, other things being equal, that the liquidator was entitled, and for practical purposes bound, to require payment from the council of the £1,374.56 due to the sub-contractors.

In general any power given by the contract to make direct payments to nominated sub-contractors is ineffective whenever the contractor becomes insolvent because it will then conflict with general principles of insolvency law.

B. Mullen & Sons (Contractors) Ltd v. Ross

COURT OF APPEAL IN NORTHERN IRELAND

(1996) 54 Con LR 163

CARSWELL LJ: Counsel for the appellant accordingly argued that the material property of the contractor which vested in the liquidators consisted of a chose in action, the right to sue the employers for payment of the moneys due to him, including that portion which he would have to pay over to the sub-contractor. It was subject to a contingency, namely that the employer might exercise his right to pay the sub-contractor direct. As Kerr J pointed out in his judgment, however, the provision in the contract permitting this was an enabling provision; the employers were entitled, but not obliged, to make such direct payments, which is a direct point of distinction between this case and *Glow Heating Ltd* v. *Eastern Health Board*. In our opinion the matter is determined by the effect upon the operation of the *pari passu* principle of such exercise of the employers' right after the commencement of the liquidation. Until the winding-up resolution was passed, the contractor's interest may have been defeasible by the exercise of the employers' right to make direct payment. Once the company went into liquidation, however, the exercise of the right would remove the sum so paid from the property which should come to the hands of the liquidator, so reducing the amount divisible among the general creditors, and such a result would offend against the *pari passu* principle. It therefore is in our opinion void and the employers have not been entitled to exercise the right of direct payment to the sub-contractor since McLaughlin & Harvey plc went into liquidation.

'Delay on the part of nominated sub-contractors or nominated suppliers' in JCT 98, clause 25.4.7 (JCT 63, clause 23(g)) means a delay on the part of the nominated sub-contractor before the date of completion. It is apparent completion which matters and if the sub-contractor's work proves to be defective, the delay caused by his subsequently returning to site is not within the provision.

City of Westminster v. J. Jarvis & Sons Ltd and Peter Lind Ltd

HOUSE OF LORDS

(1970) 7 BLR 64

Jarvis contracted with Westminster for the erection of a multi-storey car park, with flats, offices, showrooms and ancillary works. The contract was in JCT 63 form. Lind were nominated sub-contractors for the piling work. Lind carried out the work and purported to complete by the sub-contract completion date. Lind left the site. Some weeks later, it was discovered that many of the piles were defective, either as a result of bad workmanship or because of poor materials. Lind carried out remedial works. The main contract works were delayed for some 21½ weeks. The main contractor (Jarvis) claimed an extension of time under clause 23(g).

Held: Jarvis was not entitled to an extension of time since the works had apparently been completed by the due date. 'Delay' within the meaning of the clause occurs only if, by the sub-contract completion date, the sub-contractor has failed to achieve such completion of his work that he cannot hand it over to the main contractor.

LORD WILBERFORCE: [What] is to be made of the clause? One thing seems clear, that 'delay' does not, as the appellant at one time contended, mean 'sloth' or 'dilatoriness' on the part of the sub-contractor. There are at least two good reasons against this meaning; in the first place, it would put an impossible burden on the architect if he were required to form an opinion that the sub-contractor had not worked as fast or diligently as he might have done, and to measure the extent to which time could have been saved had he done so. This part of the contract would, in practice, become unworkable. And, secondly, it is contractually irrelevant whether the sub-contractor could have worked faster: what matters is whether he has done what he agreed to do in the contractual time. If he has, it does not matter, for the purposes of the contract, whether he achieved his target by leisurely methods: if he has not, it does not matter with what feverish energy he set about his work. This suggested test, then, is both unworkable and irrelevant.

That leaves two alternatives. The sub-contractor contends that he is in delay whenever in any respect he fails to fulfil in time his contractual obligation. The employer contends that there is only delay if, by the sub-contract date, the sub-contractor fails to achieve such completion of his work that he cannot hand over to the contractor. Or, putting it negatively, that the sub-contractor is not in delay so long as, by the sub-contract date, he achieves such apparent completion that the contractor is able to take over, notwithstanding that the work so apparently completed may in reality be defective. This, on the employer's argument, may involve a breach of contract, but does not involve delay.

The sub-contractor's contention has at least the merit of simplicity and comprehensiveness: its counsel admitted, indeed claimed, that it brings within the clause any breach of contract, any failure timeously to perform its total obligation, whether by repudiation, or insolvency, or departure from the site, provided always of course that such breach or failure led to delay in the progress of the main contract works. So it was in delay here on and after 20 June 1966, because it had not constructed sound piles by that date. This contention, it was admitted, might seem a little extreme, or even in some cases absurd, but the extremity or absurdity lay in the inclusion of the clause in the contract and this once swallowed the consequences should not be strained at.

My Lords, if such an interpretation were imposed by the words used, it would have to be accepted whatever (short of completely frustrating the contract) the consequences might be. Within the limit I have mentioned the parties must abide by what they have agreed to and it is not for the courts to make a sensible bargain for them. But the words used do not suggest that this is the right meaning: for if it were, why should the word 'delay' be used? Why not frankly exonerate the contractor for any delay in completion due to any breach of contract or failure, *eo nomine*, of the sub-contractor? Add to this difficulty the unattractive consequences which follow and the objections become overwhelming: for the clause would amount to little better than a charter for bad work, wilful failure and default of every kind. The worse the sub-contractor's conduct of his sub-contract and the greater impact it has upon his timeous fulfilment and that of the contractor, the more complete is his escape from liability, the more firmly the loss so occasioned is made to rest on the helpless employer.

The employer's construction at least avoids the worst of these consequences, for it relates the impact of the clause to such default as prevents the contractor from taking over and proceeding with his work. It fits in reasonably well with those provisions in the contract which distinguish between such completion as enables the contract to proceed, and such final and verified completion as enables certificates to be given, final payment to be made and the party cleared of all obligation under the contract.

It is not without its difficulties . . . It is only necessary to point to the fact that if the defects in the piles had been discovered before the sub-contract completion date, and work had been at once put in hand to remedy them – thereby producing a similar period of delay in the completion of the main contract, the clause would, it seems, have applied, but it does not do so if the work was 'complete' (though defective) on that same date so that the contractor could take over. One must set against this the advantage that, if the sub-contract work is apparently completed and handed over, and some defects appear very much later, but before the contract date, as they might in a large contract, this would not, on the employer's construction, be a case of delay, though it might be so on the sub-contractor's. But even so the first type of difficulty is a very grave defect and a serious reflection on the clause: indeed, I cannot believe that the professional body, realising how defective this clause is, will allow it to remain in its present form. But in my opinion, though it is never agreeable to have to choose the lesser of two incongruities, we have to do so here and I find the employer's version qualifies for this not very flattering description.

So finally, how does this construction fit the facts? It is common ground that the sub-contractor 'purported to complete' the piling works by 20 June 1966, and left the site. We know, too, that 'further construction work on the site was then carried out by the contractor'. On 13 July 1966, the architect wrote to the contractor a letter in which he said that 'further to the completion of the [piling] work' the sub-contractor might request the release of retention money against a guarantee. He did, it is true, go on to record suspicion as to the soundness of the piles and to suggest an increase in the guarantee, but the letter was consistent only with his belief that the sub-contract work was 'completed'. The sub-contractor had recorded that this was its position on 9 June 1966. On 16 August 1966, after defects had been revealed, the architect wrote again to the contractor stating that 'as the piling works . . . have not been completed to our satisfaction within the period of the sub-contract' they certified that they ought reasonably to have been completed within that time. This I do not consider alters the view of the matter which all concerned held on 20 June 1966, that the work was completed – though it might turn out to be defective: it is saying no more than that it is now seen to be defective and that the architect was not satisfied with it. The learned judge made no explicit finding but his judgment proceeded on the basis that the sub-contractor had achieved apparent completion, and handed over to the contractor on 20 June 1966. I think that this was correct in fact and in law.

JCT 63 does not deal explicitly with the question of re-nomination where a nominated sub-contractor repudiates the sub-contract. It has been held however that it is implicit in the contract that the employer comes under a duty to re-nominate. JCT 98, clause 35.24 imposes an express duty of re-nomination should a nominated sub-contractor fail.

North-West Metropolitan Regional Hospital Board v. T.A. Bickerton & Son Ltd

HOUSE OF LORDS

[1970] 1 All ER 1039

The facts are set out in the speech of Lord Reid.

LORD REID: My Lords, the appellants employed the respondents as contractors to erect certain buildings at their hospital at Abbotts Langley, and they nominated a company called Speediwarm Ltd as the sub-contractor who was to provide the heating system. Speediwarm then contracted with the contractor but very soon went into liquidation and the liquidator refused to carry out the contract. Then, without prejudice to their respective rights, the employers and contractor arranged that the contractor should do this work. It did so and there is no complaint about its work. But the work cost substantially more than the price at which Speediwarm had agreed to do it. The employers contend that the contractor is only entitled to be paid Speediwarm's price. The contractor contends that, when Speediwarm fell out, the employers ought to have nominated another sub-contractor and paid its price: and that, as they did not do so, they must pay the contractors on the basis of *quantum meruit*.

I must first explain the nature of the contract between the employers and the contractors. The employers put out for tender bills of quantities and drawings. With regard to the greater part of the work the bills were detailed and the contractor filled in its prices. But certain parts of the work were reserved for sub-contractors to be nominated by the employers. With regard to these parts no details were given, and sums known as prime cost sums were inserted by the employers as estimates of what those parts were likely to cost. So the tendering contractor had no concern either with the details of this work or the price to be paid for it. The work was to be part of the contract work and so the contractor's tender was made up of the sums for which it offered to do its part of the work together with the prime cost sums settled by the employers. The employers obtained tenders from specialists, selected by them for the prime cost work, and then, when they had made their contract with the contractor, they instructed the contractor to enter into a contract with the sub-contractor whom they nominated in terms which they dictated, having settled those terms with the nominated sub-contractor. Then, if the sum to be paid to the nominated sub-contractor differed from the estimated prime cost sum, as it almost certainly would, the contractor's contract price for the whole work was adjusted to take account of this difference.

The contractor's contract with the employers included bills of quantities, contract drawings and the conditions contained in the RIBA form of contract – a form published by the Royal Institute of British Architects. So the question at issue in this case must be answered by construing those conditions. Condition 1(1) is in these terms:

> The Contractor shall upon and subject to these Conditions carry out and complete the Works shown upon the Contract Drawings and described by or referred to in the Contract

> Bills and in these Conditions in every respect to the reasonable satisfaction of the Architect/Supervising Officer.

The works there referred to clearly include prime cost work to be done by nominated sub-contractors so the principal contractor is in breach of its contract with the employers if these works are not duly carried out and completed. But it will be able to sue the nominated sub-contractor for breach of the sub-contract. Admittedly the principal contractor is liable if the prime cost works are defective. How much farther its obligation goes must depend on the true interpretation of later conditions.

The main condition with regard to nominated sub-contractors is condition 27. The relevant provisions are as follows:

> The following provisions of this Condition shall apply where prime cost sums are included in the Contract Bills, or arise as a result of Architect's/Supervising Officer's instructions given in regard to the expenditure of provisional sums, in respect of persons to be nominated by the Architect/Supervising Officer to supply and fix materials or goods or to execute work.
>
> (a) Such sums shall be deemed to include 2½ per cent cash discount and shall be expended in favour of such persons as the Architect/Supervising Officer shall instruct, and all specialists or others who are nominated by the Architect/Supervising Officer are hereby declared to be sub-contractors employed by the Contractor and are referred to in these Conditions as 'nominated sub-contractors'. Provided that the Architect/Supervising Officer shall not nominate any person as a sub-contractor against whom the contractor shall make reasonable objection, or (save where the Architect/Supervising Officer and Contractor shall otherwise agree) who will not enter into a sub-contract which provides (*inter alia*): –
>
> (i) That the nominated sub-contractor shall carry out and complete the sub-contract Works in every respect to the reasonable satisfaction of the Contractor and of the Architect/Supervising Officer and in conformity with all the reasonable directions and requirements of the Contractor.
>
> (ii) That the nominated sub-contractor shall observe, perform and comply with all the provisions of this Contract on the part of the Contractor to be observed, performed and complied with (other than clause 20[A] of these Conditions, if applicable) so far as they relate and apply to the sub-contract Works or to any portion of the same.
>
> (iii) That the nominated sub-contractor shall indemnify the Contractor against the same liabilities in respect of the sub-contract Works as those for which the Contractor is liable to indemnify the Employer under this Contract.

Conditions 11(3) and 30 also throw some light on the matter. But I do not think that they affect the proper construction of condition 27, and I need not set them out. The crucial problem is to discover the meaning of the first part of condition 27(a). 'Such sums' must refer back to 'prime cost sums' in the preceding sentence; ' . . . such persons as the Architect . . . shall instruct' must be the same as 'persons to be nominated by the Architect . . . to execute work', i.e. nominated sub-contractors. And no light is thrown on the problem by the initial words of condition 27(a), 'Such sums shall be deemed to include 2½ per cent cash discount'. I did not understand any of that to be disputed in argument. So I can rewrite the crucial words – 'prime cost sums shall be expended in favour of persons nominated by the Architect to execute work'. This is meaningless if 'prime cost sums' has its ordinary meaning of sums entered or provided in the bills of quantities for work to be executed by nominated sub-contractors. Such sums are never expended; they are only estimates of the sums which will later appear in the sub-contract between the contractor and nominated sub-contractors as the prices to be paid to the nominated sub-contractors for the prime cost work to be done by them. Once these sub-contracts are made, the prime cost sums have no further part to play except in accounting as provided in condition 30(5)(c).

In order to give this sentence an intelligible meaning, either 'prime cost sums' or 'expended' must be given an unusual meaning. It was argued for the employers that 'expended' could mean 'allocated'. But that would not help, because prime cost sums are never allocated to a nominated sub-contractor or to anybody else, and in the end counsel for the employers were unable to suggest any word or phrase the substitution of which for the word 'expended' would make sense of this sentence. On the other hand, if 'prime cost sums' can be read as meaning the sums which become payable for prime cost work then the meaning of the sentence becomes clear. And that does not appear to me to be doing too great violence to the words in the clause. I would, therefore, read this clause as directing that sums payable in respect of prime cost work 'shall' be expended in favour of nominated sub-contractors and no one else. 'Expended in favour of' is a rather odd expression but it is, I think, accounted for by the fact that payments for work done by the nominated sub-contractor are paid in the first instance by the employers to the principal contractor, but they are earmarked and are then paid by the contractor to the nominated sub-contractor.

The RIBA form of contract makes no express provision for an event which cannot be very uncommon – that for one reason or another the contract between the principal contractor and the nominated sub-contractor is terminated before the sub-contract work is completed. The problem is what is to happen in that event. There appear to me to be three possibilities, and no more were suggested in argument. The first is that the employers must then make a new nomination, or, if they do not wish the work to be continued, give an instruction to that effect by way of variation. If they do make a new nomination, then as in the case of the first nomination they must make a bargain with the new nominees and then instruct the principal contractor to make a sub-contract in terms of that bargain. Then, if the price in the second sub-contract exceeds the price in the first sub-contract which has come to an end, the employers must bear the loss. That is the contention of the contractor. The second possibility is that, when the contract with the nominated sub-contractor is terminated, it becomes the right and duty of the principal contractor to do the prime cost work itself at the price fixed in the sub-contract which has been terminated. The third possibility is that the principal contractor must be held to have undertaken that the nominated sub-contractor will complete the work so that, when, by reason of the termination of the sub-contract, it becomes impossible for that obligation to be performed, there is an irremediable breach of contract by the principal contractor. The employers argued that one or other of these interpretations is correct.

It appears to me that there are insuperable objections to both of the employers' contentions. The scheme for nominated sub-contractors is an ingenious method of achieving two objects which at first sight might seem incompatible. The employers want to choose who is to do the prime cost work and to settle the terms on which it is to be done, and at the same time to avoid the hazards and difficulties which might arise if they entered into a contract with the person they have chosen to do the work. The scheme creates a chain of responsibility. Subject to a very limited right to object, the principal contractor is bound to enter into a contract with the employers' nominee, but it has no concern with the terms of that contract, for those terms are settled by the employers and their nominee. I can find nothing anywhere to indicate that the principal contractor can ever have in any event either the right or the duty to do any of the prime cost work itself. That would, I think, be contrary to the whole purpose of the scheme, and it would be strange if the contractor could have to do work for which it never tendered and at a price which it never agreed. Moreover, if I have correctly construed condition 27, its provision that payment for prime cost work shall be made in favour of nominated sub-contractors and, therefore, cannot be made to the contractor on its own account necessarily involves the conclusion that the contractor is not to do prime cost work itself. So I reject the employers' first contention.

The employers' alternative contention involves similar difficulties. If they could establish that, under their contract, there can never be any duty on the employers to nominate a second sub-contractor if the first nominee drops out, then it could be said that, when the principal contractor undertook to complete the prime cost work – as it did in condition 1 – it undertook that it would be completed by the first nominated sub-contractor, for it was not entitled to do the work itself. But if, on a proper construction of the contract, the employers are bound to re-nominate if the first nominee drops out, then there is no question of guaranteeing the first nominated sub-contractor. The contractor's obligation is simply duly to hand over the completed work. If it does not do so, it is in breach of contract. If the fault is its own, it bears the loss, but if the fault is that of the nominated sub-contractor it can sue the sub-contractor for the damages it has to pay to the employers.

There is a further objection to this contention of the employers. If the contractor is bound to complete the prime cost work but cannot do it itself under the contract and it cannot be done by a nominated sub-contractor because the first has dropped out and there is no provision for a second, then there is deadlock. There is no way in which the work can be done under the contract. It is said that the contractor could then go in and do the work itself in order to mitigate damages. But then the work would not be done under the contract and the employers and their architect would lose all control over it. That cannot have been the intention, and I gravely doubt whether the contractor could have any such right. So, unless the contractor's contention is correct, there will be grave and perhaps insuperable difficulties when any nominated sub-contractor refuses to complete its contract.

But I see no great difficulty in holding that the contract requires a second nomination if the original nominated sub-contractor drops out. It is said that there is no express provision for this. But then neither is there any express provision for the first nomination. Indeed, the absence of any such provision has led to a view, which has attracted considerable support, that the employers have no duty to make any nomination but only have a right or option to do so. But that cannot be right. The contract provides that the prime cost work shall be done, and it may be that the contractor cannot do or at least cannot finish its own work until it is done. The prime cost work is not even defined until the nominated sub-contractor is brought in. No one suggests that the principal contractor has any concern with prime cost work until it is required to make a contract with a nominated sub-contractor. It would be a clear breach of contract by the employers if their failure to nominate a sub-contractor impeded the contractor in the execution of its own work.

Once it is accepted that the principal contractor has no right or duty to do the work itself when the nominated sub-contractor drops out any more than it had before the sub-contractor was nominated, then equally it must be the duty of the employers to make a new nomination when a nominated sub-contractor does drop out. For otherwise the contract work cannot be completed. Moreover, condition 27 requires that payment for prime cost work shall be expended in favour of nominated sub-contractors. So if the first drops out, no payment for the prime cost work still to be done can be made under the contract unless that work is done by a second nominated sub-contractor.

Perhaps I should add that there was an argument that, if my view is right, there will be grave difficulty if a contractor wrongfully terminates the sub-contract – it may be because it thinks erroneously that the sub-contractor is in fundamental breach of the sub-contract. Then, in my view, the contractor would be in breach of his contract with the employers. A new sub-contractor would have to be nominated. But the contractor would have to pay damages for its breach of contract including any loss caused to the employers by that breach.

Although I have come to a clear conclusion that there was in this case a duty to re-nominate, the provisions of the RIBA form of contract are so confused and obscure that no conclusion can be reached without a long and complicated chain of reasoning. The RIBA form of conditions sponsored by the Institute is in very common use. It has been amended from time to time. For a long time it has been well known that the question at issue in the present case has given rise to doubt and controversy. It could have been set at rest by a small amendment of these conditions. But the Institute have chosen not to do that, and they have thereby caused the long and expensive litigation in the present case.

I would dismiss this appeal.

The reasoning in the *Bickerton* case rests on the holding that the contractor is neither bound nor entitled to do the nominated sub-contract work himself. On the other hand it is clear that, in principle, the contractor is liable to the employer for bad work by the nominated sub-contractor and himself has an action against the nominated sub-contractor. In *Bickerton* the nominated sub-contractor did no work at all. Is the position different where the nominated sub-contractor withdraws in mid-contract? It has been held that in this case, the employer remains bound to re-nominate and is therefore responsible for any loss due to delay in re-nomination, but the loss arising from the nominated sub-contractor's withdrawal falls on the contractor. See also *Fairclough Building Ltd* v. *Rhuddlan Borough Council* (1985), p. 427.

Percy Bilton Ltd v. Greater London Council

HOUSE OF LORDS

(1982) 20 BLR 1

The facts of this case are set out on p. 328.

The case arose from the withdrawal of a nominated sub-contractor. The parties agreed that the sub-contractor's withdrawal had caused both direct delay and also led to further delay arising from the GLC's delay in making an effective re-nomination.

Held: Although the second category was the employer's responsibility, the first was the responsibility of the contractor.

LORD FRASER OF TULLYBELTON [having stated the general position as set out on p. 329, continued]: These principles do not, in my opinion, operate in any way harshly or unfairly against the appellant. The so-called 'dropping out' of the nominated sub-contractor was not merely unilateral action by him. The mechanics of the matter were that on 26 July 1978 a firm of building and quantity surveyors, acting on the instructions of the receiver and manager of Lowdells, wrote to the appellant intimating that their labour would be withdrawn by 28 July 1978. That was a notice of intention to repudiate the sub-contract. It was in effect, accepted by the appellant in a reply dated 31 July, in which it purported to 'determine' the subcontract under clause 20(a)(i) which entitled it to determine the sub-contract if the sub-contractor 'wholly suspends the carrying out of the Sub-Contract Works' and continues in such default for ten days. But the clause was not applicable because the appellant's letter of 31 July was sent before the default had lasted for ten days. It must, in my opinion, be read simply as an acceptance of the repudiatory breach by the receiver on behalf of the sub-contractors. Accordingly, the sub-contract was terminated by the appellant's acceptance of repudiation, which it was under no obligation to give. If it had withheld acceptance of the repudiation, the sub-contract would have remained alive; it is possible that the appellant might then have been entitled to claim an extension of time under clause 23(g) on account of the sub-contractor's 'delay'. But this point was not fully argued and I express no concluded view on it. There was also another course open to the appellant. It could have exercised its right of 'reasonable objection' under clause 27(a) to prevent the nomination of any new sub-contractor who did not offer to complete his part of the work within the overall completion period for the contract as a whole. In fact, the appellant accepted the nomination, and relied on the probability that it would be allowed an extension of time. Its reliance was not misplaced, for, as already mentioned, it was allowed 14 weeks for the delay arising from the withdrawal of the sub-contractor . . .

It is common ground between the parties that the delay which followed the dropping out of Lowdells should be divided into two parts – first, the part arising directly from the withdrawal, and secondly, that arising from the failure of the respondent to nominate a replacement with reasonable promptness. The respondent was clearly responsible for the second part, and there is no doubt that, if it had been the only delay, the appellant would have been entitled to a reasonable extension of time to allow for it in accordance with clause 23(f). The dispute centres on the consequence of the first period of the delay which, as parties are agreed, does not fall within any of the provisions of clause 23. The appellant contends that the loss directly caused by the withdrawal of the nominated sub-contractor must fall on the respondent, on the ground that it has a responsibility not only to nominate the original sub-contractor and any necessary replacement, but to maintain a sub-contractor in the field so long as work of the kind allotted to him needs to be done. This is said to flow from the decision of your Lordships' House in *Bickerton* (1970). What was actually decided in that case was that, where the original nominated sub-contractor has gone into liquidation and dropped out, the main contractor had neither the right nor the duty to do any of the sub-contractor's work himself, and that it was the duty of the employer to make a new nomination. Consequently (so it was argued for the appellant), if the nominated sub-contractor withdraws at a time when his withdrawal must inevitably cause delay, the main contractor is disabled from performing his obligations for want of a sub-contractor whom only the employer can provide, and the main contractor is thus 'impeded' from working . . . in these circumstances, it was said that the contractual time limit ceases to apply, the time for completion becomes at large, and the employer cannot rely on the provisions for liquidated damages . . . I cannot accept the argument. I respectfully agree with the passage in the judgment of the Court of Appeal to the effect that:

> insofar as delay was caused by the departure of Lowdells . . . it was a delay which is not within any of the provisions of clause 23 and therefore the plaintiff was not entitled to any extension in respect of it, with the result not that time became at large but that, at that stage, the date for completion remained unaffected by that period of delay. The delay that followed, a delay caused by the failure of the employer for several months to re-nominate, [was] quite separate and clearly within the provisions of clause 23(f), per Stephenson LJ.

The appellant submitted further that clause 23(f) did not apply even to the second part of the delay, because it can only apply where the main contractor makes his application for instructions at a time when the employer, through the architect, could by timeous nomination have avoided the delay altogether. Clause 23(f) applies to delay caused by the contractor not having received instructions 'in due time', and the argument was that those words meant 'in time to avoid delay'. I do not agree. In my opinion, the words mean 'in a reasonable time' and they are therefore applicable in a case such as the present where some delay is inevitable but it has been increased by the employer's fault. Accordingly, I agree with the Court of Appeal that the second part of the delay does fall within clause 23(f).

The problems on re-nomination illustrated in the two previous cases were again considered in the following case, which discusses the grounds on which the main contractor is entitled to reject a re-nomination and the question of what is a reasonable time. It also establishes that the main contractor is not liable for defects in prime cost work which arise before completion where the original nominated sub-contractor drops out.

Fairclough Building Ltd v. Rhuddlan Borough Council

COURT OF APPEAL

(1985) 3 Con LR 38

Fairclough contracted with the Council in JCT 63 terms to construct a complex of swimming pools, a theatre and other amenities at Rhyl. Fairclough took possession of the site on 5 January 1976, and the original completion date was 2 May 1977. The contract sum was just short of £3m. Gunite (Swimpool) Ltd were nominated for specialist sprayed concrete work for the sum of £96,700. Fairclough determined the sub-contract for delay on 29 September 1977 and on the same day wrote requesting the architect to make a renomination of a sub-contractor 'to complete the sprayed concrete pool construction . . . including making good . . . any defects in . . . Gunite's work'.

The architect did not renominate for some months. On 20 January 1978 he issued an instruction requiring Fairclough to 'enter into a sub-contract with M. for the sprayed concrete completion work' but the parties agreed that this did not constitute a valid re-nomination. By an instruction of 13 February 1978 the architect required Fairclough 'to investigate and make good (Gunite's) defective workmanship'. The validity of this instruction was disputed by Fairclough.

On 24 February 1978 the architect required Fairclough to sub-contract with M. for the completion but not the remedial work. Fairclough rejected the nomination on the basis that (a) they were entitled to an extension of time to accommodate M.'s programme and (b) that the employer's duty was 'to renominate a sub-contractor to carry out all work necessary (whether remedial or not) to complete the work covered by the original sub-contract'. On 23 May 1978, without prejudice, the architect nominated another company to complete Gunite's work. Gunite were in liquidation. The trial judge found in favour of Fairclough and the Court of Appeal dismissed the Council's appeal.

Held: (1) The architect was entitled to have regard to the employer's interests by seeking lump sum tenders from three possible replacements and was justified in holding to that course until reasonably satisfied that a lump sum was not forthcoming. The architect was not at fault in not re-nominating before 24 February 1978.

(2) Fairclough were entitled to reject the nomination on the ground that the proposed nominee's programme overran its own, in the absence of an undertaking from the architect that the main contract time-limits would be made compatible with the sub-contract. Fairclough were also entitled to so reject the re-nomination because of the architect's failure to require M. to do the remedial as well as the completion work.

(3) Fairclough were not under any liability in respect of defects in Gunite's work which arose before completion, whether those defects were the result of faulty design, bad workmanship or otherwise.

(4) Fairclough were entitled to an extension of time in respect of all such delay as resulted from the employer's failure validly to re-nominate on 24 February 1978.

PARKER LJ:

(a) Did the time provided for in the proposed sub-contract entitle the contractors to refuse the nomination?

By way of preface it is necessary to state that when the contractors objected to the nomination in February 1978, *Percy Bilton Ltd* v. *Greater London Council* had not been decided, much less reported. This fact needs to be taken into account when considering contentions raised by the contractors in contemporary correspondence on which Mr Keating for the employers sought to base an argument.

The employers' case on this issue is founded on clause 27(a)(ii) of the contract and on the admitted facts (i) that at the time of AI 137 the date for completion of the main contract was fixed at 10 May 1978; (ii) that William Mulcaster Ltd would not enter into a contract which provided for completion of the swimming pool work by that date but required until 1 September 1978 for the completion of such work.

Clause 27(a) (ii), so far as material, provides:

> 27 The following provisions of this Condition, shall apply where prime cost sums are included in the Contract Bills . . . in respect of persons to be nominated by the Architect . . . to supply and fix materials or goods or to execute work.
>
> (a) . . . all specialists . . . who are nominated . . . are hereby declared to be sub-contractors employed by the Contractor and are referred to in the conditions as 'nominated sub-contractors'. Provided that the Architect . . . shall not nominate any person as a sub-contractor against whom the Contractor shall make reasonable objection or (save where the Architect . . . and Contractor shall otherwise agree) who will not enter into a sub-contract which provides (inter alia):
>
> (i) . . .
>
> (ii) That the nominated sub-contractor shall observe, perform and comply with all the provisions of this Contract on the part of the Contractor to be observed, performed and complied with . . . so far as they relate and apply to the sub-contract works or to any portion of the same.

The contractors' case is simple. They say that they were under a contractual obligation to complete all the contract work, which included the sub-contract work, by 10 May 1978. Accordingly, unless William Mulcaster Ltd would enter into a contract to perform the sub-contract work by that date, which it is common ground they would not, the architect was not entitled to nominate them and his purported nomination was invalid.

On the plain wording of clause 27(a)(ii) and the admitted facts this contention, even unsupported by authority, appears, at any rate at first sight to be unanswerable. It is, moreover, supported by *Bilton* v. *GLC*.

The facts of that case are somewhat complicated but so far as material for present purposes may be stated in simple form. The main contract was entered into on 25 October 1976 and was in the same form as the contract presently under consideration. The original completion date was 24 January 1979, but it had been extended to 16 March 1979, when on 31 July 1978 a nominated sub-contractor's contract was determined, as in this case, owing to his withdrawal from the site. There then followed a series of abortive attempts to find a new sub-contractor but no possibly effective re-nomination was made until, at the earliest, 31 October 1978. The new sub-contractor at that time required until November 1979 to complete the sub-contract work but the contractor nevertheless accepted the nomination and entered into a sub-contract which eventually provided for completion in 53 weeks from commencement of work, which was specified to be 22 January 1979. The main contract completion date remained at 16 March 1979 until on 9 May it was extended to 21 June 1979. Thereafter further extensions were granted, finally to 1 February 1980, and on 4 February 1980 the architect issued a certificate that the contract ought reasonably to have been completed by 1 February 1980. In March 1980 the GLC deducted the sum of £24,000 odd from monies otherwise due to the contractor in respect of liquidated damages calculated as from 1 February 1980. This deduction was challenged successfully before Judge Stabb QC on grounds which we need not rehearse, and the GLC appealed to this court with partial success. Again, apart from one matter we need not go into the reasons.

That matter and its importance cannot be better shown than by quotation of a passage from the judgment of Sir David Cairns (with whom Stephenson and Dunn LJJ agreed) in the Court of Appeal. He said:

> A quite separate argument by Mr Garland is what he described as his 'overshoot' submission; that is to say that, at the time of the application for re-nomination, the new sub-contractor's date for completion was later than the plaintiffs' date for completion and that, since this would make it impossible for the plaintiffs both to accept the new sub-contractor and to comply with the provision in their own contract as to the time for completion, therefore the time provision must go completely, time will be at large, and the right to liquidated damages will disappear.
>
> I do not accept this argument. The contractor, faced with a sub-contract with such a provision as to completion, would be entitled to refuse to accept that sub-contractor under clause 27; or what the contractor could do would be to say that he would not agree to accept that sub-contract unless at the same time the employer would agree to an extension of time for the completion of the main contract. Or he could simply accept the nomination and rely on the probability that extensions which would later be given would cover the whole of the period which was required by the nominated sub-contractor for completion of his own work. That is what seems to have happened in this case, and with good reason, because already at that time it was probably in everybody's mind that there were going to be such extensions as would make the ultimate completion date for the main contract later than the completion date in the sub-contractor's contract.

Whether, as we think, this is part of the *ratio* or not, the contractors' contention that the nomination in AI 137 was invalid thus has the unanimous support of all three members of the court.

On appeal to the House of Lords the decision of this court was affirmed. The speech of Lord Fraser of Tullybelton (with whose speech the other members of the House of Lords agreed) contains the following passage:

> There was also another course open to the appellant. It could have exercised its right of 'reasonable objection' under clause 27(a) to prevent the nomination of any new sub-contractor who did not offer to complete his part of the work within the overall completion period for the contract.

The mention in this passage of the right of 'reasonable objection' appears to be a reference to the alternative ground for refusing nomination under clause 27 rather than that presently under consideration unless, as may well be the case, Lord Fraser regarded an objection on such ground as being necessarily reasonable. Be that as it may, it is clear that the contractors' contention not only has the unanimous support of three members of this Court but also the equally unanimous support of five members of the House of Lords. This is hardly surprising for, whatever may be said of the shortcomings in respect of clarity of other provisions of the contract under consideration, the wording in this particular instance is so clear that no other conclusion appears possible.

(b) Was the nomination invalid by reason of the fact that the proposed sub-contract did not cover remedial work?

The contractor's objection on this ground is based on the decision of the House of Lords in *North West Metropolitan Regional Hospital Board* v. *T. A. Bickerton & Son Ltd* (1970). In that case a nominated sub-contractor had gone into liquidation and the liquidator elected not to perform the remaining work. The employer although requested to issue a variation order or nominate a new sub-contractor refused to do so but called upon the main contractor to complete the sub-contract work, maintaining that he was bound to do so. The point for determination was whether the employer was bound to re-nominate. It was held that he was obliged to do so on the true construction of clause 27 of the contract, which was in the same terms as clause 27 of the contract here under consideration. It was so held on the specific ground that under that clause the main contractor was neither bound nor entitled to do any of the sub-contract work himself.

In reaching his conclusion that the contractors' objection to the nomination on the ground now under consideration was valid, the learned judge relied on the following passages in the speeches in their Lordships' House:

Lord Reid:

> I can find nothing anywhere to indicate that the principal contractor can ever have in any event either the right or the duty to do any of the prime cost work himself.

Lord Guest:

> Upon a proper construction of the conditions of the contract, the contractor is not entitled to do work by himself for which a prime cost sum is allocated in the contract.

Viscount Dilhorne:

> The effect of a sub-contractor repudiating his contract and ceasing to act is that, unless there is a fresh nomination, there is no one left who can carry out the prime cost work. The

> contract gives to the contractor no right to do it. In the case of ordinary sub-contractors the contractor may be able to remedy the breach before the date for completion but, when the breach is by a nominated sub-contractor he cannot do that as he is debarred from doing that sub-contract work.

To those we would add one further quotation wherein Lord Reid sets out the contention which was advanced by the contractors and which was upheld:

> The RIBA form of contract makes no express provision for an event which cannot be very uncommon – that for one reason or another the contract between the principal contractor and the nominated sub-contractor is terminated before the sub-contract work is completed. The problem is what is to happen in that event. There appear to me to be three possibilities and no more were suggested in argument.
>
> The first is that the employer must then make a new nomination, or, if he does not wish the work to be continued, give an instruction to that effect by way of variation. If he does make a new nomination, then as in the case of the first nomination he must make a bargain with the new nominee and then instruct the principal contractor to make a sub-contract in terms of that bargain. Then if the price in the second sub-contract exceeds the price in the first sub-contract which has come to an end, the employer must bear the loss. That is the contention of the respondents.

It is true that only Lord Dilhorne specifically refers to remedial work, and it was submitted that there is a distinction between remedial work and work which has not been done at all. This distinction was said to be supported by the facts: (i) that under clause 6(3) and (4) the architects can require the contractor to open up for inspection any work covered up and arrange for tests to be carried out and can also issue instructions in regard to the removal from the site of any work, materials or goods which are not in accordance with the contract; and, (ii) that under clause 15 the contractor is obliged to make good defects in all work – including sub-contract work which may appear in the defects liability period under the contract.

For present purposes we are content to accept, without deciding, that clauses 6(3) and 6(4) do apply to nominated sub-contract work, although in our view this is not necessarily so. It was common ground that clause 15 does so apply. Even so they do not avail the employers. Clause 6 does not deal with remedial work. If defects are found in main contract work the contractor will have to put it right pursuant to his overall obligation to perform the contract. So also in the case of work which he has elected to have performed by a sub-contractor. He has, however, no obligation to do nominated sub-contract or prime cost work. [Counsel] submitted that remedial work was not nominated sub-contract work but we do not follow this. The nominated sub-contractor is obliged to produce work which satisfies the sub-contract. If he does some work which does not satisfy the contract he is obliged, if he stays, to cure the position. This may require either relatively minor remedial work or, if the original work is very bad, complete removal and a fresh start. If he leaves the site with neither course taken, there is an outstanding nominated sub-contract obligation still to be performed. This obligation the contractor is neither entitled nor obliged to perform. The employer must therefore either order its omission, or issue a variation order and pay the contractor for it or negotiate a new sub-contract covering such work. In our opinion this follows inevitably from *Bickerton*. If he chooses the re-nomination route but does not provide for the new sub-contractor to carry out the remedial work he had not carried out his re-nomination obligation and the contractor can object. It was submitted that this conflicted with clause 15. In our view it does not. It accords with it. The overall intention is that the contractor shall not be liable to perform any part of the nominated sub-contractor work. That is why, under clause 27, he is entitled to demand a re-nomination if the nominated sub-contractor drops out, and why the nominated sub-contractor can be rejected if he will not enter into a contract containing the provision specified in clause 27(a)(ii). If the renomination includes remedial work and defects in such work are later discovered he can call upon the sub-contractor to remedy them. This is the plain intent of the parties. If the new sub-contract does not cover remedial work that intention is defeated. In the present case all the original sub-contract work had to be removed and the work started afresh. The submission that if the contractor did remove the old work and start afresh he would not be doing work which *Bickerton* decides he is neither obliged nor entitled to do is in our view wholly unsustainable. It is true that in the result the employer will suffer a loss, but that this might happen was specifically recognized by Lord Reid. There is no reason why he should do so if the contract contains no protective provision. It is he who has chosen the sub-contractor who first did the defective work and then quit the site, leaving the contractor no remedy for any delay resulting from the need to find a replacement so long as the delay is not unreasonable. This delay could well be much more serious financially than the cost to the employer of including remedial work in the new sub-contract. If it is, the contractor must bear the burden. That the employer should bear the burden of paying for remedial work does not justify construing remedial work as not including work which the original sub-contractor would have been obliged to do had he not left the site.

(c) Are the contractors entitled to an extension for the 8 weeks' delay incurred by Gunite before they withdrew?

Clause 23(g) of the contract included amongst causes of delay,

> delay on the part of nominated sub-contractors which the contractor has taken all practicable steps to avoid or reduce but such delay will only be considered for those reasons for which the contractor could obtain an extension of time under the contract.

The words stressed were added in type at the end of the printed provision. Their meaning is in our view plain. It is to ensure that the contractor does not get an extension for mere delay on the part of the sub-contractor but can do so only if the sub-contractor's delay is itself due to one or other of the causes of delay specified in the other sub-clauses of clause 23.

Gunite's delay was not so due. Accordingly we, like the learned judge, conclude that the contractors are not entitled to an extension in respect of it.

(d) Are the employers entitled to charge the contractor with the full cost of remedial work or only obtain the amount which they had already paid in respect of Gunite's work before their withdrawal?

It is accepted that they are entitled to be credited by the contractors with the lesser amount. Once it is concluded that the obligation to renominate includes the obligation to include remedial work in the work to be done by the renominated subcontractor and that the contractors are neither entitled nor obliged to do such work, we can see no basis upon which the employers can charge them with the cost of such work. In the course of argument [Counsel] came very near to accepting that this must be so but sought to say that there was a route by which they could be so charged. We can only say that we could not follow any route, however ingenious, which leads to the result, in effect, that the contractors must bear the cost of work the burden of which has been held to fall upon the employers, and which the contractors are not obliged or entitled to do. We therefore reject the employers' contention.

In the result we would dismiss both appeal and cross appeal and affirm the judgment appealed from on all issues.

Where the works are defective as a result of the fault of the nominated sub-contractor, the contractor will normally be liable to the employer and will have an action against the sub-contractor. However, the employer may also have a direct action against the nominated sub-contractor, either in contract, if a collateral warranty has been executed (as where the JCT Agreement NSC/2 is used), or, in the most restricted circumstances, in tort if the nominated sub-contractor has been negligent.

Independent Broadcasting Authority v. EMI Electronics and BICC Construction Ltd

HOUSE OF LORDS

(1980) 14 BLR 1

The first defendants were the main contractors and the second defendants the nominated sub-contractors for the erection of a television mast 1,250ft high erected on Emley Moor in Yorkshire. The mast was handed over in November 1966 It collapsed in March 1969. The plaintiffs sued both defendants and the first defendants sued the second defendants. The trial judge held that the cause of the collapse was primarily vortex shedding and secondarily asymmetric ice-loading and that the second defendants who had designed the mast were negligent in not having adequately taken these factors into account.

Held:
(1) EMI had impliedly undertaken that the mast was reasonably fit for the purpose for which it was erected.
(2) BICC were similarly in breach of their sub-contract with EMI.
(3) BICC were liable in tort to IBA as they had been negligent in assuring them in correspondence in 1964 that the design was satisfactory.
(4) The assurance given in this correspondence did not amount to a direct contract between BICC and IBA.

Note: These assurances were given some two years after the letting of the contracts. If they had been given before the nomination of BICC the argument that they were contractually binding would have been much stronger.

LORD FRASER OF TULLYBELTON: The contract was constituted by EMI's tender and IBA's acceptance. The main part of the tender consisted of a quotation for work which, as I have said, included the design of the mast. What IBA accepted was 'your quotation'. Responsibility by EMI for the design was thus in my opinion, expressly made part of their contract with IBA. Moreover EMI's references to 'our' line diagram drawing tends to show that they were putting forward the drawing as their own. The tender and acceptance superseded the original invitation to tender which I regard merely as part of the prior communings leading up to the contract. I therefore would agree with O'Connor J and with the Court of Appeal, both of whom held EMI were under a contractual obligation to IBA in respect of the design of the mast as well as of its supply and delivery.

If the terms of the contract alone had left room for doubt about that, I think that in a contract of this nature a condition would have been implied to the effect that EMI had accepted some responsibility for the quality of the mast, including its design, and possibly also for its fitness for the purpose for which it was intended. The extent of the responsibility was not fully explored in argument, and, having regard to the decision on negligence in the design, it does not require to be decided. It is now well recognised that in a building contract for work and materials a term is normally implied that the main contractor will accept responsibility to his employer for materials provided by nominated sub-contractors. The reason for the presumption is the practical convenience of having a chain of contractual liability from the employer to the main contractor and from the main contractor to the sub-contractor – see *Young & Marten Ltd* v. *McManus Childs Ltd* (1969). Of course, as Lord Reid pointed out in that case, 'No warranty ought to be implied in a contract unless it is in all circumstances reasonable'. In most cases the implication will work reasonably because if the main contractor is liable to the employer for defective material, he will generally have a right of redress against the person from whom he bought the material. In the present case it is accepted by BIC that, if EMI are liable in damages to IBA for the design of the mast, then BIC will be liable in turn to EMI. Accordingly, the principle that was applied in *Young & Marten Ltd* in respect of materials, ought in my opinion to be applied here in respect of the complete structure, including its design. Although EMI had no specialist knowledge of mast design, and although IBA knew that and did not rely on their skill to any extent for the design, I see nothing unreasonable in holding that EMI are responsible to IBA for the design seeing that they can in turn recover from BIC who did the actual designing. On the other hand it would seem to be very improbable that IBA would have entered into a contract of this magnitude and this degree of risk without providing for some right of recourse against the principal contractor or the sub-contractors for defects of design.

LORD SCARMAN: The finding of negligence in the design of the mast makes it unnecessary for the House to determine the extent of the contractual responsibility for the design assumed by EMI to IBA (and by BIC to EMI). As my Lord Fraser of Tullybelton observes,

> it must at the very least have been to ensure that the design would not be made negligently.

But I would not wish it to be thought that I accept that this is the extent of the design obligation assumed in this case. The extent of the obligation is, of course, to be determined as a matter of construction of the contract. But, in the absence of a clear, contractual indication to the contrary, I see no reason why one who in the course of his business contracts to design, supply, and erect a television aerial mast is not under an obligation to ensure that it is reasonably fit for the purpose for which he knows it is intended to be used. The Court of Appeal held that this was the contractual obligation in this case, and I agree with them. The critical question of fact is whether he for whom the mast was designed relied upon the skill of the supplier (i.e. his or his sub-contractor's skill) to design and supply a mast fit for the known purpose for which it was required.

Counsel for the appellants, however, submitted that, where a design, as in this case, requires the exercise of professional skill, the obligation is no more than to exercise the care and skill of the ordinarily competent member of the profession. Although it might be negligence to-day for a constructional engineer not to realise the danger to a cylindrical mast of the combined forces of vortex shedding (with lock-on) and asymmetric ice loading of the stays, he submitted that it would not have been negligence before the collapse of this mast: for the danger was not then appreciated by the profession. For the purpose of the argument, I will assume (contrary to my view) that there was no negligence in the design of the mast, in that the profession was at that time unaware of the danger. However, I do not accept that the design obligation of the supplier of an article is to be equated with the obligation of a professional man in the practice of his profession. In *Samuels* v. *Davis* (1943), the Court of Appeal held that, where a dentist undertakes for reward to make a denture for a patient, it is an implied term of the contract that the denture will be reasonably fit for its intended purpose. I would quote two passages from the judgment of du Parcq LJ:

> . . . if someone goes to a professional man . . . and says: 'Will you make me something which will fit a particular part of my body?' . . . and the professional gentleman says: 'Yes,' without qualification, he is then warranting that when he has made the article it will fit the part of the body in question.

And he added:

> If a dentist takes out a tooth or a surgeon removes an appendix, he is bound to take reasonable care and to show such skill as may be expected from a qualified practitioner. The case is entirely different where a chattel is ultimately to be delivered.

I believe the distinction drawn by du Parq LJ to be a sound one.

In the absence of any terms (express or to be implied) negativing the obligation, one who contracts to design an article for a purpose made known to him undertakes that the design is reasonably fit for the purpose. Such a design obligation is consistent with the statutory law regulating the sale of goods: see Sale of Goods Act 1893, the original section 14 and its modern substitution enacted by section 3, Supply of Goods (Implied Terms) Act 1973.

Until 1982, it was generally assumed that liability in tort for negligence on the part of the nominated sub-contractor would not extend to cases where the result of the negligence was that the building was sub-standard but not dangerous to person or property. However, in the case below, the House of Lords held that this was not so, though the decision is now very restricted in its effect.

Junior Books Ltd v. Veitchi Co Ltd

HOUSE OF LORDS

(1982) 21 BLR 66

This case came to the House of Lords on appeal from Scotland. Under the relevant Scottish procedure the legal validity of the plaintiff's claims was being argued before the trial of the facts. At this stage, therefore, it was assumed that the plaintiff's allegations could all be proved.

The respondents, building owners, engaged a building company to build a factory for them and the appellants were nominated as flooring sub-contractors. Two years after the floor had been laid it developed cracks. It was not alleged that this made the floor dangerous, but it did involve the respondents in serious additional expense and the respondents brought an action in tort against the appellants. [It was not alleged that there was any direct warranty nor, apparently, did the respondents sue the main contractors.]

Held: The appellants owed the respondents a duty of care since they knew that the respondents relied on their skill and experience to lay the floor and that financial loss to the respondents would arise directly from their failure to take such care.

> LORD ROSKILL: My Lords, to my mind in the instant case there is no physical damage to the flooring in the sense in which that phrase was used in *Dutton* v. *Bognor Regis UDC* (1972), *Batty* v. *Metropolitan Property Realizations Ltd* (1978) and *Bowen* v. *Paramount* (1977) and some of the other cases. As my noble and learned friend Lord Russell said during the argument, the question which your Lordships' House now has to decide is whether the relevant Scots and English law today extends the duty of care beyond a duty to prevent harm being done by faulty work to a duty to avoid such faults being present in the work itself. It was powerfully urged on behalf of the appellants that were your Lordships so to extend the law a pursuer in the position of the pursuer in *Donoghue* v. *Stevenson* (1932) could in addition to recovering for any personal injury suffered have also recovered for the diminished value of the offending bottle of ginger beer. Any remedy of that kind it was argued must lie in contract and not in delict or tort. My Lords, I seem to detect in that able argument reflections of the previous judicial approach to comparable problems before *Donoghue* v. *Stevenson* was decided. That approach usually resulted in the conclusion that in principle the proper remedy lay in contract and not outside it. But that approach and its concomitant philosophy ended in 1932 and for my part I should be reluctant to countenance its re-emergence some fifty years later in the instant case. I think today the proper control lies not in asking whether the proper remedy should lie in contract or instead in delict or tort, not in somewhat capricious judicial determination whether a particular case falls on one side of the line or the other, not in somewhat artificial distinctions between physical and economic or financial loss when the two sometimes go together and sometimes do not (it is sometimes overlooked that virtually all damage including physical damage is in one sense financial or economic for it is compensated by an award of damages) but in the first instance in establishing the relevant principles and then in deciding whether the particular case falls within or without those principles. To state this is to do no more than to restate what Lord Reid said in *Dorset Yacht Co Ltd* v. *Home Office* (1970) and Lord Wilberforce in *Anns* v. *Merton LBC* (1978). Lord Wilberforce in the passage I have already quoted enunciated the two tests which have to be satisfied. The first is 'sufficient relationship of proximity', the second any considerations negativing, reducing or limiting the scope of the duty or the class of person to whom it is owed or the damages to which a breach of the duty may give rise. My Lords, it is I think in the application of those two principles that the ability to control the extent of liability in delict or in negligence lies. The history of the development

of the law in the last fifty years shows that fears aroused by the 'floodgates' argument have been unfounded. Cook J in *Bowen* described the 'floodgates' argument as specious and the argument against allowing a cause of action such as was allowed in *Dutton*, *Anns* and *Bowen* as '*in terrorem* or doctrinaire'.

Turning back to the present appeal I therefore ask first whether there was the requisite degree of proximity so as to give rise to the relevant duty of care relied on by the respondents. I regard the following facts as of crucial importance in requiring an affirmative answer to that question: (1) the appellants were nominated sub-contractors; (2) the appellants were specialists in flooring; (3) the appellants knew what products were required by the respondents and their main contractors and specialised in the production of those products; (4) the appellants alone were responsible for the composition and construction of the flooring; (5) the respondents relied on the appellants' skill and experience; (6) the appellants as nominated sub-contractors must have known that the respondents relied on their skill and experience; (7) the relationship between the parties was as close as it could be short of actual privity of contract; (8) the appellants must be taken to have known that if they did the work negligently (as it must be assumed that they did) the resulting defects would at some time require remedying by the respondents expending money on the remedial measures as a consequence of which the respondents would suffer financial or economic loss.

My Lords, reverting to Lord Devlin's speech in *Hedley Byrne & Co Ltd* v. *Heller & Partners Ltd* (1963), it seems to me that all the conditions existed which give rise to the relevant duty of care owed by the appellants to the respondents.

I then turn to Lord Wilberforce's second proposition. On the facts I have just stated, I see nothing whatever to restrict the duty of care arising from the proximity of which I have spoken. During the argument it was asked what the position would be in a case where there was a relevant exclusion clause in the main contract. My Lords, that question does not arise for decision in the instant appeal, but in principle I would venture the view that such a clause according to the manner in which it was worded might in some circumstances limit the duty of care just as in the *Hedley Byrne* case the plaintiffs were ultimately defeated by the defendants' disclaimer of responsibility. But in the present case the only suggested reason for limiting the damage (*ex hypothesi* economic or financial only) recoverable for the breach of the duty of care just enunciated is that hitherto the law has not allowed such recovery and therefore ought not in the future to do so. My Lords, with all respect to those who find this a sufficient answer I do not. I think this is the next logical step forward in the development of this branch of the law. I see no reason why what was called during the argument 'damage to the pocket' *simpliciter* should be disallowed when 'damage to the pocket' coupled with physical damage has hitherto always been allowed. I do not think that this development, if development it be, will lead to untoward consequences. The concept of proximity must always involve, at least in most cases, some degree of reliance; I have already mentioned the words 'skill' and 'judgment' in the speech of Lord Morris in *Hedley Byrne*. These words seem to me to be an echo, be it conscious or unconscious, of the language of section 14(1) of the Sale of Goods Act 1893. My Lords, though the analogy is not exact, I do not find it unhelpful for I think the concept of proximity of which I have spoken and the reasoning of Lord Devlin in the *Hedley Byrne* case involve factual considerations not unlike those involved in a claim under section 14(1); and as between an ultimate purchaser and a manufacturer would not easily be found to exist in the ordinary everyday transaction of purchasing chattels when it is obvious that in truth the real reliance was on the immediate vendor and not on the manufacturer.

In later cases the courts have taken a very narrow view of the scope of the above case, as is shown by the next case.

Greater Nottingham Co-operative Society Ltd v. Cementation Piling and Foundations Ltd and Others

COURT OF APPEAL

(1988) 17 Con LR 43

The plaintiffs owned a store in Skegness and in 1979 arranged to build an extension to an adjoining site. The second defendants were the main contractors and the first defendants nominated sub-contractors for piling work. The third defendants were consultant structural engineers who had been engaged by the plaintiffs to design and supervise the piling operations. The main contract was on JCT 1963 terms (private edition with quantities, 1977 revision). The employers and nominated sub-contractors entered into a direct employer-nominated sub-contractor agreement.

The site was liable to subsidence because of the silty nature of the subsoil. The original building had been built on a raft but this had not proved entirely satisfactory. In 1971 an earlier extension had been built upon piles which had been driven into the ground. This pile driving resulted in excessive vibration and consequential damage. The plaintiffs therefore decided after consulting their professional advisers to use the auguring system of piling, which is comparatively vibration-free.

The invitation to tender, prepared by the third defendants, specified 'bored' piling and drew attention to the need for the piling to be vibration-free. The specialists invited to tender were provided with a geological report on the site of 1971 but not an earlier report of 1958 on subsidence in the main store. The first defendants submitted a tender based on piles of 450 mm diameter; a competing tender provided for piles of 300 mm (the smaller the diameter of the piles, the more are needed).

Piling work started on 17 July 1979. About midday on 19 July cracks appeared in a nearby building and piling work was suspended until October 1979, when it was restarted in a substantially modified form and completed on 20 November 1979. The plaintiffs claimed for the cost of meeting the claim by the adjoining landowner for damage to his building; for the extra cost of the piling operation; (against the first defendants) for the extra sums paid to the second defendants for delay and disruption and for its own losses arising out of late completion.

The trial judge held that:

(1) The damage to the adjoining property was due to excessive over-excavation combined with the effect of persistent attempts at penetration resulting in liquefaction of the subsoil and the straight withdrawal of the auger (without reverse screwing) producing a void of at least 1m some 7m below ground.

(2) Although experience showed that a design based on a pile of 300 mm diameter was more satisfactory, the first defendants were not negligent in putting forward the original design based on 450 mm diameter.

(3) The first defendants were negligent, however, in the conduct of the piling operations both by repeated attempts at penetration and by straight-lifting the auger without reverse screwing.

(4) The first defendants were liable to the plaintiffs for the money paid to the owners of the adjoining property to compensate them and also for the additional cost of the revised piling scheme; for the additional sums paid to the second defendants under JCT 63 clause 23 and for the plaintiffs' own losses due to late completion.

(5) In computing the first defendants' liability the period from 6 August 1979 to 3 September 1979 should be excluded, as the delay during this period was the plaintiffs' own fault.

(6) The second defendants were liable to the plaintiffs for the cost of compensating the adjoining landowner but could recover a complete indemnity from the first defendants under the Civil Liability (Contribution) Act 1978.

(7) The third defendants were not negligent.

The first defendants appealed. On the appeal counsel for the plaintiff accepted that the trial judge's finding that the first defendants were liable for breach of the direct employer/sub-contractor contract could not be supported but sought to support the judgment on the basis of liability in negligence.

Held: (1) It was not open to the first defendants to attack the trial judge's finding that the effective cause of the loss in the case was the negligence of the piling operator but rather the variation order of the architect, since causation in this case was a question of fact and appeals on questions of fact from Official Referees were not permitted under RSC Order 58 rule 4; but

(2) The relevant facts were, in substance, exactly the same as those considered by the House of Lords in *Junior Books Ltd* v. *Veitchi Co Ltd* [1983] AC 520, save only that in the present case there was a contract between the plaintiffs and the first defendants, under which the first defendants assumed responsibility for loss to the plaintiffs for lack of reasonable care and skill in the design of the pile driving operation but did not assume similar responsibility for lack of care in the pile driving operation itself. In this circumstance it would be wrong to make the first defendants liable in tort for economic loss flowing from such carelessness, and the appeal was allowed.

PURCHAS LJ: Neither exegesis, however, touches on the central point of distinction in the present appeal, namely what impact on the otherwise close relationship should the existence of the contract between Cementation and the society have on tortious liability. It is at this point that, in my judgment, the question of policy arises as counsel for Cementation submitted.

The terms of the direct contractual relationship between Cementation and the society involve the warranties already set out and no other obligations imposed on Cementation by way of a direct duty towards the society. In line with the approach of Robert Goff LJ in the *Muirhead* case and that of Bingham LJ in the *Simaan* case, in considering whether there should be a concurrent but more extensive liability in tort as between the two parties arising out of the execution of the contract it is relevant to bear in mind (a) that the parties had an actual opportunity to define their relationship by means of contract and took it and (b) that the general contractual structure as between the society, the main contractor and Cementation as well as the professional advisers provided a channel of claim which was open to the society such as Bingham LJ mentioned in the *Simaan* case as being available in that case to the sheikh.

Although this is new ground, doing the best I can to distil from the mass of authorities which have already been considered in detail in the two judgments of Robert Goff and Bingham LJJ, I do not believe that it would be in accordance with present policy to extend Junior Books rather than to restrict it. This does give rise to an apparent inconsistency, namely the effect of enhancing the close relationship on which Lord Roskill based his duty in tort in *Junior Books* by adding a direct contractual relationship does not confirm a duty to avoid economic loss but negatives that liability. But in this compartment of consideration it is not only the proximity of the relationship giving rise to reliance which is critical but also the policy of the law as to whether or not in these circumstances damages for pecuniary loss ought to be recoverable.

The argument on the other side is, of course, that the collateral contract between the society and Cementation was restricted to the specific topics therein set out. These, as will appear from the extract already cited in this judgment, placed on Cementation a contractual duty not to expose the society to claims under the main contract for extensions of time under clause 23 and reciprocal undertakings by the Society to ensure that Cementation were not prejudiced by any act on their part under the terms of the main contract in clauses 25, 27 and 30 of the main contract. If the parties intended that there should be a contractual restriction of the duty in tort owed by the one to the other it would have been expected that specific provision would be made in the terms of the contract. Bearing in mind the authorities which establish quite clearly that the duty in tort is capable of running coincidentally with liability in contract there is nothing in the terms of the collateral contract to destroy a liability which otherwise would have arisen in tort.

Thus it is said, reverting to Lord Roskill's check-list, the presence of the contract, far from restricting the ambit of the duty in tort to pay damages for pecuniary loss which otherwise would have arisen from the close proximity of the parties and the reliance of the one on the other, the existence of the contract both establishes that special relationship and was a justification for awarding pecuniary damages where these flowed according to the ordinary criteria of damages in tort from the established negligence of O'Brien. This appears to be a part of the path towards Pandora's box hitherto untrod.

PART V

BUILDING CONTRACTS AND THE LAW OF TORT

Chapter 23

Building Contracts and the Law of Tort

This book has been taken up with the contractual remedies which arise in relation to building contracts. Over the last 30 years, however, there have been many tort cases where a plaintiff claims that he can recover from a builder or from a local authority because he has bought a house that is less valuable than it would have been if the builder had built more carefully or the local authority had inspected more thoroughly.

The House of Lords upheld such a claim in Anns ***v.*** London Borough of Merton ***(see p. 389), but in the next three cases the House of Lords completely reversed direction.***

D. & F. Estates Ltd v. Church Commissioners for England

HOUSE OF LORDS

(1988) 15 Con LR 35

The Church Commissioners owned a luxury block of flats in London, built by Wates Ltd between 1963 and 1965. D & F, a company in which the plaintiffs had a controlling interest, took a lease of one of the flats for 98 years from 25 March 1963 from the Commissioners and went into occupation of the flat.

In 1980, when the plaintiffs were on holiday, part of the plaster from the ceiling of one of the bedrooms fell to the floor. Subsequent investigation revealed widespread faults in the plastering, which the trial judge found to have been caused by faulty work by the plastering sub-contractor. He further held that the respondents had acted reasonably in sub-contracting the plastering work and in selecting the sub-contractor, but they had failed adequately to supervise the work. He awarded damages to D & F for the cost of remedial work undertaken in 1980 and for the estimated cost of future remedial works, as well as damages in respect of loss of rent while the future remedial works were carried out. He awarded £500 each to the plaintiffs in respect of loss of amenity during the period when they were occupying the flat while remedial works were done.

The Court of Appeal reversed the judge's decision, primarily on the grounds that Wates, having employed competent sub-contractors to carry out the plastering work, owed no further duty of care to the plaintiffs in relation to the execution of the work by the sub-contractors. The Court of Appeal also held that Wates were not liable for

Powell-Smith and Furmston's Building Contract Casebook, Fifth Edition. Michael Furmston.

damages in tort to the plaintiffs in respect of the cost of the future remedial work not yet carried out, since it represented pure economic loss. The plaintiffs appealed.

Held (dismissing the appeal):

(1) The cost of replacing the defective plaster itself, either as carried out in 1980 or as intended to be carried out in future, was not an item of damage for which Wates as builders could possibly be held liable in negligence. To make Wates so liable would be to impose upon them for the benefit of those with whom they had no contractual relationship an obligation of one who warranted the quality of the plaster as regards materials, workmanship and fitness for purpose.

(2) Wates were not vicariously liable for the negligence of their sub-contractor nor as builders did they owe a duty to future lessees and occupiers of the building to take reasonable care that it contained no hidden defects to persons or property.

(3) Wates were not under a tortious duty to the appellants to supervise the work of their sub-contractors to ensure that the sub-contracted work was not negligently performed so as to cause such latent defects.

LORD BRIDGE OF HARWICH: I do not intend to embark on the daunting task of reviewing the wealth of other . . . authority which bears, directly or indirectly, on the question whether the cost of making good defective plaster in the instant case is irrecoverable as economic loss, which seems to me to be the most important question for determination in the present appeal. My abstention may seem pusillanimous, but it stems from a recognition that the authorities, as it seems to me, speak with such an uncertain voice that, no matter how searching the analysis to which they are subject, they yield no clear and conclusive answer. It is more profitable, I believe, to examine the issue in the light of first principles.

However, certain authorities are of prime importance and must be considered. The decision of your Lordships' House in *Junior Books Ltd* v. *Veitchi Co Ltd* (1982) has been analysed in many subsequent decisions of the Court of Appeal. I do not intend to embark on a further such analysis. The consensus of judicial opinion, with which I concur, seems to be that the decision of the majority is so far dependent upon the unique, albeit non-contractual relationship between the pursuer and the defender in that case and the unique scope of the duty of care owed by the defender to the pursuer arising from that relationship that the decision cannot be regarded as laying down any principle of general application in the law of tort or delict. The dissenting speech of Lord Brandon of Oakbrook on the other hand enunciates with cogency and clarity principles of fundamental importance which are clearly applicable to determine the scope of the duty of care owed by one party to another in the absence, as in the instant case, of either any contractual relationship or any such uniquely proximate relationship as that on which the decision of the majority in Junior Books was founded. Lord Brandon said:

> My Lords, it appears to me clear beyond doubt that, there being no contractual relationship between the respondents and the appellants in the present case, the foundation, and the only foundation, for the existence of a duty of care owed by the defenders to the pursuers, is the principle laid down in the decision of your Lordships' House in *Donoghue* v. *Stevenson* (1932). The actual decision in that case related only to the duty owed by a manufacturer of goods to their ultimate user or consumer, and can be summarised in this way: a person who manufactures goods which he intends to be used or consumed by others, is under a

> duty to exercise such reasonable care in their manufacture as to ensure that they can be used or consumed in the manner intended without causing physical damage to persons or their property.
>
> While that was the actual decision in *Donoghue* v. *Stevenson*, it was based on a much wider principle embodied in passages in the speech of Lord Atkin, which have been quoted so often that I do not find it necessary to quote them again here. Put shortly, that wider principle is that, when a person can or ought to appreciate that a careless act or omission on his part may result in physical injury to other persons or their property, he owes a duty to all such persons to exercise reasonable care to avoid such careless act or omission.
>
> It is, however, of fundamental importance to observe that the duty of care laid down in *Donoghue* v. *Stevenson* was based on the existence of a danger of physical injury to persons or their property. That this is so, is clear from the observations made by Lord Atkin with regard to the statements of law of Brett MR in *Heaven* v. *Pender* (1883). It has further, until the present case, never been doubted, so far as I know, that the relevant property for the purpose of the wider principle on which the decision in *Donoghue* v. *Stevenson* was based, was property other than the very property which gave rise to the danger of physical damage concerned.

Later, Lord Brandon, having referred to the well-known two-stage test of the existence of a duty of care propounded by Lord Wilberforce in *Anns*' case (1977), asked himself, at the second stage, the question 'whether there are any considerations which ought, inter alia, to limit the scope of the duty which exists'. He continued:

> To that second question I would answer that there are two important considerations which ought to limit the scope of the duty of care which it is common ground was owed by the appellants to the respondents on the assumed facts of the present case.
>
> The first consideration is that, in *Donoghue* v. *Stevenson* itself and in all the numerous cases in which the principle of that decision has been applied to different but analogous factual situations, it has always been either stated expressly, or taken for granted, that an essential ingredient in the cause of action relied on was the existence of danger, or the threat of danger, of physical damage to persons or their property, excluding for this purpose the very piece of property from the defective condition of which such danger, or threat of danger, arises. To dispense with that essential ingredient in a cause of action of the kind concerned in the present case would, in my view, involve a radical departure from long-established authority.
>
> The second consideration is that there is no sound policy reason for substituting the wider scope of the duty of care put forward for the respondents for the more restricted scope of such duty put forward by the appellants. The effect of accepting the respondents' contention with regard to the scope of the duty of care involved would be, in substance, to create, as between two persons who are not in any contractual relationship with each other, obligations of one of those two persons to the other which are only really appropriate as between persons who do have such a relationship between them.
>
> In the case of a manufacturer or distributor of goods, the position would be that he warranted to the ultimate user or consumer of such goods that they were as well designed, as merchantable and as fit for their contemplated purpose as the exercise of reasonable care could make them. In the case of sub-contractors such as those concerned in the present case, the position would be that they warranted to the building owner that the flooring, when laid, would be as well designed, as free from defects of any kind and as fit for its contemplated purpose as the exercise of reasonable care could make it. In my view, the

> imposition of warranties of this kind on one person in favour of another, when there is no contractual relationship between them, is contrary to any sound policy requirement.
>
> It is, I think, just worth while to consider the difficulties which would arise if the wider scope of the duty of care put forward by the respondents were accepted. In any case where complaint was made by an ultimate consumer that a product made by some persons with whom he himself had no contract was defective, by what standard or standards of quality would the question of defectiveness fall to be decided? In the case of goods bought from a retailer, it could hardly be the standard prescribed by the contract between the retailer and the wholesaler, or between the wholesaler and the distributor, or between the distributor and the manufacturer, for the terms of such contract would not even be known to the ultimate buyer. In the case of sub-contractors such as the appellants in the present case, it could hardly be the standard prescribed by the contract between the sub-contractors and the main contractors, for, although the building owner would probably be aware of those terms, he could not, since he was not a party to such contract, rely on any standard or standards prescribed in it. It follows that the question by what standard or standards alleged defects in a product complained of by its ultimate user or consumer are to be judged remains entirely at large and cannot be given any just or satisfactory answer.

The reasoning in these passages receives powerful support from the unanimous decision of the Supreme Court of the United States of America in *East River Steamship Corp* v. *Transamerica Delaval Inc* (1986). Charterers of supertankers claimed damages from turbine manufacturers resulting from alleged design and manufacturing defects which caused the supertankers to malfunction while on the high seas. The court held *inter alia*, that:

> whether stated in negligence or strict liability, no products-liability claim lies in admiralty when a commercial party alleges injury only to the product itself resulting in purely economic loss . . .

This appears to undermine the earlier American authorities referred to by Richmond P in the New Zealand case of *Bowen* v. *Paramount Builders (Hamilton) Ltd* (1977). The opinion of Lord Brandon of Oakbrook in *Junior Books Ltd* v. *Veitchi Co Ltd* (1982) and that expressed by the Supreme Court of the United States of America are entirely in line with the majority decision of the Supreme Court of Canada in *Rivtow Marine Ltd* v. *Washington Iron Works* (1973) that the damages recoverable from the manufacturer by the hirers of a crane which was found to have a defect which made it unsafe to use did not include the cost of repairing the defect.

These principles are easy enough to comprehend and probably not difficult to apply when the defect complained of is in a chattel supplied complete by a single manufacturer. If the hidden defect in the chattel is the cause of personal injury or of damage to property other than the chattel itself, the manufacturer is liable. But if the hidden defect is discovered before any such damage is caused, there is no longer any room for the application of the *Donoghue* v. *Stevenson* (1932) principle. The chattel is now defective in quality, is no longer dangerous. It may be valueless or it may be capable of economic repair. In either case the economic loss is recoverable in contract by a buyer or hirer of the chattel entitled to the benefit of a relevant warranty of quality, but is not recoverable in tort by a remote buyer or hirer of the chattel.

If the same principle applies in the field of real property to the liability of the builder of a permanent structure which is dangerously defective, that liability can only arise if the defect remains hidden until the defective structure causes personal injury or damage to property other than the structure itself. If the defect is discovered before any damage is done, the loss sustained by the owner of the structure, who has to repair or demolish it to avoid a potential source of danger to third parties, would seem to be purely economic. Thus, if I acquire a property with a dangerously defective garden wall which is attributable to the bad workmanship of the original builder, it is difficult to see any basis in principle on which I can sustain an action in tort against the builder for the cost of either repairing or demolishing the wall. No physical damage has been caused. All that has happened is that the defect in the wall has been discovered in time to prevent damage occurring. I do not find it necessary for the purpose of deciding the present appeal to express any concluded view as to how far, if at all, the *ratio decidendi* of *Anns* v. *Merton London Borough* (1977) involves a departure from this principle establishing a new cause of action in negligence against a builder when the only damage alleged to have been suffered by the plaintiff is the discovery of a defect in the very structure which the builder erected.

My example of the garden wall, however, is that of a very simple structure. I can see that more difficult questions may arise in relation to a more complex structure like a dwelling-house. One view would be that such a structure should be treated in law as a single indivisible unit. On this basis, if the unit becomes a potential source of danger when a hitherto hidden defect in construction manifests itself, the builder, as in the case of the garden wall, should not in principle be liable for the cost of remedying the defect. It is for this reason that I now question the result, as against the builder, of the decision in *Batty* v. *Metropolitan Property Realization Ltd* (1978).

However, I can see that it may well be arguable that in the case of complex structures, as indeed possibly in the case of complex chattels, one element of the structure should be regarded for the purpose of the application of the principles under discussion as distinct from another element, so that damage to one part of the structure caused by a hidden defect in another part may qualify to be treated as damage to 'other property', and whether the argument should prevail may depend on the circumstances of the case. It would be unwise and it is unnecessary for the purpose of deciding the present appeal to attempt to offer authoritative solutions to these difficult problems in the abstract. I should wish to hear fuller argument before reaching any conclusion as to how far the decision of the New Zealand Court of Appeal in *Bowen* v. *Paramount Builders (Hamilton) Ltd* (1977) should be followed as a matter of English law. I do not regard *Anns* v. *Merton London Borough* (1977) as resolving that issue.

In the instant case the only hidden defect was in the plaster. The only item pleaded as damage to other property was 'cost of cleaning carpets and other possessions damaged or dirtied by falling plaster; £50'. Once it appeared that the plaster was loose, any danger of personal injury or of further injury to other property could have been simply avoided by the timely removal of the defective plaster. The only function of plaster on walls and ceilings, unless it is itself elaborately decorative, is to serve as a smooth surface on which to place decorative paper or paint. Whatever case there may be for treating a defect in some part of the structure of a building as causing damage to 'other property' when some other part of the building is injuriously affected, as for example cracking in walls caused by defective foundations, it would seem to me entirely artificial to treat the plaster as distinct from the decorative surface placed upon it. Even if it were so treated, the only damage to 'other property' caused by the defective plaster would be the loss of value of the existing decorations occasioned by the necessity to remove loose plaster which was in danger of falling. When the loose plaster in [the flat] was first discovered in 1980, the flat was in any event being redecorated.

It seems to me clear that the cost of replacing the defective plaster itself, either as carried out in 1980 or as intended to be carried out in future, was not an item of damage for which the builder of [the flats] could possibly be made liable in negligence under the principle of *Donoghue* v. *Stevenson* or any legitimate development of that principle. To make him so liable would be to impose upon him for the benefit of those with whom he had no contractual relationship the obligation of one who warranted the quality of the plaster as regards materials, workmanship and fitness for purpose. I am glad to reach the conclusion that this is not the law, if only for the reason that a conclusion to the opposite effect would mean that the courts, in developing the common law, had gone much farther than the legislature were prepared to go in 1972, after comprehensive examination of the subject by the Law Commission, in making builders liable for defects in the quality of their work to all who subsequently acquire interests in buildings they have erected. The statutory duty imposed by the Act of 1972 was confined to dwelling-houses and limited to defects appearing within six years. The common law duty, if it existed, could not be so confined or so limited. I cannot help feeling that consumer protection is an area of law where legislation is much better left to the legislators.

It follows from these conclusions that, even if Wates themselves had been responsible for the plaster-work in [the flat], the damages recoverable from them by D & F Estates would have been a trivial sum and [the plaintiffs] could have established no claim for damages for disturbance. But, as already indicated, the Court of Appeal's primary ground for allowing Wates' appeal was that they had properly employed competent sub-contractors to do the plaster work for whose negligence they were not liable, and it is to this issue that I must now turn. The submission in support of the appeal was put in three ways which amount, as it seems to me, to three alternative formulations of what is, in essence, the same proposition of law. Expressed in summary form the three formulations are (i) that Wates were vicariously liable for the negligence of their subcontractor; (ii) that Wates as main contractors responsible for building [the flats] owed a duty to future lessees and occupiers of flats to take reasonable care that the building should contain no hidden defects of the kind which might cause injury to persons or property and that this duty could not be delegated; (iii) that Wates as main contractors owed a duty of care to future lessees and occupiers of flats to supervise their sub-contractors to ensure that the sub-contracted work was not negligently performed so as to cause such defects.

It is trite law that the employer of an independent contractor is, in general, not liable for the negligence or other torts committed by the contractor in the course of the execution of the work. To this general rule there are certain well-established exceptions or apparent exceptions. Without enumerating them it is sufficient to say that it was accepted by [counsel on behalf of the appellants] that the instant case could not be accommodated without any of the recognised and established categories by which the exceptions are classified. But it has been rightly said that the so-called exceptions

> are not true exceptions (at least so far as the theoretical nature of the employer's liability is concerned) for they are dependent upon a finding that the employer is, himself, in breach of some duty which he personally owes to the plaintiff. The liability is thus not truly a vicarious liability and is to be distinguished from the vicarious liability of a master for his servant.' (See *Clerk and Lindsell on Torts* (15th edn, 1982) paras 3–37, p. 185.)

Herein lies [counsel's] real difficulty. If Wates are to be held liable for the negligent workmanship of their sub-contractors (assumed for this purpose to result in dangerously defective work) it must first be shown that in the circumstances they had assumed a personal duty to all the world to ensure that [the flats] should be free of dangerous effects. This was the assumption on which the judge proceeded when he said: 'The duty of care itself is of course not delegable.' Whence does this nondelegable duty arise? [Counsel] submits that it is a duty undertaken by any main contractor in the building industry who contracts to erect an entire building. I cannot agree because I cannot recognise any legal principle to which such an assumption of duty can be related. Just as I may employ a building contractor to build me a house, so may the building contractor, subject to the terms of my contract with him, in turn employ another to undertake part of the work. If the mere fact of employing a contractor to undertake building work automatically involved the assumption by the employer of a duty of care to any person who may be injured by a dangerous defect in the work caused by the negligence of the contractor, this would obviously lead to absurd results. If the fact of employing a contractor does not involve the assumption of any such duty by the employer, then one who has himself contracted to erect a building assumes no such liability when he employs an apparently competent independent sub-contractor to carry out part of the work for him. The main contractor may, in the interests of the proper discharge for his own contractual obligations, exercise a greater or lesser degree of supervision over the work done by the sub-contractor. If in the course of supervision the main contractor in fact comes to know that the sub-contractor's work is being done in a defective and foreseeably dangerous way and if he condones that negligence on the part of the sub-contractor, he will no doubt make himself potentially liable for the consequences as a joint tortfeasor. But the judge made no finding against Wates of actual knowledge and his finding that they 'ought to have known' what the manufacturer's instructions were depended on and was vitiated by his earlier misdirection that Wates owed a duty of care to future lessees of [the] flats in relation to their sub-contractor's work.

[He] relied on the decision of Judge Edgar Fay QC in *Queensway Discount Warehouses* v. *Graylaw Properties Ltd* (1982) and on the decision of Judge Sir William Stabb QC in *Cynat Products Ltd* v. *Landbuild (Investment and Property) Ltd* (1984). In so far as the former decision relied on any general principle of law that a main contractor is liable to a third party who suffers damage from the negligently defective work done by his sub-contractor, I can only say, as already indicated, that I can find no basis in law to support any such principle. The relevant issue in the latter case, as in *Batty* v. *Metropolitan Property Realizations Ltd* (1978) in relation to the liability of the developer defendants, was whether the defendants' admitted contractual liability was matched by a parallel liability in tort. In both cases the issue was of importance only as bearing upon the liability of insurers to indemnify the defendants. I do not find authorities directed to that question of any assistance in determining the scope of the duty of care which one person owes to another entirely independently of any contractual relationship on the basis of the *Donoghue* v. *Stevenson* (1932) principle.

More important is the decision of the New Zealand Court of Appeal in *Mount Albert Borough Council* v. *Johnson* (1979). This was another case of the purchaser of a flat suffering damage due to the subsidence of a building erected on inadequate foundations. One of the issues was whether the plaintiff was entitled to recover damages against the development company which had employed independent contractors to erect the building Delivering the judgment of Somers J and himself Cooke J said:

> In the instant type of case a development company acquires land, subdivides it and has homes built on the lots for sale to members of the general public. The company's interest is primarily a business one. For that purpose it has buildings put up which are intended to house people for many years and it makes extensive and abiding changes in the landscape. It is not a case of a landowner having a house built for his own occupation initially – as to which we would say nothing except that Lord Wilberforce's two-stage approach to duties of care in *Anns* may prove of guidance on questions of non-delegable duty also. There appears to be no authority directly in point on the duty of such a development company. We would hold that it is a duty to see that proper care and skill are exercised in the building of the houses and that it cannot be avoided by delegation to an independent contractor.

As a matter of social policy this conclusion may be entirely admirable. Indeed, it corresponds almost precisely to the policy underlying the Law Commission's recommendations in para 26 of the report 'Civil Liability of Vendors and Lessors for Defective Premises' (Law Com No 40) to which I have already referred and which was implemented by section 1(1) and (4) of the 1972 Act. As a matter of legal principle, however, I can discover no basis on which it is open to the court to embody this policy in the law without the assistance of the legislature and it is again, in my opinion, a dangerous course for the common law to embark upon the adoption of novel policies which it sees as instruments of social justice but to which, unlike the legislature, it is unable to set carefully defined limitations.

The conclusion I reach is that Wates were under no liability to the plaintiffs for damage attributable to the negligence of their plastering sub-contractor in failing to follow the instructions of the manufacturer of the plaster they were using, but that in any event such damage could not have included the cost of renewing the plaster.

Murphy v. Brentwood District Council

HOUSE OF LORDS

(1990) 21 Con LR 1

LORD KEITH OF KINKEL: Consideration of the nature of the loss suffered in this category of cases is closely tied up with the question of when the cause of action arises. Lord Wilberforce in *Anns* (1978) regarded it as arising when the state of the building is such that there is present an imminent danger to the health or safety of persons occupying it. That state of affairs may exist when there is no actual physical damage to the building itself, though Lord Wilberforce had earlier referred to the relevant damage being material physical damage. So his meaning may have been that there must be a concurrence of material physical damage and also present or imminent danger to the health or safety of occupants. On that view there would be no cause of action where the building had suffered no damage (or possibly, having regard to the word 'material', only very slight damage) but a structural survey had revealed an underlying defect, presenting imminent danger. Such a discovery would inevitably cause a fall in the value of the building, resulting in economic loss to the owner. That such is the nature of the loss is made clear in cases where the owner abandons the building as incapable of being put in a safe condition (as in *Batty*), or where he chooses to sell it at the lower value rather than undertake remedial works. In *Pirelli General Cable Works Ltd* v. *Oscar Faber & Partners* (1983) it was held that the cause of action in tort against consulting engineers who had negligently approved a defective design for a chimney arose when damage to the chimney caused by the defective design first occurred, not when the damage was discovered or with reasonable diligence might have been discovered. The defendants there had in relation to the design been in contractual relations with the plaintiffs, but it was common ground that a claim in contract was time-barred. If the plaintiffs had happened to discover the defect before any damage had occurred there would seem to be no good reason for holding that they would not have had a cause of action in tort at that stage, without having to wait until some damage had occurred. They would have suffered economic loss through having a defective chimney upon which they required to expend money for the purpose of removing the defect. It would seem that in a case such as *Pirelli* where the tortious liability arose out of a contractual relationship with professional people, the duty extended to take reasonable care not to cause economic loss to the client by the advice given. The plaintiffs built the chimney as they did in reliance on that advice. The case would accordingly fall within the principle of *Hedley Byrne & Co Ltd* v. *Heller*

& Partners Ltd (1964). I regard *Junior Books Ltd* v. *Veitchi Co Ltd* (1983) as being an application of that principle.

In my opinion it must now be recognised that, although the damage in *Anns* was characterised as physical damage by Lord Wilberforce, it was purely economic loss.

In *D & F Estates Ltd* v. *Church Commissioners for England* (1989) both Lord Bridge of Harwich and Lord Oliver of Aylmerton expressed themselves as having difficulty in reconciling the decision in *Anns* with pre-existing principle and as being uncertain as to the nature and scope of such new principle as it introduced. Lord Bridge suggested that in the case of a complex structure such as a building one element of the structure might be regarded for *Donoghue* v. *Stevenson* purposes as distinct from another element, so that damage to one part of the structure caused by a hidden defect in another part might qualify to be treated as damage to 'other property'. I think that it would be unrealistic to take this view as regards a building the whole of which had been erected and equipped by the same contractor. In that situation the whole package provided by the contractor would, in my opinion, fall to be regarded as one unit rendered unsound as such by a defect in the particular part. On the other hand where, for example, the electric wiring had been installed by a sub-contractor and due to a defect caused by lack of care a fire occurred which destroyed the building, it might not be stretching ordinary principles too far to hold the electrical sub-contractor liable for the damage. If in the *East River* case the defective turbine had caused the loss of the ship the manufacturer of it could consistently with normal principles, I would think, properly have been held liable for that loss. But even if Lord Bridge's theory were to be held acceptable, it would not seem to extend to the founding of liability upon a local authority, considering that the purposes of the Act of 1936 are concerned with averting danger to health and safety, not danger or damage to property. Further, it would not cover the situation which might arise through discovery, before any damage had occurred, of a defect likely to give rise to damage in the future

Liability under the *Anns* decision is postulated upon the existence of a present or imminent danger to health or safety. But considering that the loss involved in incurring expenditure to avert the danger is pure economic loss, there would seem to be no logic in confining the remedy to cases where such danger exists. There is likewise no logic in confining it to cases where some damage (perhaps comparatively slight) has been caused to the building, but refusing it where the existence of the danger has come to light in some other way, for example through a structural survey which happens to have been carried out, or where the danger inherent in some particular component or material has been revealed through failure in some other building. Then there is the question whether the remedy is available where the defect is rectified, not in order to avert danger to an inhabitant occupier himself, but in order to enable an occupier, who may be a corporation, to continue to occupy the building through its employees without putting those employees at risk.

In my opinion it is clear that *Anns* did not proceed upon any basis of established principle, but introduced a new species of liability governed by a principle indeterminate in character but having the potentiality of covering a wide range of situations, involving chattels as well as real property, in which it had never hitherto been thought that the law of negligence had any proper place.

In my opinion there can be no doubt that *Anns* has for long been widely regarded as an unsatisfactory decision. In relation to the scope of the duty owed by a local authority it proceeded upon what must, with due respect to its source, be regarded as a somewhat superficial examination of principle and there has been extreme difficulty, highlighted most recently by the speeches in *D & F Estates*, in ascertaining upon exactly what basis of principle it did proceed. I think it must now be recognised that it did not proceed on any basis of principle at all, but constituted a remarkable example of judicial legislation. It has engendered a vast spate of litigation, and each of the cases in the field which have reached this House has been distinguished. Others have been distinguished in the Court of Appeal. The result has been to keep the effect of the decision within reasonable bounds, but that has been achieved only by applying strictly the words of Lord Wilberforce and by refusing to accept the logical implications of the decision itself. These logical implications show that the case properly considered has potentiality for collision with long-established principles regarding liability in the tort of negligence for economic loss. There can be no doubt that to depart from the decision would reestablish a degree of certainty in this field of law which it has done a remarkable amount to upset.

LORD BRIDGE OF HARWICH: There may, of course, be situations where, even in the absence of contract, there is a special relationship of proximity between builder and building owner which is sufficiently akin to contract to introduce the element of reliance so that the scope of the duty of care owed by the builder to the owner is wide enough to embrace purely economic loss. The decision in *Junior Books Ltd* v. *Veitchi Co. Ltd* (1983) can, I believe, only be understood on this basis.

Department of the Environment v. Thomas Bates and Sons Ltd and others

HOUSE OF LORDS

(1990) 21 Con LR 54

LORD KEITH: The foundation of the plaintiffs' case is *Anns* v. *Merton London Borough Council* (1978). That decision was concerned directly only with the liability in negligence of a local authority in respect of its functions in regard to securing compliance with building byelaws and regulations. The position of the builder as regards liability towards a remote purchaser of a building which suffered from defects due to carelessness in construction was touched on very briefly. However, it has since been generally accepted that similar principles govern the liability both of the local authority and of the builder.

It has been held by this House in *Murphy* v. *Brentwood District Council* that *Anns* was wrongly decided and should be departed from, by reason of the erroneous views there expressed as to the scope of any duty of care owed to purchasers of houses by local authorities when exercising the powers conferred upon them for the purpose of securing compliance with building regulations. The process of reasoning by which the House reached its conclusion necessarily included close examination of the position of the builder who was primarily responsible, through lack of care in the construction process, for the presence of defects in the building. It was the unanimous view that, while the builder would be liable under the principle of *Donoghue* v. *Stevenson* (1932) in the event of the defect, before it had been discovered, causing physical injury to persons or damage to property other than the building itself, there was no sound basis in principle for holding him liable for the pure economic loss suffered by a purchaser who discovered the defect, however such discovery might come about, and required to expend money in order to make the building safe and suitable for its intended purpose.

In the present case it is clear that the loss suffered by the plaintiffs is pure economic loss. At the time the plaintiffs carried out the remedial work on the concrete pillars the building was not unsafe by reason of the defective construction of these pillars. It did, however, suffer from a defect of quality which made the plaintiffs' lease less valuable than it would otherwise have been, in respect that the building could not be loaded up to its design capacity unless any occupier who wished so to load it had incurred the expenditure necessary for the strengthening of the pillars. It was wholly uncertain whether during the currency of their lease the plaintiffs themselves would ever be likely to require to load the building up to its design capacity, but a purchaser from them might well have wanted to do so. Such a purchaser, faced with the need to strengthen the pillars, would obviously have paid less for the lease than if they had been sound. This underlines the purely economic character of the plaintiffs' loss. To hold in favour of the plaintiffs would involve a very significant extension of the doctrine of *Anns* so as to cover the situation where there existed no damage to the building and no imminent danger to personal safety or health. If *Anns* were correctly decided, such an extension could reasonably be regarded as entirely logical. The undesirability of such an extension, for the reasons stated in *Murphy* v. *Brentwood District Council*, formed an important part of the grounds which led to the conclusion that *Anns* was not correctly decided. That conclusion must lead inevitably to the result that the plaintiffs' claim fails.

I would dismiss the appeal.

However, these decisions have not been accepted by the highest courts of major Commonwealth countries as a persuasive analysis of the problem.

Winnipeg Condominium Corp No. 36 v. Bird Construction Co Ltd

SUPREME COURT OF CANADA

(1995) 50 Con LR 124

In 1972 a Winnipeg land developer entered into a contract with the defendant (Bird Construction) for the erection of a 15-storey, 94-unit apartment. The plans were prepared by the interveners, an architectural firm (Smith Carter). The building was completed in December 1974. In 1978 ownership of the building was transferred to the plaintiff (Winnipeg Condominium). In 1982 Winnipeg Condominium became concerned about the exterior cladding which consisted of 4-inch thick slabs of stone. They consulted Smith Carter and a firm of structural engineers who reported that the cladding was structurally sound and recommended minor works, which were carried out.

On 8 May 1989 a storey-high section of cladding, approximately 20 feet in length, fell from the ninth storey. Winnipeg Condominium spent in excess of $1.5m removing and replacing the entire cladding.

Winnipeg Condominium commenced an action in negligence against Bird Construction, Smith Carter and the cladding sub-contractors (Kornovski and Keller). Bird and Kornovski and Keller moved to strike out the claim as disclosing no reasonable cause of action. This motion was dismissed by Galanchuk J. Bird appealed and the Manitoba Court of Appeal allowed the appeal. Winnipeg Condominium appealed to the Supreme Court of Canada.

Held (allowing the appeal):

Contractors who take part in the construction and design of a building owe a duty in tort to subsequent purchasers of the building if it is foreseeable that failure to take reasonable care in constructing the building will create defects that pose a substantial danger to the health and safety of the occupants. *D & F Estates Ltd* v. *Church Comrs for England* (1988) 15 Con LR 35 and *Murphy* v. *Brentwood DC* (1990) 21 Con LR 1 not followed. *Anns* v. *Merton London Borough* [1977] 2 All ER 492; *City of Kamloops* v. *Nielson* [1984] 2 SCR 2 followed.

Per curiam:

(1) Allowing recovery against contractors in tort for the cost of repair of dangerous defects serves an important preventative function by encouraging building owners to repair.
(2) There is a policy difference between work which is dangerously defective and work which is merely shoddy and substandard.

LA FOREST J: This case gives this court the opportunity once again to address the question of recoverability in tort for economic loss. In *Canadian National Rly Co* v. *Norsk Pacific Steamship Co* [1992] 1 SCR 1021 at 1049, 91 DLR (4th) 289 at 299, I made reference to an article by Professor Feldthusen, 'Economic Loss in the Supreme Court of Canada: Yesterday and Tomorrow' (1990–91) 17 *Can Bus LJ* 356 at 357–358, in which he outlined five different categories of cases where the question of recoverability in tort for economic loss has arisen, namely: 1. The independent liability of statutory public authorities; 2. Negligent misrepresentation; 3. Negligent performance of a service; 4. Negligent supply of shoddy goods or structures; and 5. Relational economic loss. I stressed in *Canadian National Rly Co* v. *Norsk Pacific Steamship Co* that the question of recoverability for economic loss must be approached with reference to the unique and distinct policy issues raised in each of these categories. That is because ultimately the issues concerning recovery for economic loss are concerned with determining the proper ambit of the law of tort, an exercise that must take account of the various situations where that question may arise. This case raises issues different from that in *Canadian National Rly Co* v. *Norsk Pacific Steamship Co*, which fell within the fifth category. The present case, which involves the alleged negligent construction of a building, falls partially within the fourth category, although subject to an important caveat. The negligently supplied structure in this case was not merely shoddy; it was dangerous. In my view, this is important because the degree of danger to persons and other property created by the negligent construction of a building is a cornerstone of the policy analysis that must take place in determining whether the cost of repair of the building is recoverable in tort. As I will attempt to show, a distinction can be drawn on a policy level between 'dangerous' defects in buildings and merely 'shoddy' construction in buildings and that, at least with respect to dangerous defects, compelling policy reasons exist for the imposition upon contractors of tortious liability for the cost of repair of these defects . . .

My conclusion that the type of economic loss claimed by the Condominium Corp is recoverable in tort is therefore based in large part upon what seem to me to be compelling policy considerations. I shall elaborate in more detail upon these later in my reasons. However, before doing so, I think it important to clarify why the *D & F Estates Ltd* case should not, in my view, be seen as having strong persuasive authority in Canadian tort law as that law is currently developing. My reasons for coming to this conclusion are twofold: first, to the extent that the decision of the House of Lords in *D & F Estates Ltd* rests upon the assumption that liability in tort for the cost of repair of defective houses represents an unjustifiable intrusion of tort into the contractual sphere, it is inconsistent with recent Canadian decisions recognising the possibility of concurrent contractual and tortious duties; second, to the extent that the *D & F Estates Ltd* decision formed part of a line of English cases leading ultimately to the rejection of the *Anns* case, it is inconsistent with this court's continued application of the principles established in *Anns* . . .

Bryan v. Maloney

HIGH COURT OF AUSTRALIA

(1995) 51 Con LR 29

The appellant/defendant (Bryan) built a house in Launceston in 1979 for Mrs Manion. Mrs Manion in due course sold the house to a Mr and Mrs Quittenden and in 1986 Mr and Mrs Quittenden sold the house to the plaintiff/respondent, Mrs Maloney (Maloney). Before the High Court of Australia, it was common ground that Bryan was negligent in building the house with inadequate footings; that the damage suffered by Mrs Maloney was the loss involved in the decrease in the value of the house resulting from the inadequacy of the footings and that that damage was sustained by Mrs Maloney when the inadequacy of the footings first became manifest by reason of the cracks appearing in the walls of the house some six months after she had purchased it. It was accepted that that damage was a foreseeable consequence of Bryan's negligence. In these circumstances, the trial judge and the Full Court of the Supreme Court of Tasmania held that Mrs Maloney could successfully maintain an action against Bryan, who appealed to the High Court of Australia.

Held (Brennan J dissenting):

Foreseeability of loss flowing from the defendant's negligence was not sufficient for the plaintiff to establish liability where the consequential loss was pure economic loss. The builder of a permanent house could readily foresee that the house was likely to be owned and occupied by a number of families during its lifetime and that second and subsequent purchasers would rely on the house having been properly built, particularly as to the matter such as foundations which they could not readily check for themselves. So the factors of reliance and assumption of responsibility which were usually present when courts imposed liability for pure economic loss were present and the factors which normally pointed to rejection of such liability, such as avoidance of indefinite and indeterminate liability and the desire to protect the legitimate pursuit of personal advantage, were absent or weak in the circumstances of cases such as this. There was, therefore, a sufficient degree of proximity between the builder and the subsequent purchaser to justify holding that there was a duty of care, especially since the relevant policy considerations pointed in favour of imposing rather than negativing liability.

Per Brennan J (dissenting). A remedy in tort for economic loss would be available only to a plaintiff who actually incurred expense in removing dangerous defects which posed a substantial risk of physical damage to person or property.

MASON CJ, DEANE AND GAUDRON JJ: Moreover, the policy considerations underlying the reluctance of the courts to recognise a relationship of proximity and a consequent duty of care in cases of mere economic loss are inapplicable to a relationship of the kind which existed between Mr Bryan and Mrs Manion as regards the kind of economic loss sustained by Mrs Maloney. Thus, there is no basis for thinking that recognition of a relationship of proximity between builder and first owner with respect to that particular kind of economic loss would give rise to the type of liability 'in an indeterminate amount for an indeterminate time to an indeterminate class' which the courts are reluctant to recognise. Again, in circumstances where the builder is, in any event, under a duty of care to the first owner to avoid physical injury to that owner's person or property by reason of inadequacy of the footings, there can be no real question of inconsistency between the existence of a relationship of proximity with respect to that particular kind of economic loss and the legitimate pursuit by the builder of his or her own financial interests. Nor, as has been seen, is it legitimate to assert that, as a matter of policy, the sanctity of contract or the compartmentalisation of the law dictates that liability under the ordinary principles of negligence in respect of either damages generally or a particular kind of damage must be excluded as between parties in a contractual relationship notwithstanding the absence of any actual agreement between the parties to that effect. Whatever may have been or may be the position in other times or in other places, the law of this country knows no such policy.

On the other hand, there are strong reasons for acknowledging the existence of a relevant relationship of proximity between a builder such as Mr Bryan and a first owner such as Mrs Manion with respect to the kind of economic loss sustained by Mrs Maloney. In particular, the ordinary relationship between a builder of a house and the first owner with respect to that kind of economic loss is characterised by the kind of assumption of responsibility on the one part (i.e. the builder) and known reliance on the other (i.e. the building owner) which commonly exists in the special categories of case in which a relationship of proximity and a consequent duty of care exists in respect of pure economic loss. There is nothing to suggest that the relationship between Mr Bryan and Mrs Manion was not characterised by such an assumption of responsibility and such reliance . . .

It is in the context of the above-mentioned relationships of proximity that one must determine whether the relationship which exists between a professional builder of a house, such as Mr Bryan, and a subsequent owner, such as Mrs Maloney, possesses the requisite degree of proximity to give rise to a duty to take reasonable care on the part of the builder to avoid the kind of economic loss sustained by Mrs Maloney in the present case. It is likely that the only connection between such a builder and such a subsequent owner will be the house itself. None the less, the relationship between them is marked by proximity in a number of important respects. The connecting link of the house is itself a substantial one. It is a permanent structure to be used indefinitely and, in this country, is likely to represent one of the most significant, and possibly the most significant, investment which the subsequent owner will make during his or her lifetime. It is obviously foreseeable by such a builder that the negligent construction of the house with inadequate footings is likely to cause economic loss, of the kind sustained by Mrs Maloney, to the owner of the house at the time when the inadequacy of the footings first becomes manifest. When such economic loss is eventually sustained and there is no intervening negligence or other causative event, the causal proximity between the loss and the builder's lack of reasonable care is unextinguished by either lapse of time or change of ownership.

The only factor which arguably precludes the recognition of a relevant relationship of proximity between builder and subsequent owner for the purposes of the present case is the kind of damage involved, namely, mere economic loss. As has been seen, a relevant relationship of proximity would have existed between the builder and Mrs Maloney with respect to ordinary physical injury to her person or other property caused by a partial collapse of the house due to its inadequate footings even if she had not been the owner. Here again, it is important to bear in mind the particular kind of economic loss involved. As has been said, the distinction between that kind of economic loss and ordinary physical damage to property is an essentially technical one. Indeed, the economic loss sustained by the owner of a house by reason of diminution in value when the inadequacy of the footings first becomes manifest by consequent damage to the fabric of the house is, at least arguably, less remote and more readily foreseeable than ordinary physical damage to other property of the owner which might be caused by an actual collapse of part of the house as a result of the inadequacy of those footings. Again, the policy considerations underlying the reluctance of the courts to recognise a relationship of proximity and a consequential duty of care in cases of mere economic loss are largely inapplicable to the relationship between builder and subsequent owner as regards that particular kind of economic loss. There can be no question of inconsistency with the builder's legitimate pursuit of his or her own financial interests since, as has been seen, the builder owed a duty of care to the first owner with respect to such loss. In circumstances where the particular kind of economic loss is that sustained by an owner of the house on the occasion when the inadequacy of the footings first becomes manifest, there is no basis for thinking that recognition of a relevant relationship of proximity between builder and that owner would be more likely to give rise to liability 'in an indeterminate amount . . . to an indeterminate class' than does recognition of such an element of proximity in the relationship between builder and first owner. It is true that, in so far as 'an indeterminate time' is concerned, the time span in which liability to a subsequent owner might arise could be greater than if liability were restricted to the first owner. None the less, the extent of that time span would be limited by the element of reasonableness both in the requirement that damage be foreseeable and in the context of the duty of care: compare *Askin* v. *Knox* [1989] 1 NZLR 248. In any event, it would *prima facie* correspond with that applicable to the relationship of proximity which clearly exists as regards physical injury to person or other property. Moreover, any difference in duration between liability to the first owner and liability to a subsequent owner is likely to do no more than reflect the chance element of whether and when the first owner disposes of the house.

Upon analysis, the relationship between builder and subsequent owner with respect to the particular kind of economic loss is, like that between the builder and first owner, marked by the kind of assumption of responsibility and a relationship of proximity exists with respect to pure economic loss. In ordinary circumstances, the builder of a house undertakes the responsibility of erecting a structure on the basis that its footings are adequate to support it for a period during which it is likely that there will be one or more subsequent owners. Such a subsequent owner will ordinarily have no greater, and will often have less, opportunity to inspect and test the footings of the house than the first owner. Such a subsequent owner is likely to be unskilled in building matters and inexperienced in the niceties of real property investment. Any builder should be aware that such a subsequent owner will be likely, if inadequacy of the footings has not become manifest, to assume that the house has been competently built and that the footings are in fact adequate.

Ultimately, it seems to us that, from the point of view of proximity, the similarities between the relationship between the builder and first owner and the relationship between the builder and subsequent owner as regards the particular kind of economic loss are of much greater significance than the differences to which attention has been drawn, namely, the absence of direct contact or dealing and the possibly extended time in which liability might arise. Both relationships are characterised, to a comparable extent, by assumption of responsibility on the part of the builder and likely reliance on the part of the owner. No distinction can be drawn between the two relationships in so far as the foreseeability that that loss will be sustained by whichever of the first or subsequent owners happens to be the owner at the time when the inadequacy of the footings becomes manifest. In the absence of competing or intervening negligence or other causative event, the causal proximity between negligence on the part of the builder in constructing the footings and consequent economic loss on the part of the owner when the inadequacy of the footings becomes manifest is the same regardless of whether the owner in question is the first owner or a subsequent owner. In the case of both relationships, the policy considerations which ordinarily militate against the recognition of a relationship of proximity and a consequent duty of care with respect to pure economic loss are insignificant. Moreover, there are persuasive policy reasons supporting the recognition of a relationship of proximity between the builder and a subsequent owner of an ordinary dwelling house with respect to the particular kind of economic loss. As Wright J pointed out, at first instance, a number of those policy considerations were identified in the judgment of the Supreme Court of New Hampshire, delivered by Thayer J, in *Lempke* v. *Dagenais* (1988) 547 A 2d 290 at 294–295. They include the consideration that, by virtue of superior knowledge, skill and experience in the construction of houses, it is likely that a builder will be better qualified and positioned to avoid, evaluate and guard against the financial risk posed by latent defect in the structure of a house: see *Lempke* v. *Dagenais* (1988) 547 A 2d 290 at 295, quoting Whichard J in *George* v. *Veach* (1984) 313 SE 2d 920 at 923. See also Sir Robin Cooke, 'An Impossible Distinction' (1991) 107 *LQR* 46 at 61–63. In all the circumstances, the relationship between builder and subsequent owner as regards the particular kind of economic loss should be accepted as possessing a comparable degree of proximity to that possessed by the relationship between builder and first owner and as giving rise to a duty to take reasonable care on the part of the builder to avoid such loss.

The conclusion that a relationship of proximity existed between Mr Bryan, as the builder, and Mrs Maloney, as subsequent owner, with respect to the particular kind of economic loss is also supported by an analogy with the relationship which would have existed between Mr Bryan, as the builder, and any person who suffered physical injury to person or property in the event that the house or part of the house had collapsed at the time when the inadequacy of the foundations first became manifest. It is difficult to see why, as a matter of principle, policy or common sense, a negligent builder should be liable for ordinary physical injury caused to any person or to other property by reason of the collapse of a building by reason of the inadequacy of the foundations but be not liable to the owner of the building for the cost of remedial work necessary to remedy that inadequacy and to avert such damage. Indeed, there is obvious force in the view expressed by Lord Denning MR in *Dutton* v. *Bognor Regis United Building Co Ltd* [1972] 1 All ER 462 at 474, [1972] 1 QB 373 at 396, that, as a rational basis for differentiating between circumstances of liability and circumstances of no liability, such a distinction is an 'impossible' one . . .

Invercargill City Council v. Hamlin

PRIVY COUNCIL

(1996) 50 Con LR 105

In 1972 the plaintiff contracted with a firm of builders in New Zealand whereby they sold him a plot of land and agreed to construct a house on it. In the course of construction a building inspector employed by the defendant council inspected and approved the work in accordance with the council's building byelaws. In particular, the inspector approved foundations to a depth of 15 in to sit on firm clay. In fact, on one side of the house the foundations were only taken down to a depth of 7 or 8 in and over the years a number of cracks and minor defects appeared. In 1989 the plaintiff was advised by another builder that the cause of the problem was the foundations. In 1990 the plaintiff, who was unable to sue the original builder as it was no longer in business, commenced proceedings against the council. The judge found that the inspector could have discovered without difficulty that the foundations had not been carried down to firm clay and that a reasonably prudent home owner would not have discovered the cause of the cracks before 1989. On those findings, the judge held that the inspector had been negligent in carrying out his inspection and awarded the plaintiff damages.

The council appealed to the New Zealand Court of Appeal, which, following a line of New Zealand authority in preference to contrary House of Lords authority, held that the council owed the plaintiff a duty of care and that the cause of action was not time-barred because it did not accrue until the damage could with reasonable diligence have been discovered. The court accordingly dismissed the appeal. The council appealed to the Privy Council, contending that the Board ought to apply the House of Lords authorities on the duty of care point, to the effect that the loss was economic rather than physical loss and there was insufficient proximity between a local authority's building inspector and a house owner to give rise to a duty of care.

Held the appeal would be dismissed because:

The New Zealand Court of Appeal was entitled to develop the common law of New Zealand according to local policy considerations in areas of the common law which were developing, not settled. The law of negligence in relation to a local authority's liability for the negligence of a building inspector was particularly unsuited to a single solution applicable in all common law jurisdictions regardless of differing local circumstances. The perception in New Zealand was that community standards and expectations demanded the imposition of a duty of care on local authorities and builders alike to ensure compliance with local byelaws and the Court of Appeal had, in common with other common law jurisdictions, built up a line of authority based on the linked concepts of control by the local authority of building works through the enforcement of its byelaws and reliance on that control by purchasers. The present case had been decided in accordance with that line of authority and therefore on the duty of care issue the Board would indorse in relation to New Zealand the approach taken by the New Zealand courts, notwithstanding House of Lords authority to the contrary; dictum of Lord Diplock in *Cassell & Co Ltd* v. *Broome* [1972] 1 All ER 801 at 871 applied; *Bowen* v. *Paramount Builders (Hamilton) Ltd* [1977] 1 NZLR 394 and *City of Kamloops* v. *Nielson* [1984] 2 SCR 2 followed; *D & F Estates Ltd* v. *Church Comrs for England* (1988) 15 Con LR 35 and *Murphy* v. *Brentwood DC* (1990) 21 Con LR 1 not followed.

LORD LLOYD OF BERWICK: Their Lordships cite these judgments in other common law jurisdictions not to cast any doubt on *Murphy*'s case, but rather to illustrate the point that in this branch of the law more than one view is possible: there is no single correct answer. In *Bryan* v. *Maloney* the majority decision was based on the twin concepts of assumption of responsibility and reliance by the subsequent purchaser. If that be a possible and indeed respectable view, it cannot be said that the decision of the Court of Appeal in the present case, based as it was on the same or very similar twin concepts, was reached by a process of faulty reasoning, or that the decision was based on some misconception: see *Australian Consolidated Press Ltd* v. *Uren* [1967] 3 All ER 523, [1969] 1 AC 590.

In truth, the explanation for divergent views in different common law jurisdictions (or within different jurisdictions of the United States of America) is not far to seek. The decision whether to hold a local authority liable for the negligence of a building inspector is bound to be based at least in part on policy considerations. As Mason CJ said in *Bryan* v. *Maloney* (1995) 128 ALR 163 at 166:

> Inevitably, the policy considerations which are legitimately taken into account in determining whether sufficient proximity exists in a novel category will be influenced by the court's assessment of community standards and demands.

In a succession of cases in New Zealand over the last 20 years it has been decided that community standards and expectations demand the imposition of a duty of care on local authorities and builders alike to ensure compliance with local byelaws. New Zealand judges are in a much better position to decide on such matters than the Board. Whether circumstances are in fact so very different in England and New Zealand may not matter greatly. What matters is the perception. Both Richardson J and McKay J in their judgments in the court below stress that to change New Zealand law so as to make it comply with *Murphy*'s case would have 'significant community implications' and would require a 'major attitudinal shift'. It would be rash for the Board to ignore those views . . .

Woolcock Street Investments Pty Ltd v. CDG Pty Ltd

HIGH COURT OF AUSTRALIA

(2004) 205 ALR 522

CDG designed the foundations of a warehouse and office complex for the then owner of the premises. Woolcock subsequently bought the premises. It then became apparent that the building was suffering substantial structural distress as a result of settlement of the foundations.

Woolcock alleged that CDG owed it a duty of care and were in breach. A case was stated for the Queensland Court of Appeal asking whether Woolcock's statement of claim disclosed a cause of action in negligence.

The Queensland Court of Appeal said that it didn't and the High Court of Australia (Kirby J dissenting) agreed.

GLEESON CJ, GUMMOW, HAYNE AND HEYDON JJ:

Criticisms of *Bryan* v. *Maloney*

16. The decision in *Bryan* v. *Maloney* has not escaped criticism. Some of those criticisms found reflection in the series of questions posed by Brooking JA in *Zumpano* v. *Montagnese*. It is not necessary, in this case, to attempt to deal with all of those criticisms, or to attempt to answer all of the questions posed in *Zumpano*. Rather, two points should be made.

17. First, for the reasons given earlier, it may be doubted that the decision in *Bryan* v. *Maloney* should be understood as depending upon drawing a bright line between cases concerning the construction of dwellings and cases concerning the construction of other buildings. If it were to be understood as attempting to draw such a line, it would turn out to be far from bright, straight, clearly defined, or even clearly definable. As has been pointed out subsequently, some buildings are used for mixed purposes: shop and dwelling; dwelling and commercial art gallery; general practitioner's surgery and residence. Some high-rise apartment blocks are built in ways not very different from high-rise office towers. The original owner of a high-rise apartment block may be a large commercial enterprise. The list of difficulties in distinguishing between dwellings and other buildings could be extended.

18. Secondly, the decision in *Bryan* v. *Maloney* depended upon the view that 'the overriding requirement of a relationship of proximity represents the conceptual determinant and the unifying theme of the categories of case in which the common law of negligence recognises the existence of a duty to take reasonable care to avoid a reasonably foreseeable risk of injury to another'. It was the application of this 'conceptual determinant' of proximity that was seen as both permitting and requiring the equation of the duty owed to the first owner with the duty owed to the subsequent purchaser. Decisions of the court after *Bryan* v. *Maloney* reveal that proximity is no longer seen as the 'conceptual determinant' in this area.

Economic loss

19. The damage for which the appellant seeks a remedy in this case is the economic loss it alleges it has suffered as a result of buying a building which is defective. Circumstances can be imagined in which, had the defects not been discovered, some damage to person or property might have resulted from those defects. But that is not what has happened. The defects have been identified. Steps can be taken to prevent damage to person or property.

20. A view was adopted for a time in England that, because there was physical damage to the building, a claim of the kind made by the appellant was not solely for economic loss. That view was questioned in *Sutherland Shire Council* v. *Heyman* and rejected in *Bryan* v. *Maloney*. It was subsequently also rejected by the House of Lords in *Murphy* v. *Brentwood District Council*. There is no reason now to reopen that debate and neither side in the present matter sought to do so. The damage which the appellant alleges it has suffered is pure economic loss.

21. Claims for damages for pure economic loss present peculiar difficulty. Competition is the hallmark of most forms of commercial activity in Australia. As Brennan J said in *Bryan* v. *Maloney*:

> If liability were to be imposed for the doing of anything which caused pure economic loss that was foreseeable, the tort of negligence would destroy commercial competition, sterilise many contracts and, in the well-known dictum of Chief Judge Cardozo, expose defendants to potential liability 'in an indeterminate amount for an indeterminate time to an indeterminate class'.

That is why damages for pure economic loss are not recoverable if all that is shown is that the defendant's negligence was a cause of the loss and the loss was reasonably foreseeable.

22. In *Caltex Oil (Aust) Pty Ltd* v. *The Dredge 'Willemstad'*, the court held that there were circumstances in which damages for economic loss were recoverable. In *Caltex Oil*, cases for recovery of economic loss were seen as being exceptions to a general rule, said to have been established in *Cattle* v. *Stockton Waterworks*, that even if the loss was foreseeable, damages are not recoverable for economic loss which was not consequential upon injury to person or property. In *Caltex Oil*, Stephen J isolated a number of 'salient features' which combined to

constitute a sufficiently close relationship to give rise to a duty of care owed to Caltex for breach of which it might recover its purely economic loss. Chief among those features was the defendant's knowledge that to damage the pipeline which was damaged was inherently likely to produce economic loss.

23. Since *Caltex Oil*, and most notably in *Perre* v. *Apand Pty Ltd*, the vulnerability of the plaintiff has emerged as an important requirement in cases where a duty of care to avoid economic loss has been held to have been owed. 'Vulnerability', in this context, is not to be understood as meaning only that the plaintiff was likely to suffer damage if reasonable care was not taken. Rather, 'vulnerability' is to be understood as a reference to the plaintiff's inability to protect itself from the consequences of a defendant's want of reasonable care, either entirely or at least in a way which would cast the consequences of loss on the defendant. So, in *Perre*, the plaintiffs could do nothing to protect themselves from the economic consequences to them of the defendant's negligence in sowing a crop which caused the quarantining of the plaintiffs' land. In *Hill* v. *Van Erp*, the intended beneficiary depended entirely upon the solicitor performing the client's retainer properly and the beneficiary could do nothing to ensure that this was done. But in *Esanda Finance Corp Ltd* v. *Peat Marwick Hungerfords*, the financier could itself have made inquiries about the financial position of the company to which it was to lend money, rather than depend upon the auditor's certification of the accounts of the company.

Linklaters Business Services v. Sir Robert McAlpine Ltd and another (No. 2)

QUEEN'S BENCH DIVISIN

(2010) 133 Con LR 211, [2010] EWHC 2931 (TCC)

The claimants (Linklaters) were the service company of the well-known solicitors. In late 1996 they moved into premises consisting of two buildings, at 1 Silk Street, EC2, on a 25-year lease. The premises had been the subject of a major redevelopment in which Sir Robert McAlpine Ltd (McAlpine) were the main contractors, How Engineering services Ltd (How) were the mechanical and electrical sub-contractors and Southern Insulation (Medway) Ltd (Southern) were sub-sub-contractors for the installation of the insulation works. The buildings were fully air conditioned, which involved insulated chilled water pipework throughout the building.

In June 2006 a leak was noticed from one of the chilled water rise pipes. Investigation revealed extensive corrosion throughout the chilled water pipework. In due course Linklaters replaced the corroded pipework throughout the building. McAlpine and How had each given collateral warranties to Linklaters that they had performed the main contract and sub-contract respectively. Linklaters made claims against McAlpine and How in contract. How made claims against Southern in tort. How and McAlpine accepted that if McAlpine were liable in contract it was entitled to 100 per cent indemnity from How.

AKENHEAD J:

Essentially, the arguments between the parties can be summarised as follows. How argues that there is a duty of care which extends to damage to the steel pipework caused by careless insulation work. Southern says that it does not so extend and that one cannot and should not differentiate between two components (insulation and pipework) which go to make up one installation (the insulated chilled water pipework); thus, damage to the pipework is damage 'to the thing itself'. It is uncontroversial that, in the case of physical damage negligence, damage to the carelessly manufactured designed or constructed 'thing itself' does not found a cause of action.

There is very little direct, non-obiter, authority in this country on the topic. As indicated in my judgment of 21 May 2010 ([2010] EWHC 1145 (TCC), (2010) 130 Con LR 111), there are some obiter dicta, particularly in *Murphy* v. *Brentwood DC* (1990) 21 Con LR 1, [1991] 1 AC 398 which give examples of what might give rise to a cause of action in negligence, such a carelessly installed electrical wiring causing fire damage to the building per Lord Keith of Kinkel). Another example was given in *D & F Estates Ltd* v. *Church Comrs for England* (1988) 15 Con LR 35 at 61-62, [1989] AC 177 at 211, by Lord Oliver of Aylmerton:

> If I buy a secondhand car to which there has been fitted a pneumatic tyre which, as a result of carelessness in manufacture, is dangerously defective and which bursts, causing injury to me or the car, no doubt the negligent manufacturer is liable in tort on the ordinary application of [*Donoghue (or M'Alister)* v. *Stevenson* [1932] AC 562 [1932] All ER Rep 1]...

These examples however do not satisfactorily deal with components of the offending installation such as the reinforcement within a tyre or a valve within a boiler, causing damage only to the tyre or the boiler. The question therefore arises as to whether in English law as a matter of authority or policy (if different) there is or should be a distinction. I would be slow to suggest that the *obiter dicta* of eminent members of the House of Lords were wrong as a matter of principle but it is clear that they were not applying their minds to a carelessly installed or manufactured component causing damage to the thing or installation of which it was merely a component.

Southern's counsel have assiduously researched relatively recent authorities in the United States of America. The reason for doing so was the approving references in *Murphy*'s case to the United States Supreme Court decision in *East River Steamship Corp* v. *Transamerica Delaval Inc* (1986) 476 US 858. That case involved claims in tort by the charterers of ships against the manufacturers of turbine engines within the ships said to have been defective. The court regarded each turbine as a single unit, noting with approval *Northern Power and Engineering Corp* v. *Caterpillar Tractor Co* (1981) 623 P 2d 324 at 330:

> since all but the very simplest of machines have component parts, [a contrary] holding would require a finding of 'property damage', in virtually every case where a producer damages itself such a holding would eliminate the distinction between warranty and strict product liability.

This approach has been followed in a number of other American cases such as *Shipco 2295 Inc* v. *Avondale Shipyards Inc* (1987) 825 F 2d 925. Of course, one must appreciate that American law in general and the law of individual states is or may be different to English law. The legislative background is or may be different, an example being that in many states there is a product liability law which may in certain circumstances give rise to statutory remedies. There does however seem to be logic as well as common sense in this line of authority.

Having now heard and understood the evidence, I have formed the view that the insulated chilled water pipework was essentially one 'thing' for the purposes of tort. One would simply never have chilled water pipework without insulation because the chilled water would not remain chilled and it would corrode. The insulation is a key component but a component nonetheless. It would follow that no cause of action arises in tort as between Southern and Linklaters. That is not at all unreasonable in any way because Linklaters or people in their position can protect themselves, as Linklaters did, with the securing of contractual warranties from relevant parties such as the key contractors in any given development.

If I had found that Southern had been materially negligent, this would not have been a case in which what would be called the usual apportionment of responsibility would have been made. In most cases in which a negligent sub-contractor has been responsible for providing the offending work, an apportionment as between it and the careless supervising main contractor would be about 75:25. However, in this case, if any carelessness on the part of Southern had materially contributed to the excessive corrosion, How would in any event have had to bear the lion's share. That would be because How was responsible for the design or management decisions in relation to the insulation around the wooden supports and Victaulic joints. As the large part of the corrosion was attributable to the deficiencies in those areas, the apportionment would have been closer to 10:90 in favour of Southern.

Robinson v. PE Jones (Contractors) Ltd

COURT OF APPEAL, CIVIL DIVISION

(2011) 134 Con LR 26, [2011] EWCA Civ 9

In December 1999 the claimant and his wife agreed to buy from the defendant a house in Prestbury. The contract incorporated the standard NHBC agreement. In September 2004 it was discovered that the flue for one of the gas fires was defectively built and that the remedial work would be substantial. The claimant brought proceedings claiming damages in contract and tort. The trial of preliminary issues was ordered.

Held

(1) There was no reason in principle why there should not be concurrent liability in contract and tort in respect of defective building work.

(2) The building contract provided that the defendant's liability was limited to that covered by the NHBC conditions under which the builder had extensive liability for defects for two years after completion and more limited liability for the next eight years. This limitation of liability was entirely reasonable within the scope of the Unfair Contract Terms Act 1977.

(3) The defendant did not owe the claimant a duty of care in tort in relation to the economic loss arising from the defective flue.

JACKSON LJ:

In the present case I see nothing to suggest that the defendant 'assumed responsibility' to the claimant in the *Hedley Byrne* sense. The parties entered into a normal contract whereby the defendant would complete the construction of a house for the claimant to an agreed specification and the claimant would pay the purchase price. The defendant's warranties of quality were set out and the claimant's remedies in the event of breach of warranty were also set out. The parties were not in a professional relationship whereby, for example, the claimant was paying the defendant to give advice or to prepare reports or plans upon which the claimant would act.

Even if the agreement did not contain cll 8 and 10 of the building conditions, I would be disinclined to find that the defendant owed to the claimant the duty of care which is alleged in this case. To my mind, however, cll 8 and 10 of the building conditions put the matter beyond doubt. Those clauses limit the defendant's liability for building defects to the first two years, after which different provision is made for dealing with defects. For the reasons set out in part 5 above, those two clauses satisfy the test of reasonableness in UCTA. It would be inconsistent with the whole scheme of this contract, if the law were to impose upon the defendant duties of care in tort far exceeding the defendant's contractual liabilities. Finally, cl 10 of the building conditions is relevant in another way. The parties expressly agreed that the defendant's only liability to the claimant should be that arising from the NHBC agreement. The parties were thereby expressly agreeing to exclude any liability in negligence which might otherwise arise.

Bole and another v. Huntsbuild Ltd and another

COURT OF APPEAL, CIVIL DIVISION

(2009) 127 Con LR 154, [2009] EWCA Civ 1146

By a contract made on 19 September 2001 the first defendant (Huntsbuild) agreed to construct and sell a house to the claimants (Mr and Mrs Bole). The second defendant (RMA) were a firm of structural engineers engaged by Huntsbuild to carry out site investigations and to advise on and design the foundations for the house. It was accepted that RMA owed a duty to the claimants under s. 1 of the Defective Premises Act 1972.

His Honour Judge Toulmin CMG QC held (2009) 124 Con LR 1 that RMA were in breach of the 1972 Act. RMA appealed.

Held

(1) It was a relevant factor in deciding whether the house was fit for habitation that the remedial works would require the claimants to vacate the house for at least six months (the defendants' expert) or perhaps as much as 12 months (the claimants' expert). This was not necessarily decisive but it was likely to be highly relevant.

(2) In deciding whether the house was fit for habitation it was not right to consider each of the defects in isolation. It was appropriate to consider whether the dwelling as a whole was unfit. There was a fundamental defect in the inadequate foundations.

(3) The judge was entitled to hold that the cost of remedying all the defects attributable to defective foundations was a foreseeable consequence of the breach of s. 1.

DYSON LJ:

As for the other alleged error in the judge's reasoning, it might have been better if the judge had explained what he meant by a dwelling being unfit for its purpose. It is perhaps understandable that he used the phrase 'unsuitable for purpose' in the light of para 34 of the Law Commission report. I cannot, however, accept that there is any material error here. In my view, it is clear that the judge meant that defects which render a dwelling unfit for its purpose, are defects which render it unfit for habitation. The obvious purpose of a dwelling is for it to be occupied and inhabited safely and without inconvenience. That was clearly the view of the judge. What other purpose could he have had in mind when he said (at [38] (ii)) 6a unfitness for habitation extends to defects of quality, rendering the dwelling unsuitable for its purpose, as well as to 'dangerous defects' and again, when he said (at [3S](iii)) that unfitness for habitation relates to defects rendering the dwelling dangerous or unsuitable for its purpose?

The judge was not obliged to approach the question of whether there had been a breach of s. 1 in this case, by considering each effect individually and asking whether that defect, or indeed, that defect taken in conjunction with other individual defects, rendered the dwelling unfit for habitation. He was entitled to ask himself whether the dwelling as a whole was unfit for habitation. The judge found that it suffered from a fundamental defect, namely inadequate foundations. There was what the judge described (at [169]) as 'widespread cracking' which was the result of ground heave, which in turn was attributable to the inadequacy of the foundations. Some cracks, when viewed as individual cracks, were more serious than others. The cracks referred to in paras [157]–[162] were, of themselves, quite serious and even when considered on their own, arguably rendered the dwelling unfit for habitation. Other cracks and defects were in themselves less serious, nevertheless, all the defects, including all the cracks as well as warped doors et cetera, were the result of the ground heave and ultimately inadequate foundations. In reaching his conclusion that the dwelling was unfit for habitation, the judge took into account all the defects and the fact that they were caused by a fundamental defect, namely, the inadequacy of the foundations.

Table of Cases

Page numbers in **heavy type** indicate a verbatim reporting of the case.

Powell-Smith and Furmston's Building Contract Casebook, Fifth Edition. Michael Furmston.

INDEX